AF411365

Killing Cancer

Killing Cancer: Discovery and Selection of New Target Molecules

Editor

Tuula Kallunki

MDPI • Basel • Beijing • Wuhan • Barcelona • Belgrade • Manchester • Tokyo • Cluj • Tianjin

Editor
Tuula Kallunki
Danish Cancer Society Research Center
Denmark

Editorial Office
MDPI
St. Alban-Anlage 66
4052 Basel, Switzerland

This is a reprint of articles from the Special Issue published online in the open access journal *Cells* (ISSN 2073-4409) (available at: https://www.mdpi.com/journal/cells/special_issues/killing_cancer).

For citation purposes, cite each article independently as indicated on the article page online and as indicated below:

LastName, A.A.; LastName, B.B.; LastName, C.C. Article Title. *Journal Name* **Year**, *Article Number*, Page Range.

ISBN 978-3-03943-440-4 (Hbk)
ISBN 978-3-03943-441-1 (PDF)

Cover image courtesy of Tuula Kallunki.

Contents

About the Editor

Tuula Kallunki is a group leader in the Unit of Cell Death and Metabolism at the Danish Cancer Society Research Center and Associate Professor at the Department of Drug Design and Pharmacology of the University of Copenhagen. Dr. Kallunki did her PhD on biochemistry and molecular biology at the University of Oulu, Finland; prior to joining the Danish Cancer Society with a Marie Curie fellowship of the European Union, she did her postdoctoral training at the Department of Pharmacology, University of California, San Diego. Her research focuses on the role of lysosomes in the aggressiveness of breast and ovarian cancer. She is an author of 48 peer-reviewed publications with over 5600 citations; her h-index is 29. In addition to her research career, she has been active in organizational work as a member of the Federation of European Biochemical Societies (FEBS) Science and Society Committee, member of the National Executive Committee of the International Council of Science (ICSU) Denmark, and Chairperson of the Danish Society for Biochemistry and Molecular Biology (DSBMB); prior to that, she was a delegate of DSBMB to the FEBS council meetings and the International Union of Biochemistry and Molecular Biology (IUBMB) general assemblies. Tuula Kallunki has extensive experience as a grant evaluator for several international foundations and organizations, and she has been a co-organizer for numerous scientific meetings in Copenhagen.

Editorial

Seeking and Exploring Efficient Ways to Target Cancer

Tuula Kallunki [1,2]

1   The Invasion and Signaling Group, Cell Death and Metabolism, Center for Autophagy, Recycling and Disease, Danish Cancer Society Research Center, Strandboulevarden 49, 2100 Copenhagen, Denmark; tk@cancer.dk; Tel.: +45-35-257-746
2   Department of Drug Design and Pharmacology, Faculty of Health and Medical Sciences, University of Copenhagen, 2200 Copenhagen, Denmark

Received: 15 September 2020; Accepted: 16 September 2020; Published: 17 September 2020

Anti-cancer treatments have never been so numerous and so efficient. As a result, the number of cancer survivors has been steadily increasing. At the same time, we are facing the challenge of a decade with increasing cancer incidence, which is expected to increase by 80% in middle and low-income countries and 40% in high-income countries from 2008 to 2030 [1]. In 2008, 7.6 million people died of cancer, and with the current trend, it is estimated that by 2030 the annual cancer death rate will reach 13 million. It is unlikely that this raise will stop there. Now is the time to focus on efforts to stop this development; obviously not only more research, but also more focused cancer research is needed to discover and develop even better anti-cancer treatments. All the "easy" drugs have been discovered and thus all innovative, "out of the mainstream" scientific ideas to target cancer should be thoroughly studied. For this, we need co-operation between cancer researchers, politicians, patients and their families, and at the same we need to retain open and curious minds to explore new ideas and to find new and better ways to revisit the old ones. To obtain optimal individual benefit, cancer treatments would need to be personalized with the most suitable drugs, drug combinations and doses, which requires more research, resulting in an increased understanding of cancer biology. This will allow us to find new therapeutic target molecules and refine our understanding of the old ones. In this issue, we show insights into some promising as well as some possible drug/treatment targets, as well as the means to target them, and address some novel and unconventional ideas and approaches.

Immunotherapy, and especially different ways to boost patient immune systems to detect and fight cancer, has become an efficient way to treat cancer, showing great promise and potential against often otherwise non-curable cancers. In the review by Lichtenstern et al. [2], the authors discuss the development of the immunotherapy options for colorectal cancer (CRC), which is one of the most common and most lethal cancer types and is especially problematic to treat when detected at advanced stages. Despite scientific and clinical advances in terms of new treatment options, the five-year survival rate for metastatic CRC is still only about 14%. Immunotherapy has risen as a promising treatment against metastatic CRC due to the typically high mutational burden of this cancer type. An especially promising approach is immune checkpoint inhibition (ICI) when administered to mismatch repair deficient and microsatellite instability high (dMMR/MSI-H) CRC tumors. However not all dMMR/MSI-H tumors respond to ICI, and as of yet, no response is seen in mismatch repair proficient and microsatellite instability low (pMMR/MSI-L) tumors. An additional challenge is the gut microbiome, which has been implicated as a cause of variation in response rates to ICI depending on the microbe strain and cancer types. The poor treatment response to date has led to the necessity for the identification of new targets that could be used in combinatorial treatments such as KRAS(G12C) via its specific inhibitor AMG 510.

It is known that solid tumors can often be difficult to treat with targeted therapeutic intervention strategies such as antibody–drug conjugates and immunotherapy. This is especially true for tumors with a low mutation burden, which are often not antigenic enough to be targeted by immunotherapy.

In their review article, Khazamipour et al. [3] point out how good anti-cancer treatment targets in solid tumors need to be differentially over-expressed in primary tumors and metastases in contrast to healthy organs. With this respect, they set their focus to oncofetal chondroitin sulfate, which is an excellent anti-cancer treatment target due to its virtue of being a cancer-specific secondary glycosaminoglycan modification to proteoglycans that are often expressed in various solid tumors and metastases. This is discussed as an opportunity to curb childhood solid tumors.

Cancer cell metabolism differs from that of normal cells. Tumors commonly activate metabolic pathways that upregulate nutrient synthesis and intake. Magaway et al. [4] discuss in their review the targetability of one of the central signaling metabolic pathways upregulated in tumors, the mammalian target of rapamycin (mTOR) pathway. By sensing the intracellular nutrient status, mTOR controls metabolic reprogramming via nutrient uptake and flux through various additional metabolic pathways. This makes it a promising target for anti-cancer therapy. Numerous clinical trials are ongoing to evaluate the efficacy of mTOR inhibition for cancer treatment and analogs of rapamycin, a natural mTOR inhibitor, have been approved to treat specific types of cancer. Since rapamycin does not fully inhibit mTOR kinase activity, new compounds have been engineered to better inhibit the catalytic activity of mTOR in order to more potently block its functions. Early clinical trial results of these second generation mTOR inhibitors point towards increased toxicity combined with modest antitumor activity. Magaway et al. further discuss how one of the problems encountered in these studies is the plasticity of metabolic processes. Thus, identifying metabolic vulnerabilities in different types of tumors could present opportunities for rational therapeutic strategies. A novel application of mTOR inhibitors is in the possible improvement of immunotherapeutic strategies. Using mTOR inhibitors to improve cancer vaccination strategies can also have a major impact towards their further development.

Extracellular signal-regulated kinases (Erks) encompass another kinase family that is often activated in cancers. Erks possess unique features that make them differ from other eukaryotic protein kinases. Unlike others, Erks do not autoactivate and they manifest no basal activity. They are activated as unique targets of the receptor tyrosine kinases (RTKs)–Ras–Raf–MEK signaling cascade, which controls numerous physiological processes and is mutated in most cancers. Smorodinsky-Atias et al. [5] discuss in their review how Erks have been long considered to be immune to activating mutations. Nevertheless, several such mutations have been generated in laboratory conditions and the number of mutations identified in Erks has dramatically increased following the development of Erk-specific pharmacological inhibitors and the identification of mutations that cause resistance to these compounds. Several Erk mutations have also been recently found in cancer patients. In their review Smorodinsky-Atias et al. summarize the impressive number of mutations identified in Erks so far, describe their properties, and discuss their possible mechanism of action. They discuss the possibilities to develop isoform-specific inhibitors that would specifically target Erk1 and its mutated versions. In future precision medicine, Erk1 mutations that cause drug resistance could already be taken into consideration when a therapeutic strategy is planned, and then specific drugs could be applied according to the mutation that appears in the tumor.

Multidrug resistance is a serious problem in cancer and targeting multidrug resistance by re-sensitizing resistant cancer cells is one of the big challenges in cancer biology. Among the key multidrug resistance mediating proteins are the ATP-binding cassette (ABC) transporters and the breast cancer resistance protein (BCRP). ABC transporters are plasma membrane-bound proteins that transport nutrients into cells and unwanted toxic metabolites out of cells. Cancer cells can utilize this function for transporting cancer drugs out of the cells. Several attempts to target ABC transporters to gain control in cancer have been reported, but thus far none of the inhibitors developed have been clinically approved. Ambjorner et al. [6] describe their studies on a novel BCRP inhibitor, SCO-201, which directly binds BCRP and thus prevents its function. In this study, the authors provide evidence of the specific, potent and non-toxic effects of SCO-201 in reversing the BCRP-mediated resistance in cancer cells and bringing hope to its use in the clinic.

Another membrane bound protein family with important function in cancer is the vacuolar H^+ ATPase (V-ATPase) that is located at lysosomal membranes, but can also be found at the plasma membranes of cancer cells that exhibit increased metabolic acid production, which makes them dependent on increased net acid extrusion. An acidic microenvironment favors cancer cell proliferation and survival and promotes their invasion. V-ATPase consists of at least 13 subunits, of which Flinck et al. [7] has identified ATP6V0a3 (one of the six V0 transmembrane subunits) as a negative regulator of migration and invasion of pancreatic ductal adenocarcinoma cells, where a3 is mainly found in lysosomal membranes and consistent with its mainly lysosomal role, a3 knockdown does not decrease net acid extrusion. Interestingly, Flinck et al. demonstrated that ATP6V0a3, but not the whole V-ATPase, was upregulated in pancreatic adenocarcinoma cells in comparison to pancreatic ductal epithelial cells, suggesting an additional role for it in cancer and presenting a possibility that its upregulation could be used to inhibit the invasion of these cancer cells.

In their review, Peulen et al. [8] discuss ferlins, which are phospholipid-interacting proteins involved in membrane processes such as fusion, recycling, endocytosis and exocytosis. The expression of several ferlin genes is described as altered in several tumor tissues; specifically, myoferlin, otoferlin and Fer1L4 expression have been negatively correlated with patient survival in some cancer types. Targeting myoferlin using pharmacological compounds, gene transfer technology, or interfering RNA could be considered as a novel therapeutic anti-cancer strategy, since several correlations link ferlins (and most particularly myoferlin) to cancer prognosis. However, further investigations are still needed to discover the direct link between myoferlin and cancer biology. Encouragingly, many indications suggest that myoferlin depletion can interfere with growth factor exocytosis, surface receptor fate determination, exosome composition, and metabolism.

The vast majority of cancer deaths are caused by the primary tumor metastasizing into other organs. Invasion is a prerequisite for metastasis formation, and for this reason, the inhibition of invasion could efficiently prevent metastasis formation. For this, targeting the molecules regulating invasion could be useful. One of these is an oncogenic transcription factor, myeloid zinc finger 1 (MZF1), as reviewed by Brix et al. [9]. P21 activating kinase 4 (PAK4) is another kinase that is often activated in cancers. It is also a kinase that can activate MZF1 by phosphorylating it in response to ErbB2 activation. PAK4 is considered to be a good target for the treatment of a variety of solid cancers, including breast cancer, where its inhibition for this purpose has been patented, and PAK4 inhibitors have reached clinical trials. The identification of MZF1 as an oncogenic target of PAK4, whose activity is important for the invasiveness of ErbB2 positive breast cancer cells, suggests that PAK4 inhibitors might be useful for the treatment of cancers whose aggressiveness depends on MZF1. In general, more research is still needed to increase our understanding of the detailed function of MZF1 in cancer, of the cellular cancer-promoting programs it regulates, the cancers where its inhibition would be most beneficial, and how it should be achieved.

Nuclear protein localization protein 4 (NPL4) functions as an essential chaperone that regulates microtubule structures when a cell re-enters interphase. Majera et al. [10] provide evidence of targeting of NPL4 by disulfiram (tetraethylthiuram disulfide, DSF), a drug commercially known as Antabus and commonly used for alcohol-aversion. DFS metabolizes rapidly to a diethyldithiocarbamate–copper complex, CuET, which is highly toxic to cancer cells via a mechanism that is likely to involve the immobilization and inactivation of NPL4 via its CuET-induced aggregation. CuET is especially toxic to cells that are lacking functional breast cancer genes 1 and 2 (BRCA1 and BRCA2), interfering with their DNA replication and causing replication stress. While the exact mechanism of how DSF kills BRCA1 and 2 deficient cells is not known, it is likely to involve increased NPL4–CuET-induced, supra-threshold DNA replication stress and the concomitant CuET-induced activation of the ATR-Chk1 signaling pathway.

Cancer stem cells (CSCs) are often responsible for therapeutic resistance. The study by Choi et al. [11] presents sulconazole—an antifungal medicine in the imidazole class—as a way to inhibit cell proliferation, tumor growth, and CSC formation. Part of this effect could be via the transcription factor nuclear factor kappa-light-chain-enhancer of activated B cells (NF-κB), whose expression level is decreased upon sulconazole treatment, leading to the decreased expression of interleukin-8 (IL-8). Sulconazole treatment also reduces the number of cells expressing CSC markers (CD44^{high}/CD24^{low}) as well as aldehyde dehydrogenase (ALDH), among others, suggesting that additional factors can be targeted by sulconazole. NF-κB/IL-8 signaling is important for CSC formation and may be an important therapeutic target for treatment targeting breast cancer stem cells.

The formation of three-dimensional (3D) multicellular spheroids (MCS) in microgravity, mimicking tissue culture conditions, was used by Melnik et al. [12] as a method for growing the metastatic follicular thyroid carcinoma cell line (FTC)-133. For this study, they utilized a random positioning machine with which the cells can be induced to detach from an already established cellular network to form 3D spheroids, which were then used as an ex vivo model system to mimic micro-metastases formation. In these conditions, the authors provide evidence that dexamethasone can prevent spheroid formation, which has previously been reported to involve NF-κB, but actually most likely functions via several different unidentified target molecules, suggesting that instead of one target, several targets should be inhibited.

A study by Gruber et al. [13] on the other hand was conducted on 278 primary tumor samples from patients with resectable esophageal cancer. This study identifies ALL1 fused gene from chromosome 1q (AF1q), a cofactor for the transcription factor 7 (TCF7), as a nuclear protein that links two important oncogenic signaling pathways activated in many cancers: WNT and signal transducer and activator of transcription 3 (STAT3). Its expression is increased in cancer progression and it is connected to the increased expression of WNT and STAT3 targets cluster of differentiation 44 (CD44) and tyrosine 705 phosphorylated STAT3 (pYSTAT3) in esophageal cancer. Patients with AF1q-positive esophageal cancer relapsed and died earlier than those with AF1q-negative disease, suggesting that AF1q could act as a cofactor to boost the transcription of CD44 and pYSTAT3, and thus implicating its involvement in the regulation of the aggressiveness of esophageal cancer and identifying it as a potential novel target molecule.

Funding: The Danish Cancer Society Scientific Committee (KBVU) (R124-A7854-15-S2 and R56-A3108-12-S2), Novo Nordisk Foundation (NNF15OC0017324) and the Danish National Research Foundation (DNRF125).

Conflicts of Interest: The authors declare no conflict of interest.

References

1. World Health Organization. Key Statistics. Available online: https://www.who.int/cancer/resources/keyfacts/en/ (accessed on 17 September 2020).

2. Lichtenstern, C.R.; Ngu, R.K.; Shalapour, S.; Karin, M. Immunotherapy, Inflammation and Colorectal Cancer. *Cells* **2020**, *9*, 618. [CrossRef] [PubMed]

3. Khazamipour, N.; Al-Nakouzi, N.; Oo, H.Z.; Orum-Madsen, M.; Steino, A.; Sorensen, P.H.; Daugaard, M. Oncofetal Chondroitin Sulfate: A Putative Therapeutic Target in Adult and Pediatric Solid Tumors. *Cells* **2020**, *9*, 818. [CrossRef] [PubMed]

4. Magaway, C.; Kim, E.; Jacinto, E. Targeting mTOR and Metabolism in Cancer: Lessons and Innovations. *Cells* **2019**, *8*, 1584. [CrossRef] [PubMed]

5. Smorodinsky-Atias, K.; Soudah, N.; Engelberg, D. Mutations That Confer Drug-Resistance, Oncogenicity and Intrinsic Activity on the ERK MAP Kinases-Current State of the Art. *Cells* **2020**, *9*, 129. [CrossRef] [PubMed]

6. Ambjorner, S.E.B.; Wiese, M.; Kohler, S.C.; Svindt, J.; Lund, X.L.; Gajhede, M.; Saaby, L.; Brodin, B.; Rump, S.; Weigt, H.; et al. The Pyrazolo [3,4-d] pyrimidine Derivative, SCO-201, Reverses Multidrug Resistance Mediated by ABCG2/BCRP. *Cells* **2020**, *9*, 613. [CrossRef] [PubMed]

7. Flinck, M.; Hagelund, S.; Gorbatenko, A.; Severin, M.; Pedraz-Cuesta, E.; Novak, I.; Stock, C.; Pedersen, S.F. The Vacuolar H(+) ATPase alpha3 Subunit Negatively Regulates Migration and Invasion of Human Pancreatic Ductal Adenocarcinoma Cells. *Cells* **2020**, *9*, 465. [CrossRef] [PubMed]

8. Peulen, O.; Rademaker, G.; Anania, S.; Turtoi, A.; Bellahcene, A.; Castronovo, V. Ferlin Overview: From Membrane to Cancer Biology. *Cells* **2019**, *8*, 954. [CrossRef]

9.   Brix, D.M.; Bundgaard Clemmensen, K.K.; Kallunki, T. Zinc Finger Transcription Factor MZF1-A Specific Regulator of Cancer Invasion. *Cells* **2020**, *9*, 223. [CrossRef]

10. Majera, D.; Skrott, Z.; Chroma, K.; Merchut-Maya, J.M.; Mistrik, M.; Bartek, J. Targeting the NPL4 Adaptor of p97/VCP Segregase by Disulfiram as an Emerging Cancer Vulnerability Evokes Replication Stress and DNA Damage while Silencing the ATR Pathway. *Cells* **2020**, *9*, 469. [CrossRef] [PubMed]

11. Choi, H.S.; Kim, J.H.; Kim, S.L.; Lee, D.S. Disruption of the NF-kappaB/IL-8 Signaling Axis by Sulconazole Inhibits Human Breast Cancer Stem Cell Formation. *Cells* **2019**, *8*, 1007. [CrossRef] [PubMed]

12. Melnik, D.; Sahana, J.; Corydon, T.J.; Kopp, S.; Nassef, M.Z.; Wehland, M.; Infanger, M.; Grimm, D.; Kruger, M. Dexamethasone Inhibits Spheroid Formation of Thyroid Cancer Cells Exposed to Simulated Microgravity. *Cells* **2020**, *9*, 367. [CrossRef] [PubMed]

13. Gruber, E.S.; Oberhuber, G.; Birner, P.; Schlederer, M.; Kenn, M.; Schreiner, W.; Jomrich, G.; Schoppmann, S.F.; Gnant, M.; Tse, W.; et al. The Oncogene AF1Q is Associated with WNT and STAT Signaling and Offers a Novel Independent Prognostic Marker in Patients with Resectable Esophageal Cancer. *Cells* **2019**, *8*, 1357. [CrossRef] [PubMed]

Review

Immunotherapy, Inflammation and Colorectal Cancer

Charles Robert Lichtenstern [1,2], Rachael Katie Ngu [1,2], Shabnam Shalapour [1,2,*] and Michael Karin [1,2,3]

1   Department of Pharmacology, School of Medicine, University of California, San Diego, La Jolla, CA 92093, USA; karinoffice@health.ucsd.edu
2   Laboratory of Gene Regulation and Signal Transduction, Department of Pharmacology, School of Medicine, University of California, San Diego, La Jolla, CA 92093, USA
3   Moores Cancer Center, University of California, San Diego, La Jolla, CA 92093, USA
*   Correspondence: sshalapour@health.ucsd.edu

Received: 30 January 2020; Accepted: 3 March 2020; Published: 4 March 2020

Abstract: Colorectal cancer (CRC) is the third most common cancer type, and third highest in mortality rates among cancer-related deaths in the United States. Originating from intestinal epithelial cells in the colon and rectum, that are impacted by numerous factors including genetics, environment and chronic, lingering inflammation, CRC can be a problematic malignancy to treat when detected at advanced stages. Chemotherapeutic agents serve as the historical first line of defense in the treatment of metastatic CRC. In recent years, however, combinational treatment with targeted therapies, such as vascular endothelial growth factor, or epidermal growth factor receptor inhibitors, has proven to be quite effective in patients with specific CRC subtypes. While scientific and clinical advances have uncovered promising new treatment options, the five-year survival rate for metastatic CRC is still low at about 14%. Current research into the efficacy of immunotherapy, particularly immune checkpoint inhibitor therapy (ICI) in mismatch repair deficient and microsatellite instability high (dMMR–MSI-H) CRC tumors have shown promising results, but its use in other CRC subtypes has been either unsuccessful, or not extensively explored. This Review will focus on the current status of immunotherapies, including ICI, vaccination and adoptive T cell therapy (ATC) in the treatment of CRC and its potential use, not only in dMMR–MSI-H CRC, but also in mismatch repair proficient and microsatellite instability low (pMMR-MSI-L).

Keywords: colorectal cancer; immunotherapy; inflammation; microsatellite instability

1. Introduction

Colorectal cancer (CRC) is the third most common cancer type and a leading cause of mortality among cancer-related deaths in the United States [1]. While scientific and clinical advances in early detection and surgery have led to five-year survival rates of 90% and 71% for localized and regionalized CRCs, respectively, the five-year survival rate for metastatic CRC is low, remaining at around 14% [2]. Moreover, 25% of CRC patients display metastasis at diagnosis, and roughly 50% of those treated will eventually develop metastasis during their lifetime [3]. These alarming statistics can most likely be attributed to the ineffectiveness of standard treatment regimens, and thus indicates an urgent need for the development of more effective treatment options. Immunotherapy, a treatment option that takes advantage of the body's own immune system to attack cancer, has shown promise in the treatment of certain cancers [4–7]. Whereas some cancers, such as melanoma and lung cancer, respond well to immune checkpoint inhibitor therapy (ICI), others do not.

More recently, ICIs were found effective in a specific subset of CRC that is mismatch-repair-deficient (dMMR) and microsatellite instability-high (MSI-H) (referred to as dMMR-MSI-H tumors) and ineffective in subsets that are mismatch-repair-proficient (pMMR) and microsatellite instability-low

(MSI-L) (referred to as pMMR-MSI-L tumors) [8]. This Review will serve to discuss recent findings in the effectiveness of immunotherapies in the treatment of CRC, both localized and metastatic, from clinical trials and experimental models, and its potential use in pMMR-MSI-L tumors and other CRC subsets.

2. Origins of CRC

CRC can originate from a multitude of intrinsic and extrinsic factors, including an accumulation of new mutations, pre-existing mutations, and susceptibility alleles associated with family history, or chronic, lingering inflammation, as described in Figure 1. The majority (75%) of CRCs are sporadic, meaning family history is not involved in their pathogenesis [9]. Common mutations in tumor suppressor genes and oncogenes that give rise to CRC include adenomatous polyposis coli (*APC*), tumor protein 53 (*TP53*), and Kirsten rat sarcoma (*KRAS*), which are present in 81%, 60% and 43% of the cases of sporadic CRCs, respectively [10]. The role of these genetic alterations in the pathogenesis of CRC has been extensively reviewed [11–13]. Most CRC-inducing mutations act in a particular order, controlling the adenoma–carcinoma sequence, which describes the progression of a normal intestinal epithelia to an adenoma, invasive carcinoma, and eventual metastatic tumor [14,15].

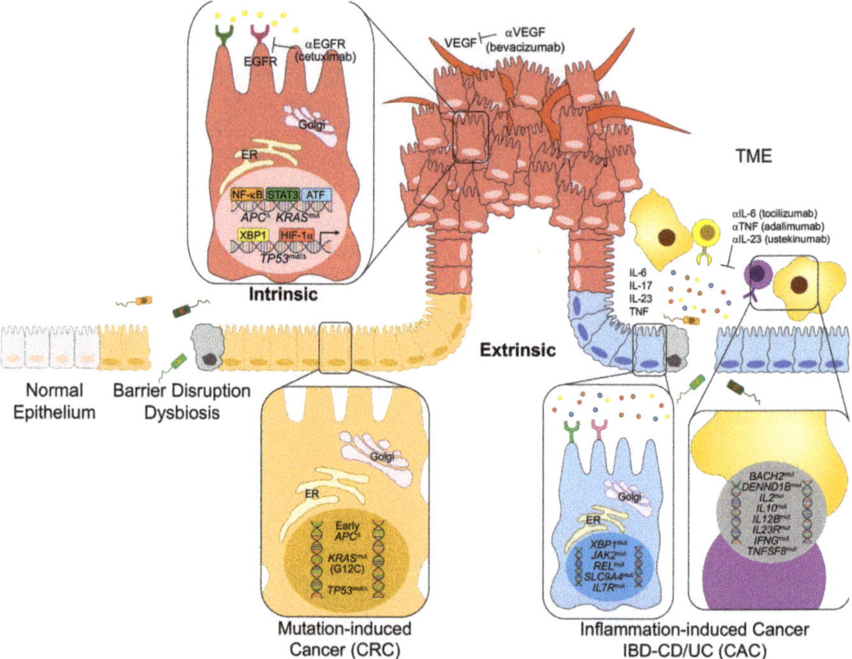

Figure 1. Intrinsic and extrinsic factors contributing to the pathogenesis of colorectal cancer (CRC). CRC can develop from a multitude of both intrinsic and extrinsic factors. Extrinsic factors, including inflammation from hyperactivated immune cells, the release of proinflammatory cytokines, or gut dysbiosis, can lead to an inflammatory and possibly premalignant environment. Intrinsic factors include sporadic mutations, such as those leading to mutation-induced CRC (sporadic CRC). Similarly, precancerous mutations, or mutations induced by prior inflammation, can lead to colitis-associated cancer (CAC), a specific subset of CRC stemming from chronic inflammation caused by inflammatory bowel disease (IBD), specifically ulcerative colitis (UC) or Crohn's disease (CD).

Family history is implicated in approximately 10–30% of CRCs [16,17]. For example, familial adenomatous polyposis (FAP) and hereditary nonpolyposis colorectal cancer (Lynch syndrome) are the most commonly inherited CRC syndromes, and account for 2–4% and 1% of CRC cases, respectively [17].

Although 96% of all CRCs do not develop in the context of pre-existing inflammation, the roles of chronic inflammation, tumor-elicited inflammation, the tumor microenvironment (TME), and partially adaptive immune cells in CRC development, have been established, particularly in the context of their interaction with gut dysbiosis [18–23]. Colitis-associated cancer (CAC) is a specific subset of CRC characterized by its implication with inflammation that accounts for 1%–2% of all CRCs [24]. CAC, originating from either the chronic inflammation in both the colon and the small intestine, or solely the colon, as is the case of Crohn's disease (CD) or ulcerative colitis (UC), respectively, is classified by the excessive activation and recruitment of immune cells that produce inflammatory cytokines, such as TNF, IL-17, IL-23 and IL-6, that lead to the propagation of an inflammatory and possibly premalignant environment [25]. Mutations involved in inflammatory bowel disease (IBD) development include genes that regulate immune activation and the subsequent response, such as *IL12B, IL2, IFNG, IL10, TNFSF8, TNFSF15, IL7R, DENND1B, JAK2* and those that also regulate ER stress, glucose, bile salt transfer and organic ion transporter, including *XBP1, SLC9A4, SLC22A5* and *SCL11A1*, as shown in Figure 1 [26]. Both CRC and CAC exhibit inflammatory microenvironments, but the order in which inflammation and tumorigenesis occur seems to be different. In CRC, inflammation follows tumorigenesis. Mutations due to environmental factors initiate tumor development in CRCs, and the subsequent activation of inflammatory cells can induce further DNA damage through the production of reactive oxygen species (ROS) and reactive nitrogen intermediates (RNIs) [25,27]. On the other hand, inflammation precedes tumorigenesis in CAC. Inflammation induced by the activation of immune cells and their release of proinflammatory cytokines can induce DNA damage and mutations in CAC [25]. Correspondingly, both CRC and CAC may entail similar mutations, but the timing and order of these mutations are different, as displayed by early *APC* and late *TP53* mutations in CRC, and early *TP53* and late *APC* mutations in CAC [28–30]. Another important contributor to CRC emergence is so-called tumor-elicited inflammation driven by the loss of normal barrier function as a result of *APC* inactivation [18].

3. Mismatch Repair Deficiency and Microsatellite Instability in CRC

dMMR or MSI-H exists in about 15% of all cases of CRC, but only in 4% of metastatic CRC, as opposed to pMMR or MSI-L, which is present in roughly 85% of all cases of CRC. MSI occurs in both spontaneous CRC and IBD-induced CAC, although the rates and timing at which MSI occurs are similar in both malignancies [31].

Microsatellites are repetitive DNA sequences that can experience a sudden and prolonged change in size, due to errors during DNA replication, such as the formation of small loops in the DNA strands, leading to MSI-H [32]. These errors are combated by the mismatch repair (MMR) system, an ancient mechanism used to correct insertions, deletions, or mismatched bases that are generated by the erroneous loops that form during DNA replication [32–34]. However, if there is a dysfunction or mutation in the MMR system, referred to as dMMR, these errors are left uncorrected, allowing them to be integrated into the DNA permanently [32]. Thus, MSI-H tumors have varied lengths of microsatellites (compared to MSI-L) due to errors in the MMR system, as shown in Figure 2.

The MMR system relies on the DNA repair genes *MLH1, MSH2, PMS1, MSH6, PMS2* and *MSH3*, all of which are involved in correcting mismatched or wrongly inserted or deleted bases in DNA [32,35]. Loss, inactivation, or the silencing of any one of these genes, classifies a patient as dMMR. More importantly, errors in this repair system lead to a high mutational profile, which explains why dMMR tumors have an average mutational profile of 1782, compared to 73 for pMMR tumors [36].

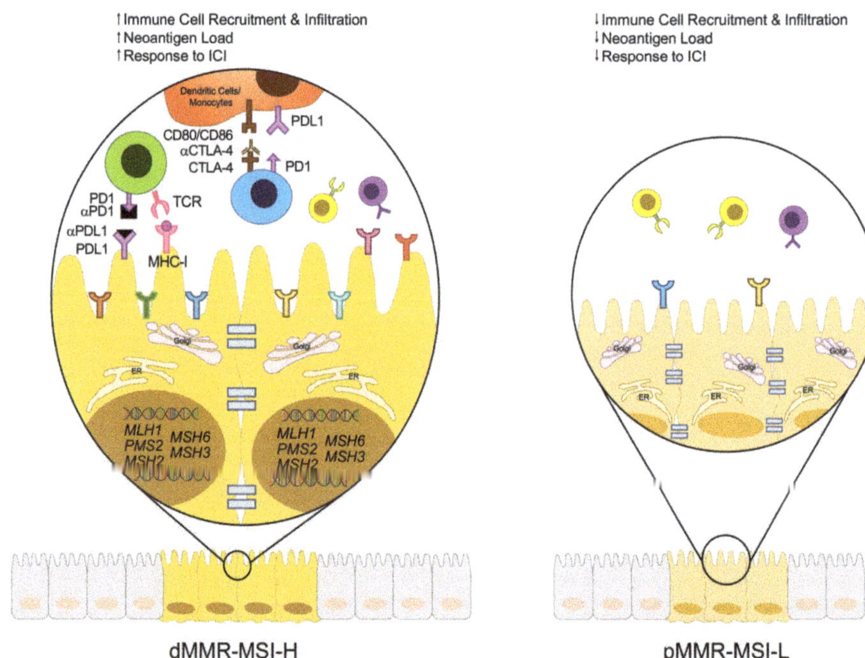

Figure 2. Immuno-landscape of dMMR-MSI-H and pMMR-MSI-L CRC. CRC can be classified into two subsets based on its MMR/MSI status. The DNA MMR system relies on key genes, such as MLH1, MSH2, MSH6, PMS2, or MSH3, that correct mismatched or wrongly inserted or deleted bases in the DNA. If this machinery fails due to defects in one or more of the repair genes, these errors are free to be integrated into the DNA permanently, forming microsatellites. Thus, dMMR-MSI-H tumors are those that have a defect in one of the major DNA repair genes (dMMR), resulting in high levels of microsatellites (MSI-H). On the other hand, pMMR-MSI-L tumors have a functional MMR system (pMMR), resulting in low or stable levels of microsatellites (MSI-L). The result of this damaged repair system in dMMR-MSI-H tumors is a higher mutational burden, which correlates with a higher expression of neoantigens on MHC-I molecules.

As the identification and classification of CRCs is necessary and crucial for proper diagnoses and treatments, methods have been practiced in order to detect MSI. Current methods include the amplification and examination of polymerase chain reaction (PCR) products from commonly affected microsatellite markers in tumors [34,37,38]. These markers include two mononucleotide repeat markers (BAT-25 and BAT-26) and three dinucleotide repeat markers (D2S123, D5S346, and D17S250) [37]. MSI-H status is classified if instability is present in two or more of the markers, whereas the MSI-L status is classified if instability is only detected in one of the markers. More recently, however, are methods that use DNA-sequencing technology for MSI detection and classification on the same markers [33,39,40]. Regardless of the screening method, albeit some more efficient and accurate than others, classification of MSI status in regard to the CRC subtype is of the upmost importance for proper treatment planning, and should be one of the primary steps when diagnosing patients.

4. Classical Treatment Options

CRC treatment can be divided into two main treatment categories: neoadjuvant and adjuvant. Neoadjuvant therapy refers to therapeutics that are given before the main cancer treatment, usually surgery, whereas adjuvant refers to that which is given after or in combination with the main cancer treatment. Neoadjuvant therapy offers many clinical benefits, in that it can potentially lessen the severity of the malignancy, through eliminating early metastatic tumors, preventing complications during surgery, and allowing for a more accurate plan for adjuvant therapy (if necessary), based on the subsequent response to neoadjuvant therapy [41–43]. Most studies have shown that neoadjuvant chemotherapy may improve overall survival, depending on the severity and stage of the disease [41,44–46].

Chemotherapy is usually the first line of defense in the treatment of CRC. 5-fluorouracil (5-FU), the most common of the chemotherapeutic agents for CRC, acts through inhibition of thymidylate synthase, which converts deoxyuridine monophosphate (dUMP) to deoxythymidine monophosphate (dTMP), causing DNA damage [47]. While it is relatively effective in early disease stages, response rates in metastatic CRC are only 10–15% [47,48]. On the other hand, combinatorial chemotherapeutic regimens consisting of 5-FU, in combination with oxaliplatin (FOLFOX) or irinotecan (FOLFIRI), have heightened response rates to 40–50% [47]. Studies into the usefulness of using MMR/MSI status as a predictor of responsiveness to chemotherapy have shown mixed results, depending on the stage of the disease and the specific type of chemotherapy, thus explaining the necessity for a more reliable and dependable treatment option for these CRC subsets [49–53].

More recently introduced are the targeted therapies, including monoclonal antibodies against epidermal growth factor receptor (EGFR) and vascular endothelial growth factor (VEGF), which inhibit cancer cell proliferation and angiogenesis, respectively. Bevacizumab, a monoclonal antibody against VEGF, was shown to improve the survival of patients with metastatic CRC in combination with 5-FU [54] and oxaliplatin-based therapies [55]. Moreover, patients with irinotecan- [56] and fluoropyrimidine- and oxaliplatin-resistant [57] CRCs were shown to have improved response rates when treated with cetuximab, a monoclonal antibody against EGFR, alone or in combination with irinotecan. Extensive research has shown *KRAS* mutational status to be a predictor of non-responsiveness to EGFR inhibitors [58–60]. It was found that patients with pMMR tumors that had mutations in *BRAF* or *KRAS*, had worse survival rates than patients with pMMR tumors free of these mutations, and patients with dMMR tumors [61]. Despite major scientific and clinical research into targeted therapies, patients that do respond to EGFR inhibitors only show improvements for 3–12 months before disease progression, suggesting that this specific therapy is not conducive to long term survival and remission [56,58,62,63]. This obstacle has paved the way for research into the efficacy of immunotherapy in the treatment of CRCs.

5. Role of Immune Cells and Tumor Microenvironment in the Classification of CRC

A positive correlation is seen between tumoral $CD3^+$ and $CD8^+$ T cell densities and the risk of recurrence, disease-free survival rate, and the overall survival rate in patients with different stages of CRC [64]. This is in accordance and supports evidence which shows that increased amounts of tumor-infiltrating lymphocytes correlate with an improved clinical outcome and prognosis [65–68]. Both dMMR-MSI-H and pMMR-MSI-L tumors have distinctly different TME makeups and distributions of immune cell populations, contributing to the variation in response rates to therapy, treatment targets and clinical prognoses [69–71]. Comparison of the makeup of the TME shows a higher expression of cytotoxic, Th1, Th2, $CD8^+$ T and follicular helper (Tfh) cell markers, in addition to macrophages and B cells in dMMR-MSI-H tumors than pMMR-MSI-L tumors [69,72]. Some of these immune cells can mediate antitumor immune responses, thus explaining why dMMR-MSI-H tumors have better response rates and clinical outcomes [73]. Higher mutational load in dMMR-MSI-H tumors correlates with the higher expression of neoantigens on major histocompatibility complex (MHC)-I molecules, thus recruiting more cytotoxic $CD8^+$ T cells for the subsequent immune response and tumor destruction, which follows the notion that frameshift mutations positively correlate with $CD8^+$ T cell infiltration in CRCs [74,75].

Since T cell infiltration is representative of a better clinical outcome in CRC patients, it is clear why dMMR-MSI-H tumors respond well to ICI, and pMMR-MSI-L tumors do not [76,77].

In addition to the wide variety of immune cells distributed throughout the TME, there are also many cytokines and other molecules secreted by these cells that have specific roles in inflammation, immunity and CRC development. These cytokines can have both antitumorigenic properties, such as interferon-gamma (IFN-γ) and granulysin, or pro-tumorigenic properties, such as IL-6, IL-23 and IL-17. IFN-γ [78] and granulysin [79] bolster and induce MHC-I antigen processing and presentation machinery, and they also recruit antigen presenting cells to stimulate tumor destruction, thus showcasing their antitumorigenic functions, and as so, are overexpressed in dMMR-MSI-H tumors [69]. Induction of proinflammatory cytokines originates as a result of NF-κB and STAT3 activation in epithelial cells, and serves an important role in supporting colorectal tumorigenesis [80–82]. IL-6 is overexpressed in CRC [83–85], and serves a pro-tumorigenic function through multiple processes, including bolstering angiogenesis through an enhanced expression of VEGF [86], protecting both healthy and malignant intestinal epithelial cells (IECs) from damage-associated molecular patterns (DAMPS), and pathogen-associated molecular patterns (PAMPS), by supporting their growth and survival [87–91], along with bolstering defects in the DNA MMR system [92]. Ablation of IL-6 in the dextran sodium sulfate/azoxymethane (DSS/AOM) mouse model of CRC resulted in diminished tumorigenesis, thus confirming its pro-tumorigenic properties [87]. Both IL-23 and IL-17 have also been implicated in the pathogenesis of CRC in human and murine models. An upregulation of IL-17 and IL-23 expression was found in tumors excised from the CPC-APC mouse model of CRC [18]. IL-23 enhances the production of IL-17, and IL-17 activates NF-κB which stimulates the proliferation and survival of IECs, resulting in accelerated colorectal tumorigenesis [18,19,80]. Correspondingly, the elevated expression of IL-6, IL-23 and IL-17 in CRC correlates with a worse prognosis and clinical outcome [93]. The role of other immunomodulatory cytokines involved in CRC has also been discussed [81,82,94]. The presence of a wide variety of immune cells and other cytokines and signaling molecules in the TME provide important topics for future research, but most importantly can serve as new possible targets for immunotherapy.

Moreover, CAC presents a different immuno-profile compared to CRC, which may increase the responsiveness to immunotherapy. However, it may also increase the risk for immunopathological side effects.

6. Why Immunotherapy?

Immunotherapy, particularly ICI, has revolutionized cancer treatment, and although response rates rarely exceed 20%, those who do respond show a durable response [95–97]. The responsiveness to ICI was suggested to depend on several key factors, including mutational load (high levels of tumor neoantigens), tumor-infiltrating lymphocytes and regulatory checkpoint receptors. ICI, a specific type of immunotherapy, functions through inhibiting negative regulatory receptors, such as cytotoxic T lymphocyte antigen 4 (CTLA4) and programmed cell death 1 (PD-1), on T cells, and thereby boosts antitumor immune responses [98–101]. T cells enable the immune system to recognize foreign antigens through an interaction between their T cell receptors (TCR) and peptide epitopes presented by MHC-I molecules on tumor cells [102,103]. Thus, it was suggested that cancers that are characterized by high mutational profiles can produce and present more neoantigens via their MHC-I molecules, and thereby lead to recognition, T cell activation and eventual self-destruction [8,104,105]. However, these effector T cells can become exhausted due to prolonged antigen stimulation, or through an interaction between their surface PD-1 with PD-L1 expressed by immune cells or tumor cells, or their surface CTLA-4 with CD80/CD86 expressed by dendritic cells, which are professional antigen-presenting cells (DC-APC) [101]. Inhibition of these interactions has been observed to partially reactivate exhausted T cells and induce tumor regression [106]. Higher response rates in non-small cell lung cancer (NSCLC) [104,107] and melanoma [108–110] have been attributed to the higher mutational loads in these tumor types [111].

However, for some tumors with lower inflammation and T cell infiltration, which could be due to defects on priming or the absence of high affinity T cells, vaccinations or more specific approaches like adoptive T cell therapy (ATC), which are specific for a particular mutated antigen, may prove favorable options in combination with ICI.

Therapeutic cancer vaccines can induce an immune response through a direct stimulation of the immune system by delivering antigens to DC-APC, which prime and activate CD4$^+$ and CD8$^+$ T cells to initiate tumor destruction [112]. Therapeutic cancer vaccines can target tumor-associated antigens (TAAs) or tumor-specific antigens (TSAs). TAAs are self-antigens expressed in both tumor and normal cells, however T cells that bind to these self-antigens can be removed from the immune system through immunotolerance mechanisms [113]. On the other hand, TSAs are unique to tumor cells, and can strongly induce an immune response through the binding and activation of T cells [113]. However, cancer vaccines comprising TSAs present a noticeable limitation, the necessity for a personalized vaccine specific to the individual's particular tumor neoantigen.

Moreover, ATC provides tumor-antigen-specific approaches that have been shown to have promising results, and may be useful when the neoantigen load is lower, or if information regarding this neoantigen load is unavailable [114]. For example, CD8 T cells targeting mutant KRAS [115] or TP53 "Hotspot" Mutations [116] have been identified. Moreover, circulating PD-1$^+$ lymphocytes have recently been shown to recognize human gastrointestinal cancer neoantigens [117].

7. Is There a Place for Immunotherapy in CRC?

Initially, ICI was not considered a viable treatment option for CRC. An initial phase II study assessed the efficacy of tremelimumab, a monoclonal antibody against CTLA4, in patients with previous treatment-refractory CRC, which resulted in no improvement post-treatment [118]. Furthermore, two phase I studies of anti-PD-1 [119] and anti-PD-L1 [120] antibodies in previously-treated CRC patients produced no responses. Unfortunately, the MMR/MSI status of the patients in both of these studies was unknown, compromising the interpretation of the results. Indeed, a subsequent phase I clinical trial of an anti-PD-1 antibody (MDX-1106) in patients with a variety of treatment-resistant tumors, including one patient with CRC, culminated in the patient achieving a durable complete response [121]. In accordance with the understanding that the response to ICI may correlate with mutational burden, Le et al. postulated that CRC tumors that are characterized by high mutational burdens due to mismatch–repair deficiencies may respond to ICI [36]. The results of the study showed that patients with dMMR-MSI-H tumors had a 40% objective response rate when treated with pembrolizumab, as compared to 0% for patients with pMMR-MSI-L tumors, and also exhibited 78% immune-related progression-free survival [36]. Importantly, these results suggested that the MMR/MSI status can be an accurate predictor of responsiveness to ICI using pembrolizumab.

Currently, a plethora of clinical trials aim to further examine ICIs in combination with a variety of other therapeutics in the treatment of CRC. Progress has led to United States Food and Drug Administration (FDA) approval of pembrolizumab and nivolumab in patients with dMMR-MSI-H CRC. Approval of pembrolizumab followed the results of the aforementioned study, being the first FDA approval based on a genetic biomarker of a particular tumor type [36]. Approval of nivolumab in patients with dMMR-MSI-H CRC followed the results of CheckMate-142, which showed a 31% objective response rate and 73% twelve month overall survival rate in treatment-resistant dMMR-MSI-H CRC [122]. This same trial also examined the efficacy of the combination of nivolumab and ipilimumab in treatment-resistant dMMR-MSI-H CRC, resulting in a 55% objective response rate and 85% twelve month overall survival rate [123]. The results of this study paved the way for FDA approval of that ICI combination in treatment-resistant dMMR-MSI-H CRC.

As the responsiveness to immunotherapy is generally associated with mutational load, as discussed previously, and dMMR-MSI-H patients comprise high mutational profiles, vaccinations targeting individuals' unique neoantigens may prove to be effective, specifically in dMMR-MSI-H patients. In a murine model of induced dMMR by knockout of MLH1, vaccination extended overall survival and reduced the tumor burden, proving that vaccination can be a viable option for treatment in mouse models of dMMR [124].

Similarly, human clinical trials of therapeutic cancer vaccines have shown promising results depending on MSI status [125,126]. Ultimately, the main question to be determined is whether the combination of ICI and vaccination may prove to be more efficacious than ICI alone in dMMR-MSI-H CRC, or have the ability to elicit a response in pMMR-MSI-L CRC, which is unresponsive to ICI alone.

8. ICI-Resistance in pMMR-MSI-L CRC

Despite its effectiveness in dMMR-MSI-H CRC, ICI is not effective in pMMR-MSI-L CRC. The lack of response of pMMR-MSI-L tumors to ICI has been suggested to trace back to the diminished antitumor immune response, due to the inability for recognition by immune cells as a result of the low mutational profile of these tumors. This lack of response was also shown to be consistent in mouse models, as mice injected with MSI-H CRC experienced greater tumor regression and T cell infiltration than MSI-L or MSI-intermediate CRC when treated with anti-PD-1 therapy [127].

Although MSI-L tumors do not respond to ICI, higher T cell infiltration in MSI-L CRC is correlated with better disease free survival, indicating that some of these tumors can be recognized by T cells [64]. Thus, the main question to be discussed is whether MSI-L CRC utilizes other mechanisms to escape immunorecognition. Perhaps those patients with higher T cell infiltration can be selected for responsiveness to ICI.

One phase 3 trial examined the combination of cobimetinib, an MEK inhibitor, with atezolizumab, an anti-PD-L1 monoclonal antibody, in patients with metastatic CRC [128]. MEK inhibition resulted in increased amounts of tumor-infiltrating $CD8^+$ T cells, and the combination with anti-PD-L1 treatment potentiated tumor regression in mouse models [129]. Despite the promising data in mouse models, the phase 3 trial failed to reach improved response or survival [128], leading to the conclusion that even when combined with MEK inhibitors, anti-PD-(L)1 is not effective in low immunoscore tumors, such as pMMR-MSI-L.

9. Conclusion: Thoughts, Obstacles, and Future Possibilities

CRC is a highly multifaceted and complex disease with an extensive mutational signature and an intricate TME. Just as complex as the disease itself, are the therapies used to combat it. Despite ICI's initial effectiveness in patients exhibiting dMMR-MSI-H tumors, not all dMMR-MSI-H responds to ICI, and as of yet, no response is seen in pMMR-MSI-L. This has led to the necessity for new combinatorial targets that can be used to further bolster the response or lack of response to ICI in these two CRC subsets, as described in Figure 3. Recent advances in the development of new CRC therapeutics include AMG 510, a KRAS(G12C) inhibitor [130]. Analysis of AMG 510 in mouse models with $KRAS^{G12C}$-injected tumors resulted in tumor regression, and combining this molecule with chemotherapy (carboplatin) or ICI (anti-PD-1) resulted in a further increase in tumor regression [130]. Analysis of these tumors showcased increased amounts of $CD8^+$ T cells, macrophages and DC-APC in both the AMG 510 alone, and combination with anti-PD-1 treatment groups [130]. Clinical trials with AMG 510 in four patients with NSCLC resulted in objective partial responses and stable disease in two patients each [130]. The two partial responders were unresponsive to previous chemotherapy and ICI treatment, but exhibited tumor reduction of 34% and 67% when treated with AMG 510 [130]. Overall, this data suggests that AMG 510 may have the ability to induce T cell recruitment, and thus potentiate antitumor immunity. Whether AMG 510 can be combined with ICI in the treatment of CRC remains to be seen.

Cancer vaccines are a rapidly expanding immunotherapeutic approach that also seeks to exploit the body's immune system to fight cancer. Therapeutic cancer vaccines can stimulate and activate T cells to initiate an immune response through the detection of TAAs or TSAs specific to the individuals' tumors. Cancer vaccinations have shown mixed results in different stages of CRC, and more research is needed to truly uncover benefits [131–133]. Moreover, studies have shown that vaccinations may be efficacious in dMMR-MSI-H tumors, but not pMMR-MSI-L [126]. More interestingly, recent studies suggest that the combination of both ICI and cancer vaccinations may result in an improved response in some cancers, but not others [134–144]. Expanding on the possible immunotherapeutic options available, ATC may also provide a possible route in treating specific pMMR-MSI-L CRCs comprising KRAS mutations [145]. However, more research is warranted to determine its effectiveness in CRC, particularly dMMR-MSI-H and pMMR-MSI-L CRC in combination with ICIs. Just as CRC followed melanoma and NSCLC in its application in ICI therapy, CRC may be the next poster boy for vaccination and ICI combination-based therapy.

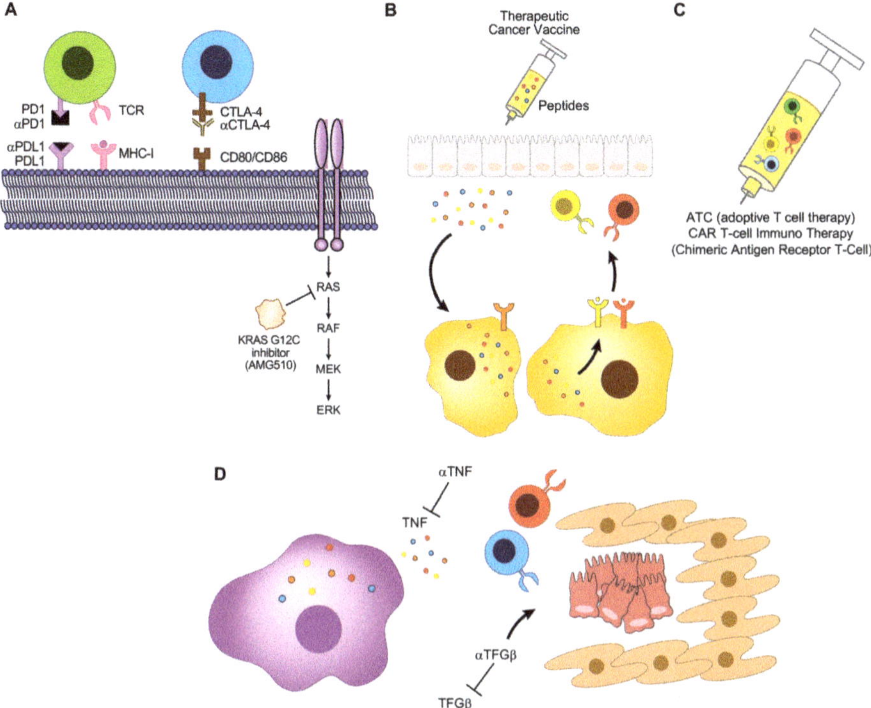

Figure 3. The future of CRC therapy: combinatorial agents. The current status of the use of inhibitor therapy (ICI) in the treatment of CRC has shown promising results, despite the lack of a complete response in dMMR-MSI-H tumors, and no response in pMMR-MSI-L. This obstacle has paved the way for insight and research into plausible combinatorial agents that can overcome this scientific impediment. (**A**) ICI in combination with AMG 510, a KRAS (G12C) inhibitor, or (**B**) therapeutic cancer vaccines, or (**C**) adoptive T cell therapy, or (**D**) TNF and TGFβ inhibitors may serve as the next candidates for combinatorial therapy with ICI.

There are also a number of other factors that can modulate the response to ICI. The gut microbiome has been implicated in variations in response rates to ICI. The presence of particular microbiota seems to be correlated with a heightened response to ICI, depending on the strain and cancer type [146–150].

A possible adverse side effect common with ICI is immune-related colitis, which is often treated with antitumor necrosis factor α (TNF) antibodies. Such treatment in combination with ICI has been shown to improve antitumor immune responses and the severity of colitis in mouse models [151,152].

Furthermore, transforming growth factor β (TGFβ) signaling has been shown to cause resistance to ICI, and inhibition of TGFβ signaling in combination with ICI led to greater tumor regression, as opposed to ICI alone in mouse models, by inhibiting the cancer-associated fibroblast, and increasing the accessibility of cancer cells to T cells [153,154] (Figure 3D). These are just three further examples of possible factors that may be utilized to produce a better response to ICI in CRC.

Immunotherapy serves as a groundbreaking step towards new and more rational treatment options, and lay the groundwork for new combinatorial agents. Further research should be conducted to investigate new combinations of treatments that can be used to produce an improved response to ICI in dMMR-MSI-H CRC, and furthermore, a response that has not yet been obtained in pMMR-MSI-L CRC.

Funding: S.S. was supported by SCRC for ALPD and Cirrhosis funded by the NIAAA (P50 AA011999). Work in M.K. laboratory was supported by grants from the NIH (R01 AI043477, R01 CA211794) and Tower Cancer Research Foundation. Additional support came from U01 AA027681 to S.S. and M.K., P01 CA128814 to M.K./Ze'ev Ronai and Padres Pedal the Cause C3 award #PTC2018.

Conflicts of Interest: The authors declare no conflicts of interest.

References

1. Siegel, R.; DeSantis, C.; Jemal, A. Colorectal cancer statistics, 2014. *Cancer J. Clin.* **2014**, *64*, 104–117. [CrossRef]

2. American Cancer Society. Cancer Facts & Figures, American Cancer Society, Atlanta, Georgia, 2019. Available online: https://www.cancer.org/content/dam/cancer-org/research/cancer-facts-and-statistics/annual-cancer-facts-and-figures/2019/cancer-facts-and-figures-2019.pdf (accessed on 4 January 2020).

3. Vatandoust, S.; Price, T.J.; Karapetis, C.S. Colorectal cancer: Metastases to a single organ. *World J. Gastroenterol.* **2015**, *21*, 11767–11776. [CrossRef]

4. Eggermont, A.M.M.; Blank, C.U.; Mandala, M.; Long, G.V.; Atkinson, V.; Dalle, S.; Haydon, A.; Lichinitser, M.; Khattak, A.; Carlino, M.S.; et al. Adjuvant Pembrolizumab versus Placebo in Resected Stage III Melanoma. *N. Engl. J. Med.* **2018**, *378*, 1789–1801. [CrossRef]

5. Gandhi, L.; Rodríguez-Abreu, D.; Gadgeel, S.; Esteban, E.; Felip, E.; De Angelis, F.; Domine, M.; Clingan, P.; Hochmair, M.J.; Powell, S.F.; et al. Pembrolizumab plus Chemotherapy in Metastatic Non-Small-Cell Lung Cancer. *N. Engl. J. Med.* **2018**, *378*, 2078–2092. [CrossRef]

6. Hugo, W.; Zaretsky, J.M.; Sun, L.; Song, C.; Moreno, B.H.; Hu-Lieskovan, S.; Berent-Maoz, B.; Pang, J.; Chmielowski, B.; Cherry, G.; et al. Genomic and Transcriptomic Features of Response to Anti-PD-1 Therapy in Metastatic Melanoma. *Cell* **2016**, *165*, 35–44. [CrossRef]

7. Schachter, J.; Ribas, A.; Long, G.V.; Arance, A.; Grob, J.-J.; Mortier, L.; Daud, A.; Carlino, M.S.; McNeil, C.; Lotem, M.; et al. Pembrolizumab versus ipilimumab for advanced melanoma: Final overall survival results of a multicentre, randomised, open-label phase 3 study (KEYNOTE-006). *Lancet* **2017**, *390*, 1853–1862. [CrossRef]

8. Ganesh, K.; Stadler, Z.K.; Cercek, A.; Mendelsohn, R.B.; Shia, J.; Segal, N.H.; Diaz, L.A. Immunotherapy in colorectal cancer: Rationale, challenges and potential. *Nat. Rev. Gastroenterol Hepatol.* **2019**, *16*, 361–375. [CrossRef]

9. Yamagishi, H.; Kuroda, H.; Imai, Y.; Hiraishi, H. Molecular pathogenesis of sporadic colorectal cancers. *Chin. J. Cancer* **2016**, *35*, 4. [CrossRef]

10. Robles, A.I.; Traverso, G.; Zhang, M.; Roberts, N.J.; Khan, M.A.; Joseph, C.; Lauwers, G.Y.; Selaru, F.M.; Popoli, M.; Pittman, M.E.; et al. Whole-Exome Sequencing Analyses of Inflammatory Bowel Disease-Associated Colorectal Cancers. *Gastroenterology* **2016**, *150*, 931–943. [CrossRef]

11. Schell, M.J.; Yang, M.; Teer, J.K.; Lo, F.Y.; Madan, A.; Coppola, D.; Monteiro, A.N.A.; Nebozhyn, M.V.; Yue, B.; Loboda, A.; et al. A multigene mutation classification of 468 colorectal cancers reveals a prognostic role for APC. *Nat. Commun.* **2016**, *7*, 1–12. [CrossRef]

12. Pandurangan, A.k.; Divya, T.; Kumar, K.; Dineshbabu, V.; Velavan, B.; Sudhandiran, G. Colorectal carcinogenesis: Insights into the cell death and signal transduction pathways: A review. *World J. Gastrointest Oncol.* **2018**, *10*, 244–259. [CrossRef] [PubMed]

13. Fearon, E.R. Molecular Genetics of Colorectal Cancer. *Annu. Rev. Pathol. Mech. Dis.* **2011**, *6*, 479–507. [CrossRef]

14. Leslie, A.; Carey, F.A.; Pratt, N.R.; Steele, R.J.C. The colorectal adenoma–carcinoma sequence. *Br. J. Surg.* **2002**, *89*, 845–860. [CrossRef]

15. Taylor, D.P.; Burt, R.W.; Williams, M.S.; Haug, P.J.; Cannon-Albright, L.A. Population-based family history-specific risks for colorectal cancer: A constellation approach. *Gastroenterology* **2010**, *138*, 877–885. [CrossRef]

16. Kerber, R.A.; Neklason, D.W.; Samowitz, W.S.; Burt, R.W. Frequency of Familial Colon Cancer and Hereditary Nonpolyposis Colorectal Cancer (Lynch Syndrome) in a Large Population Database. *Familial. Cancer* **2005**, *4*, 239–244. [CrossRef]

17. Stoffel, E.M.; Kastrinos, F. Familial colorectal cancer, beyond Lynch syndrome. *Clin. Gastroenterol. Hepatol.* **2014**, *12*, 1059–1068. [CrossRef]

18. Grivennikov, S.I.; Wang, K.; Mucida, D.; Stewart, C.A.; Schnabl, B.; Jauch, D.; Taniguchi, K.; Yu, G.-Y.; Österreicher, C.H.; Hung, K.E.; et al. Adenoma-linked barrier defects and microbial products drive IL-23/IL-17-mediated tumour growth. *Nature* **2012**, *491*, 254–258. [CrossRef]

19. Wang, K.; Kim, M.K.; Di Caro, G.; Wong, J.; Shalapour, S.; Wan, J.; Zhang, W.; Zhong, Z.; Sanchez-Lopez, E.; Wu, L.-W.; et al. Interleukin-17 Receptor A Signaling in Transformed Enterocytes Promotes Early Colorectal Tumorigenesis. *Immunity* **2014**, *41*, 1052–1063. [CrossRef]

20. Dmitrieva-Posocco, O.; Dzutsev, A.; Posocco, D.F.; Hou, V.; Yuan, W.; Thovarai, V.; Mufazalov, I.A.; Gunzer, M.; Shilovskiy, I.P.; Khaitov, M.R.; et al. Cell-Type-Specific Responses to Interleukin-1 Control Microbial Invasion and Tumor-Elicited Inflammation in Colorectal Cancer. *Immunity* **2019**, *50*, 166–180.e7. [CrossRef]

21. Greten, F.R.; Grivennikov, S.I. Inflammation and Cancer: Triggers, Mechanisms, and Consequences. *Immunity* **2019**, *51*, 27–41. [CrossRef]

22. Ziegler, P.K.; Bollrath, J.; Pallangyo, C.K.; Matsutani, T.; Canli, Ö.; De Oliveira, T.; Diamanti, M.A.; Müller, N.; Gamrekelashvili, J.; Putoczki, T.; et al. Mitophagy in Intestinal Epithelial Cells Triggers Adaptive Immunity during Tumorigenesis. *Cell* **2018**, *174*, 88–101.e16. [CrossRef] [PubMed]

23. Goldszmid, R.S.; Dzutsev, A.; Viaud, S.; Zitvogel, L.; Restifo, N.P.; Trinchieri, G. Microbiota modulation of myeloid cells in cancer therapy. *Cancer Immunol. Res.* **2015**, *3*, 103–109. [CrossRef] [PubMed]

24. Zhen, Y.; Luo, C.; Zhang, H. Early detection of ulcerative colitis-associated colorectal cancer. *Gastroenterol. Rep. (Oxf.)* **2018**, *6*, 83–92. [CrossRef]

25. Terzić, J.; Grivennikov, S.; Karin, E.; Karin, M. Inflammation and Colon Cancer. *Gastroenterology* **2010**, *138*, 2101–2114.e5. [CrossRef]

26. Khor, B.; Gardet, A.; Xavier, R.J. Genetics and pathogenesis of inflammatory bowel disease. *Nature* **2011**, *474*, 307–317. [CrossRef]

27. Shaked, H.; Hofseth, L.J.; Chumanevich, A.; Chumanevich, A.A.; Wang, J.; Wang, Y.; Taniguchi, K.; Guma, M.; Shenouda, S.; Clevers, H.; et al. Chronic epithelial NF- B activation accelerates APC loss and intestinal tumor initiation through iNOS up-regulation. *PNAS* **2012**, *109*, 14007–14012. [CrossRef]

28. Levin, B.; Lieberman, D.A.; McFarland, B.; Andrews, K.S.; Brooks, D.; Bond, J.; Dash, C.; Giardiello, F.M.; Glick, S.; Johnson, D.; et al. Screening and Surveillance for the Early Detection of Colorectal Cancer and Adenomatous Polyps, 2008: A Joint Guideline From the American Cancer Society, the US Multi-Society Task Force on Colorectal Cancer, and the American College of Radiology. *Gastroenterology* **2008**, *134*, 1570–1595. [CrossRef]

29. Carethers, J.M.; Jung, B.H. Genetics and Genetic Biomarkers in Sporadic Colorectal Cancer. *Gastroenterology* **2015**, *149*, 1177–1190.e3. [CrossRef]

30. Kameyama, H.; Nagahashi, M.; Shimada, Y.; Tajima, Y.; Ichikawa, H.; Nakano, M.; Sakata, J.; Kobayashi, T.; Narayanan, S.; Takabe, K.; et al. Genomic characterization of colitis-associated colorectal cancer. *World J. Surg. Oncol.* **2018**, *16*. [CrossRef]

31. Fleisher, A.S.; Esteller, M.; Harpaz, N.; Leytin, A.; Rashid, A.; Xu, Y.; Liang, J.; Stine, O.C.; Yin, J.; Zou, T.-T.; et al. Microsatellite Instability in Inflammatory Bowel Disease-associated Neoplastic Lesions Is Associated with Hypermethylation and Diminished Expression of the DNA Mismatch Repair Gene, hMLH1. *Cancer Res.* **2000**, *60*, 4864–4868. [PubMed]

32. Wheeler, J.M.D.; Bodmer, W.F.; Wheeler, J.M.D.; Mortensen, N.J.M. DNA mismatch repair genes and colorectal cancer. *Gut* **2000**, *47*, 148–153. [CrossRef] [PubMed]

33. Hause, R.J.; Pritchard, C.C.; Shendure, J.; Salipante, S.J. Classification and characterization of microsatellite instability across 18 cancer types. *Nat. Med.* **2016**, *22*, 1342–1350. [CrossRef] [PubMed]

34. De la Chapelle, A.; Hampel, H. Clinical relevance of microsatellite instability in colorectal cancer. *J. Clin. Oncol.* **2010**, *28*, 3380–3387. [CrossRef] [PubMed]

35. Chen, W.; Swanson, B.J.; Frankel, W.L. Molecular genetics of microsatellite-unstable colorectal cancer for pathologists. *Diagn. Pathol.* **2017**, *12*. [CrossRef]

36. Le, D.T.; Uram, J.N.; Wang, H.; Bartlett, B.R.; Kemberling, H.; Eyring, A.D.; Skora, A.D.; Luber, B.S.; Azad, N.S.; Laheru, D.; et al. PD-1 Blockade in Tumors with Mismatch-Repair Deficiency. *N. Engl. J. Med.* **2015**, *372*, 2509–2520. [CrossRef]

37. Boland, C.R.; Thibodeau, S.N.; Hamilton, S.R.; Sidransky, D.; Eshleman, J.R.; Burt, R.W.; Meltzer, S.J.; Rodriguez-Bigas, M.A.; Fodde, R.; Ranzani, G.N.; et al. A National Cancer Institute Workshop on Microsatellite Instability for Cancer Detection and Familial Predisposition: Development of International Criteria for the Determination of Microsatellite Instability in Colorectal Cancer. *Cancer Res.* **1998**, *58*, 5248. [PubMed]

38. Bacher, J.W.; Flanagan, L.A.; Smalley, R.L.; Nassif, N.A.; Burgart, L.J.; Halberg, R.B.; Megid, W.M.A.; Thibodeau, S.N. Development of a fluorescent multiplex assay for detection of MSI-High tumors. *Dis. Markers* **2004**, *20*, 237–250. [CrossRef]

39. Lu, Y.; Soong, T.D.; Elemento, O. A novel approach for characterizing microsatellite instability in cancer cells. *PLoS ONE* **2013**, *8*, e63056. [CrossRef]

40. Huang, M.N.; McPherson, J.R.; Cutcutache, I.; Teh, B.T.; Tan, P.; Rozen, S.G. MSIseq: Software for Assessing Microsatellite Instability from Catalogs of Somatic Mutations. *Sci. Rep.* **2015**, *5*, 13321. [CrossRef]

41. Dehal, A.; Graff-Baker, A.N.; Vuong, B.; Fischer, T.; Klempner, S.J.; Chang, S.-C.; Grunkemeier, G.L.; Bilchik, A.J.; Goldfarb, M. Neoadjuvant Chemotherapy Improves Survival in Patients with Clinical T4b Colon Cancer. *J. Gastrointest Surg.* **2018**, *22*, 242–249. [CrossRef]

42. Denoya, P.; Wang, H.; Sands, D.; Nogueras, J.; Weiss, E.; Wexner, S.D. Short-term outcomes of laparoscopic total mesorectal excision following neoadjuvant chemoradiotherapy. *Surg. Endosc.* **2010**, *24*, 933–938. [CrossRef] [PubMed]

43. Sauer, R.; Becker, H.; Hohenberger, W.; Rödel, C.; Wittekind, C.; Fietkau, R.; Martus, P.; Tschmelitsch, J.; Hager, E.; Hess, C.F.; et al. Preoperative versus Postoperative Chemoradiotherapy for Rectal Cancer. *N. Engl. J. Med.* **2004**, *351*, 1731–1740. [CrossRef] [PubMed]

44. Seymour, M.T.; Morton, D. FOxTROT: An international randomised controlled trial in 1052 patients (pts) evaluating neoadjuvant chemotherapy (NAC) for colon cancer. *J. Clin. Oncol.* **2019**, *37*, 3504. [CrossRef]

45. De Gooyer, J.-M.; Verstegen, M.G.; 't Lam-Boer, J.; Radema, S.A.; Verhoeven, R.H.A.; Verhoef, C.; Schreinemakers, J.M.J.; de Wilt, J.H.W. Neoadjuvant Chemotherapy for Locally Advanced T4 Colon Cancer: A Nationwide Propensity-Score Matched Cohort Analysis. *Dig. Surg.* **2019**, 1–10. [CrossRef] [PubMed]

46. Miyamoto, R.; Kikuchi, K.; Uchida, A.; Ozawa, M.; Sano, N.; Tadano, S.; Inagawa, S.; Oda, T.; Ohkohchi, N. Pathological complete response after preoperative chemotherapy including FOLFOX plus bevacizumab for locally advanced rectal cancer: A case report and literature review. *Int. J. Surg. Case. Rep.* **2019**, *62*, 85–88. [CrossRef] [PubMed]

47. Longley, D.B.; Harkin, D.P.; Johnston, P.G. 5-Fluorouracil: Mechanisms of action and clinical strategies. *Nat. Rev. Cancer* **2003**, *3*, 330–338. [CrossRef] [PubMed]

48. Giacchetti, S.; Perpoint, B.; Zidani, R.; Le Bail, N.; Faggiuolo, R.; Focan, C.; Chollet, P.; Llory, J.f.; Letourneau, Y.; Coudert, B.; et al. Phase III Multicenter Randomized Trial of Oxaliplatin Added to Chronomodulated Fluorouracil–Leucovorin as First-Line Treatment of Metastatic Colorectal Cancer. *J. Clin. Oncol.* **2000**, *18*, 136. [CrossRef] [PubMed]

49. Ribic, C.M.; Sargent, D.J.; Moore, M.J.; Thibodeau, S.N.; French, A.J.; Goldberg, R.M.; Hamilton, S.R.; Laurent-Puig, P.; Gryfe, R.; Shepherd, L.E.; et al. Tumor Microsatellite-Instability Status as a Predictor of Benefit from Fluorouracil-Based Adjuvant Chemotherapy for Colon Cancer. *N. Engl. J. Med.* **2003**, *349*, 247–257. [CrossRef] [PubMed]

50. Vilar, E.; Gruber, S.B. Microsatellite instability in colorectal cancer—The stable evidence. *Nat. Rev. Clin. Oncol.* **2010**, *7*, 153–162. [CrossRef] [PubMed]

51. Sinicrope, F.A.; Foster, N.R.; Thibodeau, S.N.; Marsoni, S.; Monges, G.; Labianca, R.; Yothers, G.; Allegra, C.; Moore, M.J.; Gallinger, S.; et al. DNA Mismatch Repair Status and Colon Cancer Recurrence and Survival in Clinical Trials of 5-Fluorouracil-Based Adjuvant Therapy. *J. Natl. Cancer Inst.* **2011**, *103*, 863–875. [CrossRef] [PubMed]

52. Jover, R.; Zapater, P.; Castells, A.; Llor, X.; Andreu, M.; Cubiella, J.; Piñol, V.; Xicola, R.M.; Bujanda, L.; Reñé, J.M.; et al. Mismatch repair status in the prediction of benefit from adjuvant fluorouracil chemotherapy in colorectal cancer. *Gut* **2006**, *55*, 848–855. [CrossRef] [PubMed]

53. Bertagnolli, M.M.; Niedzwiecki, D.; Compton, C.C.; Hahn, H.P.; Hall, M.; Damas, B.; Jewell, S.D.; Mayer, R.J.; Goldberg, R.M.; Saltz, L.B.; et al. Microsatellite Instability Predicts Improved Response to Adjuvant Therapy With Irinotecan, Fluorouracil, and Leucovorin in Stage III Colon Cancer: Cancer and Leukemia Group B Protocol 89803. *J. Clin. Oncol.* **2009**, *27*, 1814–1821. [CrossRef]

54. Hurwitz, H.; Fehrenbacher, L.; Novotny, W.; Cartwright, T.; Hainsworth, J.; Heim, W.; Berlin, J.; Baron, A.; Griffing, S.; Holmgren, E.; et al. Bevacizumab plus Irinotecan, Fluorouracil, and Leucovorin for Metastatic Colorectal Cancer. *N. Engl. J. Med.* **2004**, *350*, 2335–2342. [CrossRef]

55. Saltz, L.B.; Clarke, S.; Díaz-Rubio, E.; Scheithauer, W.; Figer, A.; Wong, R.; Koski, S.; Lichinitser, M.; Yang, T.-S.; Rivera, F.; et al. Bevacizumab in combination with oxaliplatin-based chemotherapy as first-line therapy in metastatic colorectal cancer: A randomized phase III study. *J. Clin. Oncol.* **2008**, *26*, 2013–2019. [CrossRef]

56. Cunningham, D.; Humblet, Y.; Siena, S.; Khayat, D.; Bleiberg, H.; Santoro, A.; Bets, D.; Mueser, M.; Harstrick, A.; Verslype, C.; et al. Cetuximab Monotherapy and Cetuximab plus Irinotecan in Irinotecan-Refractory Metastatic Colorectal Cancer. *N. Engl. J. Med.* **2004**, *351*, 337–345. [CrossRef]

57. Sobrero, A.F.; Maurel, J.; Fehrenbacher, L.; Scheithauer, W.; Abubakr, Y.A.; Lutz, M.P.; Vega-Villegas, M.E.; Eng, C.; Steinhauer, E.U.; Prausova, J.; et al. EPIC: Phase III Trial of Cetuximab Plus Irinotecan After Fluoropyrimidine and Oxaliplatin Failure in Patients With Metastatic Colorectal Cancer. *J. Clin. Oncol.* **2008**, *26*, 2311–2319. [CrossRef]

58. Lièvre, A.; Bachet, J.-B.; Corre, D.L.; Boige, V.; Landi, B.; Emile, J.-F.; Côté, J.-F.; Tomasic, G.; Penna, C.; Ducreux, M.; et al. KRAS Mutation Status Is Predictive of Response to Cetuximab Therapy in Colorectal Cancer. *Cancer Res.* **2006**, *66*, 3992–3995. [CrossRef]

59. Cunningham, D.; Atkin, W.; Lenz, H.-J.; Lynch, H.T.; Minsky, B.; Nordlinger, B.; Starling, N. Colorectal cancer. *Lancet* **2010**, *375*, 1030–1047. [CrossRef]

60. Misale, S.; Yaeger, R.; Hobor, S.; Scala, E.; Janakiraman, M.; Liska, D.; Valtorta, E.; Schiavo, R.; Buscarino, M.; Siravegna, G.; et al. Emergence of KRAS mutations and acquired resistance to anti-EGFR therapy in colorectal cancer. *Nature* **2012**, *486*, 532–536. [CrossRef] [PubMed]

61. Sinicrope, F.A.; Shi, Q.; Smyrk, T.C.; Thibodeau, S.N.; Dienstmann, R.; Guinney, J.; Bot, B.M.; Tejpar, S.; Delorenzi, M.; Goldberg, R.M.; et al. Molecular markers identify subtypes of stage III colon cancer associated with patient outcomes. *Gastroenterology* **2015**, *148*, 88–99. [CrossRef] [PubMed]

62. Van Emburgh, B.O.; Sartore-Bianchi, A.; Di Nicolantonio, F.; Siena, S.; Bardelli, A. Acquired resistance to EGFR-targeted therapies in colorectal cancer. *Mol. Oncol.* **2014**, *8*, 1084–1094. [CrossRef] [PubMed]

63. Van Cutsem, E.; Peeters, M.; Siena, S.; Humblet, Y.; Hendlisz, A.; Neyns, B.; Canon, J.-L.; Van Laethem, J.-L.; Maurel, J.; Richardson, G.; et al. Open-Label Phase III Trial of Panitumumab Plus Best Supportive Care Compared With Best Supportive Care Alone in Patients With Chemotherapy-Refractory Metastatic Colorectal Cancer. *J. Clin. Oncol.* **2007**, *25*, 1658–1664. [CrossRef] [PubMed]

64. Pagès, F.; Mlecnik, B.; Marliot, F.; Bindea, G.; Ou, F.-S.; Bifulco, C.; Lugli, A.; Zlobec, I.; Rau, T.T.; Berger, M.D.; et al. International validation of the consensus Immunoscore for the classification of colon cancer: A prognostic and accuracy study. *Lancet* **2018**, *391*, 2128–2139. [CrossRef]

65. Galon, J.; Costes, A.; Sanchez-Cabo, F.; Kirilovsky, A.; Mlecnik, B.; Lagorce-Pagès, C.; Tosolini, M.; Camus, M.; Berger, A.; Wind, P.; et al. Type, Density, and Location of Immune Cells Within Human Colorectal Tumors Predict Clinical Outcome. *Science* **2006**, *313*, 1960–1964. [CrossRef]

66. Mlecnik, B.; Tosolini, M.; Kirilovsky, A.; Berger, A.; Bindea, G.; Meatchi, T.; Bruneval, P.; Trajanoski, Z.; Fridman, W.-H.; Pagès, F.; et al. Histopathologic-Based Prognostic Factors of Colorectal Cancers Are Associated With the State of the Local Immune Reaction. *J. Clin. Oncol.* **2011**, *29*, 610–618. [CrossRef]

67. Bindea, G.; Mlecnik, B.; Tosolini, M.; Kirilovsky, A.; Waldner, M.; Obenauf, A.C.; Angell, H.; Fredriksen, T.; Lafontaine, L.; Berger, A.; et al. Spatiotemporal Dynamics of Intratumoral Immune Cells Reveal the Immune Landscape in Human Cancer. *Immunity* **2013**, *39*, 782–795. [CrossRef]

68. Galon, J.; Angell, H.K.; Bedognetti, D.; Marincola, F.M. The continuum of cancer immunosurveillance: Prognostic, predictive, and mechanistic signatures. *Immunity* **2013**, *39*, 11–26. [CrossRef]

69. Mlecnik, B.; Bindea, G.; Angell, H.K.; Maby, P.; Angelova, M.; Tougeron, D.; Church, S.E.; Lafontaine, L.; Fischer, M.; Fredriksen, T.; et al. Integrative Analyses of Colorectal Cancer Show Immunoscore Is a Stronger Predictor of Patient Survival Than Microsatellite Instability. *Immunity* **2016**, *44*, 698–711. [CrossRef]

70. Ogino, S.; Nosho, K.; Irahara, N.; Meyerhardt, J.A.; Baba, Y.; Shima, K.; Glickman, J.N.; Ferrone, C.R.; Mino-Kenudson, M.; Tanaka, N.; et al. Lymphocytic Reaction to Colorectal Cancer is Associated with Longer Survival, Independent of Lymph Node Count, MSI and CpG Island Methylator Phenotype. *Clin. Cancer Res.* **2009**, *15*, 6412–6420. [CrossRef]

71. Popat, S.; Hubner, R.; Houlston, R.S. Systematic Review of Microsatellite Instability and Colorectal Cancer Prognosis. *J. Clin. Oncol.* **2005**, *23*, 609–618. [CrossRef]

72. Narayanan, S.; Kawaguchi, T.; Peng, X.; Qi, Q.; Liu, S.; Yan, L.; Takabe, K. Tumor Infiltrating Lymphocytes and Macrophages Improve Survival in Microsatellite Unstable Colorectal Cancer. *Sci. Rep.* **2019**, *9*, 1–10. [CrossRef] [PubMed]

73. Tosolini, M.; Kirilovsky, A.; Mlecnik, B.; Fredriksen, T.; Mauger, S.; Bindea, G.; Berger, A.; Bruneval, P.; Fridman, W.-H.; Pagès, F.; et al. Clinical Impact of Different Classes of Infiltrating T Cytotoxic and Helper Cells (Th1, Th2, Treg, Th17) in Patients with Colorectal Cancer. *Cancer Res.* **2011**, *71*, 1263–1271. [CrossRef] [PubMed]

74. Maby, P.; Tougeron, D.; Hamieh, M.; Mlecnik, B.; Kora, H.; Bindea, G.; Angell, H.K.; Fredriksen, T.; Elie, N.; Fauquembergue, E.; et al. Correlation between Density of CD8+ T-cell Infiltrate in Microsatellite Unstable Colorectal Cancers and Frameshift Mutations: A Rationale for Personalized Immunotherapy. *Cancer Res.* **2015**, *75*, 3446–3455. [CrossRef]

75. Zhao, P.; Li, L.; Jiang, X.; Li, Q. Mismatch repair deficiency/microsatellite instability-high as a predictor for anti-PD-1/PD-L1 immunotherapy efficacy. *J. Hematol. Oncol.* **2019**, *12*, 54. [CrossRef]

76. Pagès, F.; Berger, A.; Camus, M.; Sanchez-Cabo, F.; Costes, A.; Molidor, R.; Mlecnik, B.; Kirilovsky, A.; Nilsson, M.; Damotte, D.; et al. Effector memory T cells, early metastasis, and survival in colorectal cancer. *N. Engl. J. Med.* **2005**, *353*, 2654–2666. [CrossRef]

77. Tang, H.; Wang, Y.; Chlewicki, L.K.; Zhang, Y.; Guo, J.; Liang, W.; Wang, J.; Wang, X.; Fu, Y.-X. Facilitating T cell infiltration in tumor microenvironment overcomes resistance to PD-L1 blockade. *Cancer Cell* **2016**, *29*, 285–296. [CrossRef]

78. Zhou, F. Molecular Mechanisms of IFN-γ to Up-Regulate MHC Class I Antigen Processing and Presentation. *Int. Rev. Immunol.* **2009**, *28*, 239–260. [CrossRef] [PubMed]

79. Tewary, P.; Yang, D.; de la Rosa, G.; Li, Y.; Finn, M.W.; Krensky, A.M.; Clayberger, C.; Oppenheim, J.J. Granulysin activates antigen-presenting cells through TLR4 and acts as an immune alarmin. *Blood* **2010**, *116*, 3465–3474. [CrossRef]

80. Taniguchi, K.; Karin, M. NF-κB, inflammation, immunity and cancer: Coming of age. *Nat. Rev. Immunol.* **2018**, *18*, 309–324. [CrossRef]

81. Lasry, A.; Zinger, A.; Ben-Neriah, Y. Inflammatory networks underlying colorectal cancer. *Nat. Immunol.* **2016**, *17*, 230–240. [CrossRef] [PubMed]

82. West, N.R.; McCuaig, S.; Franchini, F.; Powrie, F. Emerging cytokine networks in colorectal cancer. *Nat. Rev. Immunol.* **2015**, *15*, 615–629. [CrossRef]

83. Zeng, J.; Tang, Z.-H.; Liu, S.; Guo, S.-S. Clinicopathological significance of overexpression of interleukin-6 in colorectal cancer. *World J. Gastroenterol.* **2017**, *23*, 1780–1786. [CrossRef]

84. Chung, Y.-C.; Chang, Y.-F. Serum interleukin-6 levels reflect the disease status of colorectal cancer. *J. Surg. Oncol.* **2003**, *83*, 222–226. [CrossRef] [PubMed]

85. Knüpfer, H.; Preiss, R. Serum interleukin-6 levels in colorectal cancer patients—A summary of published results. *Int. J. Colorectal. Dis.* **2010**, *25*, 135–140. [CrossRef] [PubMed]

86. Nagasaki, T.; Hara, M.; Nakanishi, H.; Takahashi, H.; Sato, M.; Takeyama, H. Interleukin-6 released by colon cancer-associated fibroblasts is critical for tumour angiogenesis: Anti-interleukin-6 receptor antibody suppressed angiogenesis and inhibited tumour–stroma interaction. *Br. J. Cancer* **2014**, *110*, 469–478. [CrossRef] [PubMed]

87. Grivennikov, S.; Karin, E.; Terzic, J.; Mucida, D.; Yu, G.-Y.; Vallabhapurapu, S.; Scheller, J.; Rose-John, S.; Cheroutre, H.; Eckmann, L.; et al. IL-6 and Stat3 Are Required for Survival of Intestinal Epithelial Cells and Development of Colitis-Associated Cancer. *Cancer Cell* **2009**, *15*, 103–113. [CrossRef] [PubMed]

88. Corvinus, F.M.; Orth, C.; Moriggl, R.; Tsareva, S.A.; Wagner, S.; Pfitzner, E.B.; Baus, D.; Kaufmann, R.; Huber, L.A.; Zatloukal, K.; et al. Persistent STAT3 activation in colon cancer is associated with enhanced cell proliferation and tumor growth. *Neoplasia* **2005**, *7*, 545–555. [CrossRef] [PubMed]

89. Bollrath, J.; Phesse, T.J.; von Burstin, V.A.; Putoczki, T.; Bennecke, M.; Bateman, T.; Nebelsiek, T.; Lundgren-May, T.; Canli, Ö.; Schwitalla, S.; et al. gp130-Mediated Stat3 Activation in Enterocytes Regulates Cell Survival and Cell-Cycle Progression during Colitis-Associated Tumorigenesis. *Cancer Cell* **2009**, *15*, 91–102. [CrossRef] [PubMed]

90. Taniguchi, K.; Wu, L.-W.; Grivennikov, S.I.; de Jong, P.R.; Lian, I.; Yu, F.-X.; Wang, K.; Ho, S.B.; Boland, B.S.; Chang, J.T.; et al. A gp130–Src–YAP module links inflammation to epithelial regeneration. *Nature* **2015**, *519*, 57–62. [CrossRef]

91. Taniguchi, K.; Karin, M. IL-6 and related cytokines as the critical lynchpins between inflammation and cancer. *Semin. Immunol.* **2014**, *26*, 54–74. [CrossRef]

92. Tseng-Rogenski, S.; Hamaya, Y.; Choi, D.Y.; Carethers, J.M. Interleukin 6 Alters Localization of hMSH3, Leading to DNA Mismatch Repair Defects in Colorectal Cancer Cells. *Gastroenterology* **2015**, *148*, 579–589. [CrossRef] [PubMed]

93. Schetter, A.J.; Nguyen, G.H.; Bowman, E.D.; Mathé, E.A.; Yuen, S.T.; Hawkes, J.E.; Croce, C.M.; Leung, S.Y.; Harris, C.C. Association of Inflammation-Related and microRNA Gene Expression with Cancer-Specific Mortality of Colon Adenocarcinoma. *Clin. Cancer Res.* **2009**, *15*, 5878–5887. [CrossRef] [PubMed]

94. Wang, K.; Karin, M. Chapter Five—Tumor-Elicited Inflammation and Colorectal Cancer. In *Advances in Cancer Research*; Wang, X.-Y., Fisher, P.B., Eds.; Academic Press: Cambridge, MA, USA, 2015; Volume 128, pp. 173–196, ISBN 0065-230X.

95. Couzin-Frankel, J. Cancer Immunotherapy. *Science* **2013**, *342*, 1432–1433. [CrossRef] [PubMed]

96. Zugazagoitia, J.; Guedes, C.; Ponce, S.; Ferrer, I.; Molina-Pinelo, S.; Paz-Ares, L. Current Challenges in Cancer Treatment. *Clinical. Therapeutics* **2016**, *38*, 1551–1566. [CrossRef] [PubMed]

97. Franke, A.J.; Skelton, W.P., IV; Starr, J.S.; Parekh, H.; Lee, J.J.; Overman, M.J.; Allegra, C.; George, T.J. Immunotherapy for Colorectal Cancer: A Review of Current and Novel Therapeutic Approaches. *J. Natl. Cancer Inst.* **2019**, *111*, 1131–1141. [CrossRef]

98. Sharma, P.; Allison, J.P. The future of immune checkpoint therapy. *Science* **2015**, *348*, 56–61. [CrossRef]

99. Topalian, S.L.; Drake, C.G.; Pardoll, D.M. Immune Checkpoint Blockade: A Common Denominator Approach to Cancer Therapy. *Cancer Cell* **2015**, *27*, 450–461. [CrossRef]

100. Pardoll, D.M. The blockade of immune checkpoints in cancer immunotherapy. *Nat. Rev. Cancer* **2012**, *12*, 252–264. [CrossRef]

101. Wei, S.C.; Duffy, C.R.; Allison, J.P. Fundamental Mechanisms of Immune Checkpoint Blockade Therapy. *Cancer Discov.* **2018**, *8*, 1069–1086. [CrossRef]

102. Khalil, D.N.; Smith, E.L.; Brentjens, R.J.; Wolchok, J.D. The future of cancer treatment: Immunomodulation, CARs and combination immunotherapy. *Nat. Rev. Clin. Oncol.* **2016**, *13*, 273–290. [CrossRef]

103. Khalil, D.N.; Budhu, S.; Gasmi, B.; Zappasodi, R.; Hirschhorn-Cymerman, D.; Plitt, T.; De Henau, O.; Zamarin, D.; Holmgaard, R.B.; Murphy, J.T.; et al. Chapter One—The New Era of Cancer Immunotherapy: Manipulating T-Cell Activity to Overcome Malignancy. In *Advances in Cancer Research*; Wang, X.-Y., Fisher, P.B., Eds.; Academic Press: Cambridge, MA, USA, 2015; Volume 128, pp. 1–68. ISBN 0065-230X.

104. Rizvi, N.A.; Hellmann, M.D.; Snyder, A.; Kvistborg, P.; Makarov, V.; Havel, J.J.; Lee, W.; Yuan, J.; Wong, P.; Ho, T.S.; et al. Cancer immunology. Mutational landscape determines sensitivity to PD-1 blockade in non-small cell lung cancer. *Science* **2015**, *348*, 124–128. [CrossRef] [PubMed]

105. Schumacher, T.N.; Schreiber, R.D. Neoantigens in cancer immunotherapy. *Science* **2015**, *348*, 69–74. [CrossRef] [PubMed]

106. Keir, M.E.; Butte, M.J.; Freeman, G.J.; Sharpe, A.H. PD-1 and Its Ligands in Tolerance and Immunity. *Annu. Rev. Immunol.* **2008**, *26*, 677–704. [CrossRef] [PubMed]

107. Garon, E.B.; Rizvi, N.A.; Hui, R.; Leighl, N.; Balmanoukian, A.S.; Eder, J.P.; Patnaik, A.; Aggarwal, C.; Gubens, M.; Horn, L.; et al. Pembrolizumab for the Treatment of Non-Small-Cell Lung Cancer. *N. Engl. J. Med.* **2015**, *372*, 2018–2028. [CrossRef]

108. Snyder, A.; Makarov, V.; Merghoub, T.; Yuan, J.; Zaretsky, J.M.; Desrichard, A.; Walsh, L.A.; Postow, M.A.; Wong, P.; Ho, T.S.; et al. Genetic Basis for Clinical Response to CTLA-4 Blockade in Melanoma. *N. Engl. J. Med.* **2014**, *371*, 2189–2199. [CrossRef]

109. Van Allen, E.M.; Miao, D.; Schilling, B.; Shukla, S.A.; Blank, C.; Zimmer, L.; Sucker, A.; Hillen, U.; Geukes Foppen, M.H.; Goldinger, S.M.; et al. Genomic correlates of response to CTLA-4 blockade in metastatic melanoma. *Science* **2015**, *350*, 207. [CrossRef]

110. Hodi, F.S.; O'Day, S.J.; McDermott, D.F.; Weber, R.W.; Sosman, J.A.; Haanen, J.B.; Gonzalez, R.; Robert, C.; Schadendorf, D.; Hassel, J.C.; et al. Improved Survival with Ipilimumab in Patients with Metastatic Melanoma. *N. Engl. J. Med.* **2010**, *363*, 711–723. [CrossRef]

111. Alexandrov, L.B.; Nik-Zainal, S.; Wedge, D.C.; Aparicio, S.A.J.R.; Behjati, S.; Biankin, A.V.; Bignell, G.R.; Bolli, N.; Borg, A.; Børresen-Dale, A.-L.; et al. Signatures of mutational processes in human cancer. *Nature* **2013**, *500*, 415–421. [CrossRef]

112. Sahin, U.; Türeci, Ö. Personalized vaccines for cancer immunotherapy. *Science* **2018**, *359*, 1355. [CrossRef]

113. Hollingsworth, R.E.; Jansen, K. Turning the corner on therapeutic cancer vaccines. *NPJ Vaccines* **2019**, *4*, 7. [CrossRef]

114. Blankenstein, T.; Leisegang, M.; Uckert, W.; Schreiber, H. Targeting cancer-specific mutations by T cell receptor gene therapy. *Curr. Opin. Immunol.* **2015**, *33*, 112–119. [CrossRef] [PubMed]

115. Tran, E.; Robbins, P.F.; Lu, Y.-C.; Prickett, T.D.; Gartner, J.J.; Jia, L.; Pasetto, A.; Zheng, Z.; Ray, S.; Groh, E.M.; et al. T-Cell Transfer Therapy Targeting Mutant KRAS in Cancer. *N. Engl. J. Med.* **2016**, *375*, 2255–2262. [CrossRef] [PubMed]

116. Malekzadeh, P.; Pasetto, A.; Robbins, P.F.; Parkhurst, M.R.; Paria, B.C.; Jia, L.; Gartner, J.J.; Hill, V.; Yu, Z.; Restifo, N.P.; et al. Neoantigen screening identifies broad TP53 mutant immunogenicity in patients with epithelial cancers. *J. Clin. Invest.* **2019**, *129*, 1109–1114. [CrossRef] [PubMed]

117. Gros, A.; Tran, E.; Parkhurst, M.R.; Ilyas, S.; Pasetto, A.; Groh, E.M.; Robbins, P.F.; Yossef, R.; Garcia-Garijo, A.; Fajardo, C.A.; et al. Recognition of human gastrointestinal cancer neoantigens by circulating PD-1+ lymphocytes. *J. Clin. Invest.* **2019**, *129*, 4992–5004. [CrossRef] [PubMed]

118. Chung, K.Y.; Gore, I.; Fong, L.; Venook, A.; Beck, S.B.; Dorazio, P.; Criscitiello, P.J.; Healey, D.I.; Huang, B.; Gomez-Navarro, J.; et al. Phase II Study of the Anti-Cytotoxic T-Lymphocyte–Associated Antigen 4 Monoclonal Antibody, Tremelimumab, in Patients With Refractory Metastatic Colorectal Cancer. *J. Clin. Oncol.* **2010**, *28*, 3485–3490. [CrossRef] [PubMed]

119. Topalian, S.L.; Hodi, F.S.; Brahmer, J.R.; Gettinger, S.N.; Smith, D.C.; McDermott, D.F.; Powderly, J.D.; Carvajal, R.D.; Sosman, J.A.; Atkins, M.B.; et al. Safety, Activity, and Immune Correlates of Anti–PD-1 Antibody in Cancer. *N. Engl. J. Med.* **2012**, *366*, 2443–2454. [CrossRef]

120. Brahmer, J.R.; Tykodi, S.S.; Chow, L.Q.M.; Hwu, W.-J.; Topalian, S.L.; Hwu, P.; Drake, C.G.; Camacho, L.H.; Kauh, J.; Odunsi, K.; et al. Safety and Activity of Anti–PD-L1 Antibody in Patients with Advanced Cancer. *N. Engl. J. Med.* **2012**, *366*, 2455–2465. [CrossRef]

121. Brahmer, J.R.; Drake, C.G.; Wollner, I.; Powderly, J.D.; Picus, J.; Sharfman, W.H.; Stankevich, E.; Pons, A.; Salay, T.M.; McMiller, T.L.; et al. Phase I study of single-agent anti-programmed death-1 (MDX-1106) in refractory solid tumors: Safety, clinical activity, pharmacodynamics, and immunologic correlates. *J. Clin. Oncol.* **2010**, *28*, 3167–3175. [CrossRef]

122. Overman, M.J.; McDermott, R.; Leach, J.L.; Lonardi, S.; Lenz, H.-J.; Morse, M.A.; Desai, J.; Hill, A.; Axelson, M.; Moss, R.A.; et al. Nivolumab in patients with metastatic DNA mismatch repair-deficient or microsatellite instability-high colorectal cancer (CheckMate 142): An open-label, multicentre, phase 2 study. *Lancet Oncol.* **2017**, *18*, 1182–1191. [CrossRef]

123. Overman, M.J.; Lonardi, S.; Wong, K.Y.M.; Lenz, H.-J.; Gelsomino, F.; Aglietta, M.; Morse, M.A.; Van Cutsem, E.; McDermott, R.; Hill, A.; et al. Durable Clinical Benefit With Nivolumab Plus Ipilimumab in DNA Mismatch Repair-Deficient/Microsatellite Instability-High Metastatic Colorectal Cancer. *J. Clin. Oncol.* **2018**, *36*, 773–779. [CrossRef]

124. Maletzki, C.; Gladbach, Y.S.; Hamed, M.; Fuellen, G.; Semmler, M.-L.; Stenzel, J.; Linnebacher, M. Cellular vaccination of MLH1(-/-) mice—An immunotherapeutic proof of concept study. *Oncoimmunology* **2017**, *7*, e1408748. [CrossRef] [PubMed]

125. Kloor, M.; Reuschenbach, M.; Karbach, J.; Rafiyan, M.; Al-Batran, S.-E.; Pauligk, C.; Jaeger, E.; von Knebel Doeberitz, M. Vaccination of MSI-H colorectal cancer patients with frameshift peptide antigens: A phase I/IIa clinical trial. *J. Clin. Oncol.* **2015**, *33*, 3020. [CrossRef]

126. De Weger, V.A.; Turksma, A.W.; Voorham, Q.J.M.; Euler, Z.; Bril, H.; van den Eertwegh, A.J.; Bloemena, E.; Pinedo, H.M.; Vermorken, J.B.; van Tinteren, H.; et al. Clinical effects of adjuvant active specific immunotherapy differ between patients with microsatellite-stable and microsatellite-instable colon cancer. *Clin. Cancer Res.* **2012**, *18*, 882–889. [CrossRef]

127. Mandal, R.; Samstein, R.M.; Lee, K.-W.; Havel, J.J.; Wang, H.; Krishna, C.; Sabio, E.Y.; Makarov, V.; Kuo, F.; Blecua, P.; et al. Genetic diversity of tumors with mismatch repair deficiency influences anti-PD-1 immunotherapy response. *Science* **2019**, *364*, 485. [CrossRef]

128. Eng, C.; Kim, T.W.; Bendell, J.; Argilés, G.; Tebbutt, N.C.; Di Bartolomeo, M.; Falcone, A.; Fakih, M.; Kozloff, M.; Segal, N.H.; et al. Atezolizumab with or without cobimetinib versus regorafenib in previously treated metastatic colorectal cancer (IMblaze370): A multicentre, open-label, phase 3, randomised, controlled trial. *Lancet Oncol.* **2019**, *20*, 849–861. [CrossRef]

129. Ebert, P.J.R.; Cheung, J.; Yang, Y.; McNamara, E.; Hong, R.; Moskalenko, M.; Gould, S.E.; Maecker, H.; Irving, B.A.; Kim, J.M.; et al. MAP Kinase Inhibition Promotes T Cell and Anti-tumor Activity in Combination with PD-L1 Checkpoint Blockade. *Immunity* **2016**, *44*, 609–621. [CrossRef]

130. Canon, J.; Rex, K.; Saiki, A.Y.; Mohr, C.; Cooke, K.; Bagal, D.; Gaida, K.; Holt, T.; Knutson, C.G.; Koppada, N.; et al. The clinical KRAS(G12C) inhibitor AMG 510 drives anti-tumour immunity. *Nature* **2019**, *575*, 217–223. [CrossRef]

131. Hanna, M.G.; Hoover, H.C.; Vermorken, J.B.; Harris, J.E.; Pinedo, H.M. Adjuvant active specific immunotherapy of stage II and stage III colon cancer with an autologous tumor cell vaccine: First randomized phase III trials show promise. *Vaccine* **2001**, *19*, 2576–2582. [CrossRef]

132. Harris, J.E.; Ryan, L.; Hoover, H.C.; Stuart, R.K.; Oken, M.M.; Benson, A.B.; Mansour, E.; Haller, D.G.; Manola, J.; Hanna, M.G. Adjuvant active specific immunotherapy for stage II and III colon cancer with an autologous tumor cell vaccine: Eastern Cooperative Oncology Group Study E5283. *J. Clin. Oncol.* **2000**, *18*, 148–157. [CrossRef]

133. Ockert, D.; Schirrmacher, V.; Beck, N.; Stoelben, E.; Ahlert, T.; Flechtenmacher, J.; Hagmüller, E.; Buchcik, R.; Nagel, M.; Saeger, H.D. Newcastle disease virus-infected intact autologous tumor cell vaccine for adjuvant active specific immunotherapy of resected colorectal carcinoma. *Clin. Cancer Res.* **1996**, *2*, 21–28. [PubMed]

134. Puzanov, I.; Milhem, M.M.; Minor, D.; Hamid, O.; Li, A.; Chen, L.; Chastain, M.; Gorski, K.S.; Anderson, A.; Chou, J.; et al. Talimogene Laherparepvec in Combination With Ipilimumab in Previously Untreated, Unresectable Stage IIIB-IV Melanoma. *J. Clin. Oncol.* **2016**, *34*, 2619–2626. [CrossRef] [PubMed]

135. Chesney, J.; Puzanov, I.; Collichio, F.; Singh, P.; Milhem, M.M.; Glaspy, J.; Hamid, O.; Ross, M.; Friedlander, P.; Garbe, C.; et al. Randomized, Open-Label Phase II Study Evaluating the Efficacy and Safety of Talimogene Laherparepvec in Combination With Ipilimumab Versus Ipilimumab Alone in Patients With Advanced, Unresectable Melanoma. *J. Clin. Oncol.* **2018**, *36*, 1658–1667. [CrossRef] [PubMed]

136. Scholz, M.; Yep, S.; Chancey, M.; Kelly, C.; Chau, K.; Turner, J.; Lam, R.; Drake, C.G. Phase I clinical trial of sipuleucel-T combined with escalating doses of ipilimumab in progressive metastatic castrate-resistant prostate cancer. *Immunotargets Ther.* **2017**, *6*, 11–16. [CrossRef] [PubMed]

137. Ku, J.; Wilenius, K.; Larsen, C.; De Guzman, K.; Yoshinaga, S.; Turner, J.S.; Lam, R.Y.; Scholz, M.C. Survival after sipuleucel-T (SIP-T) and low-dose ipilimumab (IPI) in men with metastatic, progressive, castrate-resistant prostate cancer (M-CRPC). *J. Clin. Oncol.* **2018**, *36*, 368. [CrossRef]

138. Singh, H.; Madan, R.A.; Dahut, W.L.; O'Sullivan Coyne, G.H.; Rauckhorst, M.; McMahon, S.; Heery, C.R.; Schlom, J.; Gulley, J.L. Combining active immunotherapy and immune checkpoint inhibitors in prostate cancer. *J. Clin. Oncol.* **2015**, *33*, e14008. [CrossRef]

139. Jochems, C.; Tucker, J.A.; Tsang, K.-Y.; Madan, R.A.; Dahut, W.L.; Liewehr, D.J.; Steinberg, S.M.; Gulley, J.L.; Schlom, J. A combination trial of vaccine plus ipilimumab in metastatic castration-resistant prostate cancer patients: Immune correlates. *Cancer Immunol. Immunother.* **2014**, *63*, 407–418. [CrossRef]

140. Long, G.V.; Dummer, R.; Ribas, A.; Puzanov, I.; Michielin, O.; VanderWalde, A.; Andtbacka, R.H.; Cebon, J.; Fernandez, E.; Malvehy, J.; et al. A Phase I/III, multicenter, open-label trial of talimogene laherparepvec (T-VEC) in combination with pembrolizumab for the treatment of unresected, stage IIIb-IV melanoma (MASTERKEY-265). *J. Immuno. Ther. Cancer* **2015**, *3*, P181. [CrossRef]

141. Ribas, A.; Dummer, R.; Puzanov, I.; VanderWalde, A.; Andtbacka, R.H.I.; Michielin, O.; Olszanski, A.J.; Malvehy, J.; Cebon, J.; Fernandez, E.; et al. Oncolytic Virotherapy Promotes Intratumoral T Cell Infiltration and Improves Anti-PD-1 Immunotherapy. *Cell* **2017**, *170*, 1109–1119.e10. [CrossRef]

142. Weber, J.S.; Kudchadkar, R.R.; Yu, B.; Gallenstein, D.; Horak, C.E.; Inzunza, H.D.; Zhao, X.; Martinez, A.J.; Wang, W.; Gibney, G.; et al. Safety, efficacy, and biomarkers of nivolumab with vaccine in ipilimumab-refractory or -naive melanoma. *J. Clin. Oncol.* **2013**, *31*, 4311–4318. [CrossRef]

143. Gibney, G.T.; Kudchadkar, R.R.; DeConti, R.C.; Thebeau, M.S.; Czupryn, M.P.; Tetteh, L.; Eysmans, C.; Richards, A.; Schell, M.J.; Fisher, K.J.; et al. Safety, correlative markers, and clinical results of adjuvant nivolumab in combination with vaccine in resected high-risk metastatic melanoma. *Clin. Cancer Res.* **2015**, *21*, 712–720. [CrossRef]

144. Ribas, A.; Medina, T.; Kummar, S.; Amin, A.; Kalbasi, A.; Drabick, J.J.; Barve, M.; Daniels, G.A.; Wong, D.J.; Schmidt, E.V.; et al. SD-101 in Combination with Pembrolizumab in Advanced Melanoma: Results of a Phase Ib, Multicenter Study. *Cancer Discov.* **2018**, *8*, 1250–1257. [CrossRef] [PubMed]

145. Chatani, P.D.; Yang, J.C. Mutated RAS: Targeting the "Untargetable" with T-cells. *Clin. Cancer Res.* **2019**. [CrossRef] [PubMed]

146. Gopalakrishnan, V.; Spencer, C.N.; Nezi, L.; Reuben, A.; Andrews, M.C.; Karpinets, T.V.; Prieto, P.A.; Vicente, D.; Hoffman, K.; Wei, S.C.; et al. Gut microbiome modulates response to anti-PD-1 immunotherapy in melanoma patients. *Science* **2018**, *359*, 97–103. [CrossRef] [PubMed]

147. Chaput, N.; Lepage, P.; Coutzac, C.; Soularue, E.; Le Roux, K.; Monot, C.; Boselli, L.; Routier, E.; Cassard, L.; Collins, M.; et al. Baseline gut microbiota predicts clinical response and colitis in metastatic melanoma patients treated with ipilimumab. *Ann. Oncol.* **2017**, *28*, 1368–1379. [CrossRef]

148. Routy, B.; Chatelier, E.L.; Derosa, L.; Duong, C.P.M.; Alou, M.T.; Daillère, R.; Fluckiger, A.; Messaoudene, M.; Rauber, C.; Roberti, M.P.; et al. Gut microbiome influences efficacy of PD-1-based immunotherapy against epithelial tumors. *Science* **2018**, *359*, 91–97. [CrossRef]

149. Matson, V.; Fessler, J.; Bao, R.; Chongsuwat, T.; Zha, Y.; Alegre, M.-L.; Luke, J.J.; Gajewski, T.F. The commensal microbiome is associated with anti-PD-1 efficacy in metastatic melanoma patients. *Science* **2018**, *359*, 104–108. [CrossRef]

150. Routy, B.; Gopalakrishnan, V.; Daillère, R.; Zitvogel, L.; Wargo, J.A.; Kroemer, G. The gut microbiota influences anticancer immunosurveillance and general health. *Nat. Rev. Clin. Oncol.* **2018**, *15*, 382–396. [CrossRef]

151. Bertrand, F.; Montfort, A.; Marcheteau, E.; Imbert, C.; Gilhodes, J.; Filleron, T.; Rochaix, P.; Andrieu-Abadie, N.; Levade, T.; Meyer, N.; et al. TNFα blockade overcomes resistance to anti-PD-1 in experimental melanoma. *Nat. Commun.* **2017**, *8*, 2256. [CrossRef] [PubMed]

152. Perez-Ruiz, E.; Minute, L.; Otano, I.; Alvarez, M.; Ochoa, M.C.; Belsue, V.; de Andrea, C.; Rodriguez-Ruiz, M.E.; Perez-Gracia, J.L.; Marquez-Rodas, I.; et al. Prophylactic TNF blockade uncouples efficacy and toxicity in dual CTLA-4 and PD-1 immunotherapy. *Nature* **2019**, *569*, 428–432. [CrossRef]

153. Mariathasan, S.; Turley, S.J.; Nickles, D.; Castiglioni, A.; Yuen, K.; Wang, Y.; Kadel, E.E., III; Koeppen, H.; Astarita, J.L.; Cubas, R.; et al. TGFβ attenuates tumour response to PD-L1 blockade by contributing to exclusion of T cells. *Nature* **2018**, *554*, 544–548. [CrossRef]

154. Tauriello, D.V.F.; Palomo-Ponce, S.; Stork, D.; Berenguer-Llergo, A.; Badia-Ramentol, J.; Iglesias, M.; Sevillano, M.; Ibiza, S.; Cañellas, A.; Hernando-Momblona, X.; et al. TGFβ drives immune evasion in genetically reconstituted colon cancer metastasis. *Nature* **2018**, *554*, 538–543. [CrossRef] [PubMed]

Review

Oncofetal Chondroitin Sulfate: A Putative Therapeutic Target in Adult and Pediatric Solid Tumors

Nastaran Khazamipour [1,2], Nader Al-Nakouzi [1,2], Htoo Zarni Oo [1,2], Maj Ørum-Madsen [1,2], Anne Steino [2,3], Poul H Sorensen [3,4] and Mads Daugaard [1,2,*]

[1] Department of Urologic Sciences, University of British Columbia, Vancouver, BC V5Z 1M9, Canada; nkhazamipour@prostatecentre.com (N.K.); nalnakouzi@prostatecentre.com (N.A.-N.); hoo@prostatecentre.com (H.Z.O.); moerum@prostatecentre.com (M.Ø.-M.)
[2] Vancouver Prostate Centre, Vancouver, BC V6H 3Z6, Canada; asteino@bccrc.ca
[3] Department of Pathology and Laboratory Medicine, University of British Columbia, Vancouver, BC V5Z 1M9, Canada; psor@mail.ubc.ca
[4] Department of Molecular Oncology, British Columbia Cancer Research Centre, Vancouver, BC V5Z1L3, Canada
* Correspondence: mads.daugaard@ubc.ca; Tel.: +1-604-875-4111

Received: 10 February 2020; Accepted: 26 March 2020; Published: 28 March 2020

Abstract: Solid tumors remain a major challenge for targeted therapeutic intervention strategies such as antibody-drug conjugates and immunotherapy. At a minimum, clear and actionable solid tumor targets have to comply with the key biological requirement of being differentially over-expressed in solid tumors and metastasis, in contrast to healthy organs. Oncofetal chondroitin sulfate is a cancer-specific secondary glycosaminoglycan modification to proteoglycans expressed in a variety of solid tumors and metastasis. Normally, this modification is found to be exclusively expressed in the placenta, where it is thought to facilitate normal placental implantation during pregnancy. Informed by this biology, oncofetal chondroitin sulfate is currently under investigation as a broad and specific target in solid tumors. Here, we discuss oncofetal chondroitin sulfate as a potential therapeutic target in childhood solid tumors in the context of current knowhow obtained over the past five years in adult cancers.

Keywords: oncofetal chondroitin sulfate; chondroitin sulfate; cancer; solid tumors; target; pediatric cancer; VAR2

---

1. Oncofetal Similarities between the Fetal and Tumor Tissue Compartments

The placenta, an organ that develops during pregnancy, behaves in many ways like a tumor. In just 40 weeks, the placenta has to grow to a mass of ~500 grams, invade neighboring tissue, establish an elaborate vasculature, and escape the immune system, all key features of solid tumor development [1]. Moreover, similarities between placenta and cancer at the molecular level have been frequently observed. Several proto-oncogenes involved in malignant transformation and cancer progression, including *c-erbB1* family (*HER1*, *ERBB1* or *EGFR*), *c-myc*, *Fos* and *c-ras*, are preferentially expressed by trophoblast cells during the first week of pregnancy when the proliferative, migratory and invasive properties of these cells are at their peak [2,3]. For instance, *c-erbB1* is expressed exclusively by the cytotrophoblast in four- to five-week placentas and pre-dominantly in the syncytiotrophoblast compartment after six weeks of gestation [4–6]. It is also involved in the pathogenesis of numerous malignancies, including breast cancer [7] and some types of childhood cancer [8]. The *c-myc* (MYC) proto-oncogene displays strong expression in early placenta [9] and is also frequently increased in human cancers [10,11]. Hyperactivation of Ras signaling by mutations or overexpression of the *Ras* oncogenes is a powerful driver of solid tumor formation [12,13], and the *c-ras* proto-oncogene, a key player in signaling pathways that regulate cellular proliferation [14], is expressed in early villous trophoblasts [15,16]. Similarly, overexpression of the *Fos* proto-oncogene stimulates trophoblast invasion during placental implementation [17], while contributing to tumor metastasis in several types of cancer [18–20].

In addition to the expression of proto-oncogenes, a number of oncofetal proteins are also shared between placenta, tumors and fetal tissue, including pregnancy-associated plasma protein A (PAPP-A), PEG10, alpha-fetoprotein (AFP), carcinoembryonic antigen (CEA), trophoblast glycoprotein precursor (TPBG) and immature laminin receptor protein (iLRP). Based on their oncofetal properties, some of these proteins have since been pursued as potential therapeutic targets in solid tumors. For example, PAPP-A, which is produced by placental syncytiotrophoblasts and is essential for normal fetal development [21], has been shown to facilitate tumor growth and invasion in various malignancies [22]. Notably, PAPP-A has been investigated as a potent immunotherapeutic target in Ewing sarcoma [23]. Likewise, PEG10, an RNA splice factor that is crucial for placental and embryonic development [24], is reported to play a role in the progression of several types of human cancers, including leukemia, breast cancer, prostate cancer and hepatocellular carcinoma [25–27], and has been proposed as a therapeutic target for prostate cancer [26–28].

AFP is produced by the embryo during fetal development and is found in both fetal serum and amniotic fluid and is currently the most widely used prognostic marker in hepatocellular carcinoma [29,30]. Additionally, CEA produced during embryonal and fetal development is one of the most widely used tumor markers worldwide, especially in colorectal malignancies where it is used to detect and inform on the presence of liver metastasis [31]. In addition, TPBG is used as a prognostic tool in a broad spectrum of malignancies, including colorectal, ovarian and gastric cancers [32–34]. It is also the target of the cancer vaccine TroVax, currently in clinical trials for the treatment several solid tumor types [35–38]. iLRP, which is highly expressed in early fetal development, is re-expressed in many tumor types and has been associated with tumor progression and metastasis [39,40]. Moreover, iLRP has been investigated as a therapeutic target for patients with leukemic diseases and against metastatic spread of solid tumors [41]. There are thus numerous examples of oncofetal proteins that can be utilized as tumor targets.

To qualify as a tumor target, a protein must be differentially expressed between malignant and normal tissues. Inadequate differential expression of potential target proteins is a major concern for all targeted therapy approaches and there is therefore a high demand for discovery of new molecular targets, differentially expressed in malignant versus normal tissue. Post-translational modifications (PTMs) of proteins, including phosphorylation, glycosylation, ubiquitination, nitrosylation, methylation, acetylation, lipidation and proteolysis, increase the diversity of the proteome and influence almost all aspects of cell biology and pathogenesis [42]. Protein glycosylation has major effects on protein folding, conformation, distribution, stability and activity [43–47]. Given its critical role in expanding protein functionality and diversity, glycosylation is an attractive candidate source of molecular targets in cancer. Indeed, targeting the glycosylation component of a protein rather than the protein itself has clear advantages. Firstly, targeting of tumor-specific protein glycoforms could be a solution for increasing anti-tumor specificity while limiting off-target effects. Secondly, a specific glycosylation moiety or pattern can be present on several different proteins simultaneously across cell populations, including tumor stem cells, which may overcome challenges related to tumor heterogeneity and dormancy. Lastly, proteins that are not normally glycosylated may be subject to disease-specific glycosylations, thereby increasing the available tumor target reservoir [48–50].

2. Chondroitin Sulfate

Among the glycosylation components that play a critical role in protein functionality are glycosaminoglycans (GAGs). GAGs are large, linear, negatively-charged polysaccharides consisting of repeating disaccharide units that can be sulfated at different positions and to different extents [51,52]. Five GAG chains have been identified to date: Heparan sulfate (HS), chondroitin sulfate (CS), dermatan sulfate

(DS), and keratan sulfate, as well as the non-sulfated hyaluronic acid [51,52]. GAGs are expressed on virtually all mammalian cells and are usually covalently attached to proteins, forming proteoglycans (PG).

CS is the second most heterogenous GAG group after HS and functionally presented as CS proteoglycans (CSPGs) in the pericellular matrix, as well as the intracellular milieu and the extracellular matrix (ECM) [53–56]. CS interacts with multiple ligands, both soluble and insoluble, and modulates important roles in many physiological and pathophysiological processes [57,58]. CS consists of repeating N-acetylgalactosamine (GalNAc)-glucuronic acid (GlcA) disaccharide units. A complex biosynthetic machinery in the Golgi apparatus is responsible for the production and structure of CS chains [59]. Five enzymes catalyze a tetrasaccharide-linker region attached to a serine residue of the core protein and six additional CS enzymes produce the polymeric backbone. During elongation of the CS chain, the sulfation of hydroxyl groups in different positions can occur. CS may contain sulfate groups in both the carbon 4 (C4) and C6 positions of the GalNAc unit (CSE), but may also be predominantly C4-sulfated (CSA) or C6-sulfated (CSC). Four CS carbohydrate sulfotransferases (CHSTs: CHST11, CHST12, CHST13 and CHST14) can catalyze the 4-O-sulfation of GalNAc in CS [60]. The CHSTs involved in 6-O-sulfation of GalNAc include (CHST3, CHST7, CHST15). The GlcA unit can also be sulfated at the C2 position, giving rise to DS also known as CSB (4-sulfated GalNAc and 2-sulfated GlcA) and CSD (6-sulfated GalNAc and 2-sulfated GlcA) [61]. The role of CS modifications in cancer progression has been under investigation for decades. In solid tumors, CS participate in cell–cell and cell–ECM interactions that promote tumor cell adhesion and migration, thereby facilitating aggressive and metastatic behavior of malignant cells [62–65]. Increased production of CS is found in transformed fibroblasts and mammary carcinoma cells, where these polysaccharides contribute to cell proliferation, adhesion and migration [64,66,67]. Similarly during embryonic development, CS in the context of CSPGs has important morphogenetic functions, especially in relation to epithelial morphogenesis, cell migration and cell division rates [68–71]. Moreover, CS is indispensable for pluripotency and differentiation of embryonic stem cells [72]. The ECM of human placentas contain high levels of CSPGs [73]. Placental CSPGs are mainly located on trophoblast cells in the ECM surrounding the expanding syncytium [63], where they are involved in a number of physiological processes. For example, they are part of a glycocalyx double-barrier that prevents the migration of immune cells through the placenta, from the mother to the offspring [72,74].

3. Oncofetal Chondroitin Sulfate in Placenta

In pregnancy-associated malaria pathogenesis, CSPGs in the placenta mediate the sequestration of infected red blood cells (IRBCs) to the intervillous spaces of the placenta [63]. Upon infection and during the replication phase inside IRBCs, the malaria parasite *Plasmodium falciparum* expresses a specific lectin, VAR2CSA, on the surface of the IRBCs. VAR2CSA subsequently binds to CS chains expressed in the placental syncytium, thereby enabling *P. falciparum* IRBCs to exit blood circulation and avoid filtration and destruction in the spleen of the infected host [75–77]. The specific form of CS recognized by VAR2CSA is a type of CSA [78,79] presented as a PTM on PGs such as syndecan-1 [63]. Evident by the fact that VAR2CSA-positive *P. falciparum* IRBCs sequester to the placenta as the only organ in the human host, placental CSA is thought to be distinct from CSA found in other tissues. Perhaps due to the phenotypical similarities between the placenta and tumors, placental-type CSA is also found in the vast majority of solid tumors as a secondary oncofetal CS (ofCS) PTM to PGs [80]. While the exact structure and composition of ofCS is as yet poorly understood, it is clear that the ofCS GAG chain is highly sulfated on C4 of the vast majority of GalNAc residues [80], and this specific sulfation pattern is unique to CSPGs in placenta and solid tumor tissue [80]. Since ofCS is not found in other normal tissues but the placenta, this PTM constitutes an attractive tumor target.

4. Expression of Oncofetal Chondroitin Sulfate Proteoglycans in Adult Solid Tumors

Over the past five years, ofCS modifications of PGs have been described in multiple solid tumor indications [18,80–82]. Through binding and regulation of a large number of ligands, ofCS chains collaborate with other PG components to modulate cell behaviors such as proliferation, differentiation, migration and adhesion [63,80]. Although malignant tumors have individual CSPG profiles, they generally display strong ofCS expression [63]. Indeed, ~90% of breast tumors, 80% of melanomas [80], and 92% of bladder cancers [82], express ofCS-modified CSPGs on the cell surface and/or in the tumor stroma. Moreover, ofCS alterations are often linked to disease progression and outcome in cancer patients. For example, expression of ofCS in melanoma tumors is significantly increased in advanced tumors, Clark level 2–5 compared to level 1, and in metastatic/recurrent disease compared to newly diagnosed disease [80]. In non-small cell lung cancer, high expression of ofCS correlates with poor relapse-free survival [80]. In addition, high ofCS expression is correlated with advanced tumor stage, cisplatin resistance and poor overall survival of muscle-invasive bladder cancer (MIBC) patients [82]. In breast cancer, CHST11 is over-expressed in tumors as compared to normal tissues [62]. Also, high expression of the CHST11 predicts poor disease-free survival of lung, breast and colorectal cancer patients [80]. Contrarily, other studies have reported that expression of C4-S sulfotransferases including CHST11 seems to be downregulated in colorectal cancers [83]. This discrepancy between different cancers highlights a lack of knowledge about the regulation and maturation of CS chains, which is further complicated by tissue-specific expression patterns and redundancy among CS enzymes. Nevertheless, ofCS expression is currently being evaluated as a potential therapeutic target for several adult tumor types, including bladder cancer [82], prostate cancer, breast cancer and non-Hodgkin's lymphoma [80].

5. Expression of Oncofetal Chondroitin Sulfate Proteoglycans in Pediatric Solid Tumors

While the expression of ofCS and its correlation with disease progression and outcome has been demonstrated in a variety of adult tumors, the potential for utilizing ofCS expression as a therapeutic target in childhood tumors has been less explored. Pediatric solid tumors are non-hematologic malignancies that occur during childhood. This heterogeneous group of tumors represents approximately 40%–50% of all pediatric cancers [84]. The tumor distribution of malignant pediatric solid tumors in adolescents is different compared with that of younger children, in whom embryonal or developmental cancers, such as retinoblastoma, neuroblastoma, or hepatoblastoma, are more prevalent. The most common malignant solid tumors in adolescents are extracranial germ cell tumors, bone and soft tissue sarcomas, melanoma, and thyroid cancer [85]. Generally, the outcome for pediatric solid tumors depends on location of the specific disease and risk group such as histological finding, tumor stage and metastatic status.

Similar to adult tumors, childhood solid tumors express various CSPGs with diverse functions related to disease progression (Table 1). In osteosarcoma, versican upregulation promotes cell motility and correlates with disease progression [39]. In neuroblastoma (NB), the CSPG NCAN is highly expressed in the tumor ECM where it facilitates growth of NB cells and promotes disease progression [82]. Exogenous NCAN expression transforms adherent NB cells into spheroids with high malignancy potential both in vitro (anchorage-independent growth and chemoresistance) and in vivo (xenograft tumor growth) [82]. CSPG4 is a cell surface PG commonly modified with ofCS that has been exploited as a tumor target in several tumor indications [86–89]. High levels of CSPG4 are found on a variety of adult and pediatric solid tumors including melanoma [90,91], osteosarcoma [87], rhabdomyosarcoma [88] and some brain tumors [86,92]. The CSPG4 expression levels differ depending on tumor type but is often present in both high-grade and lower-grade pediatric brain tumors [93]. PTPRZ1 plays a key role in cell migration, and is a potential tumor target in glioblastoma multiforme (GBM) [94]. In Ewing sarcoma, overexpression of APLP2 results in lower sensitivity to radiotherapy-induced apoptosis and immunologic cell death [95].

Proteoglycans can harbor different and multiple GAGs at the same time. For instance, syndecans and glypicans are PGs containing both CS and HS chains [96]. Altered expression of these PGs has been reported in multiple cancers including pediatric tumors [97]. Glypican 3, for example, plays an important role in cellular growth and differentiation. It is absent or only minimally expressed in most adult tissues but highly expressed in a variety of non-central nervous system (CNS) pediatric tumors, including hepatoblastoma, Wilms tumor, rhabdomyosarcoma, and in atypical teratoid rhabdoid tumors [98,99]. Glypican 5 is expressed in rhabdomyosarcoma where it facilitates growth factor signaling, in particular FGF signaling [100]. High syndecan-1 levels are found in glioma, where it correlates with advanced clinicopathological features and poor patient survival [101]. Sarcomas commonly express ofCS chains in 50%–100% of cases, depending on subtypes. Overall, ~80% of bone sarcomas, and ~85% of soft-tissue sarcomas are positive for ofCS [80]. Pediatric sarcoma cell lines generally express high levels of ofCS, and ofCS is required for migration and invasion capacity of osteosarcoma and rhabdomyosarcoma cells [80,102]. Indeed, ofCS has also been found on pediatric glioma cells and circulating tumor cells (CTCs) from GBM patients [89], hinting that ofCS might be exploited for liquid diagnostic applications in pediatric brain cancers. Also, ofCS allows for EpCAM-independent detection of CTCs [81], which might provide access to circulating sarcoma cells. Combined, the broad expression of CSPGs and ofCS across multiple pediatric tumor indications, promotes ofCS as putative and attractive therapeutic target in pediatric solid tumors.

Table 1. Chondroitin sulfate proteoglycan (CSPG) expression in childhood solid tumors.

CS-Modified PG	Cancer Type	Function
NCAN	Neuroblastoma	Promotes cell division, undifferentiated state and malignant phenotypes [82] Provides a growth advantage to cancer cells [82]
Versican	Osteosarcoma [103] Glioblastoma multiforme (GBM) [89]	Involves in TGFß - induced cell migration and invasion [103] Relevant marker of osteosarcoma progression [103] Potential target in cancer treatment [103] Function is unknown in GBM [89]
Decorin	Osteosarcoma [104]	Necessary for MG63 cell migration [104] Counteracts the growth-limiting effects of TGF-β2 [104]
CSPG4	Osteosarcoma [87] Rhabdomyosarcomas (RMS) [88] Medulloblastoma [105] Neuroblastoma [105] Childhood diffuse intrinsic pontine glioma [86] GBM [86,89] Dysembryoplastic neuroepithelial tumors (DNETs) [86,93]	Correlates with shorter survival in osteosarcoma [87] Therapeutic option for the combination treatment of RMS [88] Potential target for immunotherapy [87,89,105] Impairs terminal differentiation [86] Increases the invasive and migratory capabilities of glioma cells by facilitating interactions with extracellular matrix proteins [86] Facilitates angiogenesis by sequestering angiostatin [86] Increases tumor growth [86] Potential therapeutic target for treating childhood CNS cancers [86,89]
CD44	GBM [89,106]	High CD44 expression identifies GBM with particular poor survival chance Promotes aggressive GBM growth [106]
PTPRZ1	GBM [89,94]	Potential anti-cancer targets in GBM [89,94] Plays critical role in GBM cell migration [94]
APLP2	GBM [89] Ewing sarcoma [95]	Function is unknown in GBM [89] Anti-apoptotic function within Ewing sarcoma cells [95]
Syndecan-1	Glioma [89,101]	Correlates with the advanced clinicopathological features and lower survival rate [101]
Glypican 3	Hepatoblastoma Wilms tumor Rhabdomyosarcoma Atypical teratoid rhabdoid tumors	Potential candidate for targeted therapies [98]
Glypican 5	Rhabdomyosarcoma	Facilitates growth factor signaling Increases cell proliferation Potential target for therapeutic approaches [100]
Testican-1	GBM [89]	Unknown
Neuropilin-1	Osteosarcoma [107] Neuroblastoma [108] GBM [89]	Regulates metastasis potency [107] Correlates with poor response to chemotherapy [107] Correlates with poor prognosis for osteosarcoma patients [107] Regulates angiogenesis [107,108] Increases tumor growth [108] Function is unknown in GBM [89]

6. Oncofetal Chondroitin Sulfate as a Therapeutic Target in Solid Tumors

As outlined above, ofCS has emerged as an attractive tumor target for both therapeutic and diagnostic applications [18,80–82]. VAR2CSA specifically recognizes and binds ofCS, and recombinant VAR2CSA (rVAR2) proteins have been utilized to probe and access the ofCS chain expressed in solid tumors [80,82]. rVAR2 has also been exploited as a delivery system to shuttle cytotoxic drugs directly into ofCS-expressing tumor cells. For example, rVAR2-DT, a recombinant protein drug consisting of the cytotoxic domain of diphtheria toxin (DT388) fused to rVAR2, is able to eliminate both epithelial and mesenchymal tumor cells without any deleterious effect to normal primary human endothelial cells (HUVEC) in vitro [80]. Moreover, rVAR2-DT can inhibit prostate tumor growth in xenograft mouse models [80]. However, because DT-fusion drugs historically have shown adverse toxicity in human clinical trials [80], other strategies for delivery of drugs to ofCS-positive tumors have been pursued, including a rVAR2 drug-conjugate, VDC886. VDC886 is comprised of a 72 kDa rVAR2 polypeptide conjugated with the hemiasterlin toxin analog KT886, derived from the marine sponge *Hemiasterella minor*. VDC886 contains an average payload of three KT886 toxins per rVAR2 protein and displays strong toxicity towards diverse tumor cell lines of both adult and pediatric origin [80]. In vivo, VDC886 significantly inhibits tumor growth and metastasis in non-Hodgkin's lymphoma, prostate cancer, and breast cancer xenograft models with no sign of adverse effects [80]. In a different study, VDC886 successfully targeted ofCS on cisplatin-resistant MIBC cells and suppressed tumor growth of MIBC in vivo [82]. Immunohistochemical analysis of two independent cohorts of matched pre- and post-neoadjuvant chemotherapy-treated MIBC patients, revealed that cisplatin-resistant residual tumors had elevated levels of ofCS expression, supporting ofCS as a marker for disease progression [82].

In summary, the broad expression of CSPGs across solid tumors, and of ofCS in particular, promotes ofCS as an attractive target for therapeutic intervention. Historically, targeted biologics-based therapies have been less successful in pediatric solid tumors as compared to adult cancers, largely due to low mutational burden and limited number of neoantigens [109]. Hence, targeting cancer-specific PTMs, such as ofCS, constitutes a novel opportunity to curb childhood solid tumors. Indeed, the ability of VDCs to target ofCS-positive solid tumors supports a rational for exploring additional ofCS-targeting strategies, such as chimeric antigen receptor (CAR) T cells and bi-specific immune-engagers (BiTEs).

Author Contributions: N.K., N.A.-N. and H.Z.O. performed the literature search and wrote the first draft of the manuscript. M.Ø.-M., A.S., P.H.S. and M.D. edited the manuscript and approved the content. All authors have read and agreed to the published version of the manuscript.

Funding: This work was funded by the St. Baldrick's Foundation and Stand Up 2 Cancer.

Conflicts of Interest: The authors declare no conflicts of interest. M.D. as the corresponding author certifies that all conflicts of interest, including specific financial interests and relationships and affiliations relevant to the subject matter or materials discussed in the manuscript (e.g., employment/affiliation, grants or funding, consultancies, honoraria, stock ownership or options, expert testimony, royalties, or patents filed, received, or pending), are the following: M.D. and P.H.S. are co-founders of, and shareholders in, VAR2 Pharmaceuticals. N.A.N. and M.Ø.M. are consultants for VAR2 Pharmaceuticals. VAR2 Pharmaceuticals is a biotechnology company that specializes in therapeutic development of the VAR2CSA technology (www.var2pharma.com).

References

1. Holtan, S.G.; Creedon, D.J.; Haluska, P.; Markovic, S.N. Cancer and pregnancy: Parallels in growth, invasion, and immune modulation and implications for cancer therapeutic agents. *Mayo Clin. Proc.* **2009**, *84*, 985–1000. [CrossRef]

2. Quenby, S.; Brazeau, C.; Drakeley, A.; I Lewis-Jones, D.; Vince, G. Oncogene and tumour suppressor gene products during trophoblast differentiation in the first trimester. *Mol. Hum. Reprod.* **1998**, *4*, 477–481. [CrossRef] [PubMed]

3. Ferretti, C.; Bruni, L.; Dangles-Marie, V.; Pecking, A.; Bellet, D. Molecular circuits shared by placental and cancer cells, and their implications in the proliferative, invasive and migratory capacities of trophoblasts. *Hum. Reprod. Updat.* **2006**, *13*, 121–141. [CrossRef] [PubMed]

4. Maruo, T.; Mochizuki, M. Immunohistochemical localization of epidermal growth factor receptor and myc oncogene product in human placenta: Implication for trophoblast proliferation and differentiation. *Am. J. Obstet. Gynecol.* **1987**, *156*, 721–727. [CrossRef]

5. Maruo, T.; Matsuo, H.; Otani, T.; Mochizuki, M. Role of epidermal growth factor (EGF) and its receptor in the development of the human placenta. *Reprod. Fertil. Dev.* **1995**, *7*, 1465–1470. [CrossRef] [PubMed]

6. Sugawara, T.; Maruo, T.; Otani, T.; Mochizuki, M. Increase in the Expression of C-Erb-a and C-Erb-B Messenger-Rnas in the Human Placenta in Early Gestation—Their Roles in Trophoblast Proliferation and Differentiation. *Endoc. J.* **1994**, *41*, S127–S133. [CrossRef]

7. Chen, S.; Qiu, Y.; Guo, P.; Pu, T.; Feng, Y.; Bu, H. FGFR1 and HER1 or HER2 co-amplification in breast cancer indicate poor prognosis. *Oncol. Lett.* **2018**, *15*, 8206–8214. [CrossRef]

8. Bodey, B.; E Kaiser, H.; E Siegel, S. Epidermal growth factor receptor (EGFR) expression in childhood brain tumors. *In Vivo* **2005**, *19*, 931–941.

9. Pfeifer-Ohlsson, S.; Goustin, A.S.; Rydnert, J.; Wahlström, T.; Bjersing, L.; Stéhelin, D.; Ohlsson, R. Spatial and temporal pattern of cellular myc oncogene expression in developing human placenta: Implications for embryonic cell proliferation. *Cell* **1984**, *38*, 585–596. [CrossRef]

10. Dang, C.V.; O'Donnell, K.A.; Juopperi, T. The great MYC escape in tumorigenesis. *Cancer Cell* **2005**, *8*, 177–178. [CrossRef]

11. Schaub, F.X.; Trivedi, M.; Richardson, A.B.; Shaw, R.; Zhao, W.; Zhang, X.; Ventura, A.; Ayer, D.; Hurlin, P.J.; Eisenman, R.N.; et al. Pan-cancer Alterations of the MYC Oncogene and Its Proximal Network across the Cancer Genome Atlas. *Cell Syst.* **2018**, *6*, 282–300. [CrossRef] [PubMed]

12. Akao, Y.; Kumazaki, M.; Shinohara, H.; Sugito, N.; Kuranaga, Y.; Tsujino, T.; Yoshikawa, Y.; Kitade, Y. Impairment of K-Ras signaling networks and increased efficacy of epidermal growth factor receptor inhibitors by a novel synthetic miR-143. *Cancer Sci.* **2018**, *109*, 1455–1467. [CrossRef] [PubMed]

13. Díaz, R.; Lopez-Barcons, L.; Ahn, D.; Yoon, A.; Matthews, J.; Mangues, R.; Pellicer, A.; Garcia-España, A.; Perez-Soler, R. Complex effects of Ras proto-oncogenes in tumorigenesis. *Carcinogenesis* **2003**, *25*, 535–539. [CrossRef] [PubMed]

14. Yu, Q.; Ciemerych, M.A.; Sicinski, P. Ras and Myc can drive oncogenic cell proliferation through individual D-cyclins. *Oncogene* **2005**, *24*, 7114–7119. [CrossRef]

15. Kohorn, E.; Sarkar, S.; Kacinski, B.; Merino, M.; Carter, D.; Blakemore, K.; Summers, W. Demonstration of myc and ras oncogene expression by in situ hybridization in hydatidiform mole and in the choriocarcinoma cell line BeWo. *Gynecol. Oncol.* **1986**, *23*, 245. [CrossRef]

16. Lu, C.-W.; Yabuuchi, A.; Chen, L.; Viswanathan, S.; Kim, K.; Daley, G.Q. Ras-MAPK signaling promotes trophectoderm formation from embryonic stem cells and mouse embryos. *Nat. Genet.* **2008**, *40*, 921–926. [CrossRef]

17. Bischof, P. Endocrine, paracrine and autocrine regulation of trophoblastic metalloproteinases. *Early Pregnancy* **2001**, *5*, 30–31.

18. Ding, Y.; Hao, K.; Li, Z.; Ma, R.; Zhou, Y.; Zhou, Z.; Wei, M.; Liao, Y.; Dai, Y.; Yang, Y.; et al. c-Fos separation from Lamin A/C by GDF15 promotes colon cancer invasion and metastasis in inflammatory microenvironment. *J. Cell. Physiol.* **2019**, *235*, 4407–4421. [CrossRef]

19. Qu, X.; Yan, X.; Kong, C.; Zhu, Y.; Li, H.; Pan, D.; Zhang, X.; Liu, Y.; Yin, F.; Qin, H. c-Myb promotes growth and metastasis of colorectal cancer through c-fos-induced epithelial-mesenchymal transition. *Cancer Sci.* **2019**, *110*, 3183–3196. [CrossRef]

20. Weekes, D.; Kashima, T.G.; Zandueta, C.; Perurena, N.; Thomas, D.P.; Sunters, A.; Vuillier, C.; Bozec, A.; El-Emir, E.; Miletich, I.; et al. Regulation of osteosarcoma cell lung metastasis by the c-Fos/AP-1 target FGFR1. *Oncogene* **2016**, *35*, 2948. [CrossRef]

21. Kalousova, M.; Muravská, A.; Zima, T. Pregnancy-associated plasma protein A (PAPP-A) and preeclampsia. *Adv. Clin. Chem.* **2014**, *63*, 169–209. [PubMed]

22. Guo, Y.; Bao, Y.; Guo, D.; Yang, W. Pregnancy-associated plasma protein a in cancer: Expression, oncogenic functions and regulation. *Am. J. Cancer Res.* **2018**, *8*, 955–963. [PubMed]

23. Heitzeneder, S.; Sotillo, E.; Shern, J.F.; Sindiri, S.; Xu, P.; Jones, R.C.; Pollak, M.; Noer, P.R.; Lorette, J.; Fazli, L.; et al. Pregnancy-Associated Plasma Protein-A (PAPP-A) in Ewing Sarcoma: Role in Tumor Growth and Immune Evasion. *J. Natl. Cancer Inst.* **2019**, *111*, 970–982. [CrossRef] [PubMed]

24. Chen, H.; Sun, M.; Zhao, G.; Liu, J.; Gao, W.; Si, S.; Meng, T. Elevated expression of PEG10 in human placentas from preeclamptic pregnancies. *Acta Histochem.* **2012**, *114*, 589–593. [CrossRef]

25. Kainz, B.; Shehata, M.; Bilban, M.; Kienle, D.; Heintel, D.; Krömer-Holzinger, E.; Le, T.; Kröber, A.; Heller, G.; Schwarzinger, I.; et al. Overexpression of the paternally expressed gene10 (PEG10) from the imprinted locus on chromosome 7q21 in high-risk B-cell chronic lymphocytic leukemia. *Int. J. Cancer* **2007**, *121*, 1984–1993. [CrossRef]

26. Ip, W.-K.; Lai, P.B.; Wong, N.L.-Y.; Sy, M.H.; Beheshti, B.; Squire, J.; Wong, N. Identification of PEG10 as a progression related biomarker for hepatocellular carcinoma. *Cancer Lett.* **2007**, *250*, 284–291. [CrossRef]

27. Akamatsu, S.; Wyatt, A.W.; Lin, N.; Lysakowski, S.; Zhang, F.; Kim, S.; Tse, C.; Wang, K.; Mo, F.; Haegert, A.; et al. The Placental Gene PEG10 Promotes Progression of Neuroendocrine Prostate Cancer. *Cell Rep.* **2015**, *12*, 922–936. [CrossRef]

28. Kim, S.; Thaper, D.; Bidnur, S.; Toren, P.; Akamatsu, S.; Bishop, J.L.; Colins, C.; Vahid, S.; Zoubeidi, A. PEG10 is associated with treatment-induced neuroendocrine prostate cancer. *J. Mol. Endocrinol.* **2019**, *63*, 39–49. [CrossRef]

29. Bai, D.-S.; Zhang, C.; Chen, P.; Jin, S.-J.; Jiang, G.-Q. The prognostic correlation of AFP level at diagnosis with pathological grade, progression, and survival of patients with hepatocellular carcinoma. *Sci. Rep.* **2017**, *7*, 12870. [CrossRef]

30. Galle, P.R.; Foerster, F.; Kudo, M.; Chan, S.L.; Llovet, J.M.; Qin, S.; Schelman, W.R.; Chintharlapalli, S.; Abada, P.B.; Sherman, M.; et al. Biology and significance of alpha-fetoprotein in hepatocellular carcinoma. *Liver Int.* **2019**, *39*, 2214–2229. [CrossRef]

31. Peng, S.; Huang, P.; Yu, H.; Wen, Y.; Luo, Y.; Wang, X.; Zhou, J.; Qin, S.; Li, T.; Chen, Y.; et al. Prognostic value of carcinoembryonic antigen level in patients with colorectal cancer liver metastasis treated with percutaneous microwave ablation under ultrasound guidance. *Medicine* **2018**, *97*, e0044. [CrossRef] [PubMed]

32. Starzyńska, T.; Marsh, P.; Schofield, P.; Roberts, S.; Myers, K.; Stern, P. Prognostic significance of 5T4 oncofetal antigen expression in colorectal carcinoma. *Br. J. Cancer* **1994**, *69*, 899–902. [CrossRef] [PubMed]

33. Naganuma, H.; Kono, K.; Mori, Y.; Takayoshi, S.; Stern, P.L.; Tasaka, K.; Matsumoto, Y. Oncofetal antigen 5T4 expression as a prognostic factor in patients with gastric cancer. *Anticancer Res.* **2002**, *22*, 1033–1038. [PubMed]

34. Wrigley, E.; McGown, A.T.; Rennison, J.; Swindell, R.; Crowther, D.; Starzyńska, T.; Stern, P.L. 5T4 oncofetal antigen expression in ovarian carcinoma. *Int. J. Gynecol. Cancer* **1995**, *5*, 269–274. [CrossRef] [PubMed]

35. Harrop, R.; Connolly, N.; Redchenko, I.; Valle, J.W.; Saunders, M.; Ryan, M.G.; Myers, K.A.; Drury, N.; Kingsman, S.M.; Hawkins, R.E.; et al. Vaccination of Colorectal Cancer Patients with Modified Vaccinia Ankara Delivering the Tumor Antigen 5T4 (TroVax) Induces Immune Responses which Correlate with Disease Control: A Phase I/II Trial. *Clin. Cancer Res.* **2006**, *12*, 3416–3424. [CrossRef] [PubMed]

36. Elkord, E.; Dangoor, A.; Drury, N.L.; Harrop, R.; Burt, D.J.; Drijfhout, J.W.; Hamer, C.; Andrews, D.; Naylor, S.; Sherlock, D.; et al. An MVA-based Vaccine Targeting the Oncofetal Antigen 5T4 in Patients Undergoing Surgical Resection of Colorectal Cancer Liver Metastases. *J. Immunother.* **2008**, *31*, 820–829. [CrossRef]

37. Amato, R.J.; Drury, N.; Naylor, S.; Jac, J.; Saxena, S.; Cao, A.; et al. Vaccination of prostate cancer patients with modified vaccinia ankara delivering the tumor antigen 5T4 (TroVax): A phase 2 trial. *J. Immunother.* **2008**, *31*, 577–585. [CrossRef]

38. Amato, R.J.; Hawkins, R.E.; Kaufman, H.L.; Thompson, J.A.; Tomczak, P.; Szczylik, C.; McDonald, M.; Eastty, S.; Shingler, W.H.; De Belin, J.; et al. Vaccination of Metastatic Renal Cancer Patients with MVA-5T4: A Randomized, Double-Blind, Placebo-Controlled Phase III Study. *Clin. Cancer Res.* **2010**, *16*, 5539–5547. [CrossRef]

39. Barsoum, A.L.; Schwarzenberger, P.O. Oncofetal antigen/immature laminin receptor protein in pregnancy and cancer. *Cell. Mol. Boil. Lett.* **2014**, *19*, 393–406. [CrossRef]

40. Song, T.; Choi, C.H.; Cho, Y.J.; Sung, C.O.; Song, S.Y.; Kim, T.-J.; Bae, D.-S.; Lee, J.-W.; Kim, B.-G. Expression of 67-kDa laminin receptor was associated with tumor progression and poor prognosis in epithelial ovarian cancer. *Gynecol. Oncol.* **2012**, *125*, 427–432. [CrossRef]

41. McClintock, S.D.; Warner, R.L.; Ali, S.; Chekuri, A.; Dame, M.K.; Attili, D.; Knibbs, R.K.; Aslam, M.N.; Sinkule, J.; Morgan, A.C.; et al. Monoclonal antibodies specific for oncofetal antigen–immature laminin receptor protein: Effects on tumor growth and spread in two murine models. *Cancer Boil. Ther.* **2015**, *16*, 724–732. [CrossRef] [PubMed]

42. Duan, G.; Walther, D. The Roles of Post-translational Modifications in the Context of Protein Interaction Networks. *PLoS Comput. Boil.* **2015**, *11*, e1004049. [CrossRef] [PubMed]

43. Pol-Fachin, L.; Verli, H.; Lins, R.D. Extension and validation of the GROMOS 53A6glycparameter set for glycoproteins. *J. Comput. Chem.* **2014**, *35*, 2087–2095. [CrossRef] [PubMed]

44. Mitra, N.; Sharon, N.; Surolia, A. Role of N-Linked Glycan in the Unfolding Pathway ofErythrina corallodendron Lectin. *Biochemistry* **2003**, *42*, 12208–12216. [CrossRef]

45. Brandner, B.; Kurkela, R.; Vihko, P.; Kungl, A.J. Investigating the effect of VEGF glycosylation on glycosaminoglycan binding and protein unfolding. *Biochem. Biophys. Res. Commun.* **2006**, *340*, 836–839. [CrossRef]

46. Solá, R.J.; Griebenow, K. Effects of glycosylation on the stability of protein pharmaceuticals. *J. Pharm. Sci.* **2009**, *98*, 1223–1245. [CrossRef]

47. Dumez, M.-E.; Teller, N.; Mercier, F.; Tanaka, T.; Vandenberghe, I.; Vandenbranden, M.; Devreese, B.; Luxen, A.; Frère, J.-M.; Matagne, A.; et al. Activation mechanism of recombinant Der p 3 allergen zymogen: Contribution of cysteine protease Der p 1 and effect of propeptide glycosylation. *J. Boil. Chem.* **2008**, *283*, 30606–30617. [CrossRef]

48. Rossig, C.; Kailayangiri, S.; Jamitzky, S.; Altvater, B. Carbohydrate Targets for CAR T Cells in Solid Childhood Cancers. *Front. Oncol.* **2018**, *8*. [CrossRef]

49. Mereiter, S.; Balmaña, M.; Campos, D.; Gomes, J.; Reis, C.A. Glycosylation in the Era of Cancer-Targeted Therapy: Where Are We Heading? *Cancer Cell* **2019**, *36*, 6–16. [CrossRef]

50. Steentoft, C.; Migliorini, D.; King, T.R.; Mandel, U.; June, C.H.; Posey, A.D. Glycan-directed CAR-T cells. *Glycobiology* **2018**, *28*, 656–669. [CrossRef]

51. Esko, J.D.; Kimata, K.; Lindahl, U. *Proteoglycans and Sulfated Glycosaminoglycans*; Varki, A., Cummings, R.D., Esko, J.D., Freeze, H.H., Stanley, P., Eds.; Essentials of Glycobiology: Cold Spring Harbor, NY, USA, 2009.

52. Pomin, V.H.; Mulloy, B. Glycosaminoglycans and Proteoglycans. *Pharmaceuticals* **2018**, *27*, 11.

53. Wang, Q.G.; El Haj, A.J.; Kuiper, N.J. Glycosaminoglycans in the pericellular matrix of chondrons and chondrocytes. *J. Anat.* **2008**, *213*, 266–273. [CrossRef] [PubMed]

54. Hedman, K.; Johansson, S.; Vartio, T.; Kjellén, L.; Vaheri, A.; Höök, M. Structure of the pericellular matrix: Association of heparan and chondroitin sulfates with fibronectin-procollagen fibers. *Cell* **1982**, *28*, 663–671. [CrossRef]

55. Munakata, H.; Takagaki, K.; Majima, M.; Endo, M. Interaction between collagens and glycosaminoglycans investigated using a surface plasmon resonance biosensor. *Glycobiology* **1999**, *9*, 1023–1027. [CrossRef] [PubMed]

56. Wu, Y.J.; La Pierre, D.P.; Wu, J.; Yee, A.J.; Yang, B.B. The interaction of versican with its binding partners. *Cell Res.* **2005**, *15*, 483–494. [CrossRef] [PubMed]

57. Zhou, Z.-H.; Karnaukhova, E.; Rajabi, M.; Reeder, K.; Chen, T.; Dhawan, S.; Kozlowski, S. Oversulfated Chondroitin Sulfate Binds to Chemokines and Inhibits Stromal Cell-Derived Factor-1 Mediated Signaling in Activated T Cells. *PLoS ONE* **2014**, *9*, e94402. [CrossRef] [PubMed]

58. García-Suárez, O.; Garcia, B.; Fernandez-Vega, I.; Astudillo, A.; Quiros-Fernandez, L.M. Neuroendocrine Tumors Show Altered Expression of Chondroitin Sulfate, Glypican 1, Glypican 5, and Syndecan 2 Depending on Their Differentiation Grade. *Front. Oncol.* **2014**, *4*, 15. [CrossRef]

59. Prabhakar, V.; Sasisekharan, R. The Biosynthesis and Catabolism of Galactosaminoglycans. *Adv. Pharmacol.* **2006**, *53*, 69–115.

60. Mikami, T.; Kitagawa, H. Biosynthesis and function of chondroitin sulfate. *Biochim. Biophys. Acta Gen. Subj.* **2013**, *1830*, 4719–4733. [CrossRef]

61. Da Costa, D.S.; Reis, R.L.; Pashkuleva, I. Sulfation of Glycosaminoglycans and Its Implications in Human Health and Disorders. *Annu. Rev. Biomed. Eng.* **2017**, *19*, 1–26. [CrossRef]

62. Cooney, C.; Jousheghany, F.; Yao-Borengasser, A.; Phanavanh, B.; Gomes, T.; Kieber-Emmons, A.M.; Siegel, E.R.; Suva, L.J.; Ferrone, S.; Kieber-Emmons, T.; et al. Chondroitin sulfates play a major role in breast cancer metastasis: A role for CSPG4 and CHST11gene expression in forming surface P-selectin ligands in aggressive breast cancer cells. *Breast Cancer Res.* **2011**, *13*, R58. [CrossRef] [PubMed]

63. Pereira, M.M.B.A.; Clausen, T.M.; Pehrson, C.; Mao, Y.; Resende, M.; Daugaard, M.; Kristensen, A.R.; Spliid, C.; Mathiesen, L.; Knudsen, L.E.; et al. Placental Sequestration of Plasmodium falciparum Malaria Parasites Is Mediated by the Interaction Between VAR2CSA and Chondroitin Sulfate A on Syndecan-1. *PLoS Pathog.* **2016**, *12*, e1005831.

64. Nadanaka, S.; Kinouchi, H.; Kitagawa, H. Chondroitin sulfate-mediated N-cadherin/beta-catenin signaling is associated with basal-like breast cancer cell invasion. *J. Biol. Chem.* **2018**, *293*, 444–465. [CrossRef] [PubMed]

65. Pudełko, A.; Wisowski, G.; Olczyk, K.; Koźma, E.M. The dual role of the glycosaminoglycan chondroitin-6-sulfate in the development, progression and metastasis of cancer. *FEBS J.* **2019**, *286*, 1815–1837. [CrossRef]

66. Fthenou, E.; Zong, F.; Zafiropoulos, A.; Dobra, K.; Hjerpe, A.; Tzanakakis, G.N. Chondroitin sulfate A regulates fibrosarcoma cell adhesion, motility and migration through JNK and tyrosine kinase signaling pathways. *In Vivo* **2009**, *23*.

67. Chiarugi, V.P.; Dietrich, C.P. Sulfated mucopolysaccharides from normal and virus transformed rodent fibroblasts. *J. Cell. Physiol.* **1979**, *99*, 201–206. [CrossRef]

68. Kramer, K.L. Specific sides to multifaceted glycosaminoglycans are observed in embryonic development. *Semin. Cell Dev. Boil.* **2010**, *21*, 631–637. [CrossRef]

69. Shannon, J.M.; McCormick-Shannon, K.; Burhans, M.S.; Shangguan, X.; Srivastava, K.; Hyatt, B.A. Chondroitin sulfate proteoglycans are required for lung growth and morphogenesis in vitro. *Am. J. Physiol. Cell. Mol. Physiol.* **2003**, *285*, L1323–L1336. [CrossRef]

70. Long, K.R.; Huttner, W.B. How the extracellular matrix shapes neural development. *Open Boil.* **2019**, *9*, 180216. [CrossRef]

71. Djerbal, L.; Lortat-Jacob, H.; Kwok, J. Chondroitin sulfates and their binding molecules in the central nervous system. *Glycoconj. J.* **2017**, *34*, 363–376. [CrossRef]

72. Izumikawa, T.; Sato, B.; Kitagawa, H. Chondroitin Sulfate Is Indispensable for Pluripotency and Differentiation of Mouse Embryonic Stem Cells. *Sci. Rep.* **2014**, *4*, 3701. [CrossRef] [PubMed]

73. Lee, T.-Y.; Jamieson, A.M.; A Schafer, I. Changes in the Composition and Structure of Glycosaminoglycans in the Human Placenta during Development. *Pediatr. Res.* **1973**, *7*, 965–977. [CrossRef] [PubMed]

74. Blois, S.; Dveksler, G.; Vasta, G.R.; Freitag, N.; Blanchard, V.; Barrientos, G. Pregnancy Galectinology: Insights Into a Complex Network of Glycan Binding Proteins. *Front. Immunol.* **2019**, *10*, 1166. [CrossRef] [PubMed]

75. Salanti, A.; Dahlbaäck, M.; Turner, L.; Nielsen, M.A.; Barfod, L.; Magistrado, P.; Jensen, A.R.; Lavstsen, T.; Ofori, M.F.; Marsh, K.; et al. Evidence for the Involvement of VAR2CSA in Pregnancy-associated Malaria. *J. Exp. Med.* **2004**, *200*, 1197–1203. [CrossRef]

76. Clausen, T.; Christoffersen, S.; Dahlbäck, M.; Langkilde, A.E.; Jensen, K.E.; Resende, M.; Agerbæk, M.Ø.; Andersen, D.; Berisha, B.; Ditlev, S.B.; et al. Structural and Functional Insight into How the Plasmodium falciparum VAR2CSA Protein Mediates Binding to Chondroitin Sulfate A in Placental Malaria*. *J. Boil. Chem.* **2012**, *287*, 23332–23345. [CrossRef]

77. Fried, M.; Duffy, P.E. Adherence of Plasmodium falciparum to Chondroitin Sulfate A in the Human Placenta. *Science* **1996**, *272*, 1502–1504. [CrossRef]

78. Resende, M.; Nielsen, M.A.; Dahlbäck, M.; Ditlev, S.B.; Andersen, P.; Sander, A.F.; Ndam, N.T.; Theander, T.G.; Salanti, A. Identification of glycosaminoglycan binding regions in the Plasmodium falciparum encoded placental sequestration ligand, VAR2CSA. *Malar. J.* **2008**, *7*, 104. [CrossRef]

79. Gangnard, S.; Chêne, A.; Dechavanne, S.; Srivastava, A.; Avril, M.; Smith, J.D.; Gamain, B. VAR2CSA binding phenotype has ancient origin and arose before Plasmodium falciparum crossed to humans: Implications in placental malaria vaccine design. *Sci. Rep.* **2019**, *9*, 1–10. [CrossRef]

80. Salanti, A.; Clausen, T.M.; Agerbæk, M.Ø.; Al Nakouzi, N.; Dahlbäck, M.; Oo, H.Z.; Lee, S.; Gustavsson, T.; Rich, J.R.; Hedberg, B.J.; et al. Targeting Human Cancer by a Glycosaminoglycan Binding Malaria Protein. *Cancer Cell* **2015**, *28*, 500–514. [CrossRef]

81. Agerbæk, M.Ø.; Bang-Christensen, S.R.; Yang, M.-H.; Clausen, T.M.; Pereira, M.M.B.A.; Sharma, S.; Ditlev, S.B.; Nielsen, M.A.; Choudhary, S.; Gustavsson, T.; et al. The VAR2CSA malaria protein efficiently retrieves circulating tumor cells in an EpCAM-independent manner. *Nat. Commun.* **2018**, *9*, 3279.

82. Seiler, R.; Oo, H.Z.; Tortora, D.; Clausen, T.; Wang, C.K.; Kumar, G.; Pereira, M.M.B.A.; Ørum-Madsen, M.S.; Agerbæk, M.Ø.; Gustavsson, T.; et al. An Oncofetal Glycosaminoglycan Modification Provides Therapeutic Access to Cisplatin-resistant Bladder Cancer. *Eur. Urol.* **2017**, *72*, 142–150. [CrossRef] [PubMed]

83. Fernandez-Vega, I.; García-Suarez, O.; García, B.; Crespo, A.; Astudillo, A.; Quiros-Fernandez, L.M. Heparan sulfate proteoglycans undergo differential expression alterations in right sided colorectal cancer, depending on their metastatic character. *BMC Cancer* **2015**, *15*, 742. [CrossRef] [PubMed]

84. Allen-Rhoades, W.; Whittle, S.B.; Rainusso, N. Pediatric Solid Tumors of Infancy: An Overview. *Pediatr. Rev.* **2018**, *39*, 57–67. [CrossRef] [PubMed]

85. Allen-Rhoades, W.; Whittle, S.B.; Rainusso, N. Pediatric Solid Tumors in Children and Adolescents: An Overview. *Pediatr. Rev.* **2018**, *39*, 444–453. [CrossRef] [PubMed]

86. Yadavilli, S.; Hwang, E.I.; Packer, R.J.; Nazarian, J. The Role of NG2 Proteoglycan in Glioma. *Transl. Oncol.* **2016**, *9*, 57–63. [CrossRef]

87. Riccardo, F.; Tarone, L.; Iussich, S.; Giacobino, D.; Arigoni, M.; Sammartano, F.; Morello, E.; Martano, M.; Gattino, F.; De Maria, R.; et al. Identification of CSPG4 as a promising target for translational combinatorial approaches in osteosarcoma. *Ther. Adv. Med Oncol.* **2019**, *11*. [CrossRef]

88. Brehm, H.; Niesen, J.; Mladenov, R.; Stein, C.; Pardo, A.; Fey, G.; Helfrich, W.; Fischer, R.; Gattenlöhner, S.; Barth, S. A CSPG4-specific immunotoxin kills rhabdomyosarcoma cells and binds to primary tumor tissues. *Cancer Lett.* **2014**, *352*, 228–235. [CrossRef]

89. Bang-Christensen, S.R.; Pedersen, R.S.; Pereira, M.M.B.A.; Clausen, T.; Løppke, C.; Sand, N.T.; Ahrens, T.D.; Jørgensen, A.M.; Lim, Y.C.; Goksøyr, L.; et al. Capture and Detection of Circulating Glioma Cells Using the Recombinant VAR2CSA Malaria Protein. *Cells* **2019**, *8*, 998. [CrossRef]

90. Price, M.A.; Wanshura, L.E.C.; Yang, J.; Carlson, J.; Xiang, B.; Li, G.; Ferrone, S.; Dudek, A.Z.; Turley, E.A.; McCarthy, J.B. CSPG4, a potential therapeutic target, facilitates malignant progression of melanoma. *Pigment. Cell Melanoma Res.* **2011**, *24*, 1148–1157. [CrossRef]

91. Rolih, V.; Barutello, G.; Iussich, S.; De Maria, R.; Quaglino, E.; Buracco, P.; Cavallo, F.; Riccardo, F. CSPG4: A prototype oncoantigen for translational immunotherapy studies. *J. Transl. Med.* **2017**, *15*, 151. [CrossRef]

92. Sood, D.; Tang-Schomer, M.; Pouli, D.; Mizzoni, C.; Raia, N.; Tai, A.; Arkun, K.; Wu, J.; Black, L.D.; Scheffler, B.; et al. 3D extracellular matrix microenvironment in bioengineered tissue models of primary pediatric and adult brain tumors. *Nat. Commun.* **2019**, *10*, 1–14. [CrossRef] [PubMed]

93. Higgins, S.C.; Bolteus, A.J.; Donovan, L.K.; Hasegawa, H.; Doey, L.; Al-Sarraj, S.; King, A.; Ashkan, K.; Roncaroli, F.; Fillmore, H.L.; et al. Expression of the chondroitin sulphate proteoglycan, NG2, in paediatric brain tumors. *Anticancer Res.* **2014**, *34*.

94. Müller, S.; Kunkel, P.; Lamszus, K.; Ulbricht, U.; Lorente, G.A.; Nelson, A.M.; von Schack, D.; Chin, D.J.; Lohr, S.C.; Westphal, M.; et al. A role for receptor tyrosine phosphatase zeta in glioma cell migration. *Oncogene* **2003**, *22*, 6661–6668. [CrossRef] [PubMed]

95. Peters, H.L.; Yan, Y.; Nordgren, T.; Cutucache, C.; Joshi, S.S.; Solheim, J.C. Amyloid precursor-like protein 2 suppresses irradiation-induced apoptosis in Ewing sarcoma cells and is elevated in immune-evasive Ewing sarcoma cells. *Cancer Boil. Ther.* **2013**, *14*, 752–760. [CrossRef]

96. Nikitovic, D.; Berdiaki, A.; Spyridaki, I.; Krasanakis, T.; Aristidis, T.; Tzanakakis, G.N. Proteoglycans—Biomarkers and Targets in Cancer Therapy. *Front. Endocrinol.* **2018**, *9*. [CrossRef]

97. Iozzo, R.V.; Sanderson, R.D. Proteoglycans in cancer biology, tumour microenvironment and angiogenesis. *J. Cell. Mol. Med.* **2011**, *15*, 1013–1031. [CrossRef]

98. Ortiz, M.; Roberts, S.S.; Bender, J.G.; Shukla, N.; Wexler, L.H. Immunotherapeutic Targeting of GPC3 in Pediatric Solid Embryonal Tumors. *Front. Oncol.* **2019**, *9*, 108. [CrossRef]

99. Chan, E.S.; Pawel, B.R.; A Corao, D.; Venneti, S.; Russo, P.; Santi, M.; Sullivan, L. Immunohistochemical Expression of Glypican-3 in Pediatric Tumors: An Analysis of 414 Cases. *Pediatr. Dev. Pathol.* **2013**, *16*, 272–277. [CrossRef]

100. Williamson, D.; Selfe, J.; Gordon, T.; Lu, Y.-J.; Pritchard-Jones, K.; Murai, K.; Jones, P.H.; Workman, P.; Shipley, J.M. Role for Amplification and Expression of Glypican-5 in Rhabdomyosarcoma. *Cancer Res.* **2007**, *67*, 57–65. [CrossRef]

101. Xu, Y.; Yuan, J.; Zhang, Z.; Lin, L.; Xu, S. Syndecan-1 expression in human glioma is correlated with advanced tumor progression and poor prognosis. *Mol. Boil. Rep.* **2012**, *39*, 8979–8985. [CrossRef]

102. Clausen, T.M.; Pereira, M.M.B.A.; Al Nakouzi, N.; Oo, H.Z.; Agerbæk, M.Ø.; Lee, S.; Ørum-Madsen, M.S.; Kristensen, A.R.; El-Naggar, A.; Grandgenett, P.M.; et al. Oncofetal Chondroitin Sulfate Glycosaminoglycans Are Key Players in Integrin Signaling and Tumor Cell Motility. *Mol. Cancer Res.* **2016**, *14*, 1288–1299. [CrossRef] [PubMed]

103. Li, F.; Li, S.; Cheng, T. TGF-beta1 promotes osteosarcoma cell migration and invasion through the miR-143-versican pathway. *Cell. Phys. Biochem.* **2014**, *34*, 2169–2179. [CrossRef] [PubMed]

104. Zafiropoulos, A.; Nikitovic, D.; Katonis, P.; Aristidis, T.; Karamanos, N.K.; Tzanakakis, G.N. Decorin-Induced Growth Inhibition Is Overcome through Protracted Expression and Activation of Epidermal Growth Factor Receptors in Osteosarcoma Cells. *Mol. Cancer Res.* **2008**, *6*, 785–794. [CrossRef] [PubMed]

105. Rota, C.M.; Tschernia, N.; Feldman, S.; Mackall, C.; Lee, D.W. Abstract 3151: T cells engineered to express a chimeric antigen receptor targeting chondroitin sulfate proteoglycan 4 (CSPG4) specifically kill medulloblastoma and produce inflammatory cytokines. *Immunology* **2015**, *75*, 3151.

106. Pietras, E.J.; Katz, A.M.; Ekström, E.J.; Wee, B.; Halliday, J.J.; Pitter, K.L.; Werbeck, J.L.; Amankulor, N.M.; Huse, J.T.; Holland, E.C. Osteopontin-CD44 signaling in the glioma perivascular niche enhances cancer stem cell phenotypes and promotes aggressive tumor growth. *Cell Stem Cell* **2014**, *14*, 357–369. [CrossRef] [PubMed]

107. Zhu, H.; Cai, H.; Tang, M.; Tang, J. Neuropilin-1 is overexpressed in osteosarcoma and contributes to tumor progression and poor prognosis. *Clin. Transl. Oncol.* **2013**, *16*, 732–738. [CrossRef] [PubMed]

108. Marcus, K.; Johnson, M.; Adam, R.; O'Reilly, M.S.; Donovan, M.; Atala, A.; Freeman, M.R.; Soker, S. Tumor cell-associated neuropilin-1 and vascular endothelial growth factor expression as determinants of tumor growth in neuroblastoma. *Neuropathology* **2005**, *25*, 178–187. [CrossRef]

109. Downing, J.R.; Wilson, R.K.; Zhang, J.; Mardis, E.R.; Pui, C.-H.; Ding, L.; Ley, T.J.; Evans, W.E. The Pediatric Cancer Genome Project. *Nat. Genet.* **2012**, *44*, 619–622. [CrossRef]

Review

Targeting mTOR and Metabolism in Cancer: Lessons and Innovations

Cedric Magaway, Eugene Kim and Estela Jacinto *

Department of Biochemistry and Molecular Biology, Rutgers-Robert Wood Johnson Medical School, Piscataway, NJ 08854, USA; cgm95@rwjms.rutgers.edu (C.M.); eugene.jae.kim@gmail.com (E.K.)
* Correspondence: jacintes@rwjms.rutgers.edu; Tel.: +1-(732)-235-4476

Received: 28 October 2019; Accepted: 3 December 2019; Published: 6 December 2019

Abstract: Cancer cells support their growth and proliferation by reprogramming their metabolism in order to gain access to nutrients. Despite the heterogeneity in genetic mutations that lead to tumorigenesis, a common alteration in tumors occurs in pathways that upregulate nutrient acquisition. A central signaling pathway that controls metabolic processes is the mTOR pathway. The elucidation of the regulation and functions of mTOR can be traced to the discovery of the natural compound, rapamycin. Studies using rapamycin have unraveled the role of mTOR in the control of cell growth and metabolism. By sensing the intracellular nutrient status, mTOR orchestrates metabolic reprogramming by controlling nutrient uptake and flux through various metabolic pathways. The central role of mTOR in metabolic rewiring makes it a promising target for cancer therapy. Numerous clinical trials are ongoing to evaluate the efficacy of mTOR inhibition for cancer treatment. Rapamycin analogs have been approved to treat specific types of cancer. Since rapamycin does not fully inhibit mTOR activity, new compounds have been engineered to inhibit the catalytic activity of mTOR to more potently block its functions. Despite highly promising pre-clinical studies, early clinical trial results of these second generation mTOR inhibitors revealed increased toxicity and modest antitumor activity. The plasticity of metabolic processes and seemingly enormous capacity of malignant cells to salvage nutrients through various mechanisms make cancer therapy extremely challenging. Therefore, identifying metabolic vulnerabilities in different types of tumors would present opportunities for rational therapeutic strategies. Understanding how the different sources of nutrients are metabolized not just by the growing tumor but also by other cells from the microenvironment, in particular, immune cells, will also facilitate the design of more sophisticated and effective therapeutic regimen. In this review, we discuss the functions of mTOR in cancer metabolism that have been illuminated from pre-clinical studies. We then review key findings from clinical trials that target mTOR and the lessons we have learned from both pre-clinical and clinical studies that could provide insights on innovative therapeutic strategies, including immunotherapy to target mTOR signaling and the metabolic network in cancer.

Keywords: mTORC1; mTORC2; metabolism; rapalogs; mTOR inhibitors; cancer metabolism; mTOR in immunotherapy; nutrient metabolism; kinase inhibitors; mTOR signaling

---

1. Introduction

In 1956, Otto Warburg wrote that if we know how cancer cells have "damaged respiration and excessive fermentation", then we understand the origin of cancer cells [1]. Warburg was referring to the abnormal metabolism of glucose in cancer. Both normal and cancer cells metabolize nutrients, primarily glucose, to produce energy in the form of ATP. There are two ways for cells to generate ATP. First, through the more evolutionarily primitive means via glycolysis, a process that converts glucose to pyruvate without the need for oxygen (anaerobic). This process is also referred to as fermentation, as lower organisms ferment glucose to ethanol, whereas higher organisms can convert pyruvate to lactate in the cytoplasm. Secondly, ATP is also produced via respiration in the mitochondria. Pyruvate from glycolysis enters the mitochondria, resulting in the production of abundant ATP in the presence of oxygen (aerobic), hence the term respiration. However, Warburg observed that cancer cells increase their fermentation, referring to the anaerobic process of glucose metabolism despite the presence of oxygen. This has been termed as aerobic glycolysis or the Warburg effect. The increased fermentation favors the conversion of pyruvate to lactate over pyruvate metabolism in the mitochondria. Warburg proposed that the "driving force of the increased fermentation is the energy deficiency" in order to generate sufficient ATP in lieu of the defective respiration process. How this happens in cancer cells was then mysterious, but more modern approaches to study cancer metabolism have now provided clues that may help us understand the origin of cancer and how we can eradicate them more effectively.

More than a decade after Warburg's observations, an expedition in the island of Rapa Nui to collect soil bacteria for the purpose of isolating active natural compounds unearthed a drug that would later prove instrumental in understanding the anomalous metabolism of cancer cells. More importantly, this drug, named rapamycin, has served as a prototype for the development of more effective compounds to treat cancer as well as other metabolism-related disorders. Rapamycin forms a complex with the prolyl isomerase FKBP12 and together binds to mTOR, a protein kinase that plays a central role in controlling growth and metabolism [2]. Since the identification of mTOR, numerous studies have unraveled how mTOR signaling reprograms metabolism to acquire nutrients from the environment and intracellular sources in order to generate ATP as well as intermediates for macromolecule synthesis [3]. This central function of mTOR in metabolic reprogramming could be targeted using mTOR inhibitors and thus prevent tumor growth and malignancy [4]. The first generation mTOR inhibitors, rapamycin analogs (rapalogs), have shown promising results in pre-clinical studies but with modest results in the clinic [5]. Nevertheless, some of the rapalogs have now been approved by the FDA for the treatment of specific types of cancer. A limitation of rapamycin is that it is only an allosteric mTOR inhibitor and does not fully block its activity. Second generation mTOR inhibitors that target the catalytic site of mTOR to block mTOR activity have been developed [6]. Pre-clinical studies with these inhibitors have confirmed their potency in preventing cell growth and proliferation. However, results from clinical trials have been sobering due to their toxic nature. Hence, there is a need to improve therapeutic strategies in order to gain benefit from these compounds. Studies from the use of these mTOR inhibitors have also unraveled that tumors exhibit such metabolic plasticity and that other mTOR-independent mechanisms can be triggered to compensate for the block in mTOR activity, thus allowing cells to acquire nutrients and metabolize them for growth and proliferation. Here, we discuss the metabolic functions of mTOR that have emerged from pre-clinical studies using cells and animal models. We then review the results from studies utilizing different strategies of mTOR inhibition both from pre-clinical and clinical studies and discuss the lessons from the successes and failures gleaned from these studies. We mention rational therapeutic approaches to more specifically target tumors that will have high sensitivity to mTOR inhibition. Lastly, we discuss recent innovative strategies in manipulating the tumor microenvironment using mTOR inhibitors to target the growing tumor or to improve immunotherapy for cancer treatment.

2. mTOR Signaling and its Role in Metabolism

Early studies in yeast unraveled that TOR is involved in the regulation of growth or cell mass accumulation in response to the presence of nutrients [7,8]. Studies in other organisms such as *Drosophila* and mammals further corroborated the critical role of mTOR in promoting not only cell growth but also organismal growth [9]. The elucidation of the function of mTOR in protein synthesis and autophagy provided clues on its role in nutrient sensing and anabolic metabolism [10,11]. Genome-wide screening further uncovered the effect of rapamycin on metabolic genes, revealing that TOR/mTOR mediates the expression of genes involved in nutrient metabolism [12–15]. mTOR is part of two structurally distinct complexes, mTORC1 and mTORC2. The conserved components of mTORC1 include mTOR, raptor and mLST8 whereas mTORC2 consists of mTOR, rictor, SIN1 and mLST8 (Figure 1). Genetic studies that ablated components of the mTOR complexes in a tissue-specific manner also provided support on the role of mTOR on glucose, amino acid, lipid, nucleotide metabolism and other biosynthetic pathways [16–18]. In addition to promoting anabolic metabolism, mTOR also functions to negatively regulate catabolic processes such as autophagy. Altogether, these findings unraveled how mTOR controls cell growth via its central role in metabolism.

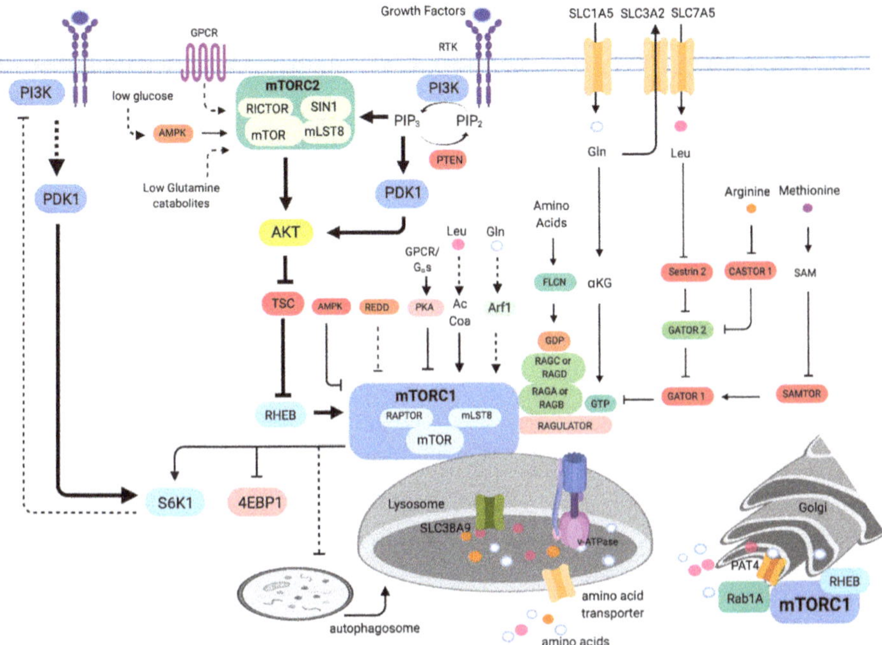

Figure 1. mTOR Signaling. mTORC1 activation is modulated by the presence of nutrients such as amino acids at the membrane surface of organelles such as the lysosomes and Golgi. Signaling to mTORC1 is potentiated by growth factor/PI3K signaling via Akt. mTORC2 activation is enhanced by the presence of growth factors and also occurs on membrane subcellular compartments. It is also augmented by G-protein coupled receptor (GPCR) signaling and by nutrient-limiting conditions. The bold lines indicate signals from growth factor signaling. The dashed lines indicate indirect modulation.

2.1. Signaling to mTOR

mTOR as part of mTORC1 is active in the presence of nutrients such as amino acids [19]. Several amino acid transporters, including the transporters for glutamine (SLC1A5/ASCT2) and leucine (SLC7A5/LAT1, which imports Leu in exchange for Gln efflux by SLC3A2/CD98/4F2hc), have been linked to mTORC1 activation and their overexpression is often associated with malignancies [20–23]. The activation of mTORC1 occurs via recruitment to the surface of the lysosomes, a major hub for the degradation and recycling of macromolecules. When nutrients are abundant, mTORC1 is activated via the Ras-related GTP binding proteins (Rags) [24,25]. RagA/B is bound to GTP while RagC/D is GDP-bound under amino acid sufficiency. The Rag heterodimers then interact with raptor and facilitate translocation of mTORC1 to the lysosomal surface. Amino acids such as leucine and arginine activate mTORC1 robustly via Rag-dependent mechanisms. In the case of leucine and arginine, proteins that bind to these amino acids, such as sestrin 2 and CASTOR1, respectively, mediate the activation of mTORC1 [26]. In the presence of leucine, the sestrin2/leucine interaction relieves the inhibition of GATOR1 by GATOR2, ultimately activating RagA/B and mTORC1

(Figure 1). Arginine binding to CASTOR1 also derepresses the inhibition of GATOR2 by CASTOR1, thereby activating mTORC1. Another Arg sensor, SLC38A9, an Arg gated amino acid transporter affects RagA/RagB nucleotide state and thus controls mTORC1 activation [27,28]. Methionine, on the other hand, is sensed as S-adenosylmethionine (SAM) via SAMTOR. The binding of SAM to SAMTOR disrupts the SAMTOR-GATOR1 association, thus activating mTORC1. Glutamine activates mTORC1 via Rag-dependent and -independent mechanisms [19,29]. Glutamine is metabolized during glutaminolysis to produce α-ketoglutarate. This process enhances mTORC1 activation via Rag. Glutamine also promotes mTORC1 translocation to the lysosomal surface via the ADP ribosylation factor (Arf1) GTPase, independent of Rag [30,31]. A lysosomal GPCR-like protein GPR137B also regulates Rag and mTORC1 localization to the lysosomes and plays a role in regulating the GTP loading state of RagA [32]. Leucine also activates mTORC1 via its catabolite acetyl-coenzyme A (AcCoA), independently of sestrin2 [33]. AcCoA promotes EP300-mediated acetylation of raptor at K1097, thus enhancing mTORC1 activity. This regulation of mTORC1 appears to be cell-type-specific, however. The activation of mTORC1 by amino acids is potentiated by signals from growth factors. The binding of growth factors such as insulin to receptor tyrosine kinases trigger phosphatidylinositol-3 kinase (PI3K) activation. PI3K signals are wired to mTORC1 via the tumor suppressor proteins tuberous sclerosis complex (TSC1 and TSC2), which acts to negatively regulate mTORC1 [34,35]. Upon PI3K activation, the TSC1/TSC2 complex is inactivated. TSC2 functions as a GTPase-activating protein (GAP) that inhibits the Rheb GTPase [36–38]. Hence, cells that lack or have inactivating mutations in TSC1 or TSC2 have increased mTORC1 activity. Rheb binds and activates mTORC1 by realignment of residues in the mTOR active site to enhance catalytic activity [37]. As discussed further below, there are a number of tumors that have upregulated mTORC1 activity due to TSC1/TSC2 inactivation. mTORC1 is also modulated by cellular energy via AMPK. AMPK modulates mTORC1 via phosphorylation of raptor and indirectly via phosphorylation of TSC2 [34,39]. Under energy-depleted conditions, AMPK becomes activated and inhibits mTORC1. Decreased oxygen also dampens mTORC1 activation via REDD1, which acts through TSC [40]. Recently, mTORC1 has been shown to also be inhibited by G-protein coupled receptor (GPCR) signaling [41]. This inhibition occurs via $G_{\alpha s}$ proteins, which increase cyclic adenosine 3′5′ monophosphate (cAMP), leading to the activation of PKA. PKA phosphorylates Ser791 of raptor, leading to diminished mTORC1 activity. In addition to mTORC1 activation on the lysosomal surface, mTORC1 is also activated by amino acids at the surface of the Golgi via Rab1A, another small GTPase [42]. Rab1A activates mTORC1 via promoting the interaction of mTORC1 with Rheb in the Golgi. The amino acid transporter PAT4 (SLC36A4), which is mainly associated with the Golgi interacts with mTORC1 and Rab1A [20]. Further studies are needed to determine how mTORC1 could be activated in different cellular compartments by amino acids and possibly other metabolites [43].

The activation of mTORC2 is relatively less understood than that of mTORC1 [44]. mTORC2 associates with plasma membrane, mitochondria and a subpopulation of endosomal vesicles [45]. mTORC2 is also activated by signals from growth factors. The precise mechanisms as to how mTORC2 is activated remain unclear but its component SIN1 has been shown to bind phosphatidylinositol (3,4,5)-trisphosphate PIP3 and could thus promote mTORC2 activation [46,47]. Increased receptor tyrosine kinase activation enhances PI3K activity, resulting in increased PIP3. This phospholipid attracts several signaling molecules including Akt and the 3-phosphoinositide dependent kinase 1 (PDK1). Signals from PI3K are antagonized by the lipid phosphatase tumor suppressor, PTEN. Tumors with activating mutations in PI3K or inactivating mutations or deletion of PTEN have upregulated mTORC2 signaling. PI3K activation increases phosphorylation of the mTORC2 target Akt at the allosteric site Ser473 while PDK1 phosphorylates Akt at the catalytic site at Thr308. mTORC2 also promotes the phosphorylation of PKC and SGK1 at sites homologous to the Akt-phosphorylated sites [48–51]. Additionally, mTORC2 promotes the phosphorylation of Akt and PKC at the turn motif site in a manner that occurs co-translationally and is not further enhanced by addition of growth factors [52]. Consistently with this, active mTORC2 is present in the ribosomes [53]. Interestingly, mTORC2 activation is also enhanced during nutrient limitation [54,55]. Glutamine or glucose starvation can increase phosphorylation of mTORC2 targets such as Akt. The activation of mTORC2 during glucose depletion occurs via a more direct action of AMPK on mTORC2 via phosphorylation of mTOR and possibly rictor [55]. β- and α-adrenergic signaling through GPCR also modulates mTORC2. β-adrenergic stimulation of mTORC2 signaling induces lipid catabolism in brown adipocytes and glucose uptake into skeletal muscle or brown fat cells [56]. The function of rictor in lipid catabolism in brown adipocytes is independent of Akt but involves FoxO1 deacetylation via SIRT6 [57]. α-adrenergic signaling to mTORC2 also promotes glucose uptake in cardiomyocytes [58]. Hence, unlike mTORC1 that is activated by anabolic and negatively regulated by catabolic signals, mTORC2 seems to be modulated positively by both types of signals. It is likely that mTORC2 maintains a basal level of activation, as supported by constitutive phosphorylation of some of its targets, but further elevates its activity to restore metabolic homeostasis during nutrient fluctuations.

2.2. Protein Synthesis

mTOR has multiple roles in the regulation of protein synthesis (Figure 2). Much of what we know about this mTOR function relates to mTORC1. mTORC1 phosphorylates two key effectors of translation, S6K1 and eIF4E-BP (4EBP). mTORC1 phosphorylates the Thr389 residue of S6K1, a serine/threonine kinase that belongs to the AGC kinase family [59]. Thr389 resides in the hydrophobic motif of S6K1, which is a motif that is common to most AGC kinases. Phosphorylation at Thr389 enables subsequent phosphorylation of S6K1 at its activation loop by PDK1. S6K1 has several downstream targets that are involved in protein synthesis, including eIF4B, a positive regulator of the 5′cap binding eIF4F complex. It also phosphorylates PDCD4, leading to its degradation and the positive regulation of eIF4B [60]. S6K1 also enhances translation efficiency of spliced mRNAs via SKAR, a component of exon-junction complexes [61]. In addition to the above functions of S6K1, it has other targets that are involved in metabolism, as discussed further below. mTORC1 also phosphorylates 4EBP at multiple sites. Phosphorylation of 4EBP promotes 5′cap-dependent mRNA translation by dissociation of 4EBP from eIF4E, thus allowing assembly of the eIF4F complex. 4EBP is also phosphorylated by other kinases to regulate its function in an mTORC1-dependent or -independent manner [62,63].

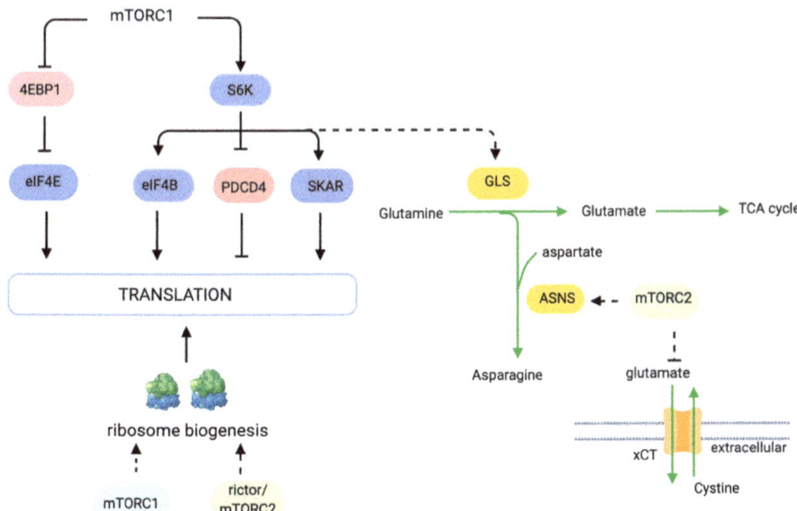

Figure 2. The role of mTOR in protein synthesis. mTOR controls several aspects of protein synthesis, including ribosome biogenesis, translation, and amino acid transport and synthesis.

In addition to regulating key effectors of protein synthesis, mTOR also regulates synthesis of amino acids, the building blocks for protein synthesis. mTOR upregulates asparagine biosynthesis via enhancing expression of the asparagine synthetase (ASNS) in colorectal cancer cells with KRAS mutations [64]. In non-small-cell lung cancer expressing oncogenic KRAS, the increased ASNS expression occurs via Akt- and Nrf2-mediated induction of ATF4. Tumor growth is prevented by inhibition of Akt together with depletion of extracellular asparagine [65]. mTOR also regulates the amounts of amino acid transporters. Genomic studies identified neutral amino acid transporters to be decreased upon rapamycin treatment [13]. mTORC2 also plays a role in regulating amino acid transporters. mTORC2 phosphorylates Ser26 of the cystine-glutamate anti-porter xCT, thereby preventing the efflux of glutamate and influx of cystine [66]. When mTORC2 activity was disrupted genetically or pharmacologically, glutamate secretion, cystine uptake and incorporation into glutathione were enhanced. This function of mTORC2 could allow highly proliferating cancer cells to utilize glutamate for TCA anaplerosis. When nutrients become limiting or when mTORC2 signals are dampened, the influx of cystine, at the expense of glutamate efflux, would allow tumor cells to relieve redox stress.

mTOR also controls the biogenesis of ribosomes, the machinery that drives protein synthesis. mTOR positively regulates several processes involved in ribosome biogenesis, including rRNA transcription as well as the synthesis of ribosomal proteins and components of ribosome assembly [67]. The mTORC2 component rictor is recruited to the nucleolar compartment during epithelial-to-mesenchymal transition (EMT)-associated ribosome biogenesis [68]. The role of rictor or mTORC2 in this compartment requires further investigation. Since previous studies demonstrated that active mTORC2 associates with ribosomes and that it could phosphorylate nascent peptides [52,53], mTORC2 could thus have multiple ribosome-associated functions. Many tumors have high rates of ribosome biogenesis to support the augmented protein synthesis necessary for growth and proliferation. Inhibition of RNA polymerase I has been used in pre-clinical studies and is also undergoing clinical trials for cancer therapy [69]. Recently, the inhibitor of Pol I-mediated rDNA transcription (CX-5461) was combined with everolimus and was shown to synergistically increase the survival of mice with a Myc-driven lymphoma [70].

2.3. Glucose Metabolism

Cancer cells enhance their rate of glucose uptake and produce pyruvate at a higher rate than can be metabolized by the mitochondria. Under this condition, the excess pyruvate is diverted from being metabolized in the mitochondria to being converted to lactate in the cytosol. This glycolytic switch can occur under aerobic conditions. Genetic mutations that lead to enhanced growth factor-PI3K/Akt/mTOR signaling can drive and/or maintain this switch. Indeed, multiple points along the glycolytic pathway are influenced by mTOR via regulation of critical transcription factors such as HIF1α and Myc (Figure 3). HIF1 (hypoxia inducible factor 1) is a heterodimer consisting of an O_2-regulated HIF1α subunit and a constitutively expressed HIF1β subunit [71]. The expression of HIF1α is dependent on mTORC1 and mTORC2 [72]. Increased HIF1α expression is sufficient to induce expression of genes whose products increase glycolytic flux [73]. While HIF1α expression is normally elevated under hypoxia, it becomes abnormally upregulated in cancer despite aerobic conditions. HIF1 expression is upregulated in many primary and metastatic human tumors [71]. Early studies in prostate cancer cells have shown that inhibiting mTOR by rapamycin blocks the growth factor- and mitogen-induced HIF1α expression [74]. Rapamycin also decreases HIF1α stabilization and transcriptional activity under hypoxic conditions [75]. mTORC1 promotes HIF1α transcription through Foxk1 [76]. By mTORC1-mediated inhibition of GSK3, Foxk1 phosphorylation is repressed and allows its accumulation in the nucleus to induce HIF1α transcription. Elevated mTORC1 activation that occurs in TSC2$^{-/-}$ cells also increases translation of HIF1α mRNA [77] while rapamycin decreases its mRNA levels [78,79]. The control of HIF1α translation involves the mTORC1 target, 4E-BP1 [14]. Thus, mTORC1 can regulate HIF1α expression at the level of transcription and translation.

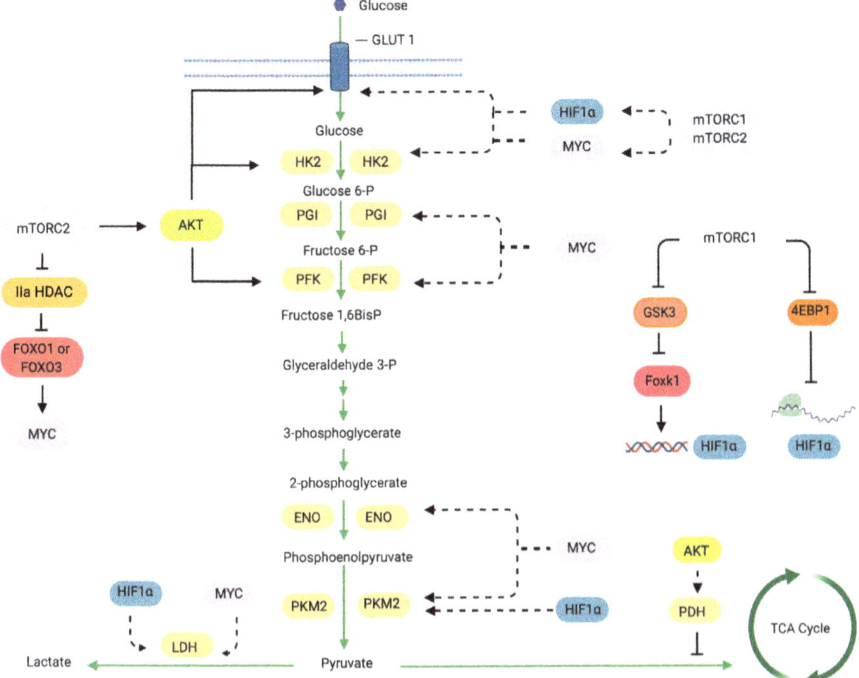

Figure 3. The role of mTOR in glucose metabolism. mTORC1 controls glycolysis via HIF1α and MYC. Through these transcription factors, it promotes the expression of genes whose products are involved in glucose metabolism including glucose transporters and glycolytic enzymes. mTORC2 also has a positive role in the regulation of glucose metabolism. The mTORC2 target, Akt, modulates glucose transport and phosphorylates glycolytic enzymes.

mTORC1 also modulates expression of the glycolytic genes via HIF1α. Transcriptional profiling of rapamycin-treated lymphocytes revealed altered glycolytic gene expression in these cells [13,80]. In prostate epithelial cells of transgenic mice expressing active Akt, rapamycin diminishes the levels of glycolytic enzyme genes controlled by HIF1α [81]. A combination of genomics, metabolomics and bioinformatics approaches further confirmed the involvement of mTORC1 in inducing a HIF1α-dependent transcriptional program to promote glycolysis [14]. Among the HIF1α-regulated genes that are transcriptionally upregulated in an mTORC1-dependent manner are glycolytic enzymes and VEGF. The expression of the rate-limiting glycolytic enzyme pyruvate kinase M2 (PKM2), which is exclusively expressed in proliferating and tumor cells, is also regulated transcriptionally by mTORC1 via HIF1α [82].

mTORC1 also controls glucose uptake via regulation of gene expression or membrane trafficking of glucose transporters. Cancer cells upregulate their consumption of glucose. This characteristic of cancer cells is exploited in fluorodeoxyglucose -positron emission tomography (FDG-PET) scan that is widely used to detect tumors. Rapamycin treatment in vivo reduces FDG uptake in kidney cancers with loss of the tumor suppressor von Hippel Landau (VHL1), supporting a role for mTORC1 in glucose uptake [79]. The sensitivity to mTOR inhibition is attributed to a block in translation of mRNA encoding HIF1α, a target of VHL. Glucose uptake, as well as the expression of glycolysis-associated genes and glucose metabolism-regulating miRNAs, were decreased by PI3K/mTOR inhibitors in lymphoma cells [83]. However, in another study using liver-specific *Tsc1* mutant mice, increased mTORC1 activation is accompanied by reduced glucose uptake [84]. This is likely due to the mTORC1/S6K1-mediated negative feedback loop that downregulates PI3K/Akt pathway, which plays a role in glucose transport [85]. The reduced glucose uptake under elevated mTORC1 activity is also in line with the findings that TSC-deficient cells are hypersensitive to glucose withdrawal [34]. It is also likely that the increased mTORC1 activation could upregulate the export or synthesis of alternative carbon or energy sources such as glutamine.

Another effector of mTORC1 that promotes glycolytic gene expression is the transcription factor Myc [86,87]. Myc stimulates transcription of genes involved in metabolism, ribosome biogenesis and mitochondrial function [88]. Using a bioinformatics approach, cis-regulatory elements among rapamycin-sensitive genes were shown to be regulated by Myc [14]. HIF1 and Myc have overlapping metabolic gene targets. One example of a common target gene is that encoding lactate dehydrogenase (LDH) [89,90]. LDH converts pyruvate to lactate and is a tetrameric enzyme composed of a combination of the subunits LDHA and LDHB. Rapamycin treatment of prostate cancer cell lines downregulates LDHA expression among other metabolic effectors [91]. On the other hand, the expression of LDHB is upregulated in an mTOR-dependent manner in murine embryonic fibroblasts that have deficiency in TSC1, TSC2 or PTEN and with activated Akt. The enhanced LDHB levels are critical for hyperactive mTOR-mediated tumorigenesis [92]. Another common target of HIF1α and Myc is PKM2. Unlike transcriptional activation of the PKM2 gene by HIF1α, Myc appears to regulate PKM2 expression in an mTORC1-dependent manner via the alternative splicing repressors, hnRNPs [82]. How mTORC1 can regulate Myc remains to be characterized.

Downstream mTORC1 substrates also mediate glycolytic metabolism. Knockdown of S6K1 in PTEN-deficient cells decreases HIF1α expression and glycolysis [93]. In these studies, targeting S6K1 in PTEN-deficient mouse model of leukemia delays leukemogenesis. Additionally, pharmacological or genetic inhibition of another mTORC1 target, 4E-BP1, is also sufficient to block Myc-driven tumorigenesis [94]. Studies using knockouts of negative regulators of mTORC1 further reveal how enhanced mTORC1 activation reprogram metabolism. Knockout of the GATOR1 component, NPRL2, in skeletal muscle increased pyruvate conversion to lactate while reducing its entry into the TCA cycle [95]. In turn, there was a compensatory increase in anaplerotic reactions accompanied by decreased amounts of amino acids such as aspartate and glutamine that are likely utilized for anaplerosis.

Although Warburg thought that respiration is defective in cancer cells, recent studies demonstrated that cancer cells also upregulate processes in the mitochondria. Mitochondrial oxidative phosphorylation (OXPHOS) via the respiratory chain is required by cancer cells for their proliferation. The mitochondrial import protein coiled-coil helix coiled-coil helix domain-containing protein (CHCHD4), which controls respiratory chain complex activity and oxygen consumption, promotes mTORC1 signaling and drives tumor cell growth [96]. mTORC1 also increases the expression of nucleus-encoded mitochondrial proteins via inhibition of 4EBP [97].

mTORC2 also plays a role in glycolysis. Early studies uncovered the role of Akt, an mTORC2 substrate, in the regulation of glycolytic enzymes such as hexokinase [98], PFK2 [99] as well as the glucose transporter GLUT1 [100,101]. In an experimental leukemia model, Akt activation was sufficient to increase the rate of glucose metabolism [102]. Furthermore, glioblastoma cells expressing constitutively active Akt have high rates of aerobic glycolysis. In prostate epithelial cells, activated Akt induces glycolytic genes via HIF1α [81]. On the other hand, Akt deficiency is sufficient to suppress tumor development in PTEN+/− mice [103]. Thus, Akt plays a crucial role in enhanced aerobic glycolysis. Studies that abrogated the mTORC2 component, rictor, highlighted the role of mTORC2 in glycolysis. For example, in a liver-specific knockout of rictor, glycolysis was impaired and the activity of glucokinase was reduced. Expression of a constitutively active Akt or glucokinase rescues glucose flux in these mice [104]. mTOR, most likely via Akt, also modulates pyruvate dehydrogenase (PDH), the gatekeeper enzyme of mitochondrial respiration [105]. Inhibiting PI3K/mTOR or Akt increased the phosphorylation of the E1a subunit of the PDH complex on Ser293, thus inhibiting its activity and reducing the oxygen consumption rate of head and neck squamous carcinoma cells. Akt may also function in glycolysis independent of mTOR. In glioblastoma cells, Akt1 promoted HIF1α translation in a manner that is insensitive to rapamycin or mTOR depletion [106].

mTORC2 also controls glycolysis via Akt-independent mechanisms. In glioblastoma mTORC2 promotes inactivating phosphorylation of Class IIa histone deacetylases, which then controls acetylation of FoxO1 and FoxO3 [107]. This in turn promotes upregulation of Myc. Furthermore, mTORC2 promotes acetylation of the histone H3K56 in glioma, which influences glycolytic gene expression due to enhanced recruitment of Sirt6 in the promoter of these genes [108].

Inhibiting mTORC2 activity such as by decreasing expression of its component, rictor, decreases glycolytic metabolism in cancers. In glioblastoma wherein mTORC2 activity is elevated, the expression of the large intergenic non-coding (Linc) RNA-RoR was attenuated. Increasing the expression of LincRNA-RoR led to decreased rictor expression, reduced mTORC2 activity, diminished expression of glycolytic effectors and diminished tumor growth [109].

2.4. Lipid Metabolism

Cancer cells undergo increased de novo lipid synthesis. Production of fatty acids and cholesterol are enhanced for biosynthesis of membranes and signaling molecules. Cell membrane lipids including phospholipids, sterols, sphingolipids and lyso-phospholipids are derived in part from acetyl CoA. A major transcriptional regulator of lipid metabolism-related genes is the sterol regulatory element binding protein (SREBP) family of transcription factors. Among the SREBP-targeted genes are ATP citrate lyase (ACLY), acetyl-CoA carboxylase 1 (ACC1), fatty acid synthase (FASN), stearoyl-CoA desaturase 1 (SCD) and fatty acid transporters. SREBP is processed and translocates to the nucleus to induce transcription of its target genes. mTOR modulates SREBP and other regulators and effectors of lipid metabolism (Figure 4). Rapamycin blocks the expression of genes involved in lipogenesis and prevents the nuclear accumulation of SREBP [110]. SREBPs are synthesized as inactive precursors that reside in the endoplasmic reticulum and translocate to the nucleus after processing from the Golgi. This active processed form induces transcription of SRE-containing genes. The processing step is thus sensitive to sterol levels and controlled by mTORC1 signaling. An enhancement of the processed forms of SREBP1 occur in the TSC-deficient cells [14]. S6K1 also regulates SREBP processing [111]. The sterol regulatory element was the most highly enriched DNA motif in a gene expression study of rapamycin-sensitive genes [14]. mTORC1 may also regulate SREBP via negative regulation of lipin1, a phosphatidic acid phosphatase that represses SREBP activity [112]. Lipin is phosphorylated by mTORC1 at multiple phosphosites including both rapamycin-sensitive and -insensitive sites [112]. When phosphorylated, lipin1 accumulates in the nucleus and represses SREBP-dependent gene transcription. mTORC1 signaling is necessary, but not sufficient to activate SREBP1 and lipid synthesis in the liver. mTORC2 is also required for SREBP1c activation and lipogenesis [113].

In B-cell non-Hodgkin lymphoma, both fatty acid synthesis and glycolysis are upregulated in a PI3K/mTOR-dependent manner. The fatty acid synthase FASN was overexpressed and thereby the lymphoma was more sensitive to the FASN inhibitor C75 than primary B cells. [114]. In hepatocellular carcinoma triggered by co-expression of Akt and c-Met, their growth was dependent on mTORC1 and fatty acid synthase (FASN) [115]. Carcinogenesis was prevented by FASN ablation. In neurofibromatosis type 2 (NF2) disorder, which is characterized by multiple tumors in the central nervous system such as schwannomas and meningiomas, the deficiency of the *Nf2* gene, which encodes the tumor suppressor Merlin, upregulated mTORC1 leading to elevated expression of key enzymes involved in lipogenesis. The inhibition or knockdown of FASN led to apoptosis of NF2-deficient cells [116]. FASN catalyzes the synthesis of palmitic acid from malonyl-CoA. When the Nf2-deficient cells were treated with compounds that blocked the production of malonyl CoA, the sensitivity to FASN inhibitors was reduced. In breast cancer, mTORC1 also promotes expression of stearoyl CoA desaturase 1 (SCD1), the rate limiting enzyme in monounsaturated fatty acid synthesis [117]. Rapamycin inhibited SCD1 promoter activity and decreased the expression of SREBP1 through a mechanism involving eIF4E.

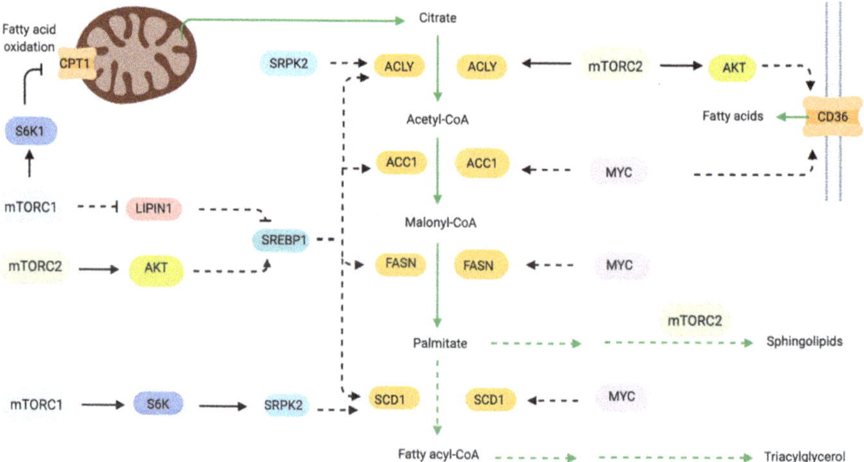

Figure 4. The role of mTOR in fatty acid and lipid synthesis. mTOR controls metabolic enzymes involved in fatty acid and lipid synthesis via SREBP1 and SRPK2. They also modulate expression of fatty acid transporters such as CD36 and carnitine palmitoyl transferase 1c (CPT1c).

In chronic myelogenous leukemia that are addicted to glucose metabolism for survival, inhibition of mTORC1 by rapamycin or S6K1 knockdown led to increased fatty acid oxidation and increased the expression of the fatty acid transporter carnitine palmitoyl transferase 1c (Cpt1c) [118]. This transporter is also linked to rapamycin resistance in human lung tumors [119]. The fatty acid binding protein 4 (FABP4), which is an adipokine for fatty acid transport increases breast cancer cell proliferation and promotes expression of fatty acid transporters such as CD36 and FABP5 [120]. Exogenous treatment of FABP4 increased Akt and MAPK signaling. The increased expression of the FA transporter CD36 in gastric cancer tissues also correlated with poor prognosis in patients [121]. CD36 mediates the palmitate acid-induced metastasis of GC via Akt. Together, these findings suggest that targeting fatty acid transport could have therapeutic benefits for cancers that rely on increased lipid metabolism for growth and proliferation.

In addition to the role of mTOR in fatty acid synthesis, it is also involved in lipid synthesis. Lipids are used not only for membrane biogenesis but also as signaling molecules. mTORC2 promotes the increased synthesis of sphingolipids and glycerosphospholipid, leading to liver steatosis and hepatocellular carcinoma [122].

2.5. Glutamine Metabolism

As with glucose metabolism, mTOR signaling impinges on multiple aspects of glutamine metabolism (Figure 5). On the other hand, mTOR is also sensitive to glutamine fluctuations and its activity is influenced by perturbations in glutamine metabolism. Glutamine is the most abundant non-essential amino acid in the plasma and is avidly used by proliferating tumor cells. It is acquired from the environment via transporters including Slc1A5 (ASCT2), Slc38A1, Slc38A2 or Slc38A5. The expression of these transporters is upregulated in many types of cancer [123]. Slc1A5 is a major transporter of glutamine in most cells. Knockdown of ASCT2 diminishes mTORC1 activity and tumor growth [124–128]. Although the absence of Slc1A5 did not significantly diminish intracellular Gln or Glu levels, it disrupted influx of leucine and diminished levels of other amino acids crucial for redox homeostasis [128]. Leucine, an essential amino acid, is acquired by the cell through counter-transport with Gln [21]. The expression of the heterodimeric glutamine antiporter Slc7A5/Slc3A2 (LAT/CD98) is associated with elevated mTORC1 activity in cancer [129].

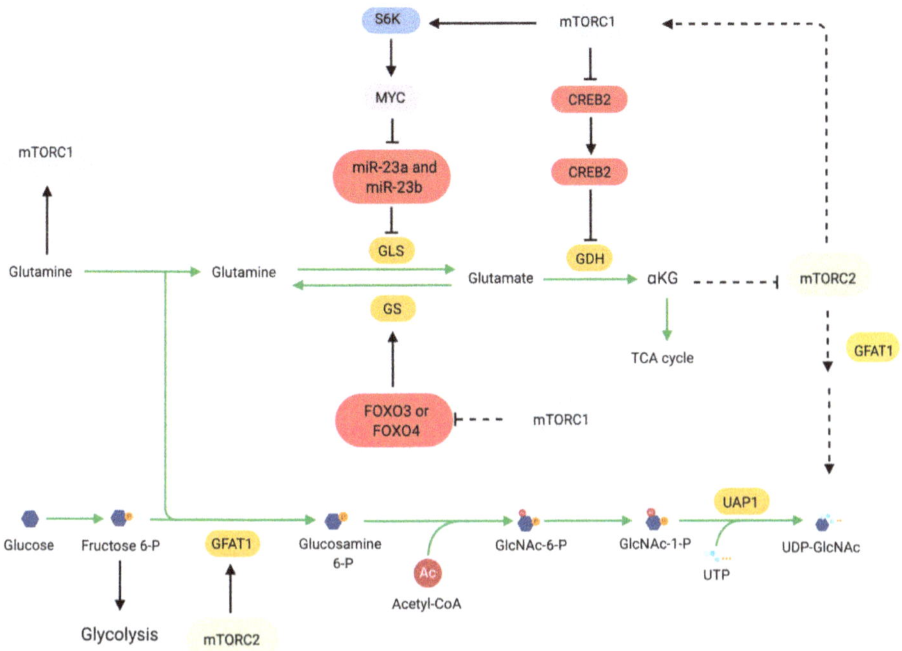

Figure 5. The role of mTOR in glutamine metabolism and hexosamine biosynthesis. mTORC1 promotes glutaminolysis and is active in the presence of sufficient glutamine. On the other hand, mTORC2 activation is enhanced when glutamine catabolite levels diminish to promote flux through the hexosamine biosynthesis pathway via control of GFAT1.

Glutamine is a versatile molecule, as it serves as an alternative carbon source for energy production and its carbon and nitrogen are also used for biosynthetic reactions [130]. Hence, cancer cells can develop addiction to glutamine and upregulate its metabolism to provide necessary building blocks for the growing tumor [131,132]. mTOR has been linked to the regulation of many of these processes. Glutamine is a precursor for α-ketoglutarate (αKG) and is used to replenish TCA intermediates (anaplerosis) in proliferating cells. Glutamine is metabolized via glutaminolysis, which consists of two steps. The first step is catalyzed by glutaminase (GLS) and converts glutamine to glutamate. The second is catalyzed by glutamate dehydrogenase (GDH) and converts glutamate to αKG. Conversely, glutamine can be generated by glutamine synthetase (GS) which uses ATP and $NH4^+$ to generate glutamine from glutamate. mTORC1 stimulates glutamine metabolism via regulation of transcription factors involved in expression of glutaminolysis-related genes. mTORC1 mediates translational upregulation of Myc via a mechanism involving the S6K1 eIF-4E [86,133]. Myc in turn represses transcription of miR-23a and miR-23b, which are both repressors of GLS [134]. Oncogenic signals such as elevated Myc levels increase glutamine uptake and metabolism through a transcriptional program that includes enhancement of expression of mitochondrial glutaminase [134,135]. Other signals independent of mTOR could also contribute to promoting glutaminolysis and could thus be exploited for co-targeting with mTOR inhibitors. In lung squamous cancer cell carcinoma that underwent chronic mTOR inhibition and suppression of glycolysis, glutaminolysis was enhanced via a mechanism involving GSK3. Increased glutaminolysis and/or increased levels of GLS have been found in a number of cancers that become resistant to other therapies. Combined treatment with the glutaminase inhibitor CB-839 and mTOR inhibitors shows efficacy in overcoming therapeutic resistance to other targeted inhibitors in these different types of cancer [136,137]. Another transcription factor that is regulated by mTORC1 to promote glutaminolysis is CREB2 (cAMP-responsive element binding 2). mTORC1 promotes the proteasome-mediated degradation of CREB2 and represses SIRT4 transcription [138]. SIRT4 negatively regulates GDH by ADP-ribosylation. Thus, mTORC1 promotes glutamine metabolism via negative regulation of CREB2, which ultimately leads to activation of GDH.

mTORC2 is also emerging to play a crucial role in glutamine metabolism. The expression of Myc in glioblastoma is mTORC2-dependent [107]. Furthermore, knockdown of rictor decreases levels of αKG that are likely derived from glutaminolysis, suggesting that mTORC2 could regulate this process as well [97]. However, inhibition of PI3K or Akt in glioma that expressed oncogenic levels of Myc did not affect glutaminolysis in these tumors [135]. A possibility is that the overexpression or deregulation of Myc uncouples it from PI3K/Akt signals.

While mTOR modulates glutamine metabolism, the latter also reciprocally modulates mTOR. Glutamine, in combination with leucine activates mTORC1 by enhancing glutaminolysis and αKG production. Glutaminolysis correlates with increased mTORC1 activity and is necessary for GTP loading of RagB and activation of mTORC1 signaling. It also promotes cell growth and inhibits autophagy via regulation of mTORC1 [29]. The import of glutamine by the transporter Slc1A5 has also been suggested to be the rate-limiting step that activates mTOR. On the other hand, glutamine depletion that can occur as an off target effect of using asparaginase, which has glutaminase activity, indirectly inhibits mTOR activity via decreased leucine uptake in AML [126]. Glutamine synthase levels also affect mTORC1 activation. In liver cancer with mutations in β-catenin, levels of GS, a target of β-catenin, are increased and its disruption prevents mTOR phosphorylation at Ser2448, suggesting downregulation of mTOR activity [139]. In some cell types, in the absence of glutamine, cells utilize ammonia as an alternative nitrogen source. This ability to utilize ammonia is linked to mTORC1 as well as GDH and AMPK [140]. mTORC2 also responds to levels of glutamine or its catabolites. It is interesting to note that glutamine depletion decreases mTORC1 activity while increasing mTORC2 activity [54,141]. This differential regulation of mTORC1 and mTORC2 is mediated by sestrin2 and allows survival of these cancer cells by maintenance of energy and redox balance [141]. However, inhibition of glutaminase or GDH did not affect Akt phosphorylation while it diminished mTORC1 activation [29]. How glutaminolysis affects mTORC2 requires further investigation.

Glutamine is also used for the production of UDP-GlcNAc, a metabolite produced by the hexosamine biosynthesis pathway (HBP), which, in turn is used for protein and lipid glycosylation [142]. During glucose or glutamine starvation, mTORC2 promotes the expression and phosphorylation of glutamine fructose-6-phosphate amidotransferase-1 (GFAT1), the rate limiting enzyme involved in the de novo hexosamine biosynthesis, in order to maintain flux through the HBP [54,143]. The expression of GFAT1 is dependent on the levels of glutamine catabolites such as α-KG. In turn, mTORC2 also promotes the generation of glutamine catabolites. Maintaining flux through the HBP via glucosamine supplementation, which bypasses the GFAT1-catalyzed reaction, during glucose starvation could rescue Akt signaling in the absence of insulin [144]. These findings revealed a feedback relationship between mTORC2 and the HBP.

2.6. Pentose Phosphate Pathway and Nucleotide Synthesis

The pentose phosphate pathway (PPP) generates reducing equivalents in the form of NADPH and ribose-5-phosphate for nucleic acid synthesis. The rate-limiting reaction in the PPP is catalyzed by Glucose-6-phosphate dehydrogenase (G6PD) [145]. PPP consists of two stages: an irreversible oxidative and a reversible nonoxidative stage (Figure 6). In the oxidative reactions, the G6PD-catalyzed reaction produces NADPH. The non-oxidative stage is a series of reversible reactions, converting glycolytic intermediates into ribose-5-phosphate, a key precursor for DNA and RNA synthesis. The PPP ultimately generate phosphoribosyl pyrophosphate (PRPP), the precursor for nucleotide synthesis. Cancer cells have increased G6PD levels and PPP activity [146–149]. Whereas pentose phosphates are essential for nucleotide production, NADPH serves as a reducing agent in several synthetic steps of fatty acid, cholesterol, and steroid hormones, along with detoxification reactions. Hence, blocking G6PD could have potent anti-tumor effects by preventing nucleotide synthesis as well as impairing redox balance. The G6PD competitive inhibitor, 6-aminonicotinamide (6-AN) is being used to target cancer cells with increased PPP activity and has shown antineoplastic effects [150,151]. Downregulating G6PD expression has been shown to decrease cell viability of bladder cancer cell lines due to accumulation of ROS [149]. mTORC1 can stimulate flux through the oxidative branch [14]. mTORC1 is involved in this pathway via transcription of genes encoding enzymes that drive the PPP. By regulating expression of these genes, mTORC1 promotes production of ribose-5-phosphates, which are used in purine and pyrimidine nucleotide synthesis and production of NADPH.

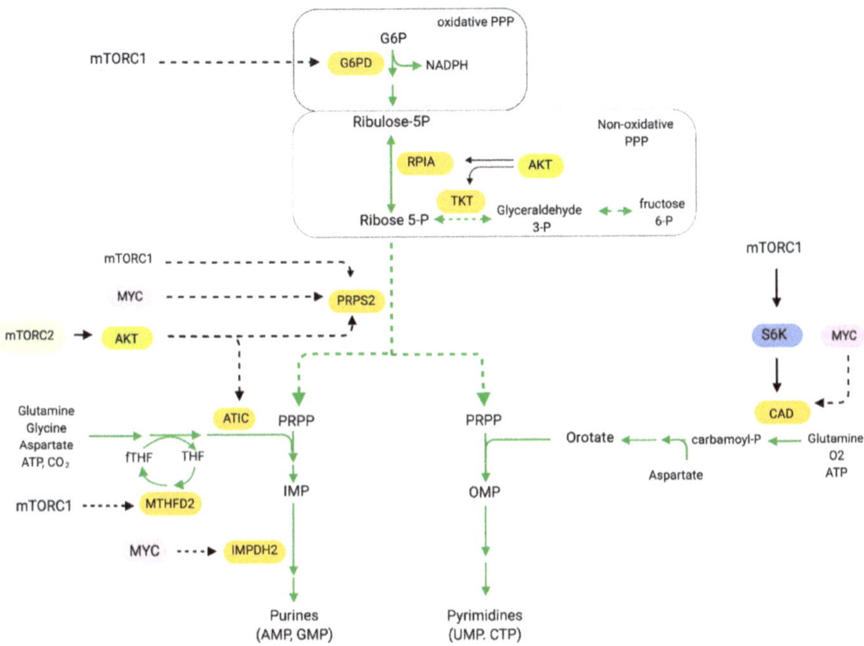

Figure 6. The role of mTOR in the pentose phosphate pathway and nucleotide synthesis. mTOR modulates the expression of the enzymes involved in the pentose phosphate pathway, purine and pyrimidine synthesis.

Cancer cells require an abundant pool of purines and pyrimidines, which serves as building blocks for DNA and RNA synthesis. Nucleotide synthesis utilizes two pathways: the de novo synthesis pathway, which assembles complex nucleotides from basic molecules and the salvage pathway, which generates nucleotides from degradation intermediates. mTORC1 is involved in the production of pyrimidines via de novo pathways [152,153]. mTORC1 promotes activation of CAD (carbamoyl-phosphate synthetase 2, aspartate transcarbamylase, and dihydro-orotase) via S6K1. S6K1 phosphorylates CAD on Ser1859 [153]. CAD catalyzes the initial steps of pyrimidine synthesis by utilizing glutamine, bicarbonate, and aspartate to generate pyrimidine rings. De novo pyrimidine synthesis is enhanced by mTORC1 and S6K in response to growth factor or amino acid stimulation, although they are not essential for de novo synthesis per se [152]. mTORC1 also promotes purine nucleotide biosynthesis. Purines are assembled on the ribose sugar PRPP, wherein carbon and nitrogen molecules are donated by non-essential amino acids and one-carbon formyl units from the tetrahydrofolate (THF) cycle. mTORC1 has been implicated in the increased expression of phosphoribosyl pyrophosphate synthase 2 (PRPS2), the rate-limiting enzyme that provides PRPP for both purine and pyrimidine nucleotides, in Myc-transformed cells [154]. mTORC1 also promotes transcription of numerous enzymes contributing to purine synthesis including the mitochondrial tetrahydrofolate cycle enzyme methylenetetrahydrofolate dehydrogenase 2 (MTHFD2). MTHFD2 expression is induced by mTORC1 via ATF4 [155].

mTORC2 is also involved in modulating the PPP. In an insulin-driven model of hepatocellular carcinoma cells, Akt drove the upregulation of the PPP through several mechanisms, including via increase of phosphate dehydrogenase and ribose 5-phosphate isomerase A expression and activity, as well as through driving glycolysis [156]. The role of mTORC2 in the PPP is further supported from studies in yeast wherein the yeast TORC2 physically interacted with proteins that play a role in the PPP [157]. Metabolic intermediates such as 6-phospho-D-gluconate (6PG) and ribose-5-phosphate are strongly downregulated in response to TOR2 inhibition, suggesting that TORC2 could post-translationally regulate PPP modulators.

mTORC2 also regulates purine synthesis via Akt [158]. The PI3K/Akt signaling axis regulates the early steps of the non-oxidative PPP at the level of PRPP synthesis and later steps by modulating the activity of aminoimidazole-carboxamide ribonucleotide transformylase IMP cyclohydrolase (ATIC) [158]. Akt also phosphorylates transketolase (TKT) on Thr382, a key enzyme of the nonoxidative PPP, leading to increased flux through this pathway and increasing purine synthesis [159].

2.7. Other Metabolic Pathways

mTORC1 also promotes serine biosynthesis and one-carbon pathway metabolism. The product of this pathway, S-adenosylmethionine (SAM), which is derived from methionine, is also sensed by mTORC1 via SAMTOR [160]. In mouse models and primary pancreatic epithelial cells, the oncogenic cooperation of KRAS activation and LKB1 loss led to mTOR-induced activation of the serine-glycine-one carbon pathway that enhances generation of S-adenosyl methionine (SAM) [161]. This was accompanied by increased expression of DNA methyltransferases that consequently modified the epigenome to promote tumorigenesis. In de novo or during therapy-induced neuroendocrine prostate cancer (NEPC), downregulation of PKCλ/ι upregulates serine biosynthesis through mTORC1 and ATF4-dependent pathway [162]. This metabolic reprogramming generates increased SAM that is utilized for epigenetic changes that promote NEPC attributes.

3. Targeting the mTOR Pathway in Cancer

Due to the critical role that mTOR plays in cell growth and metabolism and availability of the natural compound, rapamycin, to inhibit its activity, there are numerous ongoing efforts to target the mTOR signaling pathway for cancer therapy. Various studies have already shown that deregulation of the mTOR pathway is present in many cancers. For example, mutations causing an increase in mTOR activity have been found in solid tumors [163–165]. Additionally, mutations in upstream regulators of mTOR including the oncogene *PIK3CA*, which encodes PI3K and the tumor suppressor *PTEN* occur frequently in human tumors [166]. Moreover, genetic lesions can also promote the activation of the PI3K/mTOR pathway in cancer cells, such as those encoding for Ras, Akt, TSC1/2, Notch1, and receptor tyrosine kinases [167]. Mutations and genomic alterations in metabolic enzymes and other key regulators of metabolic pathways of which mTOR has been linked are also common in cancers [71,168]. The upregulation of the mTOR pathway due to one or more of these mutations make mTOR an attractive target in tumors. However, due to other signaling molecules that also play a role in the control of metabolism as well as the presence of de novo and salvage synthesis of metabolites, there is a need to design appropriate combination therapeutic strategies for more effective cancer treatment. Here, we discuss the current strategies to inhibit the mTOR pathway and combination therapy that could more effectively block critical metabolic pathways and other signaling molecules that control metabolism. The remarkable success of immunotherapy, which utilizes a patient's own immune system to combat tumors, also underscores the importance of considering not just the tumor's metabolism but also of neighboring cells, including immune cells. We include a discussion of how modulating mTOR signaling has implications in immunotherapeutic approaches.

3.1. Rapalogs: Targeting mTORC1 Activity

Rapamycin and its analogs (rapalogs) are the first generation of mTOR inhibitors, which selectively inhibit the activity of mTORC1 by binding to FKBP-12 and forming a ternary complex with mTOR. Rapamycin is an allosteric inhibitor of mTOR and it inhibits some of the functions of mTORC1, such as phosphorylation of the protein kinase S6K1. As discussed above, S6K1 has numerous downstream targets involved in protein synthesis as well as other metabolic targets [59]. The clinical use of rapamycin is limited due to poor water solubility and stability. Thus, several pharmaceutical companies have developed rapamycin analogs with improved pharmacokinetic properties (Table 1). Rapalogs differ in their chemical properties in terms of drug solubility and metabolism. For example, temsirolimus (Torisel, CCI779; Wyeth) and ridaforolimus (AP23573, deforolimus, Merck/ARIAD) are water soluble and must be administered intravenously [169,170]. Rapalogs have been undergoing clinical trials for various malignancies and have already been approved by the Food and Drug Administration (FDA) for the treatment of specific types of cancers. Everolimus (Affinitor, RAD001; Novartis) has been efficacious and FDA-approved for the treatment of advanced renal cell carcinoma that progressed after treatment with sunitinib and/or sorafenib [171,172]. Everolimus has also been approved for treatment of both subependymal giant-cell astrocytoma (SEGA) and renal angiomyolipoma with tuberous sclerosis complex [173–175]. Additionally, everolimus significantly prolonged progression-free survival (PFS) among patients with advanced pancreatic neuroendocrine tumors and was associated with a low rate of severe adverse events in a Phase III study, making everolimus the first new treatment for this type of cancer in almost 30 years [176]. Furthermore, everolimus is the first medication approved to treat progressive, nonfunctional lung and gastrointestinal tumors [177]. Temsirolimus (Torisel, CCI779; Wyeth) is another rapalog and is currently approved for advanced renal cell carcinoma [178,179] and was shown to significantly improve PFS in patients with relapsed or refractory mantle cell lymphoma [180].

There are numerous clinical trials on rapalogs as monotherapy that have yielded some promising results for the treatment of other types of cancer. Results from a Phase II study of everolimus in patients with advanced follicular -derived thyroid cancer yielded favorable results, resulting in clinically relevant antitumor activity and a relatively low toxicity profile [181]. Ridaforolimus is the newest rapalog and has been investigated in treating sarcomas. Ridaforolimus had an overall PFS rate at 6 months of 23.4% for soft tissue and bone sarcomas, including primary bone sarcomas, leiomyosarcomas, and liposarcomas in a phase II study [182]. However, an international phase III trial of ridaforolimus showed that it delayed tumor progression only to a small degree in patients with metastatic sarcoma [183]. Ridaforolimus displayed anti-tumor activity in advanced endometrial cancer patients in phase II clinical trials though with associated significant toxicity [184,185]. A phase I study of ridaforolimus in pediatric patients with advanced solid tumors had favorable toxicity and shows promise for combination therapies [186]. A phase II study of ridaforolimus in women with recurrent or metastatic endometrial cancer showed tolerability and modest activity of the drug [187].

Table 1. mTOR or mTOR complex inhibitors that are currently used in pre-clinical and clinical studies.

Drug	Target	References
Rapalogs (Rapamycin Analogs)		
Sirolimus	mTORC1/FKBP12	[188,189]
Everolimus (Affinitor, RAD001)	mTORC1/FKBP12	[171–177,181,190–194]
Temsirolimus (Torisel, CCI779)	mTORC1/FKBP12	[178–180,195]
Ridaforolimus (deforolimus, AP23573)	mTORC1/FKBP12	[182–187]

Table 1. *Cont.*

Drug	Target	References
mTOR ATP-competitive inhibitors (TORKIs)		
AZD8055	mTOR	[196–199]
AZD2014, vistusertib	mTOR	[200–203]
TAK-228; INK-128; sapanisertib	mTOR	[204,205]
OSI-027; ASP7486	mTOR	[206]
MLN0128	mTOR	[207]
CC-223	mTOR	[208]
Dual PI3K/mTOR inhibitors		
BEZ235; dactolisib	PI3K/mTOR	[209–220]
PKI-587; gedatolisib	PI3K/mTOR	[221,222]
GDC-0980, apitolisib	PI3K/mTOR	[223–225]
SAR245409, XL765; voxtalisib	PI3K/mTOR	[226]
BGT226	PI3K/mTOR	[227]
CMG002	PI3K/mTOR	[228]
GSK2126458	PI3K/mTOR	[229]
Other		
JR-AB2-011	Rictor	[230]
RapaLink1	mTOR/FKBP12	[231–233]

In many types of cancer however, rapalogs do not provide marked benefits or only stabilize the disease. For example, everolimus did not significantly improve overall survival in comparison to supportive care in previously treated gastric cancer [190]. Another study that involved patients with advanced hepatocellular carcinoma showed that everolimus did not improve overall survival when their disease progressed during or after receiving sorafenib (inhibitor of growth factor receptors and Raf) or who were intolerant of sorafenib [191]. Everolimus also failed to produce clinically relevant patient response in a phase II trial for patients with relapsed or cisplatin refractory germ cell tumors [192]. In a phase II trial of everolimus in patients with previously treated recurrent or metastatic head and neck squamous cell carcinoma, there was no efficacy likely due to toxicity [193]. In a phase II study of temsirolimus in women with platinum-refractory/resistant ovarian cancer or advanced recurrent endometrial carcinoma, the temsirolimus treatment was tolerated and a few patients had long-lasting stable disease [195].

The efficacy of rapalog treatment as a monotherapy has been modest is not surprising given results from pre-clinical studies that rapamycin only inhibits a subset of mTORC1 substrates [234]. Whereas rapamycin inhibits S6K1, it does not fully inhibit 4EBP1 phosphorylation, thus ineffective in blocking cap-dependent translation in most cell types [235]. Phosphorylated 4EBP1 inhibits the pro-oncogenic eIF4E. eIF4E-mediated translation are upregulated in tumors and blocking this pathway may be crucial to preventing tumor growth in specific cancers [236–238]. As discussed below, mTOR inhibitors that could block the catalytic activity of mTOR could more effectively inhibit mTOR functions and may have better anti-tumor activity.

Despite the limited efficacy of rapalogs as a treatment for a variety of cancers, they remain promising for particular types of cancers. In a few studies, tumors that have upregulated mTORC1 due to inactivation of TSC displayed sensitivity to rapalog treatment. Using whole genome sequencing to analyze molecular determinants of sensitivity to everolimus treatment in metastatic bladder cancer, mutations in TSC1 were uncovered [239]. In another study, treatment with oral sirolimus led to significant reductions in cardiac rhabdomyomas in infants [188]. These rhabdomyomas are associated with tuberous sclerosis complex. In a phase II trial of everolimus on patients with thyroid cancer, whole-exome sequencing of tumors prior to treatment revealed that in a responding patient, there was a mutation in TSC2 that is known to inactivate it as well as a mutation in another tumor suppressor, folliculin 1 (FLCN) that also results in increased mTORC1 signaling [194]. Another study of outlier cases in responders who had metastatic renal cell carcinoma treated with rapalogs also revealed that the strong responders had alterations in TSC1 and mTOR, leading to increased mTORC1 signaling [240]. Importantly, as mentioned above, everolimus has been approved for the treatment of SEGA and renal angiomyolipoma with tuberous sclerosis complex. Altogether, these findings suggest that tumors with increased mTORC1 activation due to TSC mutations may be vulnerable to rapalog treatment. Hence, pre-screening patients for TSC mutations may provide information on the possible benefits of rapalog therapy. However, advanced or highly metastatic tumors, despite having an upregulated mTORC1 signaling, may not necessarily be responsive to rapalog monotherapy due to additional mutations that deregulate other compensatory pathways, as further discussed below [241].

3.2. Co-targeting mTORC1 and Growth Factor Signaling

A potential hurdle in the use of rapalogs for cancer treatment is the existence of feedback mechanisms that could occur when mTORC1 signaling is downregulated. Growth factors potentiate mTORC1 signals via mechanisms that are dependent and independent of TSC. However, to prevent excessive signaling to mTORC1 that could lead to deregulated growth and proliferation, growth factor signals are subjected to negative feedback regulation by mTORC1 [85]. Activated S6K1, an mTORC1 target, reduces signals from the insulin receptor (IR) by inhibitory phosphorylation of insulin receptor substrate-1 (IRS-1), thus mitigating signals downstream of the insulin receptor (IR). Rapamycin treatment, which inactivates S6K1, can thus prevent the suppression of insulin-PI3K signaling. Since this effect has also been observed in other tyrosine kinase receptors (RTK), rapamycin could thus have the undesirable effect of upregulating growth factor/PI3K signaling. In addition, rapamycin treatment has also been shown to increase expression of growth factor receptors such as IGF1R [242]. Furthermore, RTKs not only signal to PI3K/mTOR but also transduce signals to other signaling pathways such as the Ras/MAPK pathway, that play a role in growth and proliferation [243]. Therefore, combining rapamycin treatment with inhibition of RTKs could serve as a more effective strategy for inhibiting mTOR and tumor growth.

Rapalogs have been used in combination with inhibitors of receptor tyrosine kinases (Table 2). Several clinical trials using rapalogs as combination therapy have been conducted in breast cancer, wherein HER2, a member of the epidermal growth factor receptor (EGFR), is often expressed or amplified. BOLERO-3 trial found that the addition of everolimus to trastuzumab, an anti-HER2 antibody, and vinorelbine (mitotic inhibitor) prolonged PFS in patients with trastuzumab-resistance, taxane-pretreated, and HER2-positive breast cancer [244]. A phase II study showed that ridaforolimus with trastuzumab demonstrated anti-tumor activity for patients with HER2+ trastuzumab-refractory breast cancer [245]. In a phase I study, the combination of neratinib, a small-molecule irreversible pan-HER tyrosine kinase inhibitor, and temsirolimus displayed anti-tumor activity in patients with HER2+ breast cancer resistant to trastuzumab, HER2-mutant non-small cell lung cancer, and tumors without identified mutations in the HER-PI3K-mTOR pathway [246]. In a phase I study of temsirolimus with the c-Met inhibitor tivantinib for the treatment of various advanced solid tumors, the combination treatment was well tolerated and had some clinical activity [247]. In a phase I study of combination temsirolimus with cetuximab (anti-EGFR monoclonal antibody) for treatment of advanced solid tumors, clinical activity was modest and thus not further pursued [248]. A combination of temsirolimus with the VEGF inhibitor bevacizumab and cetuximab had partial responses in head and neck squamous cell carcinoma with numerous toxicities [249]. In a phase II study of temsirolimus with bevacizumab for patients with metastatic renal cell carcinoma previously treated with other VEGFR TKI, the combination treatment resulted in modest activity and dose reductions were needed due to toxicity [250]. Another phase II study of lower dose ridaforolimus with dalotuzumab, which inhibits autophosphorylation of IGF1R in ER+ advanced breast cancer had similar efficacy but with higher incidence of adverse events compared to treatment with the aromatase inhibitor, exemestane [251].

Table 2. Other targeted or cytotoxic chemotherapy that are used in combination with mTOR inhibitors.

Drug	Target	References
RTK inhibitors		
Trastuzumab	HER2	[244,245]
Neratinib	HER2	[246]
Tivantinib	c-Met	[247]
Cetuximab	EGFR	[248,249]
Bevacizumab	VEGF	[249,250]
Dalotuzumab	IGF-1R	[251]
Cixutumumab	IGF-1R	[252]
Pazopanib	Tyrosine Kinase	[164]
Lapatinib	HER2/EGFR	[253]
Ras/MAPK Pathway Inhibitors		
Sorafenib	Raf/VEGFR	[228,254–256]
RAF265	Pan-Raf	[257]
Selumetinib	MEK1/2	[258,259]
MNKI-57; MNKI-4	Mnk1/2, Mnk2	[260]
Trametinib (GSK1120212)	MEK	[229,261]
Pimasertib	MEK	[262,263]
Cytotoxic Chemotherapy		
Carboplatin	DNA	[264–266]
Paclitaxel	Microtubules	[266–268]
Cisplatin	DNA	[268,269]
Vinorelbine	Microtubules	[244,270]
Cyclophosphamide	DNA	[271]
Doxorubicin	Topoisomerase	[272]
Vincristine	Microtubules	[273]
Docetaxel	Microtubules	[274]
Temozolomide	DNA	[275]
Oxaliplatin	DNA	[276]
Other Targeted Therapies		
Exemestane	Aromatase	[277]
Letrozole	Aromatase	[278]
Fulvestrant	Estrogen Receptor	[279]

However, the combination of cixutumumab, an anti-IGF-1R antibody, with temsirolimus did not show objective responses in pediatric and young adults with refractory or recurrent sarcoma [252]. A randomized phase II trial of ridaforolimus with the IGF1R inhibitor dalotuzumab and exemestane (R/D/E) was compared with R/E in patients with advanced breast cancer. The PFS of the R/D/E did not improve compared to R/E, likely due to lower doses of ridaforolimus in the R/D/E arm. While there was less toxicity, the efficacy of the R/D/E regimen was poorer [280]. Taken together, rapalogs in combination with growth factor receptor inhibitors have limited efficacy.

Despite the limited efficacy and increased toxicity of combined rapalog/RTK inhibitors, exceptional cases of strong responders as well as pre-clinical studies may provide insight on how this strategy can yield more positive results. In one study, tumors that may have activating mutations of mTOR may potentially be more sensitive to rapalog treatment. A mutation in the kinase domain of mTOR (E2419K) and another at the FRB domain (E2014K) activated mTORC1 signaling in a patient who had a strong response during a phase I trial of everolimus in combination with pazopanib, an inhibitor of growth factor receptors [164]. Since mTOR is also part of mTORC2, such mutation in the kinase domain may also affect the growth factor-modulated mTORC2 signaling. Hence, the molecular alterations that may be associated with vulnerability to combined rapalog/RTK inhibition warrant further investigation. In mouse models of uterine serous carcinoma, combined ridaforolimus and HER2 blockade with lapatinib/trastuzumab had a better anti-tumor activity when tumors had PIK3CA (E542K) mutations together with HER2 gene amplification rather than those without PIK3CA mutations. While lapatinib downregulated mTORC2 signaling in the former tumors, only the combined ridaforolimus/lapatinib treatment was able to abrogate S6 phosphorylation in such tumors, correlating with the anti-tumor activity [253]. These findings reveal how both mTORC1 and mTORC2 signaling downregulation is critical to achieve better efficacy when both mTOR complexes are upregulated either due to amplified RTK or mutated PIK3CA.

3.3. Targeting mTORC1 and mTORC2 with ATP-Competitive mTOR Kinase Inhibitors (TORKIs)

The second generation of mTOR inhibitors differs from rapamycin and rapalogs by the fact that they directly bind to the ATP-binding site of the kinase domain of mTOR. Furthermore, unlike rapamycin and its analogs that allosterically inhibit mTOR and block mTORC1 function, ATP inhibitors can directly inhibit the enzymatic activity of both mTORC1 and mTORC2. They are also more specific to mTOR and have an IC_{50} that is much lower than that for PI3K. However, it is worth noting that the well characterized substrates of mTOR, such as S6K1 (for mTORC1) and Akt (for mTORC2) are only allosterically regulated by mTOR via phosphorylation at their hydrophobic and/or turn motif sites [281]. The activation of these AGC protein kinases via phosphorylation at their catalytic loop occurs via PDK1. Hence, blocking mTOR catalytic activity would only partially inactivate these AGC kinases and may not fully block their functions. Nevertheless, this class of mTOR inhibitors displayed anti-proliferative and cytotoxic effects in preclinical studies [282]. Inhibition of mRNA translation and induction of cell cycle arrest, apoptosis, and autophagy was seen in the treatment of acute myeloid leukemia and T-cell acute lymphoblastic leukemia cells with ATP mTOR inhibitors [283,284]. In another study, hepatocellular carcinoma cells treated with AZD8055 (AstraZeneca) led to cell death, induction of autophagy, and activation of AMPK [196]. AZD8055 was also studied on Hep-2, a human laryngeal cancer cell line. In these cells, expression of mTOR was downregulated after treatment with AZD8055. Furthermore, pro-apoptotic factors, including Bax and Caspase3, were up-regulated with AZD8055 treatment and anti-apoptotic factor Bcl-2 was reduced. This suggests that AZD8055 can inhibit proliferation and induce apoptosis in Hep-2 cells [197].

Although pre-clinical evidence was encouraging, early clinical trials have had mixed results. Phase I trials with AZD8055 led to elevated transaminases in patients with advanced solid malignancies and lymphoma, displaying an unfavorable toxic profile. Therefore, AZD8055 is not being developed further, and a follow-up compound, AZD2014 (Vistusertib, AstraZeneca) that has no rise in transaminases is being investigated instead [198,199]. A phase I trial with AZD2014 showed a more favorable pharmacological profile overall and demonstrated efficacy as a single agent in heavily pre-treated solid tumors [200]. However, a phase II trial that studied AZD2014 versus everolimus found AZD2014 to be inferior in treating patients with VEGF-refractory renal cell carcinoma and led to early termination of the study [201]. Kinome profiling of samples from patients with advanced-stage ovarian clear cell carcinoma (OCCC) revealed increased alterations in the PI3K/Akt/mTOR pathway in about 91% of tumors [285]. The majority of the OCCC cell lines tested displayed more sensitivity to mTORC1/2 inhibition (AZD8055) than to drugs targeting the ERBB family of RTKs or to inhibitors of DNA repair signaling.

Sapanisertib (TAK-228, INK-128, Millennium Pharmaceuticals) is another ATP-competitive mTOR kinase inhibitors and has shown some promising results in phase I trials. Sapanisertib and paclitaxel with or without trastuzumab was evaluated in patients with advanced solid malignancies. The most common types of cancers in this study were lung (21%), ovarian (12%), and breast, endometrial, and esophageal (9% each). Sapanisertib was well tolerated in this study and exhibited anti-tumor activity in a range of tumor types. Out of 54 patients, eight experienced partial responses and six had stable disease lasting over 6 months [204]. Sapanisertib was also well tolerated and displayed preliminary therapeutic activity in patients with refractory multiple myeloma, non-Hodgkin's lymphoma, and Waldenström's macroglobulinemia. Almost half of the patients in the study achieved stable disease [205]. OSI-027

(ASP7486, Astellas Pharma) is another ATP-competitive mTOR kinase inhibitor currently undergoing clinical trials. A recent phase I trial of OSI-027 showed inhibition of mTORC1/mTORC2 in patients with advanced tumors. However, disturbances of renal function were common, and doses above tolerable levels in two of the tested drug schedules were required for sustained biological effects [206]. Another TORKI, MLN0128, was used in a phase II clinical trial on metastatic castration-resistant prostate cancer but had limited clinical efficacy due to dose reductions as a consequence of toxicity. There was poor inhibition of mTOR signaling targets such as Akt and 4EBP1 and compensatory increase of androgen receptor activity [207].

These early clinical trials have yielded mixed results with the negative outcome due to dose limiting toxicities. It will be important to identify predictive biomarkers to determine which patients could benefit from the TORKIs. For example, in gastric cancer and small cell lung cancer tumors that had amplification of rictor expression, there was specific sensitivity to AZD2014 [202,203]. The toxicity of the TORKIs also highlights the important role of mTOR in metabolic homeostasis of normal proliferating cells. Hence, revisions of formulations or treatment regimen may also be refined to avoid toxicities while obtaining a more durable response. More highly selective mTOR inhibitors are also currently being developed to improve specificity and metabolic stability [286].

3.4. Dual PI3K/mTOR Inhibitors: Targeting PI3K and mTOR Signaling

The PI3K signaling cascade is deregulated in many types of human cancer [287]. Upon growth factor stimulation, PI3K is activated and phosphorylates the phosphatidylinositol-4,5 bisphosphate (PIP2) to generate phosphatidylinositol-3,4,5 trisphosphate (PIP3). Increased PIP3 levels recruit a number of signaling molecules to the membrane periphery, including Akt. Signals from PI3K are antagonized by the tumor suppressor PTEN, which contains a lipid phosphatase domain that catalyzes conversion of PIP3 to PIP2. PTEN is often mutated or inactivated in several cancers. Other common mutations are those affecting the gene coding for p110α, a catalytic domain of PI3K. The p110 subunit of PI3K and catalytic domain of mTOR are structurally similar. Therefore, dual inhibitors of PI3K and mTOR that target their catalytic domains have been engineered [6]. These dual inhibitors would block not only PI3K activity but both mTORC1 and mTORC2 complexes and would thus have broader inhibitory capabilities than rapamycin. Furthermore, Akt would also be suppressed by these dual inhibitors since blocking PI3K would diminish production of PIP3, the lipid that acts as a docking site for Akt and PDK1. Inhibition of mTORC2 would also block the allosteric activation of Akt. Hence, dual targeting of PI3K and mTOR would have more extensive inhibitory effects on the PI3K/mTOR pathway.

Pre-clinical studies using cell and mouse models demonstrated the efficacy of dual PI3K/mTOR inhibitors such as BEZ235 or dactolisib (NVP-BEZ235, Novartis) in several types of cancers including breast cancer, lung adenocarcinomas, glioblastoma, [209–211]. In these models, tumors with mutations in PIK3CA (catalytic subunit) or increased PI3K/mTOR pathway activation displayed specific sensitivity to the inhibitor. In another study of HER2-amplified with or without PIK3CA mutation breast cancer cells, dactolisib induced cell death and apoptosis by activating caspase-2 and PARP cleavage [212]. In a multiple myeloma model, dactolisib prevented phosphorylation of mTORC1/2 targets and induced cell cycle arrest [213]. Gedatolisib (PKI-587; Pfizer) had better efficacy than everolimus in inhibiting proliferation of gastroenteropancreatic neuroendocrine (GEP-NENs) tumor cell lines [221]. The inhibitor promoted cell cycle arrest and induced apoptosis and prevented phosphorylation of 4EBP1. Hence, in pre-clinical studies, dual PI3K/mTOR inhibitors are more effective at promoting tumor cell death, as expected.

However, early clinical trials of dactolisib have not shown as much efficacy. In an early Phase I study, BEZ235 given to patients with advanced solid tumors led to dose limiting toxicity [214]. In one study, patients with advanced renal cell carcinoma received escalating doses of dactolisib had high incidence of significant toxicities across all dose levels tested with no objective response. Based on these results, dactolisib is not recommended for patients with renal cell carcinoma [215]. Dactolisib was also associated with a poorer tolerability profile and was not more effective than everolimus in mTOR inhibitor-naïve patients with pancreatic neuroendocrine tumors [216]. In another Phase I/IB study of BEZ235 in patients with advanced solid tumors including those with advanced breast cancer, the inhibition of PI3K/mTOR was not sufficient and adverse effects were prevalent [217]. In a phase II study of BEZ235 in patients with locally advanced or metastatic transitional cell carcinoma, clinical activity was modest and was accompanied by considerable toxicities [218]. Combination therapy utilizing dactolisib has also been disappointing. Treatment strategies utilizing dactolisib with everolimus demonstrated limited efficacy and tolerance in patients with advanced solid tumors [219]. In patients with castration-resistant prostate cancer, treatment with dactolisib and abiraterone acetate, an anti-androgen agent, also yielded poor efficacy and tolerance profile [220].

Other dual PI3K/mTOR inhibitors are also undergoing clinical trials. Apitolisib (GDC-0980, Genentech) is another dual inhibitor of PI3K/mTOR. A phase I study in patients with advanced solid tumors revealed tolerability at 30 mg, with modest but durable anti-tumor activity [223]. Apitolisib was compared against everolimus in patients with clear-cell metastatic renal cell carcinoma in a phase II trial. Not only were adverse events more frequent in patients receiving apitolisib compared with everolimus, but apitolisib was also found to be less effective [224]. In a phase II study of apitolisib for the treatment of recurrent or persistent endometrial carcinoma (MAGGIE study), the anti-tumor activity was limited by the dose toxicity particularly for patients with diabetes. However, molecular profiling of evaluable archival tumor samples revealed that the patients with confirmed response had at least one alteration in a gene involved in the PI3K pathway [225]. In a phase II trial of voxtalisib (SAR245409, XL-765, Sanofi), there was acceptable safety profile with promising efficacy in patients with follicular lymphoma but limited efficacy in patients with mantle cell lymphoma, diffuse large B-cell lymphoma, or chronic lymphocytic leukemia [226]. The limited efficacy in the aggressive lymphomas could be due to incomplete blockade of PI3Kδ. While there seemed to be no strong correlation between the presence of PI3K or PTEN mutations and the response to voxtalisib, there was one patient with a PIK3CA and KRAS mutation who had a complete response.

There have been some promising trials with dual PI3K/mTOR inhibitors. In a phase II study in patients with recurrent endometrial cancer, gedatolisib demonstrated moderate activity [222]. Another study using BGT226 (NVP-BGT226, Novartis) was generally well tolerated with only three patients out of 57 having dose-limiting toxic effects. The study also reported 30% of patients reaching stable disease with nine achieving stable disease for over 16 weeks and 53% of patients reached stable metabolic disease. However, they found inhibition of the PI3K pathway to be inconsistent, but this may have been due to low systemic exposure [227].

It is interesting to note that the dual inhibition of PI3K and mTOR in patients displayed more toxicity than combined rapalog/growth factor receptor blockade. The latter strategy should, in theory, block PI3K signaling, at least in cases wherein PI3K is not constitutively active or PTEN is inactivated. Yet, it is likely less toxic since the amplified expression of RTKs occurs in specific tissues (e.g., HER2 in breast). In contrast, increased mTOR and PI3K signaling would be pervasive not only in tumors but highly proliferating normal cells. Hence, more studies are needed to determine if dual PI3K/mTOR inhibitors would be most effective for tumors displaying hyperactive PI3K/mTOR signaling such as those harboring PTEN inactivating or PI3K- and/or mTOR-activating mutations. These types of tumors may be more sensitive to lower doses of the dual inhibitor and would thus have increased efficacy with lower toxicity.

3.5. Targeting mTORC2 Signaling

Over-activation of Akt has been found to be associated with many types of cancer, including HER2-amplified breast cancer and glioblastoma [288,289]. mTORC2 modulates Akt signaling by allosteric phosphorylation of Akt on Ser473. mTOR as part of mTORC2 requires the presence of its partners rictor and SIN1 to function. Rictor has been found to be highly expressed in certain types of cancers, such as colorectal cancer and non-small cell lung cancer [208,290]. Therefore, there has been interest in using rictor as a target for cancer therapeutics.

In HER2-enriched breast tumors, rictor expression was upregulated significantly compared to non-malignant tissues. Genetic ablation of *rictor* led to decreased cell survival and phosphorylation at S473 on Akt as well as decreased tumor formation and tumor multiplicity in a HER2/Neu mouse model of breast cancer. In the same study, it was found that there was decreased Akt activation and cell survival in multiple HER2-amplified human breast cancer cell lines with rictor loss [288]. mTOR inhibition with rapamycin was demonstrated to induce activation of Akt in human gastric and pancreatic cancer cells. Knockdown of rictor upon rapamycin treatment in these cancer cells led to diminished Akt phosphorylation and function, impaired cell motility and potentiated the anti-migratory properties of rapamycin. This suggests that simultaneous inhibition of mTORC1 and rictor could be an approach to reduce metastatic spread of tumors [291]. Another study investigated the effects of RNAi-mediated gene silencing of both rictor and EGFR in glioblastoma cells. Here, it was also demonstrated that a combined approach can reduce cell migration. Furthermore, siRNA-mediated silencing of EGFR and rictor also increased sensitivity to irinotecan, temozolomide, and vincristine in PTEN mutant human GBM cell line U251MG and increased sensitivity to vincristine and temozolomide in LN229 cell line. Silencing of both EGFR and rictor also caused complete eradication of tumors in U251MG cell line [289]. A nanoparticle-based RNAi therapeutic that was engineered to target rictor was shown to decrease breast cancer cell growth and survival via intratumoral and intravenous delivery. Furthermore, using this molecule in combination with lapatinib led to an even greater reduction of tumor growth [292]. JR-AB2-011 is a small molecule inhibitor that has been recently developed, which prevents the interaction of rictor with mTORC2. This molecule demonstrated significant anti-tumor effects in glioblastoma xenograft studies [230].

Although there have not been human clinical trials yet to test the effectiveness of rictor inhibition, patients with rictor amplification may benefit from ATP-competitive mTOR kinase inhibitors that block both mTORC1 and mTORC2. One study found that a patient with rictor-amplified non-small cell lung cancer achieved tumor stabilization for 12 months with CC-223 (Celgene). Disease rapidly progressed when treatment with CC-223 was ceased [208]. In small-cell lung cancer cell lines with rictor amplification, there was increased sensitivity to ATP-competitive mTOR kinase inhibitors [203]. Furthermore, a gastric-cancer patient derived cell line with rictor amplification was found to be most sensitive to AZD2014 [202]. The expression of SIN is also upregulated in medullary and aggressive papillary thyroid carcinomas [293]. The increased SIN1 expression was associated with enhanced Akt activation. Future studies should reveal whether increased levels of mTORC2 components could serve as predictive biomarkers for sensitivity to TORKIs or more specific mTORC2 inhibitors.

3.6. RapaLink1

Another class of mTOR inhibitor, RapaLink, has been recently developed [231]. RapaLink consists of a rapamycin-FRB binding element linked to mTOR kinase inhibitor (TORKI). This inhibitor was generated to combine the advantages while overcoming the limitations posed by each of the two types of inhibitors. While rapamycin has limited inhibitory capacity, it is more stable in cells due to its binding to FKBP12. On the other hand, the TORKI have better blockade of mTOR but has poor durability. Thus, this new bivalent inhibitor combines the durable effect of rapamycin and the superior inhibitory capacity of TORKI. Furthermore, it crosses the blood–brain barrier and was able to block in vivo glioblastoma models [232]. Follicular lymphoma, an incurable form of B cell lymphoma, with a genetic mutant of the epigenetic regulator, EZH2, was sensitive to Rapalink1 [233]. The increased mTORC1 activity due to EZH2 mutant repression of Sestrin1, which acts as a tumor suppressor, likely contributed to the sensitivity of these tumors to mTOR inhibition. Future studies should reveal whether this new class of mTOR inhibitor would have a desirable efficacy with less toxicity.

3.7. Combining mTOR Inhibition with Other Protein Kinase Inhibitors

mTOR responds to both nutrients and growth factors. Other signaling molecules could also be modulated by these signals. Furthermore, the genes encoding the proteins from these signaling pathways also undergo mutations and could drive oncogenesis. Therefore, there are numerous efforts to develop inhibitors to these molecules. One of the signaling pathways that often become deregulated in cancer is the Ras/MAPK pathway. This pathway cross-talks with the PI3K/mTOR pathway and can become highly upregulated to compensate for dampened PI3K/mTOR signals. Indeed, the Ras/MAPK and PI3K/mTOR pathways converge on regulating some common transcriptional regulators of cell growth, proliferation and metabolism [243]. Furthermore, both pathways could also be simultaneously upregulated. In kidney and endometrial carcinoma, mutant Rheb-Y35N led to constitutive activation of both mTORC1 and MEK/ERK pathway, leading to rapamycin resistance of these tumors. Hence, combined mTOR and MAPK inhibition could more effectively prevent growth of tumors that display alterations in these two critical growth-regulatory pathways [294].

The efficacy of combining mTOR inhibition with blockade of the Ras/MAPK pathway to promote cell death is supported by several pre-clinical studies. Combining rapamycin with Raf inhibitors in melanoma cell lines led to more potent growth inhibition [295,296]. The combination of PI3K/mTOR inhibitor CMG002 with sorafenib inhibited proliferation of hepatocellular carcinoma cell lines, induced apoptosis and blocked the activity of both Ras/MAPK and PI3K/mTOR pathways [228]. In thyroid cancer cell lines, BEZ235 combined with RAF265, a pan-RAF inhibitor synergistically inhibited growth of cell lines and mouse xenografts [257]. The mTORC1/2 inhibitor AZD8055 was combined with PI3K inhibitor (GDC0941) and MEK1/2 inhibitor selumetinib each at low doses in advanced stage ovarian clear cell carcinoma cell lines and patient-derived xenograft models [258]. The triple combination was more effective in preventing tumor growth and had better tolerability in PDX models. In a subset of non-small cell lung cancer that have increased rictor expression, the concomitant inhibition of mTORC1/2 and MEK1/2 had synergistic anti-tumor effects [297]. Analysis of genomic and expression data from patient samples revealed that the rictor-altered cohort had increased K-ras/MAPK axis mutations. In leukemia cells wherein Akt is not constitutively active, treatment with an inhibitor of MAPK-interacting kinases (Mnks) increased the sensitivity to rapamycin, resulting in more effective inhibition of proliferation [260]. Whereas Mnk inhibition alone was cytostatic and did not fully block 4EBP phosphorylation, combined Mnk and mTORC1 inhibition was cytotoxic and strongly inhibited 4EBP phosphorylation.

Despite the promising pre-clinical evidence, clinical trials using combined mTOR and Ras/MAPK pathway inhibition of different cancer types have had varied results. In a phase I/II study of patients with recurrent glioblastoma, combination treatment with temsirolimus and sorafenib (Raf kinase and VEGFR-2 inhibitor) led to significant toxicity and lacked efficacy [254]. In phase I studies of everolimus and sorafenib for advanced renal cell cancer, there was enhanced antitumor activity and reasonable tolerability [255,256]. In a phase Ib trial of combined everolimus and the oral MEK inhibitor trametinib (GSK1120212) for patients with advanced solid tumors, there was modest activity and poor tolerability [261]. In a phase Ib dose-escalation study of patients with advanced solid tumors treated with MEK inhibitor trametinib in combination with PI3K/mTOR inhibitor GSK2126458, there was poor tolerability and responses were minimal despite upregulation of PI3K/Ras pathway, likely due to toxicities [229]. In advanced solid tumors, a phase Ib clinical trial of combined MEK inhibitor (pimasertib) and PI3K/mTOR inhibitor (voxtalisib) had poor long-term tolerability and limited anti-tumor activity [262]. In a phase I trial of temsirolimus combined with pimasertib for patients with advanced solid tumors, there was unfavorable toxicity profile although some patients had some clinical benefit and stabilized disease [263]. A randomized phase II trial of the MEK inhibitor selumetinib in combination with temsirolimus for soft tissue sarcomas (STS) revealed that the combination treatment compared to selumetinib treatment alone did not improve progression-free survival in patients with advanced STS. However, the combination treatment seems to have better efficacy for leiomyosarcoma, thus warranting further investigation [259]. It is notable that previous studies have reported that over 70% of primary leiomyosarcoma tumors have increased Akt/mTOR signaling [298]. Hence, future studies should address whether better tolerability and enhanced efficacy would be achieved by tailoring the doses of each inhibitor depending on specific alterations in each or both of the PI3K/mTOR and Ras/MAPK pathways.

3.8. Combining mTOR Inhibition with Conventional Chemotherapies and Other Targeted Therapies

Traditional cytotoxic chemotherapy remains a staple in cancer treatment. Many of these drugs, such as alkylating agents, intercalating drugs, microtubule disruptors, and topoisomerase inhibitors, directly target the DNA of the cell [299]. Due to eventual development of resistance, these conventional chemotherapeutic agents are combined with more targeted therapy including mTOR inhibitors. Rapalogs are also largely cytostatic and would likely have more efficacy when combined with cytotoxic chemotherapy. In a phase II trial of everolimus and carboplatin, this combination displayed efficacy in treating patients with triple negative metastatic breast cancer [264]. However, the addition of everolimus to carboplatin in patients with metastatic prostate cancer had minimal clinical efficacy [265]. In a phase II clinical trial of everolimus combined with paclitaxel, which blocks the cell cycle by stabilizing the microtubules, and carboplatin as first-line treatment for metastatic large-cell neuroendocrine lung carcinoma, the treatment was well tolerated and displayed efficacy [267]. In a randomized phase II study in patients with triple negative breast cancer, addition of everolimus to the paclitaxel/cisplatin combination was associated with more adverse events without improvement in clinical response compared to paclitaxel/cisplatin treatment [268]. In a phase I/II clinical trial of everolimus combined with gemcitabine/cisplatin for metastatic triple negative breast cancer, the combination treatment did not have synergistic effects despite the majority of patients harboring PIK3CA mutations [269]. The combination of ridaforolimus with paclitaxel and carboplatin in a phase I study in patients with solid tumor cancers showed antineoplastic activity with no unanticipated toxicities [266]. In a randomized phase II study to compare the effects of monotherapy with vinorelbine, which disrupts microtubules, versus combined vinorelbine and everolimus for second-line chemotherapy in advanced HER2-negative breast cancer, the combined therapy was not superior to the monotherapy, although the treatment was well tolerated [270]. In a phase II study of everolimus in combination with CHOP (cyclophosphamide, doxorubicin, vincristine, and prednisone) as a first-line treatment for patients with peripheral T-cell lymphoma, the treatment was efficacious [271]. Immunohistochemistry revealed maintenance of PTEN expression among patients displaying a complete response. In a phase II study of temsirolimus with liposomal doxorubicin for patients with recurrent and refractory bone and soft tissue sarcomas, stable disease was achieved in more than half of the patients and therapy was well tolerated [272]. The response to treatment correlated with a decline in the highly expressing aldehyde dehydrogenase (ALDH) population of putative sarcoma stem cells, thus supporting that mTOR inhibition could sensitize this population to doxorubicin treatment. In relapsed childhood acute lymphoblastic leukemia, everolimus combined with four-drug reinduction chemotherapy (vincristine, prednisone, pegaspargase and doxorubicin) was well tolerated and had promising results [273]. In a phase I/II study of everolimus in combination with hyperCVAD chemotherapy in patients with relapsed/refractory T-ALL, the combination treatment was well tolerated and produced some favorable patient response. Interestingly, patients that had lower baseline 4EBP phosphorylation at Thr37/46 were associated with a better response to the therapy [300]. A phase I trial of temsirolimus and intensive re-induction chemotherapy in children with relapsed ALL also induced remission in about half of the patients, despite resulting in excessive toxicity at the dosages used [301]. In a study that used temsirolimus as maintenance therapy in castration-resistant prostate cancer after docetaxel induction, the regimen proved safe, and delayed the time to treatment failure to 6 months [274]. Voxtalisib plus temozolomide, an alkylating agent, with or without radiotherapy also displayed a favorable safety profile and moderate amount of PI3K/mTOR pathway inhibition in patients with high-grade glioma [275]. Based on these clinical trial results, combining mTOR inhibition with conventional cytotoxic chemotherapy warrants further investigation. These cytotoxic chemotherapies can overcome the cytostatic effects of rapalogs. As these drugs have also been used over decades, there are also better protocols available for the use of these drugs that could serve to prevent unwanted side effects.

Whether the combined mTOR inhibition and cytotoxic chemotherapy could delay the development of resistance remains to be further investigated.

Aside from the traditional cytotoxic chemotherapies, a few conventional targeted therapies have also been used in combination with mTOR inhibition. In a randomized phase II trial in post-menopausal women with hormone receptor positive, EGFR2-negative metastatic breast cancer resistant to aromatase inhibitor therapy, the combination of everolimus with fulvestrant, which downregulates the estrogen receptor, enhanced efficacy compared to fulvestrant alone [279]. In a phase III trial, everolimus combined with the aromatase inhibitor exemestane improved the progression free survival of postmenopausal hormone-receptor-positive advanced breast cancer [277]. In a phase II trial, voxtalisib was used in combination with letrozole, an aromatase inhibitor, in patients with HR+, HER2-negative metastatic breast cancer refractory to a non-steroidal aromatase inhibitor. Voxtalisib, in combination with letrozole, had an acceptable safety profile, but no objective response was observed and further investigation is not warranted [278]. These findings also reveal that combining targeted therapies with rapalogs has more efficacy than with PI3K/mTOR inhibitors. The most likely explanation for this is the increased durability of response and the lower toxicity with the use of rapalogs.

3.9. Co-targeting mTOR and Metabolism

The increased demand for nutrient-derived metabolites for macromolecule synthesis by cancer cells is a vulnerability that is often exploited for cancer therapy. The use of anti-metabolites or metabolite analogs can block the generation of critical building blocks for protein, DNA/RNA or lipid synthesis as well as production of reducing equivalents such as NADPH, thus promoting cell death. The mTOR pathway is often upregulated in cancers and promotes generation of these critical metabolic intermediates via modulating the expression or activity of metabolic enzymes or transcription factors. Thus, the dependence of cancer cells to a specific metabolic pathway makes such a pathway a viable target for inhibition and could prove to be more effective especially when combined with mTOR inhibition.

mTORC1 plays numerous roles in the pentose phosphate pathway and nucleotide synthesis. Hence, rapalogs have been used in combination with nucleotide/nucleoside analogs (Table 3). For example, 5-fluorouracil, a pyrimidine analog that has been widely used in cancer therapy has been used in combination with mTORC1 inhibition in recent clinical trials. In a phase Ib study of everolimus plus mFOLFOX-6, a combination chemotherapy regimen of 5-FU, folinic acid, and oxaliplatin, 83% of patients with metastatic gastroesophageal adenocarcinoma experienced a partial response [276]. In phase I trials of everolimus combined with capecitabine, which is a pro-drug of 5-FU, the treatment was found to be well tolerated and safe with promising clinical benefit in metastatic triple negative breast cancer and in advanced solid malignancies [302,303]. Gemcitabine is a nucleoside analogue used to treat several types of cancers. In a phase II study, 48.5% of patients with osteosarcoma were observed to have stabilized disease when treated with a combination of gemcitabine and sirolimus [304]. However, in advanced pancreatic cancer, combination treatment with temsirolimus and gemcitabine lacked clinical efficacy [305]. In a phase Ib/II study of everolimus in combination with azacitidine, a cytidine analog used in patients with relapsed/refractory acute myeloid leukemia, the treatment was tolerable and has promising clinical activity [306]. While these recent studies suggest that combined inhibition of mTORC1 and nucleotide metabolism may have efficacy in certain cancers, it may be worth evaluating mTORC1 signaling and its consequences in reprogramming nucleotide metabolism in these tumors. In a recent pre-clinical study, it was found that mTORC1 inhibition may not provide benefit for tumors with increased mTORC1 activity due to TSC deficiency. Since mTORC1 coordinately controls many anabolic processes, blocking mTORC1 activity would then generally dampen metabolic processes, thus having a cytostatic rather than cytotoxic effect. Instead, targeting one anabolic branch, in this case, inhibiting nucleotide synthesis using the IMP dehydrogenase inhibitor mizoribine, creates a metabolic imbalance, thus promoting cell death [307].

Table 3. Metabolism inhibitors or anti-metabolites and other mTOR pathway inhibitors.

Drug	Target	References
Metabolism Inhibitors		
5-Fluorouracil	Thymidylate Synthase	[276]
Capecitabine	Thymidylate Synthase	[302,303]
Gemcitabine	DNA	[304,305]
Azacitidine	DNA methyltransferase/DNA	[306]
Mizoribine	IMP dehydrogenase	[307]
L-asparaginase	Asparagine	[64]
Cerulenin	FASN	[308,309]
Metformin	Complex 1	[310–314]
Hydroxychloroquine	Lysosomes/Autophagy	[315–317]
Other mTOR Pathway Inhibitors		
NR1	Rheb	[318]
Lonafarnib	Farnesyl transferase	[319]
Perifosine	Akt	[320]
MK-2206	Akt	[321–324]
AZD5363	Akt	[325]
GSK650394	SGK1	[326,327]
SI113	SGK1	[328]
GSK2334470	PDK1	[329]

Given the role of mTOR in protein synthesis and amino acid metabolism, mTOR inhibition is also often combined with drugs that block amino acid uptake or those that prevent amino acid biosynthesis. For example, pre-clinical studies demonstrated that combined rapamycin and L-asparaginase, which depletes cells of both glutamine and asparagine, can synergistically inhibit the growth of KRAS-mutant colorectal cancer cells that have upregulated asparagine synthetase [64].

Fatty acid metabolism is also being targeted in combination with mTOR inhibitors for cancer therapy. In pre-clinical studies of ER/HER2 positive breast cancer, cerulenin, a FASN inhibitor synergized with rapamycin to induce apoptosis and inhibit tumorigenesis [308]. FASN inhibition in ovarian cancer led to cell death, which involved a caspase 2 mechanism via the mTORC1 negative regulator REDD1 [309].

Another drug that targets metabolism and has been used in combination with mTORC1 inhibition is metformin. Metformin is a first-line anti-diabetic drug that inhibits the mitochondrial respiratory chain (complex I) and activates the enzyme AMP-activated protein kinase (AMPK). Through its action on AMPK, metformin is believed to also inhibit mTOR as well as activate the tumor suppressor gene TSC2. Two different phase I clinical trials have combined temsirolimus with metformin. In one study, 56% of patients with advanced or refractory cancers were observed with stable disease after treatment of temsirolimus and metformin [310]. In another study, one of 11 patients experienced partial response while five of the remaining patients experienced stable disease. One of these patients with melanoma had stable disease for 22 months [311]. Combined everolimus and metformin in advanced solid malignancies was poorly tolerated likely due to pharmacointeractions between the two drugs since everolimus delayed and prevented the elimination of metformin [312]. In a pharmacodynamic study, patients with advanced solid tumors treated with combined sirolimus and metformin tolerated the regimen but no significant changes in mTOR inhibition or other serum pharmacodynamic biomarkers were observed [313]. The presence of predictive biomarkers of metabolism could improve the efficacy of combined metformin and mTOR inhibitor treatment. For example, in diffuse large B cell lymphoma (DLBCL) wherein about 40% are refractory to the standard combined immunotherapy (R-CHOP), the low levels of GAPDH, which predict poor response to R-CHOP, correlated with dependence on oxidative phosphorylation (OxPhos) metabolism, mTORC1 signaling and glutaminolysis. Based on such dependence, three out of four patients treated with asparaginase, temsirolimus and metformin (KTM) displayed a complete response to this combination treatment [314].

mTORC1 negatively regulates catabolic processes such as autophagy. Hence, inhibition of mTORC1 unleashes autophagy and provides salvaged metabolites to cancer cells thereby allowing their growth and proliferation. Pre-clinical studies combining mTOR and autophagy inhibition displayed synergistic cytotoxic effects [315]. In a phase I clinical trial, significant anti-tumor activity was observed in melanoma patients treated with both temsirolimus and hydroxychloroquine, an autophagy inhibitor [316]. In a phase I/II trial of everolimus with hydroxychloroquine in patients with previously treated renal cell carcinoma, the combination treatment was tolerable and achieved a >40% 6-month PFS rate [317].

As discussed in earlier sections, our knowledge of the roles of both mTOR complexes in controlling metabolic pathways is expanding. As we gain understanding of the unique metabolic needs of different tissues and how they become deregulated in cancer, we can tailor treatment strategies based on their metabolic dependencies. An intriguing concept is the integration of dietary manipulation with targeted therapies to improve patient response [330].

3.10. Other Inhibitors of the mTOR Pathway

Compounds that target key signaling molecules along the mTOR pathway have also been developed and are undergoing pre-clinical and clinical trials. A small molecule inhibitor of Rheb, NR1, binds Rheb in its switch II domain and selectively blocks mTORC1 signaling but not mTORC2 or ERK signaling in multiple cell lines [318]. The farnesyl transferase inhibitor lonafarnib, which inhibits the farnesylation of proteins, including Ras and Rheb, downregulates mTOR signaling and potentiates the apoptotic effect of the pan-Raf inhibitor sorafenib but not the Akt inhibitor on melanoma cells [319]. A phase I trial of temsirolimus with the Akt inhibitor, perifosine, for recurrent pediatric solid tumors including gliomas and medulloblastomas, was well tolerated, although partial or complete responses were not achieved [320]. It is not clear if the drugs reached their targets such as the brain. In a phase I study of temsirolimus with perifosine, the combination was generally tolerated with no dose-limiting toxicity and safe in patients with recurrent/refractory pediatric solid tumors [320]. A phase I trial of combined ridaforolimus and an Akt inhibitor (MK-2206) in patients with advanced malignancies showed promising activity in hormone-positive and -negative breast cancer with PI3K pathway dependence [321]. In a large panel of cancer cell lines, T-cell acute lymphoblastic leukemia (T-ALL) with Notch mutation were highly sensitive to Akt inhibitor (AZD5363) and the mTORC1/2 inhibitor (AZD2014) but only partially sensitive to PI3K inhibitors [325]. While combining mTOR with Akt inhibition showed promise with less toxicity, it seems that Akt inhibitor monotherapy had low clinical activity partly due to poor tolerability [322–324]. Pre-clinical studies on inhibition of SGK, another target of mTORC2, for the treatment of a variety of cancers are also showing promise [327,328,331]. SGK1 could play a role in enhancing uptake of unsaturated fatty acids in hypoxic lung cancer cell [327]. The hypoxia makes these tumors reliant on uptake rather than desaturation of saturated FA, a process that occurs in normoxia. Therefore, this vulnerability can also be exploited for more effective therapy of tumors that have deregulated fatty acid metabolism. The 3-phosphoinositide-dependent kinase 1 (PDK1) inhibitor GSK2334470 displayed antitumor activity in multiple myeloma cells. Interestingly, while it blocked phosphorylation of the mTORC1 target S6K1 (at Thr389) and the phosphorylation of Akt at Thr308, it did not affect phosphorylation of Akt at the mTORC2-targeted site Ser473. Combined treatment with GSK2334470 and the mTOR inhibitor PP242 had more potent anti-myeloma activity and led to complete inhibition of mTOR and Akt [329]. Future studies should reveal which types of tumors could benefit from this combination treatment in the clinic.

3.11. Resistance Mechanisms and Other Therapeutic Opportunities

A main challenge in cancer therapy is the development of resistance to chemotherapeutic agents and targeted therapies by malignant cells. As for mTOR inhibitors, cells acquire mutations in mTOR or its partners that could prevent drug binding or upregulate mTORC components that could enhance mTOR activity, hence limiting the efficacy of these inhibitors [231,332]. Cells can also bypass a block in mTOR by upregulating other signaling and metabolic pathways. Non-biased omics technologies have facilitated identification of such bypass mechanisms that could serve as therapeutic opportunities or predictive biomarkers for follow-up treatment. Recent efforts to gain insights into such resistance mechanisms underscore the ability of cancer cells to restore key metabolic processes that support their growth and proliferation. For example, proteomics and genomics studies that analyzed changes that occur after treatment with mTOR inhibitors revealed bypass mechanisms related to protein synthesis in Ewing sarcoma cells [333] and glioblastoma [334]. Using metabolomics studies, another resistance mechanism occurring due to PI3K/mTOR inhibition is via upregulation of the purine salvage pathway in small cell lung carcinoma [335]. Non-genetic mechanisms can also occur, as has been shown recently in a single-cell phosphoproteomics analysis of patient-derived in vivo glioblastoma model. Resistance to TORKI was monitored by single cell phosphoproteomics and revealed adaptive signaling dynamic alterations that were responsive to combinations of drugs targeting these pathways [336]. The combination of genomics, proteomics and metabolomics and improved technologies in single-cell analysis should have a tremendous impact in moving towards more personalized therapy.

3.12. Immunotherapy

mTOR plays a central role in immunity [337]. In fact, rapalogs are widely used as immunosuppressants to prevent kidney transplant rejection. Interestingly, rapamycin affects only distinct classes of immune cells [338]. Studies on the metabolic dependencies of specific T cell subsets are providing clues on the basis of rapamycin's specific effects [339]. Quiescent or naïve T cells rely on OxPhos for their metabolic needs, whereas activated highly proliferating effector T cells depend on robust glycolytic metabolism to fuel their growth and proliferation [340–342]. Inhibition of mTOR using rapalogs specifically affects these highly proliferating immune cells while being ineffective on T cells that rely on OXPHOS metabolism, such as T-regulatory (T-reg) and memory T cells. The effects of mTOR inhibition on specific T-cell subsets can be exploited to improve cancer treatment and immunotherapy. T cells are an important component of the tumor microenvironment. Inhibitors that promote cancer cell death combined with strategies to boost effector functions of T cells while repressing negative regulators of the immune responses could improve therapeutic outcome. In a phase I clinical trial for the treatment of metastatic renal cell cancer, everolimus was combined with low-dose cyclophosphamide [343]. Cyclophosphamide was administered to selectively deplete the immunosuppressive T-regs, which undergo expansion in the presence of rapamycin [344]. The results have been promising as the treatment sustained levels of CD8$^+$ T-cell population together with increased effector to suppressor ratio. The observed changes in various immune cell populations may promote antitumor immunity. These promising results have led to a phase II clinical trial. mTOR inhibition in the tumor may also have consequences that allow T cells to enhance their anti-tumor recognition. In pre-clinical studies, everolimus combined with anti-PD-L1 was more effective in decreasing tumor burden compared to individual treatment in a mouse model of renal cell carcinoma due to upregulation of PD-L1 in the tumor cells resulting in increased tumor infiltrating CD8+ lymphocytes and tumor regression [345]. Manipulating mTOR signaling in other immune cell types is also being investigated for enhancing immunotherapeutic strategies. Inhibition of mTOR using rapamycin increases cytotoxicity of Vγ4 γδ T cells towards various cancer cell lines by enhancing NKG2D expression and TNF-α expression [346].

Immunotherapeutic approaches using ex vivo cultures for adoptive transfer of genetically modified T-cells including tumor-infiltrating lymphocytes and chimeric antigen receptor-T cells (CAR-T) could also be enhanced by manipulation of mTOR activity. Akt inhibition during ex vivo expansion of tumor-infiltrating lymphocytes increased the generation of antitumor CD8+ T cells with memory cell phenotypes. This allows increased persistence following adoptive transfer, thus enhancing their antitumor activity [347]. Decreased mTORC1 activity due to IL15, which is used to expand CAR-T cells, also enhances CAR-T cell antitumor activity by preserving their stem cell memory phenotype. CAR-T cells that are less differentiated or less exhausted are more effective [348].

mTOR inhibition is also being exploited for improvement of vaccine strategies. Rapalogs can enhance the generation of $CD8^+$ memory T cells in response to vaccination. This appears to be due to the rapamycin-mediated reprogramming of metabolism to fatty acid oxidation [349,350]. T cells that have Rheb-deficiency, thereby decreased mTORC1 activity, have poor effector cells while their memory cells persisted. Conversely, T cells that are TSC2 null, thereby with hyperactive mTORC1, have increased effector phenotypes but fail to convert to memory cells [351]. mTOR inhibition also improved $CD8^+$ T cell responses to vaccinia virus vaccination in rhesus macaques [352] and enhanced immune responses to influenza vaccine in the elderly [353]. Using microparticles encapsulating rapamycin, the low dose release of rapamycin polarized vaccine-induced T cells toward central memory phenotypes, which could enhance anti-tumor immunotherapy [354]. Rapalogs could also modulate dendritic cell function thus enhancing anti-tumor effects of DNA vaccines [355,356]. A better understanding of the metabolic dependencies of various immune cells should provide more opportunities for immunotherapy.

4. Conclusions

Less than a century after Warburg's hypothesis, we now have a better understanding of how the defective metabolic processes in cancer cells are intertwined with genetic and proteomic alterations. The discovery of rapamycin and the mTOR pathway facilitated the elucidation of how cancer cells reprogram their metabolism in order to acquire nutrients that are necessary for their growth and proliferation. Perturbations in this pathway, such as oncogenic and tumor suppressor mutations that elevate mTOR signaling, lead to rewiring of metabolic pathways in ways that increase aerobic glycolysis, as Warburg reported. However, studies over the past decades unravel that cancer cells also display heterogeneous metabolic vulnerabilities that can be exploited for more effective and specific therapy. The results from rapalog clinical trials suggest how tumors that may have particular metabolic signatures due to mTORC1 activation, by virtue of mutations in TSC, are more sensitive to rapalog monotherapy. There are numerous efforts to identify predictive biomarkers and thus identify patients who would benefit most from mTOR inhibitors [239,357–359]. Such biomarkers include not only genetic, proteomic or signaling alterations but also metabolite changes. The latter is exemplified by recent studies in gliomas harboring the mutant isocitrate dehydrogenase 1 that produces the oncometabolite 2-hydroxyglutarate. These gliomas displayed sensitivity to voxtalisib, the dual PI3K/mTOR inhibitor [360]. Although other metabolites were altered upon voxtalisib treatment, the decrease in 2HG highly correlated with the increased animal survival. Thus, measurement of 2HG by magnetic resonance spectroscopy could be useful as a metabolic biomarker for mutations in isocitrate dehydrogenase 1 (IDHmut) glioma response to PI3K/mTOR inhibition. The mTOR pathway can also be activated by multiple mechanisms. In a pan-cancer proteogenomic analysis of thousands of human cancers, many of these cancers had high mTOR pathway activity despite the lack of alterations in canonical genes associated with this pathway [361]. These findings highlight the need to explore metabolomic impacts on mTOR signaling. In other studies, despite rationalizing treatment strategy based on molecular aberrations in the PI3K/mTOR and EGFR/MAPK pathways, patients did not derive significant benefits [362]. Such resistance to mTOR inhibition underscores how the metabolic plasticity of cancer cells enables the emergence of alternative mechanisms to feed the growing tumor. Other signaling pathways that converge on mTOR or metabolism could also serve as potential additional targets. For example, CDK4, which regulates the cell cycle, modulates mTORC1 via phosphorylation of the tumor suppressor FLCN [363]. This regulates mTORC1 recruitment to the lysosomal surface in response to amino acids. Epigenetic mechanisms could also contribute to metabolic reprogramming. In a phase I study using combined ridaforolimus and the histone deacetylase (HDAC) inhibitor vorinostat in advanced renal cell carcinoma, prolonged disease stabilization was observed and was tolerable at the phase II dose [364]. In a phase I study of sirolimus and vorinostat in patients with advanced malignancy, the combination treatment seemed to be safe and displayed efficacy [189].

Identification of mutations in mTOR could also inform on sensitivity to mTOR inhibitors. Indeed, mTOR mutations have been identified in human cancers [163,164,365–369]. Toxicities associated with mTOR inhibitors could also be addressed by the mode of drug delivery. Improvement of formulation and dosing could be beneficial. A twice daily 5 mg dosing instead of once-daily 10 mg regimen conducted on a randomized pharmacokinetic crossover trial of everolimus suggested that such dosing could be promising to reduce adverse toxicity while maintaining treatment efficacy [370]. Nanoparticle-based mTOR targeting would need to be improved as well in order to have more specific effects on tumors and prevent undesirable cellular perturbations [371]. Defining the metabolic conditions within a given tumor microenvironment would also improve targeting strategies and perhaps provide clues on metastatic sites that could be conducive for growth of a malignant tumor with a metabolic dependency. Furthermore, given the dynamic effects of nutrition in metabolism, gene expression and cell signaling between different tissues and individuals, a deeper understanding of the mechanisms involved in these processes should pave the way for integrating diet manipulation with targeted therapeutic strategies and more personalized therapy.

An exciting application of mTOR inhibitors is in the improvement of immunotherapeutic strategies. The use of mTOR inhibitors for the improvement of vaccination strategies also promises to have a major impact towards cancer vaccination. As we gain more understanding of the metabolic needs of different immune cell subsets as well as the metabolic vulnerabilities of the tumor in the TME, we can improve anti-tumor activity of effector T cells while preventing immunosuppressive mechanisms as well as develop more innovative targeted therapeutic strategies for more effective cancer treatment.

Funding: This work was supported by NIH grants GM079176, CA154674 and New Jersey Commission on Cancer Research Bridge Grant (DFHS18CRF008) to E.J.

Acknowledgments: We thank the members of the Jacinto lab for helpful discussions. We apologize to our colleagues in the field whose work has not been cited due to space restrictions.

Conflicts of Interest: The authors declare no conflict of interest.

References

1. Warburg, O. On the origin of cancer cells. *Science* **1956**, *123*, 309–314. [CrossRef]
2. Heitman, J.; Movva, N.R.; Hall, M.N. Targets for cell cycle arrest by the immunosuppressant rapamycin in yeast. *Science* **1991**, *253*, 905–909. [CrossRef] [PubMed]
3. Saxton, R.A.; Sabatini, D.M. mTOR Signaling in Growth, Metabolism, and Disease. *Cell* **2017**, *169*, 361–371. [CrossRef] [PubMed]
4. Mossmann, D.; Park, S.; Hall, M.N. mTOR signalling and cellular metabolism are mutual determinants in cancer. *Nat. Rev. Cancer* **2018**, *18*, 744–757. [CrossRef] [PubMed]
5. Easton, J.B.; Houghton, P.J. mTOR and cancer therapy. *Oncogene* **2006**, *25*, 6436–6446. [CrossRef]
6. Benjamin, D.; Colombi, M.; Moroni, C.; Hall, M.N. Rapamycin passes the torch: A new generation of mTOR inhibitors. *Nat. Rev. Drug Discov.* **2011**, *10*, 868–880. [CrossRef]
7. Barbet, N.C.; Schneider, U.; Helliwell, S.B.; Stansfield, I.; Tuite, M.F.; Hall, M.N. TOR controls translation initiation and early G1 progression in yeast. *Mol. Biol. Cell* **1996**, *7*, 25–42. [CrossRef]
8. Beck, T.; Hall, M.N. The TOR signalling pathway controls nuclear localization of nutrient- regulated transcription factors. *Nature* **1999**, *402*, 689–692. [CrossRef]
9. Zhang, H.; Stallock, J.P.; Ng, J.C.; Reinhard, C.; Neufeld, T.P. Regulation of cellular growth by the Drosophila target of rapamycin dTOR. *Genes Dev.* **2000**, *14*, 2712–2724. [CrossRef]
10. Dennis, P.B.; Fumagalli, S.; Thomas, G. Target of rapamycin (TOR): Balancing the opposing forces of protein synthesis and degradation. *Curr. Opin. Genet. Dev.* **1999**, *9*, 49–54. [CrossRef]
11. Noda, T.; Ohsumi, Y. Tor, a phosphatidylinositol kinase homologue, controls autophagy in yeast. *J. Biol. Chem.* **1998**, *273*, 3963–3966. [CrossRef] [PubMed]
12. Hardwick, J.S.; Kuruvilla, F.G.; Tong, J.K.; Shamji, A.F.; Schreiber, S.L. Rapamycin-modulated transcription defines the subset of nutrient- sensitive signaling pathways directly controlled by the Tor proteins. *Proc. Natl. Acad. Sci. USA* **1999**, *96*, 14866–14870. [CrossRef] [PubMed]

13. Peng, T.; Golub, T.R.; Sabatini, D.M. The immunosuppressant rapamycin mimics a starvation-like signal distinct from amino acid and glucose deprivation. *Mol. Cell Biol.* **2002**, *22*, 5575–5584. [CrossRef] [PubMed]

14. Duvel, K.; Yecies, J.L.; Menon, S.; Raman, P.; Lipovsky, A.I.; Souza, A.L.; Triantafellow, E.; Ma, Q.; Gorski, R.; Cleaver, S.; et al. Activation of a metabolic gene regulatory network downstream of mTOR complex 1. *Mol. Cell* **2010**, *39*, 171–183. [CrossRef]

15. Cardenas, M.E.; Cutler, N.S.; Lorenz, M.C.; Di Como, C.J.; Heitman, J. The TOR signaling cascade regulates gene expression in response to nutrients. *Genes Dev.* **1999**, *13*, 3271–3279. [CrossRef]

16. Lynch, T.; Moloughney, J.; Jacinto, E. The mTOR complexes in cancer cell metabolism. In *Pi3k-mTOR Cancer and Cancer Therapy*; Dey, N., De, P., Leyland-Jones, B., Eds.; Springer: New York, NY, USA, 2016; pp. 29–63.

17. Laplante, M.; Sabatini, D.M. mTOR signaling in growth control and disease. *Cell* **2012**, *149*, 274–293. [CrossRef]

18. Shimobayashi, M.; Hall, M.N. Making new contacts: The mTOR network in metabolism and signalling crosstalk. *Nat. Rev. Mol. Cell Biol.* **2014**, *15*, 155–162. [CrossRef]

19. Kim, J.; Guan, K.L. mTOR as a central hub of nutrient signalling and cell growth. *Nat. Cell Biol.* **2019**, *21*, 63–71. [CrossRef]

20. Fan, S.J.; Snell, C.; Turley, H.; Li, J.L.; McCormick, R.; Perera, S.M.; Heublein, S.; Kazi, S.; Azad, A.; Wilson, C.; et al. PAT4 levels control amino-acid sensitivity of rapamycin-resistant mTORC1 from the Golgi and affect clinical outcome in colorectal cancer. *Oncogene* **2016**, *35*, 3004–3015. [CrossRef]

21. Nicklin, P.; Bergman, P.; Zhang, B.; Triantafellow, E.; Wang, H.; Nyfeler, B.; Yang, H.; Hild, M.; Kung, C.; Wilson, C.; et al. Bidirectional transport of amino acids regulates mTOR and autophagy. *Cell* **2009**, *136*, 521–534. [CrossRef]

22. Kim, J.; Guan, K.L. Amino acid signaling in TOR activation. *Annu. Rev. Biochem.* **2011**, *80*, 1001–1032. [CrossRef] [PubMed]

23. Fuchs, B.C.; Bode, B.P. Amino acid transporters ASCT2 and LAT1 in cancer: Partners in crime? *Semin. Cancer Biol.* **2005**, *15*, 254–266. [CrossRef] [PubMed]

24. Sancak, Y.; Peterson, T.R.; Shaul, Y.D.; Lindquist, R.A.; Thoreen, C.C.; Bar-Peled, L.; Sabatini, D.M. The Rag GTPases bind raptor and mediate amino acid signaling to mTORC1. *Science* **2008**, *320*, 1496–1501. [CrossRef] [PubMed]

25. Kim, E.; Goraksha-Hicks, P.; Li, L.; Neufeld, T.P.; Guan, K.L. Regulation of TORC1 by Rag GTPases in nutrient response. *Nat. Cell Biol.* **2008**, *10*, 935–945. [CrossRef] [PubMed]

26. Wolfson, R.L.; Sabatini, D.M. The Dawn of the Age of Amino Acid Sensors for the mTORC1 Pathway. *Cell Metab.* **2017**, *26*, 301–309. [CrossRef]

27. Rebsamen, M.; Pochini, L.; Stasyk, T.; de Araujo, M.E.; Galluccio, M.; Kandasamy, R.K.; Snijder, B.; Fauster, A.; Rudashevskaya, E.L.; Bruckner, M.; et al. SLC38A9 is a component of the lysosomal amino acid sensing machinery that controls mTORC1. *Nature* **2015**, *519*, 477–481. [CrossRef]

28. Wyant, G.A.; Abu-Remaileh, M.; Wolfson, R.L.; Chen, W.W.; Freinkman, E.; Danai, L.V.; Vander Heiden, M.G.; Sabatini, D.M. mTORC1 Activator SLC38A9 Is Required to Efflux Essential Amino Acids from Lysosomes and Use Protein as a Nutrient. *Cell* **2017**, *171*, 642–654. [CrossRef]

29. Duran, R.V.; Oppliger, W.; Robitaille, A.M.; Heiserich, L.; Skendaj, R.; Gottlieb, E.; Hall, M.N. Glutaminolysis activates Rag-mTORC1 signaling. *Mol. Cell* **2012**, *47*, 349–358. [CrossRef]

30. Stracka, D.; Jozefczuk, S.; Rudroff, F.; Sauer, U.; Hall, M.N. Nitrogen source activates TOR (target of rapamycin) complex 1 via glutamine and independently of Gtr/Rag proteins. *J. Biol. Chem.* **2014**, *289*, 25010–25020. [CrossRef]

31. Jewell, J.L.; Kim, Y.C.; Russell, R.C.; Yu, F.X.; Park, H.W.; Plouffe, S.W.; Tagliabracci, V.S.; Guan, K.L. Metabolism. Differential regulation of mTORC1 by leucine and glutamine. *Science* **2015**, *347*, 194–198. [CrossRef]

32. Gan, L.; Seki, A.; Shen, K.; Iyer, H.; Han, K.; Hayer, A.; Wollman, R.; Ge, X.; Lin, J.R.; Dey, G.; et al. The lysosomal GPCR-like protein GPR137B regulates Rag and mTORC1 localization and activity. *Nat. Cell Biol.* **2019**, *21*, 614–626. [CrossRef] [PubMed]

33. Son, S.M.; Park, S.J.; Lee, H.; Siddiqi, F.; Lee, J.E.; Menzies, F.M.; Rubinsztein, D.C. Leucine Signals to mTORC1 via Its Metabolite Acetyl-Coenzyme A. *Cell Metab.* **2019**, *29*, 192–201. [CrossRef] [PubMed]

34. Inoki, K.; Zhu, T.; Guan, K.L. TSC2 mediates cellular energy response to control cell growth and survival. *Cell* **2003**, *115*, 577–590. [CrossRef]

35. Manning, B.D.; Tee, A.R.; Logsdon, M.N.; Blenis, J.; Cantley, L.C. Identification of the tuberous sclerosis complex-2 tumor suppressor gene product tuberin as a target of the phosphoinositide 3-kinase/akt pathway. *Mol. Cell* **2002**, *10*, 151–162. [CrossRef]

36. Tee, A.R.; Manning, B.D.; Roux, P.P.; Cantley, L.C.; Blenis, J. Tuberous sclerosis complex gene products, Tuberin and Hamartin, control mTOR signaling by acting as a GTPase-activating protein complex toward Rheb. *Curr. Biol.* **2003**, *13*, 1259–1268. [CrossRef]

37. Yang, H.; Jiang, X.; Li, B.; Yang, H.J.; Miller, M.; Yang, A.; Dhar, A.; Pavletich, N.P. Mechanisms of mTORC1 activation by RHEB and inhibition by PRAS40. *Nature* **2017**, *552*, 368–373. [CrossRef]

38. Garami, A.; Zwartkruis, F.J.; Nobukuni, T.; Joaquin, M.; Roccio, M.; Stocker, H.; Kozma, S.C.; Hafen, E.; Bos, J.L.; Thomas, G. Insulin activation of Rheb, a mediator of mTOR/S6K/4E-BP signaling, is inhibited by TSC1 and 2. *Mol. Cell* **2003**, *11*, 1457–1466. [CrossRef]

39. Gwinn, D.M.; Shackelford, D.B.; Egan, D.F.; Mihaylova, M.M.; Mery, A.; Vasquez, D.S.; Turk, B.E.; Shaw, R.J. AMPK phosphorylation of raptor mediates a metabolic checkpoint. *Mol. Cell* **2008**, *30*, 214–226. [CrossRef]

40. Brugarolas, J.; Lei, K.; Hurley, R.L.; Manning, B.D.; Reiling, J.H.; Hafen, E.; Witters, L.A.; Ellisen, L.W.; Kaelin, W.G., Jr. Regulation of mTOR function in response to hypoxia by REDD1 and the TSC1/TSC2 tumor suppressor complex. *Genes Dev.* **2004**, *18*, 2893–2904. [CrossRef]

41. Jewell, J.L.; Fu, V.; Hong, A.W.; Yu, F.X.; Meng, D.; Melick, C.H.; Wang, H.; Lam, W.M.; Yuan, H.X.; Taylor, S.S.; et al. GPCR signaling inhibits mTORC1 via PKA phosphorylation of Raptor. *eLife* **2019**, *8*, e43038. [CrossRef]

42. Thomas, J.D.; Zhang, Y.J.; Wei, Y.H.; Cho, J.H.; Morris, L.E.; Wang, H.Y.; Zheng, X.F. Rab1A is an mTORC1 activator and a colorectal oncogene. *Cancer Cell* **2014**, *26*, 754–769. [CrossRef] [PubMed]

43. Goberdhan, D.C.; Wilson, C.; Harris, A.L. Amino Acid Sensing by mTORC1: Intracellular Transporters Mark the Spot. *Cell Metab.* **2016**, *23*, 580–589. [CrossRef] [PubMed]

44. Xie, J.; Wang, X.; Proud, C.G. Who does TORC2 talk to? *Biochem J.* **2018**, *475*, 1721–1738. [CrossRef] [PubMed]

45. Ebner, M.; Sinkovics, B.; Szczygiel, M.; Ribeiro, D.W.; Yudushkin, I. Localization of mTORC2 activity inside cells. *J. Cell Biol.* **2017**, *216*, 343–353. [CrossRef]

46. Gan, X.; Wang, J.; Su, B.; Wu, D. Evidence for Direct Activation of mTORC2 Kinase Activity by Phosphatidylinositol 3,4,5-Trisphosphate. *J. Biol. Chem.* **2011**, *286*, 10998–11002. [CrossRef]

47. Liu, P.; Gan, W.; Chin, Y.R.; Ogura, K.; Guo, J.; Zhang, J.; Wang, B.; Blenis, J.; Cantley, L.C.; Toker, A.; et al. PtdIns(3,4,5)P3-Dependent Activation of the mTORC2 Kinase Complex. *Cancer Discov.* **2015**, *5*, 1194–1209. [CrossRef]

48. Ikenoue, T.; Inoki, K.; Yang, Q.; Zhou, X.; Guan, K.L. Essential function of TORC2 in PKC and Akt turn motif phosphorylation, maturation and signalling. *Embo J.* **2008**, *27*, 1919–1931. [CrossRef]

49. Facchinetti, V.; Ouyang, W.; Wei, H.; Soto, N.; Lazorchak, A.; Gould, C.; Lowry, C.; Newton, A.C.; Mao, Y.; Miao, R.Q.; et al. The mammalian target of rapamycin complex 2 controls folding and stability of Akt and protein kinase C. *Embo J.* **2008**, *27*, 1932–1943. [CrossRef]

50. Garcia-Martinez, J.M.; Alessi, D.R. mTOR complex 2 (mTORC2) controls hydrophobic motif phosphorylation and activation of serum- and glucocorticoid-induced protein kinase 1 (SGK1). *Biochem J.* **2008**, *416*, 375–385. [CrossRef]

51. Cameron, A.J.; Linch, M.D.; Saurin, A.T.; Escribano, C.; Parker, P.J. mTORC2 targets AGC kinases through Sin1-dependent recruitment. *Biochem J.* **2011**, *439*, 287–297. [CrossRef]

52. Oh, W.J.; Wu, C.C.; Kim, S.J.; Facchinetti, V.; Julien, L.A.; Finlan, M.; Roux, P.P.; Su, B.; Jacinto, E. mTORC2 can associate with ribosomes to promote cotranslational phosphorylation and stability of nascent Akt polypeptide. *Embo J.* **2010**, *29*, 3939–3951. [CrossRef] [PubMed]

53. Zinzalla, V.; Stracka, D.; Oppliger, W.; Hall, M.N. Activation of mTORC2 by Association with the Ribosome. *Cell* **2011**, *144*, 757–768. [CrossRef] [PubMed]

54. Moloughney, J.G.; Kim, P.K.; Vega-Cotto, N.M.; Wu, C.C.; Zhang, S.; Adlam, M.; Lynch, T.; Chou, P.C.; Rabinowitz, J.D.; Werlen, G.; et al. mTORC2 Responds to Glutamine Catabolite Levels to Modulate the Hexosamine Biosynthesis Enzyme GFAT1. *Mol. Cell* **2016**, *63*, 811–826. [CrossRef] [PubMed]

55. Kazyken, D.; Magnuson, B.; Bodur, C.; Acosta-Jaquez, H.A.; Zhang, D.; Tong, X.; Barnes, T.M.; Steinl, G.K.; Patterson, N.E.; Altheim, C.H.; et al. AMPK directly activates mTORC2 to promote cell survival during acute energetic stress. *Sci. Signal* **2019**, *12*, eaav3249. [CrossRef] [PubMed]

56. Albert, V.; Svensson, K.; Shimobayashi, M.; Colombi, M.; Munoz, S.; Jimenez, V.; Handschin, C.; Bosch, F.; Hall, M.N. mTORC2 sustains thermogenesis via Akt-induced glucose uptake and glycolysis in brown adipose tissue. *Embo Mol. Med* **2016**, *8*, 232–246. [CrossRef] [PubMed]

57. Jung, S.M.; Hung, C.M.; Hildebrand, S.R.; Sanchez-Gurmaches, J.; Martinez-Pastor, B.; Gengatharan, J.M.; Wallace, M.; Mukhopadhyay, D.; Martinez Calejman, C.; Luciano, A.K.; et al. Non-canonical mTORC2 Signaling Regulates Brown Adipocyte Lipid Catabolism through SIRT6-FoxO1. *Mol. Cell* **2019**, *75*, 807–822. [CrossRef]

58. Sato, M.; Evans, B.A.; Sandstrom, A.L.; Chia, L.Y.; Mukaida, S.; Thai, B.S.; Nguyen, A.; Lim, L.; Tan, C.Y.R.; Baltos, J.A.; et al. alpha1A-Adrenoceptors activate mTOR signalling and glucose uptake in cardiomyocytes. *Biochem. Pharm.* **2018**, *148*, 27–40. [CrossRef]

59. Tavares, M.R.; Pavan, I.C.; Amaral, C.L.; Meneguello, L.; Luchessi, A.D.; Simabuco, F.M. The S6K protein family in health and disease. *Life Sci.* **2015**, *131*, 1–10. [CrossRef]

60. Dennis, M.D.; Jefferson, L.S.; Kimball, S.R. Role of p70S6K1-mediated phosphorylation of eIF4B and PDCD4 proteins in the regulation of protein synthesis. *J. Biol. Chem.* **2012**, *287*, 42890–42899. [CrossRef]

61. Ma, X.M.; Yoon, S.O.; Richardson, C.J.; Julich, K.; Blenis, J. SKAR links pre-mRNA splicing to mTOR/S6K1-mediated enhanced translation efficiency of spliced mRNAs. *Cell* **2008**, *133*, 303–313. [CrossRef]

62. Choi, S.H.; Martinez, T.F.; Kim, S.; Donaldson, C.; Shokhirev, M.N.; Saghatelian, A.; Jones, K.A. CDK12 phosphorylates 4E-BP1 to enable mTORC1-dependent translation and mitotic genome stability. *Genes Dev.* **2019**, *33*, 418–435. [CrossRef] [PubMed]

63. Qin, X.; Jiang, B.; Zhang, Y. 4E-BP1, a multifactor regulated multifunctional protein. *Cell Cycle* **2016**, *15*, 781–786. [CrossRef] [PubMed]

64. Toda, K.; Kawada, K.; Iwamoto, M.; Inamoto, S.; Sasazuki, T.; Shirasawa, S.; Hasegawa, S.; Sakai, Y. Metabolic Alterations Caused by KRAS Mutations in Colorectal Cancer Contribute to Cell Adaptation to Glutamine Depletion by Upregulation of Asparagine Synthetase. *Neoplasia* **2016**, *18*, 654–665. [CrossRef] [PubMed]

65. Gwinn, D.M.; Lee, A.G.; Briones-Martin-Del-Campo, M.; Conn, C.S.; Simpson, D.R.; Scott, A.I.; Le, A.; Cowan, T.M.; Ruggero, D.; Sweet-Cordero, E.A. Oncogenic KRAS Regulates Amino Acid Homeostasis and Asparagine Biosynthesis via ATF4 and Alters Sensitivity to L-Asparaginase. *Cancer Cell* **2018**, *33*, 91–107. [CrossRef] [PubMed]

66. Gu, Y.; Albuquerque, C.P.; Braas, D.; Zhang, W.; Villa, G.R.; Bi, J.; Ikegami, S.; Masui, K.; Gini, B.; Yang, H.; et al. mTORC2 Regulates Amino Acid Metabolism in Cancer by Phosphorylation of the Cystine-Glutamate Antiporter xCT. *Mol. Cell* **2017**, *67*, 128–138. [CrossRef]

67. Gentilella, A.; Kozma, S.C.; Thomas, G. A liaison between mTOR signaling, ribosome biogenesis and cancer. *Biochim. Biophys Acta* **2015**, *1849*, 812–820. [CrossRef]

68. Prakash, V.; Carson, B.B.; Feenstra, J.M.; Dass, R.A.; Sekyrova, P.; Hoshino, A.; Petersen, J.; Guo, Y.; Parks, M.M.; Kurylo, C.M.; et al. Ribosome biogenesis during cell cycle arrest fuels EMT in development and disease. *Nat. Commun.* **2019**, *10*, 2110. [CrossRef]

69. Bywater, M.J.; Poortinga, G.; Sanij, E.; Hein, N.; Peck, A.; Cullinane, C.; Wall, M.; Cluse, L.; Drygin, D.; Anderes, K.; et al. Inhibition of RNA polymerase I as a therapeutic strategy to promote cancer-specific activation of p53. *Cancer Cell* **2012**, *22*, 51–65. [CrossRef]

70. Devlin, J.R.; Hannan, K.M.; Hein, N.; Cullinane, C.; Kusnadi, E.; Ng, P.Y.; George, A.J.; Shortt, J.; Bywater, M.J.; Poortinga, G.; et al. Combination Therapy Targeting Ribosome Biogenesis and mRNA Translation Synergistically Extends Survival in MYC-Driven Lymphoma. *Cancer Discov.* **2016**, *6*, 59–70. [CrossRef]

71. Semenza, G.L. HIF-1 mediates metabolic responses to intratumoral hypoxia and oncogenic mutations. *J. Clin. Investig.* **2013**, *123*, 3664–3671. [CrossRef]

72. Toschi, A.; Lee, E.; Gadir, N.; Ohh, M.; Foster, D.A. Differential dependence of hypoxia-inducible factors 1 alpha and 2 alpha on mTORC1 and mTORC2. *J. Biol. Chem.* **2008**, *283*, 34495–34499. [CrossRef] [PubMed]

73. Hu, C.J.; Wang, L.Y.; Chodosh, L.A.; Keith, B.; Simon, M.C. Differential roles of hypoxia-inducible factor 1alpha (HIF-1alpha) and HIF-2alpha in hypoxic gene regulation. *Mol. Cell Biol.* **2003**, *23*, 9361–9374. [CrossRef] [PubMed]

74. Zhong, H.; Chiles, K.; Feldser, D.; Laughner, E.; Hanrahan, C.; Georgescu, M.M.; Simons, J.W.; Semenza, G.L. Modulation of hypoxia-inducible factor 1alpha expression by the epidermal growth factor/phosphatidylinositol 3-kinase/PTEN/AKT/FRAP pathway in human prostate cancer cells: Implications for tumor angiogenesis and therapeutics. *Cancer Res.* **2000**, *60*, 1541–1545. [PubMed]

75. Hudson, C.C.; Liu, M.; Chiang, G.G.; Otterness, D.M.; Loomis, D.C.; Kaper, F.; Giaccia, A.J.; Abraham, R.T. Regulation of hypoxia-inducible factor 1alpha expression and function by the mammalian target of rapamycin. *Mol. Cell Biol.* **2002**, *22*, 7004–7014. [CrossRef]

76. He, L.; Gomes, A.P.; Wang, X.; Yoon, S.O.; Lee, G.; Nagiec, M.J.; Cho, S.; Chavez, A.; Islam, T.; Yu, Y.; et al. mTORC1 Promotes Metabolic Reprogramming by the Suppression of GSK3-Dependent Foxk1 Phosphorylation. *Mol. Cell* **2018**, *70*, 949–960. [CrossRef]

77. Brugarolas, J.B.; Vazquez, F.; Reddy, A.; Sellers, W.R.; Kaelin, W.G., Jr. TSC2 regulates VEGF through mTOR-dependent and -independent pathways. *Cancer Cell* **2003**, *4*, 147–158. [CrossRef]

78. Laughner, E.; Taghavi, P.; Chiles, K.; Mahon, P.C.; Semenza, G.L. HER2 (neu) signaling increases the rate of hypoxia-inducible factor 1alpha (HIF-1alpha) synthesis: Novel mechanism for HIF-1-mediated vascular endothelial growth factor expression. *Mol. Cell Biol.* **2001**, *21*, 3995–4004. [CrossRef]

79. Thomas, G.V.; Tran, C.; Mellinghoff, I.K.; Welsbie, D.S.; Chan, E.; Fueger, B.; Czernin, J.; Sawyers, C.L. Hypoxia-inducible factor determines sensitivity to inhibitors of mTOR in kidney cancer. *Nat. Med.* **2006**, *12*, 122–127. [CrossRef]

80. Grolleau, A.; Bowman, J.; Pradet-Balade, B.; Puravs, E.; Hanash, S.; Garcia-Sanz, J.A.; Beretta, L. Global and specific translational control by rapamycin in T cells uncovered by microarrays and proteomics. *J. Biol. Chem.* **2002**, *9*, 9. [CrossRef]

81. Majumder, P.K.; Febbo, P.G.; Bikoff, R.; Berger, R.; Xue, Q.; McMahon, L.M.; Manola, J.; Brugarolas, J.; McDonnell, T.J.; Golub, T.R.; et al. mTOR inhibition reverses Akt-dependent prostate intraepithelial neoplasia through regulation of apoptotic and HIF-1-dependent pathways. *Nat. Med.* **2004**, *10*, 594–601. [CrossRef]

82. Sun, Q.; Chen, X.; Ma, J.; Peng, H.; Wang, F.; Zha, X.; Wang, Y.; Jing, Y.; Yang, H.; Chen, R.; et al. Mammalian target of rapamycin up-regulation of pyruvate kinase isoenzyme type M2 is critical for aerobic glycolysis and tumor growth. *Proc. Natl. Acad. Sci. USA* **2011**, *108*, 4129–4134. [CrossRef] [PubMed]

83. Broecker-Preuss, M.; Becher-Boveleth, N.; Bockisch, A.; Duhrsen, U.; Muller, S. Regulation of glucose uptake in lymphoma cell lines by c-MYC- and PI3K-dependent signaling pathways and impact of glycolytic pathways on cell viability. *J. Transl. Med.* **2017**, *15*, 158. [CrossRef] [PubMed]

84. Jiang, X.; Kenerson, H.; Aicher, L.; Miyaoka, R.; Eary, J.; Bissler, J.; Yeung, R.S. The tuberous sclerosis complex regulates trafficking of glucose transporters and glucose uptake. *Am. J. Pathol.* **2008**, *172*, 1748–1756. [CrossRef] [PubMed]

85. Harrington, L.S.; Findlay, G.M.; Lamb, R.F. Restraining PI3K: mTOR signalling goes back to the membrane. *Trends Biochem. Sci.* **2005**, *30*, 35–42. [CrossRef] [PubMed]

86. West, M.J.; Stoneley, M.; Willis, A.E. Translational induction of the c-myc oncogene via activation of the FRAP/TOR signalling pathway. *Oncogene* **1998**, *17*, 769–780. [CrossRef]

87. Gordan, J.D.; Thompson, C.B.; Simon, M.C. HIF and c-Myc: Sibling rivals for control of cancer cell metabolism and proliferation. *Cancer Cell* **2007**, *12*, 108–113. [CrossRef]

88. Dang, C.V. A Time for MYC: Metabolism and Therapy. *Cold Spring Harb. Symp. Quant. Biol.* **2016**, *81*, 79–83. [CrossRef]

89. Semenza, G.L. Targeting HIF-1 for cancer therapy. *Nat. Rev. Cancer* **2003**, *3*, 721–732. [CrossRef]

90. Dang, C.V.; O'Donnell, K.A.; Zeller, K.I.; Nguyen, T.; Osthus, R.C.; Li, F. The c-Myc target gene network. *Semin. Cancer Biol.* **2006**, *16*, 253–264. [CrossRef]

91. van der Poel, H.G.; Hanrahan, C.; Zhong, H.; Simons, J.W. Rapamycin induces Smad activity in prostate cancer cell lines. *Urol. Res.* **2003**, *30*, 380–386.

92. Zha, X.; Wang, F.; Wang, Y.; He, S.; Jing, Y.; Wu, X.; Zhang, H. Lactate dehydrogenase B is critical for hyperactive mTOR-mediated tumorigenesis. *Cancer Res.* **2011**, *71*, 13–18. [CrossRef] [PubMed]

93. Tandon, P.; Gallo, C.A.; Khatri, S.; Barger, J.F.; Yepiskoposyan, H.; Plas, D.R. Requirement for ribosomal protein S6 kinase 1 to mediate glycolysis and apoptosis resistance induced by Pten deficiency. *Proc. Natl. Acad. Sci. USA* **2011**, *108*, 2361–2365. [CrossRef] [PubMed]

94. Pourdehnad, M.; Truitt, M.L.; Siddiqi, I.N.; Ducker, G.S.; Shokat, K.M.; Ruggero, D. Myc and mTOR converge on a common node in protein synthesis control that confers synthetic lethality in Myc-driven cancers. *Proc. Natl. Acad. Sci. USA* **2013**, *110*, 11988–11993. [CrossRef] [PubMed]

95. Dutchak, P.A.; Estill-Terpack, S.J.; Plec, A.A.; Zhao, X.; Yang, C.; Chen, J.; Ko, B.; Deberardinis, R.J.; Yu, Y.; Tu, B.P. Loss of a Negative Regulator of mTORC1 Induces Aerobic Glycolysis and Altered Fiber Composition in Skeletal Muscle. *Cell Rep.* **2018**, *23*, 1907–1914. [CrossRef] [PubMed]

96. Thomas, L.W.; Esposito, C.; Stephen, J.M.; Costa, A.S.H.; Frezza, C.; Blacker, T.S.; Szabadkai, G.; Ashcroft, M. CHCHD4 regulates tumour proliferation and EMT-related phenotypes, through respiratory chain-mediated metabolism. *Cancer Metab.* **2019**, *7*, 7. [CrossRef] [PubMed]

97. Morita, M.; Gravel, S.P.; Chenard, V.; Sikstrom, K.; Zheng, L.; Alain, T.; Gandin, V.; Avizonis, D.; Arguello, M.; Zakaria, C.; et al. mTORC1 controls mitochondrial activity and biogenesis through 4E-BP-dependent translational regulation. *Cell Metab.* **2013**, *18*, 698–711. [CrossRef]

98. Gottlob, K.; Majewski, N.; Kennedy, S.; Kandel, E.; Robey, R.B.; Hay, N. Inhibition of early apoptotic events by Akt/PKB is dependent on the first committed step of glycolysis and mitochondrial hexokinase. *Genes Dev.* **2001**, *15*, 1406–1418. [CrossRef]

99. Deprez, J.; Vertommen, D.; Alessi, D.R.; Hue, L.; Rider, M.H. Phosphorylation and activation of heart 6-phosphofructo-2-kinase by protein kinase B and other protein kinases of the insulin signaling cascades. *J. Biol. Chem.* **1997**, *272*, 17269–17275. [CrossRef]

100. Barthel, A.; Okino, S.T.; Liao, J.; Nakatani, K.; Li, J.; Whitlock, J.P., Jr.; Roth, R.A. Regulation of GLUT1 gene transcription by the serine/threonine kinase Akt1. *J. Biol. Chem.* **1999**, *274*, 20281–20286. [CrossRef]

101. Robey, R.B.; Hay, N. Is Akt the "Warburg kinase"?-Akt-energy metabolism interactions and oncogenesis. *Semin. Cancer Biol.* **2009**, *19*, 25–31. [CrossRef]

102. Elstrom, R.L.; Bauer, D.E.; Buzzai, M.; Karnauskas, R.; Harris, M.H.; Plas, D.R.; Zhuang, H.; Cinalli, R.M.; Alavi, A.; Rudin, C.M.; et al. Akt stimulates aerobic glycolysis in cancer cells. *Cancer Res.* **2004**, *64*, 3892–3899. [CrossRef] [PubMed]

103. Chen, M.L.; Xu, P.Z.; Peng, X.D.; Chen, W.S.; Guzman, G.; Yang, X.; Di Cristofano, A.; Pandolfi, P.P.; Hay, N. The deficiency of Akt1 is sufficient to suppress tumor development in Pten+/- mice. *Genes Dev.* **2006**, *20*, 1569–1574. [CrossRef] [PubMed]

104. Hagiwara, A.; Cornu, M.; Cybulski, N.; Polak, P.; Betz, C.; Trapani, F.; Terracciano, L.; Heim, M.H.; Ruegg, M.A.; Hall, M.N. Hepatic mTORC2 Activates Glycolysis and Lipogenesis through Akt, Glucokinase, and SREBP1c. *Cell Metab.* **2012**, *15*, 725–738. [CrossRef] [PubMed]

105. Cerniglia, G.J.; Dey, S.; Gallagher-Colombo, S.M.; Daurio, N.A.; Tuttle, S.; Busch, T.M.; Lin, A.; Sun, R.; Esipova, T.V.; Vinogradov, S.A.; et al. The PI3K/Akt Pathway Regulates Oxygen Metabolism via Pyruvate Dehydrogenase (PDH)-E1alpha Phosphorylation. *Mol. Cancer* **2015**, *14*, 1928–1938. [CrossRef]

106. Pore, N.; Jiang, Z.; Shu, H.K.; Bernhard, E.; Kao, G.D.; Maity, A. Akt1 activation can augment hypoxia-inducible factor-1alpha expression by increasing protein translation through a mammalian target of rapamycin-independent pathway. *Mol. Cancer Res.* **2006**, *4*, 471–479. [CrossRef]

107. Masui, K.; Tanaka, K.; Akhavan, D.; Babic, I.; Gini, B.; Matsutani, T.; Iwanami, A.; Liu, F.; Villa, G.R.; Gu, Y.; et al. mTOR complex 2 controls glycolytic metabolism in glioblastoma through FoxO acetylation and upregulation of c-Myc. *Cell Metab.* **2013**, *18*, 726–739. [CrossRef]

108. Vadla, R.; Haldar, D. Mammalian target of rapamycin complex 2 (mTORC2) controls glycolytic gene expression by regulating Histone H3 Lysine 56 acetylation. *Cell Cycle* **2018**, *17*, 110–123. [CrossRef]

109. Li, Y.; He, Z.C.; Liu, Q.; Zhou, K.; Shi, Y.; Yao, X.H.; Zhang, X.; Kung, H.F.; Ping, Y.F.; Bian, X.W. Large Intergenic Non-coding RNA-RoR Inhibits Aerobic Glycolysis of Glioblastoma Cells via Akt Pathway. *J. Cancer* **2018**, *9*, 880–889. [CrossRef]

110. Porstmann, T.; Santos, C.R.; Griffiths, B.; Cully, M.; Wu, M.; Leevers, S.; Griffiths, J.R.; Chung, Y.L.; Schulze, A. SREBP activity is regulated by mTORC1 and contributes to Akt-dependent cell growth. *Cell Metab.* **2008**, *8*, 224–236. [CrossRef]

111. Owen, J.L.; Zhang, Y.; Bae, S.H.; Farooqi, M.S.; Liang, G.; Hammer, R.E.; Goldstein, J.L.; Brown, M.S. Insulin stimulation of SREBP-1c processing in transgenic rat hepatocytes requires p70 S6-kinase. *Proc. Natl. Acad. Sci. USA* **2012**, *109*, 16184–16189. [CrossRef]

112. Peterson, T.R.; Sengupta, S.S.; Harris, T.E.; Carmack, A.E.; Kang, S.A.; Balderas, E.; Guertin, D.A.; Madden, K.L.; Carpenter, A.E.; Finck, B.N.; et al. mTOR complex 1 regulates lipin 1 localization to control the SREBP pathway. *Cell* **2011**, *146*, 408–420. [CrossRef] [PubMed]

113. Yecies, J.L.; Zhang, H.H.; Menon, S.; Liu, S.; Yecies, D.; Lipovsky, A.I.; Gorgun, C.; Kwiatkowski, D.J.; Hotamisligil, G.S.; Lee, C.H.; et al. Akt stimulates hepatic SREBP1c and lipogenesis through parallel mTORC1-dependent and independent pathways. *Cell Metab.* **2011**, *14*, 21–32. [CrossRef] [PubMed]

114. Bhatt, A.P.; Jacobs, S.R.; Freemerman, A.J.; Makowski, L.; Rathmell, J.C.; Dittmer, D.P.; Damania, B. Dysregulation of fatty acid synthesis and glycolysis in non-Hodgkin lymphoma. *Proc. Natl. Acad. Sci. USA* **2012**, *109*, 11818–11823. [CrossRef] [PubMed]

115. Hu, J.; Che, L.; Li, L.; Pilo, M.G.; Cigliano, A.; Ribback, S.; Li, X.; Latte, G.; Mela, M.; Evert, M.; et al. Co-activation of AKT and c-Met triggers rapid hepatocellular carcinoma development via the mTORC1/FASN pathway in mice. *Sci. Rep.* **2016**, *6*, 20484. [CrossRef] [PubMed]

116. Stepanova, D.S.; Semenova, G.; Kuo, Y.M.; Andrews, A.J.; Ammoun, S.; Hanemann, C.O.; Chernoff, J. An Essential Role for the Tumor-Suppressor Merlin in Regulating Fatty Acid Synthesis. *Cancer Res.* **2017**, *77*, 5026–5038. [CrossRef] [PubMed]

117. Luyimbazi, D.; Akcakanat, A.; McAuliffe, P.F.; Zhang, L.; Singh, G.; Gonzalez-Angulo, A.M.; Chen, H.; Do, K.A.; Zheng, Y.; Hung, M.C.; et al. Rapamycin regulates stearoyl CoA desaturase 1 expression in breast cancer. *Mol. Cancer* **2010**, *9*, 2770–2784. [CrossRef] [PubMed]

118. Barger, J.F.; Gallo, C.A.; Tandon, P.; Liu, H.; Sullivan, A.; Grimes, H.L.; Plas, D.R. S6K1 determines the metabolic requirements for BCR-ABL survival. *Oncogene* **2013**, *32*, 453–461. [CrossRef]

119. Zaugg, K.; Yao, Y.; Reilly, P.T.; Kannan, K.; Kiarash, R.; Mason, J.; Huang, P.; Sawyer, S.K.; Fuerth, B.; Faubert, B.; et al. Carnitine palmitoyltransferase 1C promotes cell survival and tumor growth under conditions of metabolic stress. *Genes Dev.* **2011**, *25*, 1041–1051. [CrossRef]

120. Guaita-Esteruelas, S.; Bosquet, A.; Saavedra, P.; Guma, J.; Girona, J.; Lam, E.W.; Amillano, K.; Borras, J.; Masana, L. Exogenous FABP4 increases breast cancer cell proliferation and activates the expression of fatty acid transport proteins. *Mol. Carcinog.* **2017**, *56*, 208–217. [CrossRef]

121. Pan, J.; Fan, Z.; Wang, Z.; Dai, Q.; Xiang, Z.; Yuan, F.; Yan, M.; Zhu, Z.; Liu, B.; Li, C. CD36 mediates palmitate acid-induced metastasis of gastric cancer via AKT/GSK-3beta/beta-catenin pathway. *J. Exp. Clin. Cancer Res.* **2019**, *38*, 52. [CrossRef]

122. Guri, Y.; Colombi, M.; Dazert, E.; Hindupur, S.K.; Roszik, J.; Moes, S.; Jenoe, P.; Heim, M.H.; Riezman, I.; Riezman, H.; et al. mTORC2 Promotes Tumorigenesis via Lipid Synthesis. *Cancer Cell* **2017**, *32*, 807–823. [CrossRef] [PubMed]

123. Bhutia, Y.D.; Ganapathy, V. Glutamine transporters in mammalian cells and their functions in physiology and cancer. *Biochim. Biophys Acta* **2016**, *1863*, 2531–2539. [CrossRef] [PubMed]

124. van Geldermalsen, M.; Wang, Q.; Nagarajah, R.; Marshall, A.D.; Thoeng, A.; Gao, D.; Ritchie, W.; Feng, Y.; Bailey, C.G.; Deng, N.; et al. ASCT2/SLC1A5 controls glutamine uptake and tumour growth in triple-negative basal-like breast cancer. *Oncogene* **2016**, *35*, 3201–3208. [CrossRef] [PubMed]

125. Wang, Q.; Hardie, R.A.; Hoy, A.J.; van Geldermalsen, M.; Gao, D.; Fazli, L.; Sadowski, M.C.; Balaban, S.; Schreuder, M.; Nagarajah, R.; et al. Targeting ASCT2-mediated glutamine uptake blocks prostate cancer growth and tumour development. *J. Pathol.* **2015**, *236*, 278–289. [CrossRef] [PubMed]

126. Willems, L.; Jacque, N.; Jacquel, A.; Neveux, N.; Maciel, T.T.; Lambert, M.; Schmitt, A.; Poulain, L.; Green, A.S.; Uzunov, M.; et al. Inhibiting glutamine uptake represents an attractive new strategy for treating acute myeloid leukemia. *Blood* **2013**, *122*, 3521–3532. [CrossRef]

127. Lu, J.; Chen, M.; Tao, Z.; Gao, S.; Li, Y.; Cao, Y.; Lu, C.; Zou, X. Effects of targeting SLC1A5 on inhibiting gastric cancer growth and tumor development in vitro and in vivo. *Oncotarget* **2017**, *8*, 76458–76467. [CrossRef]

128. Ni, F.; Yu, W.M.; Li, Z.; Graham, D.K.; Jin, L.; Kang, S.; Rossi, M.R.; Li, S.; Broxmeyer, H.E.; Qu, C.K. Critical role of ASCT2-mediated amino acid metabolism in promoting leukaemia development and progression. *Nat. Metab.* **2019**, *1*, 390–403. [CrossRef]

129. Digomann, D.; Kurth, I.; Tyutyunnykova, A.; Chen, O.; Lock, S.; Gorodetska, I.; Peitzsch, C.; Skvortsova, I.I.; Negro, G.; Aschenbrenner, B.; et al. The CD98 Heavy Chain Is a Marker and Regulator of Head and Neck Squamous Cell Carcinoma Radiosensitivity. *Clin. Cancer Res.* **2019**, *25*, 3152–3163. [CrossRef]

130. Daye, D.; Wellen, K.E. Metabolic reprogramming in cancer: Unraveling the role of glutamine in tumorigenesis. *Semin. Cell Dev. Biol.* **2012**, *23*, 362–369. [CrossRef]

131. Pusapati, R.V.; Daemen, A.; Wilson, C.; Sandoval, W.; Gao, M.; Haley, B.; Baudy, A.R.; Hatzivassiliou, G.; Evangelista, M.; Settleman, J. mTORC1-Dependent Metabolic Reprogramming Underlies Escape from Glycolysis Addiction in Cancer Cells. *Cancer Cell* **2016**, *29*, 548–562. [CrossRef]

132. Qie, S.; Yoshida, A.; Parnham, S.; Oleinik, N.; Beeson, G.C.; Beeson, C.C.; Ogretmen, B.; Bass, A.J.; Wong, K.K.; Rustgi, A.K.; et al. Targeting glutamine-addiction and overcoming CDK4/6 inhibitor resistance in human esophageal squamous cell carcinoma. *Nat. Commun.* **2019**, *10*, 1296. [CrossRef] [PubMed]

133. Csibi, A.; Lee, G.; Yoon, S.O.; Tong, H.; Ilter, D.; Elia, I.; Fendt, S.M.; Roberts, T.M.; Blenis, J. The mTORC1/S6K1 pathway regulates glutamine metabolism through the eIF4B-dependent control of c-Myc translation. *Curr. Biol.* **2014**, *24*, 2274–2280. [CrossRef] [PubMed]

134. Gao, P.; Tchernyshyov, I.; Chang, T.C.; Lee, Y.S.; Kita, K.; Ochi, T.; Zeller, K.I.; De Marzo, A.M.; Van Eyk, J.E.; Mendell, J.T.; et al. c-Myc suppression of miR-23a/b enhances mitochondrial glutaminase expression and glutamine metabolism. *Nature* **2009**, *458*, 762–765. [CrossRef] [PubMed]

135. Wise, D.R.; DeBerardinis, R.J.; Mancuso, A.; Sayed, N.; Zhang, X.Y.; Pfeiffer, H.K.; Nissim, I.; Daikhin, E.; Yudkoff, M.; McMahon, S.B.; et al. Myc regulates a transcriptional program that stimulates mitochondrial glutaminolysis and leads to glutamine addiction. *Proc. Natl. Acad. Sci. USA* **2008**, *105*, 18782–18787. [CrossRef] [PubMed]

136. Momcilovic, M.; Bailey, S.T.; Lee, J.T.; Fishbein, M.C.; Braas, D.; Go, J.; Graeber, T.G.; Parlati, F.; Demo, S.; Li, R.; et al. The GSK3 Signaling Axis Regulates Adaptive Glutamine Metabolism in Lung Squamous Cell Carcinoma. *Cancer Cell* **2018**, *33*, 905–921. [CrossRef] [PubMed]

137. Demas, D.M.; Demo, S.; Fallah, Y.; Clarke, R.; Nephew, K.P.; Althouse, S.; Sandusky, G.; He, W.; Shajahan-Haq, A.N. Glutamine Metabolism Drives Growth in Advanced Hormone Receptor Positive Breast Cancer. *Front. Oncol.* **2019**, *9*, 686. [CrossRef] [PubMed]

138. Csibi, A.; Fendt, S.M.; Li, C.; Poulogiannis, G.; Choo, A.Y.; Chapski, D.J.; Jeong, S.M.; Dempsey, J.M.; Parkhitko, A.; Morrison, T.; et al. The mTORC1 pathway stimulates glutamine metabolism and cell proliferation by repressing SIRT4. *Cell* **2013**, *153*, 840–854. [CrossRef]

139. Adebayo Michael, A.O.; Ko, S.; Tao, J.; Moghe, A.; Yang, H.; Xu, M.; Russell, J.O.; Pradhan-Sundd, T.; Liu, S.; Singh, S.; et al. Inhibiting Glutamine-Dependent mTORC1 Activation Ameliorates Liver Cancers Driven by beta-Catenin Mutations. *Cell Metab.* **2019**, *29*, 1135–1150. [CrossRef]

140. Lie, S.; Wang, T.; Forbes, B.; Proud, C.G.; Petersen, J. The ability to utilise ammonia as nitrogen source is cell type specific and intricately linked to GDH, AMPK and mTORC1. *Sci. Rep.* **2019**, *9*, 1461. [CrossRef]

141. Byun, J.K.; Choi, Y.K.; Kim, J.H.; Jeong, J.Y.; Jeon, H.J.; Kim, M.K.; Hwang, I.; Lee, S.Y.; Lee, Y.M.; Lee, I.K.; et al. A Positive Feedback Loop between Sestrin2 and mTORC2 Is Required for the Survival of Glutamine-Depleted Lung Cancer Cells. *Cell Rep.* **2017**, *20*, 586–599. [CrossRef]

142. Denzel, M.S.; Antebi, A. Hexosamine pathway and (ER) protein quality control. *Curr. Opin. Cell Biol.* **2015**, *33*, 14–18. [CrossRef] [PubMed]

143. Moloughney, J.G.; Vega-Cotto, N.M.; Liu, S.; Patel, C.; Kim, P.K.; Wu, C.C.; Albaciete, D.; Magaway, C.; Chang, A.; Rajput, S.; et al. mTORC2 modulates the amplitude and duration of GFAT1 Ser-243 phosphorylation to maintain flux through the hexosamine pathway during starvation. *J. Biol. Chem.* **2018**, *293*, 16464–16478. [CrossRef] [PubMed]

144. Jones, D.R.; Keune, W.J.; Anderson, K.E.; Stephens, L.R.; Hawkins, P.T.; Divecha, N. The hexosamine biosynthesis pathway and O-GlcNAcylation maintain insulin-stimulated PI3K-PKB phosphorylation and tumour cell growth after short-term glucose deprivation. *Febs. J.* **2014**, *281*, 3591–3608. [CrossRef] [PubMed]

145. Stanton, R.C. Glucose-6-phosphate dehydrogenase, NADPH, and cell survival. *Iubmb. Life* **2012**, *64*, 362–369. [CrossRef]

146. Wang, J.; Yuan, W.; Chen, Z.; Wu, S.; Chen, J.; Ge, J.; Hou, F.; Chen, Z. Overexpression of G6PD is associated with poor clinical outcome in gastric cancer. *Tumour Biol.* **2012**, *33*, 95–101. [CrossRef]

147. Yang, C.A.; Huang, H.Y.; Lin, C.L.; Chang, J.G. G6PD as a predictive marker for glioma risk, prognosis and chemosensitivity. *J. Neurooncol.* **2018**, *139*, 661–670. [CrossRef]

148. Benito, A.; Polat, I.H.; Noe, V.; Ciudad, C.J.; Marin, S.; Cascante, M. Glucose-6-phosphate dehydrogenase and transketolase modulate breast cancer cell metabolic reprogramming and correlate with poor patient outcome. *Oncotarget* **2017**, *8*, 106693–106706. [CrossRef]

149. Chen, X.; Xu, Z.; Zhu, Z.; Chen, A.; Fu, G.; Wang, Y.; Pan, H.; Jin, B. Modulation of G6PD affects bladder cancer via ROS accumulation and the AKT pathway in vitro. *Int. J. Oncol.* **2018**, *53*, 1703–1712. [CrossRef]

150. Best, S.A.; Ding, S.; Kersbergen, A.; Dong, X.; Song, J.Y.; Xie, Y.; Reljic, B.; Li, K.; Vince, J.E.; Rathi, V.; et al. Distinct initiating events underpin the immune and metabolic heterogeneity of KRAS-mutant lung adenocarcinoma. *Nat. Commun.* **2019**, *10*, 4190. [CrossRef]

151. Parkhitko, A.A.; Priolo, C.; Coloff, J.L.; Yun, J.; Wu, J.J.; Mizumura, K.; Xu, W.; Malinowska, I.A.; Yu, J.; Kwiatkowski, D.J.; et al. Autophagy-dependent metabolic reprogramming sensitizes TSC2-deficient cells to the antimetabolite 6-aminonicotinamide. *Mol. Cancer Res.* **2014**, *12*, 48–57. [CrossRef]

152. Ben-Sahra, I.; Howell, J.J.; Asara, J.M.; Manning, B.D. Stimulation of de novo pyrimidine synthesis by growth signaling through mTOR and S6K1. *Science* **2013**, *339*, 1323–1328. [CrossRef] [PubMed]

153. Robitaille, A.M.; Christen, S.; Shimobayashi, M.; Cornu, M.; Fava, L.L.; Moes, S.; Prescianotto-Baschong, C.; Sauer, U.; Jenoe, P.; Hall, M.N. Quantitative phosphoproteomics reveal mTORC1 activates de novo pyrimidine synthesis. *Science* **2013**, *339*, 1320–1323. [CrossRef] [PubMed]

154. Cunningham, J.T.; Moreno, M.V.; Lodi, A.; Ronen, S.M.; Ruggero, D. Protein and nucleotide biosynthesis are coupled by a single rate-limiting enzyme, PRPS2, to drive cancer. *Cell* **2014**, *157*, 1088–1103. [CrossRef] [PubMed]

155. Ben-Sahra, I.; Hoxhaj, G.; Ricoult, S.J.; Asara, J.M.; Manning, B.D. mTORC1 induces purine synthesis through control of the mitochondrial tetrahydrofolate cycle. *Science* **2016**, *351*, 728–733. [CrossRef]

156. Evert, M.; Calvisi, D.F.; Evert, K.; De Murtas, V.; Gasparetti, G.; Mattu, S.; Destefanis, G.; Ladu, S.; Zimmermann, A.; Delogu, S.; et al. V-AKT murine thymoma viral oncogene homolog/mammalian target of rapamycin activation induces a module of metabolic changes contributing to growth in insulin-induced hepatocarcinogenesis. *Hepatology* **2012**, *55*, 1473–1484. [CrossRef]

157. Kliegman, J.I.; Fiedler, D.; Ryan, C.J.; Xu, Y.F.; Su, X.Y.; Thomas, D.; Caccese, M.C.; Cheng, A.; Shales, M.; Rabinowitz, J.D.; et al. Chemical genetics of rapamycin-insensitive TORC2 in S. cerevisiae. *Cell Rep.* **2013**, *5*, 1725–1736. [CrossRef]

158. Wang, W.; Fridman, A.; Blackledge, W.; Connelly, S.; Wilson, I.A.; Pilz, R.B.; Boss, G.R. The phosphatidylinositol 3-kinase/akt cassette regulates purine nucleotide synthesis. *J. Biol. Chem.* **2009**, *284*, 3521–3528. [CrossRef]

159. Saha, A.; Connelly, S.; Jiang, J.; Zhuang, S.; Amador, D.T.; Phan, T.; Pilz, R.B.; Boss, G.R. Akt phosphorylation and regulation of transketolase is a nodal point for amino acid control of purine synthesis. *Mol. Cell* **2014**, *55*, 264–276. [CrossRef]

160. Gu, X.; Orozco, J.M.; Saxton, R.A.; Condon, K.J.; Liu, G.Y.; Krawczyk, P.A.; Scaria, S.M.; Harper, J.W.; Gygi, S.P.; Sabatini, D.M. SAMTOR is an S-adenosylmethionine sensor for the mTORC1 pathway. *Science* **2017**, *358*, 813–818. [CrossRef]

161. Kottakis, F.; Nicolay, B.N.; Roumane, A.; Karnik, R.; Gu, H.; Nagle, J.M.; Boukhali, M.; Hayward, M.C.; Li, Y.Y.; Chen, T.; et al. LKB1 loss links serine metabolism to DNA methylation and tumorigenesis. *Nature* **2016**, *539*, 390–395. [CrossRef]

162. Reina-Campos, M.; Linares, J.F.; Duran, A.; Cordes, T.; L'Hermitte, A.; Badur, M.G.; Bhangoo, M.S.; Thorson, P.K.; Richards, A.; Rooslid, T.; et al. Increased Serine and One-Carbon Pathway Metabolism by PKClambda/iota Deficiency Promotes Neuroendocrine Prostate Cancer. *Cancer Cell* **2019**, *35*, 385–400. [CrossRef] [PubMed]

163. Grabiner, B.C.; Nardi, V.; Birsoy, K.; Possemato, R.; Shen, K.; Sinha, S.; Jordan, A.; Beck, A.H.; Sabatini, D.M. A diverse array of cancer-associated MTOR mutations are hyperactivating and can predict rapamycin sensitivity. *Cancer Discov.* **2014**, *4*, 554–563. [CrossRef] [PubMed]

164. Wagle, N.; Grabiner, B.C.; Van Allen, E.M.; Hodis, E.; Jacobus, S.; Supko, J.G.; Stewart, M.; Choueiri, T.K.; Gandhi, L.; Cleary, J.M.; et al. Activating mTOR mutations in a patient with an extraordinary response on a phase I trial of everolimus and pazopanib. *Cancer Discov.* **2014**, *4*, 546–553. [CrossRef]

165. Murugan, A.K. mTOR: Role in cancer, metastasis and drug resistance. *Semin. Cancer Biol.* **2019**. [CrossRef] [PubMed]

166. Wong, K.K.; Engelman, J.A.; Cantley, L.C. Targeting the PI3K signaling pathway in cancer. *Curr. Opin. Genet. Dev.* **2010**, *20*, 87–90. [CrossRef] [PubMed]

167. Shaw, R.J.; Cantley, L.C. Ras, PI(3)K and mTOR signalling controls tumour cell growth. *Nature* **2006**, *441*, 424–430. [CrossRef]

168. Gustafson, W.C.; Weiss, W.A. Myc proteins as therapeutic targets. *Oncogene* **2010**, *29*, 1249–1259. [CrossRef]

169. Meng, L.H.; Zheng, X.F. Toward rapamycin analog (rapalog)-based precision cancer therapy. *Acta Pharm. Sin.* **2015**, *36*, 1163–1169. [CrossRef]

170. Le Tourneau, C.; Faivre, S.; Serova, M.; Raymond, E. mTORC1 inhibitors: Is temsirolimus in renal cancer telling us how they really work? *Br. J. Cancer* **2008**, *99*, 1197–1203. [CrossRef]

171. Motzer, R.J.; Escudier, B.; Oudard, S.; Hutson, T.E.; Porta, C.; Bracarda, S.; Grunwald, V.; Thompson, J.A.; Figlin, R.A.; Hollaender, N.; et al. Efficacy of everolimus in advanced renal cell carcinoma: A double-blind, randomised, placebo-controlled phase III trial. *Lancet* **2008**, *372*, 449–456. [CrossRef]

172. Buti, S.; Leonetti, A.; Dallatomasina, A.; Bersanelli, M. Everolimus in the management of metastatic renal cell carcinoma: An evidence-based review of its place in therapy. *Core Evid.* **2016**, *11*, 23–36. [CrossRef] [PubMed]

173. Krueger, D.A.; Care, M.M.; Holland, K.; Agricola, K.; Tudor, C.; Mangeshkar, P.; Wilson, K.A.; Byars, A.; Sahmoud, T.; Franz, D.N. Everolimus for subependymal giant-cell astrocytomas in tuberous sclerosis. *N. Engl. J. Med.* **2010**, *363*, 1801–1811. [CrossRef] [PubMed]

174. Bissler, J.J.; Kingswood, J.C.; Radzikowska, E.; Zonnenberg, B.A.; Frost, M.; Belousova, E.; Sauter, M.; Nonomura, N.; Brakemeier, S.; de Vries, P.J.; et al. Everolimus for angiomyolipoma associated with tuberous sclerosis complex or sporadic lymphangioleiomyomatosis (EXIST-2): A multicentre, randomised, double-blind, placebo-controlled trial. *Lancet* **2013**, *381*, 817–824. [CrossRef]

175. Franz, D.N.; Belousova, E.; Sparagana, S.; Bebin, E.M.; Frost, M.D.; Kuperman, R.; Witt, O.; Kohrman, M.H.; Flamini, J.R.; Wu, J.Y.; et al. Long-Term Use of Everolimus in Patients with Tuberous Sclerosis Complex: Final Results from the EXIST-1 Study. *PLoS ONE* **2016**, *11*, e0158476. [CrossRef] [PubMed]

176. Yao, J.C.; Shah, M.H.; Ito, T.; Bohas, C.L.; Wolin, E.M.; Van Cutsem, E.; Hobday, T.J.; Okusaka, T.; Capdevila, J.; de Vries, E.G.; et al. Everolimus for advanced pancreatic neuroendocrine tumors. *N. Engl. J. Med.* **2011**, *364*, 514–523. [CrossRef] [PubMed]

177. Yao, J.C.; Fazio, N.; Singh, S.; Buzzoni, R.; Carnaghi, C.; Wolin, E.; Tomasek, J.; Raderer, M.; Lahner, H.; Voi, M.; et al. Everolimus for the treatment of advanced, non-functional neuroendocrine tumours of the lung or gastrointestinal tract (RADIANT-4): A randomised, placebo-controlled, phase 3 study. *Lancet* **2016**, *387*, 968–977. [CrossRef]

178. Hudes, G.; Carducci, M.; Tomczak, P.; Dutcher, J.; Figlin, R.; Kapoor, A.; Staroslawska, E.; Sosman, J.; McDermott, D.; Bodrogi, I.; et al. Temsirolimus, interferon alfa, or both for advanced renal-cell carcinoma. *N. Engl. J. Med.* **2007**, *356*, 2271–2281. [CrossRef]

179. Kwitkowski, V.E.; Prowell, T.M.; Ibrahim, A.; Farrell, A.T.; Justice, R.; Mitchell, S.S.; Sridhara, R.; Pazdur, R. FDA approval summary: Temsirolimus as treatment for advanced renal cell carcinoma. *Oncologist* **2010**, *15*, 428–435. [CrossRef]

180. Hess, G.; Herbrecht, R.; Romaguera, J.; Verhoef, G.; Crump, M.; Gisselbrecht, C.; Laurell, A.; Offner, F.; Strahs, A.; Berkenblit, A.; et al. Phase III study to evaluate temsirolimus compared with investigator's choice therapy for the treatment of relapsed or refractory mantle cell lymphoma. *J. Clin. Oncol.* **2009**, *27*, 3822–3829. [CrossRef]

181. Schneider, T.C.; de Wit, D.; Links, T.P.; van Erp, N.P.; van der Hoeven, J.J.; Gelderblom, H.; Roozen, I.C.; Bos, M.; Corver, W.E.; van Wezel, T.; et al. Everolimus in Patients With Advanced Follicular-Derived Thyroid Cancer: Results of a Phase II Clinical Trial. *J. Clin. Endocrinol. Metab.* **2017**, *102*, 698–707. [CrossRef]

182. Chawla, S.P.; Staddon, A.P.; Baker, L.H.; Schuetze, S.M.; Tolcher, A.W.; D'Amato, G.Z.; Blay, J.Y.; Mita, M.M.; Sankhala, K.K.; Berk, L.; et al. Phase II study of the mammalian target of rapamycin inhibitor ridaforolimus in patients with advanced bone and soft tissue sarcomas. *J. Clin. Oncol.* **2012**, *30*, 78–84. [CrossRef] [PubMed]

183. Demetri, G.D.; Chawla, S.P.; Ray-Coquard, I.; Le Cesne, A.; Staddon, A.P.; Milhem, M.M.; Penel, N.; Riedel, R.F.; Bui-Nguyen, B.; Cranmer, L.D.; et al. Results of an international randomized phase III trial of the mammalian target of rapamycin inhibitor ridaforolimus versus placebo to control metastatic sarcomas in patients after benefit from prior chemotherapy. *J. Clin. Oncol.* **2013**, *31*, 2485–2492. [CrossRef] [PubMed]

184. Colombo, N.; McMeekin, D.S.; Schwartz, P.E.; Sessa, C.; Gehrig, P.A.; Holloway, R.; Braly, P.; Matei, D.; Morosky, A.; Dodion, P.F.; et al. Ridaforolimus as a single agent in advanced endometrial cancer: Results of a single-arm, phase 2 trial. *Br. J. Cancer* **2013**, *108*, 1021–1026. [CrossRef] [PubMed]

185. Oza, A.M.; Pignata, S.; Poveda, A.; McCormack, M.; Clamp, A.; Schwartz, B.; Cheng, J.; Li, X.; Campbell, K.; Dodion, P.; et al. Randomized Phase II Trial of Ridaforolimus in Advanced Endometrial Carcinoma. *J. Clin. Oncol.* **2015**, *33*, 3576–3582. [CrossRef]

186. Pearson, A.D.; Federico, S.M.; Aerts, I.; Hargrave, D.R.; DuBois, S.G.; Iannone, R.; Geschwindt, R.D.; Wang, R.; Haluska, F.G.; Trippett, T.M.; et al. A phase 1 study of oral ridaforolimus in pediatric patients with advanced solid tumors. *Oncotarget* **2016**, *7*, 84736–84747. [CrossRef]

187. Tsoref, D.; Welch, S.; Lau, S.; Biagi, J.; Tonkin, K.; Martin, L.A.; Ellard, S.; Ghatage, P.; Elit, L.; Mackay, H.J.; et al. Phase II study of oral ridaforolimus in women with recurrent or metastatic endometrial cancer. *Gynecol. Oncol.* **2014**, *135*, 184–189. [CrossRef]

188. Lucchesi, M.; Chiappa, E.; Giordano, F.; Mari, F.; Genitori, L.; Sardi, I. Sirolimus in Infants with Multiple Cardiac Rhabdomyomas Associated with Tuberous Sclerosis Complex. *Case Rep. Oncol.* **2018**, *11*, 425–430. [CrossRef]

189. Park, H.; Garrido-Laguna, I.; Naing, A.; Fu, S.; Falchook, G.S.; Piha-Paul, S.A.; Wheler, J.J.; Hong, D.S.; Tsimberidou, A.M.; Subbiah, V.; et al. Phase I dose-escalation study of the mTOR inhibitor sirolimus and the HDAC inhibitor vorinostat in patients with advanced malignancy. *Oncotarget* **2016**, *7*, 67521–67531. [CrossRef]

190. Ohtsu, A.; Ajani, J.A.; Bai, Y.X.; Bang, Y.J.; Chung, H.C.; Pan, H.M.; Sahmoud, T.; Shen, L.; Yeh, K.H.; Chin, K.; et al. Everolimus for previously treated advanced gastric cancer: Results of the randomized, double-blind, phase III GRANITE-1 study. *J. Clin. Oncol.* **2013**, *31*, 3935–3943. [CrossRef]

191. Zhu, A.X.; Kudo, M.; Assenat, E.; Cattan, S.; Kang, Y.K.; Lim, H.Y.; Poon, R.T.; Blanc, J.F.; Vogel, A.; Chen, C.L.; et al. Effect of everolimus on survival in advanced hepatocellular carcinoma after failure of sorafenib: The EVOLVE-1 randomized clinical trial. *JAMA* **2014**, *312*, 57–67. [CrossRef]

192. Fenner, M.; Oing, C.; Dieing, A.; Gauler, T.; Oechsle, K.; Lorch, A.; Hentrich, M.; Kopp, H.G.; Bokemeyer, C.; Honecker, F. Everolimus in patients with multiply relapsed or cisplatin refractory germ cell tumors: Results of a phase II, single-arm, open-label multicenter trial (RADIT) of the German Testicular Cancer Study Group. *J. Cancer Res. Clin. Oncol.* **2019**, *145*, 717–723. [CrossRef] [PubMed]

193. Geiger, J.L.; Bauman, J.E.; Gibson, M.K.; Gooding, W.E.; Varadarajan, P.; Kotsakis, A.; Martin, D.; Gutkind, J.S.; Hedberg, M.L.; Grandis, J.R.; et al. Phase II trial of everolimus in patients with previously treated recurrent or metastatic head and neck squamous cell carcinoma. *Head Neck* **2016**, *38*, 1759–1764. [CrossRef] [PubMed]

194. Wagle, N.; Grabiner, B.C.; Van Allen, E.M.; Amin-Mansour, A.; Taylor-Weiner, A.; Rosenberg, M.; Gray, N.; Barletta, J.A.; Guo, Y.; Swanson, S.J.; et al. Response and acquired resistance to everolimus in anaplastic thyroid cancer. *N. Engl. J. Med.* **2014**, *371*, 1426–1433. [CrossRef] [PubMed]

195. Emons, G.; Kurzeder, C.; Schmalfeldt, B.; Neuser, P.; de Gregorio, N.; Pfisterer, J.; Park-Simon, T.W.; Mahner, S.; Schroder, W.; Luck, H.J.; et al. Temsirolimus in women with platinum-refractory/resistant ovarian cancer or advanced/recurrent endometrial carcinoma. A phase II study of the AGO-study group (AGO-GYN8). *Gynecol. Oncol.* **2016**, *140*, 450–456. [CrossRef] [PubMed]

196. Hu, M.; Huang, H.; Zhao, R.; Li, P.; Li, M.; Miao, H.; Chen, N.; Chen, M. AZD8055 induces cell death associated with autophagy and activation of AMPK in hepatocellular carcinoma. *Oncol. Rep.* **2014**, *31*, 649–656. [CrossRef] [PubMed]

197. Zhao, L.; Teng, B.; Wen, L.; Feng, Q.; Wang, H.; Li, N.; Wang, Y.; Liang, Z. mTOR inhibitor AZD8055 inhibits proliferation and induces apoptosis in laryngeal carcinoma. *Int J. Clin. Exp. Med.* **2014**, *7*, 337–347.

198. Naing, A.; Aghajanian, C.; Raymond, E.; Olmos, D.; Schwartz, G.; Oelmann, E.; Grinsted, L.; Burke, W.; Taylor, R.; Kaye, S.; et al. Safety, tolerability, pharmacokinetics and pharmacodynamics of AZD8055 in advanced solid tumours and lymphoma. *Br. J. Cancer* **2012**, *107*, 1093–1099. [CrossRef]

199. Asahina, H.; Nokihara, H.; Yamamoto, N.; Yamada, Y.; Tamura, Y.; Honda, K.; Seki, Y.; Tanabe, Y.; Shimada, H.; Shi, X.; et al. Safety and tolerability of AZD8055 in Japanese patients with advanced solid tumors; a dose-finding phase I study. *Investig. New Drugs* **2013**, *31*, 677–684. [CrossRef]

200. Basu, B.; Dean, E.; Puglisi, M.; Greystoke, A.; Ong, M.; Burke, W.; Cavallin, M.; Bigley, G.; Womack, C.; Harrington, E.A.; et al. First-in-Human Pharmacokinetic and Pharmacodynamic Study of the Dual m-TORC 1/2 Inhibitor AZD2014. *Clin. Cancer Res.* **2015**, *21*, 3412–3419. [CrossRef]

201. Powles, T.; Wheater, M.; Din, O.; Geldart, T.; Boleti, E.; Stockdale, A.; Sundar, S.; Robinson, A.; Ahmed, I.; Wimalasingham, A.; et al. A Randomised Phase 2 Study of AZD2014 Versus Everolimus in Patients with VEGF-Refractory Metastatic Clear Cell Renal Cancer. *Eur. Urol.* **2016**, *69*, 450–456. [CrossRef]

202. Kim, S.T.; Kim, S.Y.; Klempner, S.J.; Yoon, J.; Kim, N.; Ahn, S.; Bang, H.; Kim, K.M.; Park, W.; Park, S.H.; et al. Rapamycin-insensitive companion of mTOR (RICTOR) amplification defines a subset of advanced gastric cancer and is sensitive to AZD2014-mediated mTORC1/2 inhibition. *Ann. Oncol.* **2017**, *28*, 547–554. [CrossRef] [PubMed]

203. Sakre, N.; Wildey, G.; Behtaj, M.; Kresak, A.; Yang, M.; Fu, P.; Dowlati, A. RICTOR amplification identifies a subgroup in small cell lung cancer and predicts response to drugs targeting mTOR. *Oncotarget* **2017**, *8*, 5992–6002. [CrossRef] [PubMed]

204. Burris, H.A., 3rd; Kurkjian, C.D.; Hart, L.; Pant, S.; Murphy, P.B.; Jones, S.F.; Neuwirth, R.; Patel, C.G.; Zohren, F.; Infante, J.R. TAK-228 (formerly MLN0128), an investigational dual TORC1/2 inhibitor plus paclitaxel, with/without trastuzumab, in patients with advanced solid malignancies. *Cancer Chemother. Pharm.* **2017**, *80*, 261–273. [CrossRef] [PubMed]

205. Ghobrial, I.M.; Siegel, D.S.; Vij, R.; Berdeja, J.G.; Richardson, P.G.; Neuwirth, R.; Patel, C.G.; Zohren, F.; Wolf, J.L. TAK-228 (formerly MLN0128), an investigational oral dual TORC1/2 inhibitor: A phase I dose escalation study in patients with relapsed or refractory multiple myeloma, non-Hodgkin lymphoma, or Waldenström's macroglobulinemia. *Am. J. Hematol.* **2016**, *91*, 400–405. [CrossRef] [PubMed]

206. Mateo, J.; Olmos, D.; Dumez, H.; Poondru, S.; Samberg, N.L.; Barr, S.; Van Tornout, J.M.; Jie, F.; Sandhu, S.; Tan, D.S.; et al. A first in man, dose-finding study of the mTORC1/mTORC2 inhibitor OSI-027 in patients with advanced solid malignancies. *Br. J. Cancer* **2016**, *114*, 889–896. [CrossRef] [PubMed]

207. Graham, L.; Banda, K.; Torres, A.; Carver, B.S.; Chen, Y.; Pisano, K.; Shelkey, G.; Curley, T.; Scher, H.I.; Lotan, T.L.; et al. A phase II study of the dual mTOR inhibitor MLN0128 in patients with metastatic castration resistant prostate cancer. *Investig. New Drugs* **2018**, *36*, 458–467. [CrossRef]

208. Cheng, H.; Zou, Y.; Ross, J.S.; Wang, K.; Liu, X.; Halmos, B.; Ali, S.M.; Liu, H.; Verma, A.; Montagna, C.; et al. RICTOR Amplification Defines a Novel Subset of Patients with Lung Cancer Who May Benefit from Treatment with mTORC1/2 Inhibitors. *Cancer Discov.* **2015**, *5*, 1262–1270. [CrossRef]

209. Engelman, J.A.; Chen, L.; Tan, X.; Crosby, K.; Guimaraes, A.R.; Upadhyay, R.; Maira, M.; McNamara, K.; Perera, S.A.; Song, Y.; et al. Effective use of PI3K and MEK inhibitors to treat mutant Kras G12D and PIK3CA H1047R murine lung cancers. *Nat. Med.* **2008**, *14*, 1351–1356. [CrossRef]

210. Maira, S.M.; Stauffer, F.; Brueggen, J.; Furet, P.; Schnell, C.; Fritsch, C.; Brachmann, S.; Chene, P.; De Pover, A.; Schoemaker, K.; et al. Identification and characterization of NVP-BEZ235, a new orally available dual phosphatidylinositol 3-kinase/mammalian target of rapamycin inhibitor with potent in vivo antitumor activity. *Mol. Cancer* **2008**, *7*, 1851–1863. [CrossRef]

211. Serra, V.; Markman, B.; Scaltriti, M.; Eichhorn, P.J.; Valero, V.; Guzman, M.; Botero, M.L.; Llonch, E.; Atzori, F.; Di Cosimo, S.; et al. NVP-BEZ235, a dual PI3K/mTOR inhibitor, prevents PI3K signaling and inhibits the growth of cancer cells with activating PI3K mutations. *Cancer Res.* **2008**, *68*, 8022–8030. [CrossRef]

212. Brachmann, S.M.; Hofmann, I.; Schnell, C.; Fritsch, C.; Wee, S.; Lane, H.; Wang, S.; Garcia-Echeverria, C.; Maira, S.M. Specific apoptosis induction by the dual PI3K/mTor inhibitor NVP-BEZ235 in HER2 amplified and PIK3CA mutant breast cancer cells. *Proc. Natl. Acad. Sci. USA* **2009**, *106*, 22299–22304. [CrossRef] [PubMed]

213. Baumann, P.; Mandl-Weber, S.; Oduncu, F.; Schmidmaier, R. The novel orally bioavailable inhibitor of phosphoinositol-3-kinase and mammalian target of rapamycin, NVP-BEZ235, inhibits growth and proliferation in multiple myeloma. *Exp. Cell Res.* **2009**, *315*, 485–497. [CrossRef] [PubMed]

214. Bendell, J.C.; Kurkjian, C.; Infante, J.R.; Bauer, T.M.; Burris, H.A., 3rd; Greco, F.A.; Shih, K.C.; Thompson, D.S.; Lane, C.M.; Finney, L.H.; et al. A phase 1 study of the sachet formulation of the oral dual PI3K/mTOR inhibitor BEZ235 given twice daily (BID) in patients with advanced solid tumors. *Investig. New Drugs* **2015**, *33*, 463–471. [CrossRef] [PubMed]

215. Carlo, M.I.; Molina, A.M.; Lakhman, Y.; Patil, S.; Woo, K.; DeLuca, J.; Lee, C.H.; Hsieh, J.J.; Feldman, D.R.; Motzer, R.J.; et al. A Phase Ib Study of BEZ235, a Dual Inhibitor of Phosphatidylinositol 3-Kinase (PI3K) and Mammalian Target of Rapamycin (mTOR), in Patients With Advanced Renal Cell Carcinoma. *Oncologist* **2016**, *21*, 787–788. [CrossRef] [PubMed]

216. Salazar, R.; Garcia-Carbonero, R.; Libutti, S.K.; Hendifar, A.E.; Custodio, A.; Guimbaud, R.; Lombard-Bohas, C.; Ricci, S.; Klumpen, H.J.; Capdevila, J.; et al. Phase II Study of BEZ235 versus Everolimus in Patients with Mammalian Target of Rapamycin Inhibitor-Naive Advanced Pancreatic Neuroendocrine Tumors. *Oncologist* **2018**, *23*, 766-e90. [CrossRef] [PubMed]

217. Rodon, J.; Perez-Fidalgo, A.; Krop, I.E.; Burris, H.; Guerrero-Zotano, A.; Britten, C.D.; Becerra, C.; Schellens, J.; Richards, D.A.; Schuler, M.; et al. Phase 1/1b dose escalation and expansion study of BEZ235, a dual PI3K/mTOR inhibitor, in patients with advanced solid tumors including patients with advanced breast cancer. *Cancer Chemother. Pharm.* **2018**, *82*, 285–298. [CrossRef]

218. Seront, E.; Rottey, S.; Filleul, B.; Glorieux, P.; Goeminne, J.C.; Verschaeve, V.; Vandenbulcke, J.M.; Sautois, B.; Boegner, P.; Gillain, A.; et al. Phase II study of dual phosphoinositol-3-kinase (PI3K) and mammalian target of rapamycin (mTOR) inhibitor BEZ235 in patients with locally advanced or metastatic transitional cell carcinoma. *Bju. Int.* **2016**, *118*, 408–415. [CrossRef]

219. Wise-Draper, T.M.; Moorthy, G.; Salkeni, M.A.; Karim, N.A.; Thomas, H.E.; Mercer, C.A.; Beg, M.S.; O'Gara, S.; Olowokure, O.; Fathallah, H.; et al. A Phase Ib Study of the Dual PI3K/mTOR Inhibitor Dactolisib (BEZ235) Combined with Everolimus in Patients with Advanced Solid Malignancies. *Target. Oncol.* **2017**, *12*, 323–332. [CrossRef]

220. Massard, C.; Chi, K.N.; Castellano, D.; de Bono, J.; Gravis, G.; Dirix, L.; Machiels, J.P.; Mita, A.; Mellado, B.; Turri, S.; et al. Phase Ib dose-finding study of abiraterone acetate plus buparlisib (BKM120) or dactolisib (BEZ235) in patients with castration-resistant prostate cancer. *Eur. J. Cancer* **2017**, *76*, 36–44. [CrossRef]

221. Freitag, H.; Christen, F.; Lewens, F.; Grass, I.; Briest, F.; Iwaszkiewicz, S.; Siegmund, B.; Grabowski, P. Inhibition of mTOR's Catalytic Site by PKI-587 Is a Promising Therapeutic Option for Gastroenteropancreatic Neuroendocrine Tumor Disease. *Neuroendocrinology* **2017**, *105*, 90–104. [CrossRef]

222. Del Campo, J.M.; Birrer, M.; Davis, C.; Fujiwara, K.; Gollerkeri, A.; Gore, M.; Houk, B.; Lau, S.; Poveda, A.; Gonzalez-Martin, A.; et al. A randomized phase II non-comparative study of PF-04691502 and gedatolisib (PF-05212384) in patients with recurrent endometrial cancer. *Gynecol. Oncol.* **2016**, *142*, 62–69. [CrossRef] [PubMed]

223. Dolly, S.O.; Wagner, A.J.; Bendell, J.C.; Kindler, H.L.; Krug, L.M.; Seiwert, T.Y.; Zauderer, M.G.; Lolkema, M.P.; Apt, D.; Yeh, R.F.; et al. Phase I Study of Apitolisib (GDC-0980), Dual Phosphatidylinositol-3-Kinase and Mammalian Target of Rapamycin Kinase Inhibitor, in Patients with Advanced Solid Tumors. *Clin. Cancer Res.* **2016**, *22*, 2874–2884. [CrossRef] [PubMed]

224. Powles, T.; Lackner, M.R.; Oudard, S.; Escudier, B.; Ralph, C.; Brown, J.E.; Hawkins, R.E.; Castellano, D.; Rini, B.I.; Staehler, M.D.; et al. Randomized Open-Label Phase II Trial of Apitolisib (GDC-0980), a Novel Inhibitor of the PI3K/Mammalian Target of Rapamycin Pathway, Versus Everolimus in Patients With Metastatic Renal Cell Carcinoma. *J. Clin. Oncol* **2016**, *34*, 1660–1668. [CrossRef] [PubMed]

225. Makker, V.; Recio, F.O.; Ma, L.; Matulonis, U.A.; Lauchle, J.O.; Parmar, H.; Gilbert, H.N.; Ware, J.A.; Zhu, R.; Lu, S.; et al. A multicenter, single-arm, open-label, phase 2 study of apitolisib (GDC-0980) for the treatment of recurrent or persistent endometrial carcinoma (MAGGIE study). *Cancer* **2016**, *122*, 3519–3528. [CrossRef] [PubMed]

226. Brown, J.R.; Hamadani, M.; Hayslip, J.; Janssens, A.; Wagner-Johnston, N.; Ottmann, O.; Arnason, J.; Tilly, H.; Millenson, M.; Offner, F.; et al. Voxtalisib (XL765) in patients with relapsed or refractory non-Hodgkin lymphoma or chronic lymphocytic leukaemia: An open-label, phase 2 trial. *Lancet Haematol.* **2018**, *5*, e170–e180. [CrossRef]

227. Markman, B.; Tabernero, J.; Krop, I.; Shapiro, G.I.; Siu, L.; Chen, L.C.; Mita, M.; Melendez Cuero, M.; Stutvoet, S.; Birle, D.; et al. Phase I safety, pharmacokinetic, and pharmacodynamic study of the oral phosphatidylinositol-3-kinase and mTOR inhibitor BGT226 in patients with advanced solid tumors. *Ann. Oncol.* **2012**, *23*, 2399–2408. [CrossRef]

228. Kim, M.N.; Lee, S.M.; Kim, J.S.; Hwang, S.G. Preclinical efficacy of a novel dual PI3K/mTOR inhibitor, CMG002, alone and in combination with sorafenib in hepatocellular carcinoma. *Cancer Chemother. Pharm.* **2019**. [CrossRef]

229. Grilley-Olson, J.E.; Bedard, P.L.; Fasolo, A.; Cornfeld, M.; Cartee, L.; Razak, A.R.; Stayner, L.A.; Wu, Y.; Greenwood, R.; Singh, R.; et al. A phase Ib dose-escalation study of the MEK inhibitor trametinib in combination with the PI3K/mTOR inhibitor GSK2126458 in patients with advanced solid tumors. *Investig. New Drugs* **2016**, *34*, 740–749. [CrossRef]

230. Benavides-Serrato, A.; Lee, J.; Holmes, B.; Landon, K.A.; Bashir, T.; Jung, M.E.; Lichtenstein, A.; Gera, J. Specific blockade of Rictor-mTOR association inhibits mTORC2 activity and is cytotoxic in glioblastoma. *PLoS ONE* **2017**, *12*, e0176599. [CrossRef]

231. Rodrik-Outmezguine, V.S.; Okaniwa, M.; Yao, Z.; Novotny, C.J.; McWhirter, C.; Banaji, A.; Won, H.; Wong, W.; Berger, M.; de Stanchina, E.; et al. Overcoming mTOR resistance mutations with a new-generation mTOR inhibitor. *Nature* **2016**, *534*, 272–276. [CrossRef]

232. Fan, Q.; Aksoy, O.; Wong, R.A.; Ilkhanizadeh, S.; Novotny, C.J.; Gustafson, W.C.; Truong, A.Y.; Cayanan, G.; Simonds, E.F.; Haas-Kogan, D.; et al. A Kinase Inhibitor Targeted to mTORC1 Drives Regression in Glioblastoma. *Cancer Cell* **2017**, *31*, 424–435. [CrossRef] [PubMed]

233. Oricchio, E.; Katanayeva, N.; Donaldson, M.C.; Sungalee, S.; Pasion, J.P.; Beguelin, W.; Battistello, E.; Sanghvi, V.R.; Jiang, M.; Jiang, Y.; et al. Genetic and epigenetic inactivation of SESTRIN1 controls mTORC1 and response to EZH2 inhibition in follicular lymphoma. *Sci. Transl. Med.* **2017**, *9*. [CrossRef] [PubMed]

234. Thoreen, C.C.; Kang, S.A.; Chang, J.W.; Liu, Q.; Zhang, J.; Gao, Y.; Reichling, L.J.; Sim, T.; Sabatini, D.M.; Gray, N.S. An ATP-competitive mammalian target of rapamycin inhibitor reveals rapamycin-resistant functions of mTORC1. *J. Biol. Chem.* **2009**, *284*, 8023–8032. [CrossRef] [PubMed]

235. Choo, A.Y.; Yoon, S.O.; Kim, S.G.; Roux, P.P.; Blenis, J. Rapamycin differentially inhibits S6Ks and 4E-BP1 to mediate cell-type-specific repression of mRNA translation. *Proc. Natl. Acad. Sci. USA* **2008**, *105*, 17414–17419. [CrossRef]

236. Hsieh, A.C.; Costa, M.; Zollo, O.; Davis, C.; Feldman, M.E.; Testa, J.R.; Meyuhas, O.; Shokat, K.M.; Ruggero, D. Genetic dissection of the oncogenic mTOR pathway reveals druggable addiction to translational control via 4EBP-eIF4E. *Cancer Cell* **2010**, *17*, 249–261. [CrossRef]

237. Mallya, S.; Fitch, B.A.; Lee, J.S.; So, L.; Janes, M.R.; Fruman, D.A. Resistance to mTOR kinase inhibitors in lymphoma cells lacking 4EBP1. *PLoS ONE* **2014**, *9*, e88865. [CrossRef]

238. Bi, C.; Zhang, X.; Lu, T.; Zhang, X.; Wang, X.; Meng, B.; Zhang, H.; Wang, P.; Vose, J.M.; Chan, W.C.; et al. Inhibition of 4EBP phosphorylation mediates the cytotoxic effect of mechanistic target of rapamycin kinase inhibitors in aggressive B-cell lymphomas. *Haematologica* **2017**, *102*, 755–764. [CrossRef]

239. Iyer, G.; Hanrahan, A.J.; Milowsky, M.I.; Al-Ahmadie, H.; Scott, S.N.; Janakiraman, M.; Pirun, M.; Sander, C.; Socci, N.D.; Ostrovnaya, I.; et al. Genome sequencing identifies a basis for everolimus sensitivity. *Science* **2012**, *338*, 221. [CrossRef]

240. Voss, M.H.; Hakimi, A.A.; Pham, C.G.; Brannon, A.R.; Chen, Y.B.; Cunha, L.F.; Akin, O.; Liu, H.; Takeda, S.; Scott, S.N.; et al. Tumor genetic analyses of patients with metastatic renal cell carcinoma and extended benefit from mTOR inhibitor therapy. *Clin. Cancer Res.* **2014**, *20*, 1955–1964. [CrossRef]

241. Gao, X.; Jegede, O.; Gray, C.; Catalano, P.J.; Novak, J.; Kwiatkowski, D.J.; McKay, R.R.; George, D.J.; Choueiri, T.K.; McDermott, D.F.; et al. Comprehensive Genomic Profiling of Metastatic Tumors in a Phase 2 Biomarker Study of Everolimus in Advanced Renal Cell Carcinoma. *Clin. Genitourin. Cancer* **2018**, *16*, 341–348. [CrossRef]

242. O'Reilly, K.E.; Rojo, F.; She, Q.B.; Solit, D.; Mills, G.B.; Smith, D.; Lane, H.; Hofmann, F.; Hicklin, D.J.; Ludwig, D.L.; et al. mTOR inhibition induces upstream receptor tyrosine kinase signaling and activates Akt. *Cancer Res.* **2006**, *66*, 1500–1508.

243. Mendoza, M.C.; Er, E.E.; Blenis, J. The Ras-ERK and PI3K-mTOR pathways: Cross-talk and compensation. *Trends Biochem Sci.* **2011**. [CrossRef] [PubMed]

244. Andre, F.; O'Regan, R.; Ozguroglu, M.; Toi, M.; Xu, B.; Jerusalem, G.; Masuda, N.; Wilks, S.; Arena, F.; Isaacs, C.; et al. Everolimus for women with trastuzumab-resistant, HER2-positive, advanced breast cancer (BOLERO-3): A randomised, double-blind, placebo-controlled phase 3 trial. *Lancet. Oncol.* **2014**, *15*, 580–591. [CrossRef]

245. Seiler, M.; Ray-Coquard, I.; Melichar, B.; Yardley, D.A.; Wang, R.X.; Dodion, P.F.; Lee, M.A. Oral ridaforolimus plus trastuzumab for patients with HER2+ trastuzumab-refractory metastatic breast cancer. *Clin. Breast Cancer* **2015**, *15*, 60–65. [CrossRef]

246. Gandhi, L.; Bahleda, R.; Tolaney, S.M.; Kwak, E.L.; Cleary, J.M.; Pandya, S.S.; Hollebecque, A.; Abbas, R.; Ananthakrishnan, R.; Berkenblit, A.; et al. Phase I study of neratinib in combination with temsirolimus in patients with human epidermal growth factor receptor 2-dependent and other solid tumors. *J. Clin. Oncol* **2014**, *32*, 68–75. [CrossRef]

247. Kyriakopoulos, C.E.; Braden, A.M.; Kolesar, J.M.; Eickhoff, J.C.; Bailey, H.H.; Heideman, J.; Liu, G.; Wisinski, K.B. A phase I study of tivantinib in combination with temsirolimus in patients with advanced solid tumors. *Investig. New Drugs* **2017**, *35*, 290–297. [CrossRef]

248. Hollebecque, A.; Bahleda, R.; Faivre, L.; Adam, J.; Poinsignon, V.; Paci, A.; Gomez-Roca, C.; Thery, J.C.; Le Deley, M.C.; Varga, A.; et al. Phase I study of temsirolimus in combination with cetuximab in patients with advanced solid tumours. *Eur. J. Cancer* **2017**, *81*, 81–89. [CrossRef]

249. Liu, X.; Kambrick, S.; Fu, S.; Naing, A.; Subbiah, V.; Blumenschein, G.R.; Glisson, B.S.; Kies, M.S.; Tsimberidou, A.M.; Wheler, J.J.; et al. Advanced malignancies treated with a combination of the VEGF inhibitor bevacizumab, anti-EGFR antibody cetuximab, and the mTOR inhibitor temsirolimus. *Oncotarget* **2016**, *7*, 23227–23238. [CrossRef]

250. Mahoney, K.M.; Jacobus, S.; Bhatt, R.S.; Song, J.; Carvo, I.; Cheng, S.C.; Simpson, M.; Fay, A.P.; Puzanov, I.; Michaelson, M.D.; et al. Phase 2 Study of Bevacizumab and Temsirolimus After VEGFR TKI in Metastatic Renal Cell Carcinoma. *Clin. Genitourin Cancer* **2016**, *14*, 304–313. [CrossRef]

251. Baselga, J.; Morales, S.M.; Awada, A.; Blum, J.L.; Tan, A.R.; Ewertz, M.; Cortes, J.; Moy, B.; Ruddy, K.J.; Haddad, T.; et al. A phase II study of combined ridaforolimus and dalotuzumab compared with exemestane in patients with estrogen receptor-positive breast cancer. *Breast Cancer Res. Treat.* **2017**, *163*, 535–544. [CrossRef]

252. Wagner, L.M.; Fouladi, M.; Ahmed, A.; Krailo, M.D.; Weigel, B.; DuBois, S.G.; Doyle, L.A.; Chen, H.; Blaney, S.M. Phase II study of cixutumumab in combination with temsirolimus in pediatric patients and young adults with recurrent or refractory sarcoma: A report from the Children's Oncology Group. *Pediatr. Blood Cancer* **2015**, *62*, 440–444. [CrossRef] [PubMed]

253. Hernandez, S.F.; Chisholm, S.; Borger, D.; Foster, R.; Rueda, B.R.; Growdon, W.B. Ridaforolimus improves the anti-tumor activity of dual HER2 blockade in uterine serous carcinoma in vivo models with HER2 gene amplification and PIK3CA mutation. *Gynecol. Oncol.* **2016**, *141*, 570–579. [CrossRef] [PubMed]

254. Schiff, D.; Jaeckle, K.A.; Anderson, S.K.; Galanis, E.; Giannini, C.; Buckner, J.C.; Stella, P.; Flynn, P.J.; Erickson, B.J.; Schwerkoske, J.F.; et al. Phase 1/2 trial of temsirolimus and sorafenib in the treatment of patients with recurrent glioblastoma: North Central Cancer Treatment Group Study/Alliance N0572. *Cancer* **2018**, *124*, 1455–1463. [CrossRef] [PubMed]

255. Harzstark, A.L.; Small, E.J.; Weinberg, V.K.; Sun, J.; Ryan, C.J.; Lin, A.M.; Fong, L.; Brocks, D.R.; Rosenberg, J.E. A phase 1 study of everolimus and sorafenib for metastatic clear cell renal cell carcinoma. *Cancer* **2011**, *117*, 4194–4200. [CrossRef] [PubMed]

256. Amato, R.J.; Flaherty, A.L.; Stepankiw, M. Phase I trial of everolimus plus sorafenib for patients with advanced renal cell cancer. *Clin. Genitourin. Cancer* **2012**, *10*, 26–31. [CrossRef] [PubMed]

257. Jin, N.; Jiang, T.; Rosen, D.M.; Nelkin, B.D.; Ball, D.W. Synergistic action of a RAF inhibitor and a dual PI3K/mTOR inhibitor in thyroid cancer. *Clin. Cancer Res.* **2011**, *17*, 6482–6489. [CrossRef] [PubMed]

258. Caumanns, J.J.; van Wijngaarden, A.; Kol, A.; Meersma, G.J.; Jalving, M.; Bernards, R.; van der Zee, A.G.J.; Wisman, G.B.A.; de Jong, S. Low-dose triple drug combination targeting the PI3K/AKT/mTOR pathway and the MAPK pathway is an effective approach in ovarian clear cell carcinoma. *Cancer Lett.* **2019**, *461*, 102–111. [CrossRef] [PubMed]

259. Eroglu, Z.; Tawbi, H.A.; Hu, J.; Guan, M.; Frankel, P.H.; Ruel, N.H.; Wilczynski, S.; Christensen, S.; Gandara, D.R.; Chow, W.A. A randomised phase II trial of selumetinib vs selumetinib plus temsirolimus for soft-tissue sarcomas. *Br. J. Cancer* **2015**, *112*, 1644–1651. [CrossRef]

260. Teo, T.; Yu, M.; Yang, Y.; Gillam, T.; Lam, F.; Sykes, M.J.; Wang, S. Pharmacologic co-inhibition of Mnks and mTORC1 synergistically suppresses proliferation and perturbs cell cycle progression in blast crisis-chronic myeloid leukemia cells. *Cancer Lett.* **2015**, *357*, 612–623. [CrossRef]

261. Tolcher, A.W.; Bendell, J.C.; Papadopoulos, K.P.; Burris, H.A., 3rd; Patnaik, A.; Jones, S.F.; Rasco, D.; Cox, D.S.; Durante, M.; Bellew, K.M.; et al. A phase IB trial of the oral MEK inhibitor trametinib (GSK1120212) in combination with everolimus in patients with advanced solid tumors. *Ann. Oncol.* **2015**, *26*, 58–64. [CrossRef]

262. Schram, A.M.; Gandhi, L.; Mita, M.M.; Damstrup, L.; Campana, F.; Hidalgo, M.; Grande, E.; Hyman, D.M.; Heist, R.S. A phase Ib dose-escalation and expansion study of the oral MEK inhibitor pimasertib and PI3K/MTOR inhibitor voxtalisib in patients with advanced solid tumours. *Br. J. Cancer* **2018**, *119*, 1471–1476. [CrossRef] [PubMed]

263. Mita, M.; Fu, S.; Piha-Paul, S.A.; Janku, F.; Mita, A.; Natale, R.; Guo, W.; Zhao, C.; Kurzrock, R.; Naing, A. Phase I trial of MEK 1/2 inhibitor pimasertib combined with mTOR inhibitor temsirolimus in patients with advanced solid tumors. *Investig. New Drugs* **2017**, *35*, 616–626. [CrossRef] [PubMed]

264. Singh, J.; Novik, Y.; Stein, S.; Volm, M.; Meyers, M.; Smith, J.; Omene, C.; Speyer, J.; Schneider, R.; Jhaveri, K.; et al. Phase 2 trial of everolimus and carboplatin combination in patients with triple negative metastatic breast cancer. *Breast Cancer Res.* **2014**, *16*, R32. [CrossRef]

265. Vaishampayan, U.; Shevrin, D.; Stein, M.; Heilbrun, L.; Land, S.; Stark, K.; Li, J.; Dickow, B.; Heath, E.; Smith, D.; et al. Phase II Trial of Carboplatin, Everolimus, and Prednisone in Metastatic Castration-resistant Prostate Cancer Pretreated With Docetaxel Chemotherapy: A Prostate Cancer Clinical Trial Consortium Study. *Urology* **2015**, *86*, 1206–1211. [CrossRef] [PubMed]

266. Chon, H.S.; Kang, S.; Lee, J.K.; Apte, S.M.; Shahzad, M.M.; Williams-Elson, I.; Wenham, R.M. Phase I study of oral ridaforolimus in combination with paclitaxel and carboplatin in patients with solid tumor cancers. *BMC Cancer* **2017**, *17*, 407. [CrossRef] [PubMed]

267. Christopoulos, P.; Engel-Riedel, W.; Grohe, C.; Kropf-Sanchen, C.; von Pawel, J.; Gutz, S.; Kollmeier, J.; Eberhardt, W.; Ukena, D.; Baum, V.; et al. Everolimus with paclitaxel and carboplatin as first-line treatment for metastatic large-cell neuroendocrine lung carcinoma: A multicenter phase II trial. *Ann. Oncol.* **2017**, *28*, 1898–1902. [CrossRef]

268. Jovanovic, B.; Mayer, I.A.; Mayer, E.L.; Abramson, V.G.; Bardia, A.; Sanders, M.E.; Kuba, M.G.; Estrada, M.V.; Beeler, J.S.; Shaver, T.M.; et al. A Randomized Phase II Neoadjuvant Study of Cisplatin, Paclitaxel With or Without Everolimus in Patients with Stage II/III Triple-Negative Breast Cancer (TNBC): Responses and Long-term Outcome Correlated with Increased Frequency of DNA Damage Response Gene Mutations, TNBC Subtype, AR Status, and Ki67. *Clin. Cancer Res.* **2017**, *23*, 4035–4045.

269. Park, I.H.; Kong, S.Y.; Kwon, Y.; Kim, M.K.; Sim, S.H.; Joo, J.; Lee, K.S. Phase I/II clinical trial of everolimus combined with gemcitabine/cisplatin for metastatic triple-negative breast cancer. *J. Cancer* **2018**, *9*, 1145–1151. [CrossRef]

270. Decker, T.; Marschner, N.; Muendlein, A.; Welt, A.; Hagen, V.; Rauh, J.; Schroder, H.; Jaehnig, P.; Potthoff, K.; Lerchenmuller, C. VicTORia: A randomised phase II study to compare vinorelbine in combination with the mTOR inhibitor everolimus versus vinorelbine monotherapy for second-line chemotherapy in advanced HER2-negative breast cancer. *Breast Cancer Res. Treat.* **2019**, *176*, 637–647. [CrossRef]

271. Kim, S.J.; Shin, D.Y.; Kim, J.S.; Yoon, D.H.; Lee, W.S.; Lee, H.; Do, Y.R.; Kang, H.J.; Eom, H.S.; Ko, Y.H.; et al. A phase II study of everolimus (RAD001), an mTOR inhibitor plus CHOP for newly diagnosed peripheral T-cell lymphomas. *Ann. Oncol.* **2016**, *27*, 712–718. [CrossRef]

272. Trucco, M.M.; Meyer, C.F.; Thornton, K.A.; Shah, P.; Chen, A.R.; Wilky, B.A.; Carrera-Haro, M.A.; Boyer, L.C.; Ferreira, M.F.; Shafique, U.; et al. A phase II study of temsirolimus and liposomal doxorubicin for patients with recurrent and refractory bone and soft tissue sarcomas. *Clin. Sarcoma Res.* **2018**, *8*, 21. [CrossRef] [PubMed]

273. Place, A.E.; Pikman, Y.; Stevenson, K.E.; Harris, M.H.; Pauly, M.; Sulis, M.L.; Hijiya, N.; Gore, L.; Cooper, T.M.; Loh, M.L.; et al. Phase I trial of the mTOR inhibitor everolimus in combination with multi-agent chemotherapy in relapsed childhood acute lymphoblastic leukemia. *Pediatr. Blood Cancer* **2018**, *65*, e27062. [CrossRef] [PubMed]

274. Emmenegger, U.; Booth, C.M.; Berry, S.; Sridhar, S.S.; Winquist, E.; Bandali, N.; Chow, A.; Lee, C.; Xu, P.; Man, S.; et al. Temsirolimus Maintenance Therapy After Docetaxel Induction in Castration-Resistant Prostate Cancer. *Oncologist* **2015**, *20*, 1351–1352. [CrossRef] [PubMed]

275. Wen, P.Y.; Omuro, A.; Ahluwalia, M.S.; Fathallah-Shaykh, H.M.; Mohile, N.; Lager, J.J.; Laird, A.D.; Tang, J.; Jiang, J.; Egile, C.; et al. Phase I dose-escalation study of the PI3K/mTOR inhibitor voxtalisib (SAR245409, XL765) plus temozolomide with or without radiotherapy in patients with high-grade glioma. *Neuro-Oncol.* **2015**, *17*, 1275–1283. [CrossRef]

276. Chung, V.; Frankel, P.; Lim, D.; Yeon, C.; Leong, L.; Chao, J.; Ruel, N.; Luevanos, E.; Koehler, S.; Chung, S.; et al. Phase Ib Trial of mFOLFOX6 and Everolimus (NSC-733504) in Patients with Metastatic Gastroesophageal Adenocarcinoma. *Oncology* **2016**, *90*, 307–312. [CrossRef]

277. Baselga, J.; Campone, M.; Piccart, M.; Burris, H.A., 3rd; Rugo, H.S.; Sahmoud, T.; Noguchi, S.; Gnant, M.; Pritchard, K.I.; Lebrun, F.; et al. Everolimus in postmenopausal hormone-receptor-positive advanced breast cancer. *N. Engl. J. Med.* **2012**, *366*, 520–529. [CrossRef]

278. Blackwell, K.; Burris, H.; Gomez, P.; Lynn Henry, N.; Isakoff, S.; Campana, F.; Gao, L.; Jiang, J.; Mace, S.; Tolaney, S.M. Phase I/II dose-escalation study of PI3K inhibitors pilaralisib or voxtalisib in combination with letrozole in patients with hormone-receptor-positive and HER2-negative metastatic breast cancer refractory to a non-steroidal aromatase inhibitor. *Breast Cancer Res. Treat.* **2015**, *154*, 287–297. [CrossRef]

279. Kornblum, N.; Zhao, F.; Manola, J.; Klein, P.; Ramaswamy, B.; Brufsky, A.; Stella, P.J.; Burnette, B.; Telli, M.; Makower, D.F.; et al. Randomized Phase II Trial of Fulvestrant Plus Everolimus or Placebo in Postmenopausal Women With Hormone Receptor-Positive, Human Epidermal Growth Factor Receptor 2-Negative Metastatic Breast Cancer Resistant to Aromatase Inhibitor Therapy: Results of PrE0102. *J. Clin. Oncol.* **2018**, *36*, 1556–1563.

280. Rugo, H.S.; Tredan, O.; Ro, J.; Morales, S.M.; Campone, M.; Musolino, A.; Afonso, N.; Ferreira, M.; Park, K.H.; Cortes, J.; et al. A randomized phase II trial of ridaforolimus, dalotuzumab, and exemestane compared with ridaforolimus and exemestane in patients with advanced breast cancer. *Breast Cancer Res. Treat.* **2017**, *165*, 601–609. [CrossRef]

281. Jacinto, E.; Lorberg, A. TOR regulation of AGC kinases in yeast and mammals. *Biochem. J.* **2008**, *410*, 19–37. [CrossRef]

282. Feldman, M.E.; Apsel, B.; Uotila, A.; Loewith, R.; Knight, Z.A.; Ruggero, D.; Shokat, K.M. Active-site inhibitors of mTOR target rapamycin-resistant outputs of mTORC1 and mTORC2. *PLoS Biol.* **2009**, *7*, e38. [CrossRef] [PubMed]

283. Willems, L.; Chapuis, N.; Puissant, A.; Maciel, T.T.; Green, A.S.; Jacque, N.; Vignon, C.; Park, S.; Guichard, S.; Herault, O.; et al. The dual mTORC1 and mTORC2 inhibitor AZD8055 has anti-tumor activity in acute myeloid leukemia. *Leukemia* **2012**, *26*, 1195–1202. [CrossRef] [PubMed]

284. Evangelisti, C.; Ricci, F.; Tazzari, P.; Tabellini, G.; Battistelli, M.; Falcieri, E.; Chiarini, F.; Bortul, R.; Melchionda, F.; Pagliaro, P.; et al. Targeted inhibition of mTORC1 and mTORC2 by active-site mTOR inhibitors has cytotoxic effects in T-cell acute lymphoblastic leukemia. *Leukemia* **2011**, leu201120. [CrossRef] [PubMed]

285. Caumanns, J.J.; Berns, K.; Wisman, G.B.A.; Fehrmann, R.S.N.; Tomar, T.; Klip, H.; Meersma, G.J.; Hijmans, E.M.; Gennissen, A.M.C.; Duiker, E.W.; et al. Integrative Kinome Profiling Identifies mTORC1/2 Inhibition as Treatment Strategy in Ovarian Clear Cell Carcinoma. *Clin. Cancer Res.* **2018**, *24*, 3928–3940. [CrossRef] [PubMed]

286. Borsari, C.; Rageot, D.; Dall'Asen, A.; Bohnacker, T.; Melone, A.; Sele, A.M.; Jackson, E.; Langlois, J.B.; Beaufils, F.; Hebeisen, P.; et al. A Conformational Restriction Strategy for the Identification of a Highly Selective Pyrimido-pyrrolo-oxazine mTOR Inhibitor. *J. Med. Chem.* **2019**. [CrossRef]

287. Yuan, T.L.; Cantley, L.C. PI3K pathway alterations in cancer: Variations on a theme. *Oncogene* **2008**, *27*, 5497–5510. [CrossRef]

288. Morrison Joly, M.; Hicks, D.J.; Jones, B.; Sanchez, V.; Estrada, M.V.; Young, C.; Williams, M.; Rexer, B.N.; Sarbassov dos, D.; Muller, W.J.; et al. Rictor/mTORC2 Drives Progression and Therapeutic Resistance of HER2-Amplified Breast Cancers. *Cancer Res.* **2016**, *76*, 4752–4764. [CrossRef]

289. Verreault, M.; Weppler, S.A.; Stegeman, A.; Warburton, C.; Strutt, D.; Masin, D.; Bally, M.B. Combined RNAi-mediated suppression of Rictor and EGFR resulted in complete tumor regression in an orthotopic glioblastoma tumor model. *PLoS ONE* **2013**, *8*, e59597. [CrossRef]

290. Wang, L.; Qi, J.; Yu, J.; Chen, H.; Zou, Z.; Lin, X.; Guo, L. Overexpression of Rictor protein in colorectal cancer is correlated with tumor progression and prognosis. *Oncol. Lett.* **2017**, *14*, 6198–6202. [CrossRef]

291. Lang, S.A.; Hackl, C.; Moser, C.; Fichtner-Feigl, S.; Koehl, G.E.; Schlitt, H.J.; Geissler, E.K.; Stoeltzing, O. Implication of RICTOR in the mTOR inhibitor-mediated induction of insulin-like growth factor-I receptor (IGF-IR) and human epidermal growth factor receptor-2 (Her2) expression in gastrointestinal cancer cells. *Biochim. Biophys. Acta* **2010**, *1803*, 435–442. [CrossRef]

292. Werfel, T.A.; Wang, S.; Jackson, M.A.; Kavanaugh, T.E.; Joly, M.M.; Lee, L.H.; Hicks, D.J.; Sanchez, V.; Ericsson, P.G.; Kilchrist, K.V.; et al. Selective mTORC2 Inhibitor Therapeutically Blocks Breast Cancer Cell Growth and Survival. *Cancer Res.* **2018**, *78*, 1845–1858. [CrossRef] [PubMed]

293. Moraitis, D.; Karanikou, M.; Liakou, C.; Dimas, K.; Tzimas, G.; Tseleni-Balafouta, S.; Patsouris, E.; Rassidakis, G.Z.; Kouvaraki, M.A. SIN1, a critical component of the mTOR-Rictor complex, is overexpressed and associated with AKT activation in medullary and aggressive papillary thyroid carcinomas. *Surgery* **2014**, *156*, 1542–1548. [CrossRef] [PubMed]

294. Wang, Y.; Hong, X.; Wang, J.; Yin, Y.; Zhang, Y.; Zhou, Y.; Piao, H.L.; Liang, Z.; Zhang, L.; Li, G.; et al. Inhibition of MAPK pathway is essential for suppressing Rheb-Y35N driven tumor growth. *Oncogene* **2017**, *36*, 756–765. [CrossRef] [PubMed]

295. Lasithiotakis, K.G.; Sinnberg, T.W.; Schittek, B.; Flaherty, K.T.; Kulms, D.; Maczey, E.; Garbe, C.; Meier, F.E. Combined inhibition of MAPK and mTOR signaling inhibits growth, induces cell death, and abrogates invasive growth of melanoma cells. *J. Investig. Derm.* **2008**, *128*, 2013–2023. [CrossRef]

296. Molhoek, K.R.; Brautigan, D.L.; Slingluff, C.L., Jr. Synergistic inhibition of human melanoma proliferation by combination treatment with B-Raf inhibitor BAY43-9006 and mTOR inhibitor Rapamycin. *J. Transl. Med.* **2005**, *3*, 39. [CrossRef]

297. Ruder, D.; Papadimitrakopoulou, V.; Shien, K.; Behrens, C.; Kalhor, N.; Chen, H.; Shen, L.; Lee, J.J.; Hong, W.K.; Tang, X.; et al. Concomitant targeting of the mTOR/MAPK pathways: Novel therapeutic strategy in subsets of RICTOR/KRAS-altered non-small cell lung cancer. *Oncotarget* **2018**, *9*, 33995–34008. [CrossRef]

298. Setsu, N.; Yamamoto, H.; Kohashi, K.; Endo, M.; Matsuda, S.; Yokoyama, R.; Nishiyama, K.; Iwamoto, Y.; Dobashi, Y.; Oda, Y. The Akt/mammalian target of rapamycin pathway is activated and associated with adverse prognosis in soft tissue leiomyosarcomas. *Cancer* **2012**, *118*, 1637–1648. [CrossRef]

299. Malhotra, V.; Perry, M.C. Classical chemotherapy: Mechanisms, toxicities and the therapeutic window. *Cancer Biol.* **2003**, *2*, S2–S4. [CrossRef]

300. Daver, N.; Boumber, Y.; Kantarjian, H.; Ravandi, F.; Cortes, J.; Rytting, M.E.; Kawedia, J.D.; Basnett, J.; Culotta, K.S.; Zeng, Z.; et al. A Phase I/II Study of the mTOR Inhibitor Everolimus in Combination with HyperCVAD Chemotherapy in Patients with Relapsed/Refractory Acute Lymphoblastic Leukemia. *Clin. Cancer Res.* **2015**, *21*, 2704–2714. [CrossRef]

301. Rheingold, S.R.; Tasian, S.K.; Whitlock, J.A.; Teachey, D.T.; Borowitz, M.J.; Liu, X.; Minard, C.G.; Fox, E.; Weigel, B.J.; Blaney, S.M. A phase 1 trial of temsirolimus and intensive re-induction chemotherapy for 2nd or greater relapse of acute lymphoblastic leukaemia: A Children's Oncology Group study (ADVL1114). *Br. J. Haematol.* **2017**, *177*, 467–474. [CrossRef]

302. Vidal, G.A.; Chen, M.; Sheth, S.; Svahn, T.; Guardino, E. Phase I Trial of Everolimus and Capecitabine in Metastatic HER2(-) Breast Cancer. *Clin. Breast Cancer* **2017**, *17*, 418–426. [CrossRef] [PubMed]

303. Deenen, M.J.; Klumpen, H.J.; Richel, D.J.; Sparidans, R.W.; Weterman, M.J.; Beijnen, J.H.; Schellens, J.H.; Wilmink, J.W. Phase I and pharmacokinetic study of capecitabine and the oral mTOR inhibitor everolimus in patients with advanced solid malignancies. *Investig. New Drugs* **2012**, *30*, 1557–1565. [CrossRef] [PubMed]

304. Martin-Broto, J.; Redondo, A.; Valverde, C.; Vaz, M.A.; Mora, J.; Garcia Del Muro, X.; Gutierrez, A.; Tous, C.; Carnero, A.; Marcilla, D.; et al. Gemcitabine plus sirolimus for relapsed and progressing osteosarcoma patients after standard chemotherapy: A multicenter, single-arm phase II trial of Spanish Group for Research on Sarcoma (GEIS). *Ann. Oncol.* **2017**, *28*, 2994–2999. [CrossRef] [PubMed]

305. Karavasilis, V.; Samantas, E.; Koliou, G.A.; Kalogera-Fountzila, A.; Pentheroudakis, G.; Varthalitis, I.; Linardou, H.; Rallis, G.; Skondra, M.; Papadopoulos, G.; et al. Gemcitabine Combined with the mTOR Inhibitor Temsirolimus in Patients with Locally Advanced or Metastatic Pancreatic Cancer. A Hellenic Cooperative Oncology Group Phase I/II Study. *Target. Oncol.* **2018**, *13*, 715–724. [CrossRef] [PubMed]

306. Tan, P.; Tiong, I.S.; Fleming, S.; Pomilio, G.; Cummings, N.; Droogleever, M.; McManus, J.; Schwarer, A.; Catalano, J.; Patil, S.; et al. The mTOR inhibitor everolimus in combination with azacitidine in patients with relapsed/refractory acute myeloid leukemia: A phase Ib/II study. *Oncotarget* **2017**, *8*, 52269–52280. [CrossRef] [PubMed]

307. Valvezan, A.J.; Turner, M.; Belaid, A.; Lam, H.C.; Miller, S.K.; McNamara, M.C.; Baglini, C.; Housden, B.E.; Perrimon, N.; Kwiatkowski, D.J.; et al. mTORC1 Couples Nucleotide Synthesis to Nucleotide Demand Resulting in a Targetable Metabolic Vulnerability. *Cancer Cell* **2017**, *32*, 624–638. [CrossRef]

308. Yan, C.; Wei, H.; Minjuan, Z.; Yan, X.; Jingyue, Y.; Wenchao, L.; Sheng, H. The mTOR inhibitor rapamycin synergizes with a fatty acid synthase inhibitor to induce cytotoxicity in ER/HER2-positive breast cancer cells. *PLoS ONE* **2014**, *9*, e97697. [CrossRef]

309. Yang, C.S.; Matsuura, K.; Huang, N.J.; Robeson, A.C.; Huang, B.; Zhang, L.; Kornbluth, S. Fatty acid synthase inhibition engages a novel caspase-2 regulatory mechanism to induce ovarian cancer cell death. *Oncogene* **2015**, *34*, 3264–3272. [CrossRef]

310. Khawaja, M.R.; Nick, A.M.; Madhusudanannair, V.; Fu, S.; Hong, D.; McQuinn, L.M.; Ng, C.S.; Piha-Paul, S.A.; Janku, F.; Subbiah, V.; et al. Phase I dose escalation study of temsirolimus in combination with metformin in patients with advanced/refractory cancers. *Cancer Chemother. Pharm.* **2016**, *77*, 973–977. [CrossRef]

311. MacKenzie, M.J.; Ernst, S.; Johnson, C.; Winquist, E. A phase I study of temsirolimus and metformin in advanced solid tumours. *Investig. New Drugs* **2012**, *30*, 647–652. [CrossRef]

312. Molenaar, R.J.; van de Venne, T.; Weterman, M.J.; Mathot, R.A.; Klumpen, H.J.; Richel, D.J.; Wilmink, J.W. A phase Ib study of everolimus combined with metformin for patients with advanced cancer. *Investig. New Drugs* **2018**, *36*, 53–61. [CrossRef] [PubMed]

313. Sehdev, A.; Karrison, T.; Zha, Y.; Janisch, L.; Turcich, M.; Cohen, E.E.W.; Maitland, M.; Polite, B.N.; Gajewski, T.F.; Salgia, R.; et al. A pharmacodynamic study of sirolimus and metformin in patients with advanced solid tumors. *Cancer Chemother. Pharm.* **2018**, *82*, 309–317. [CrossRef] [PubMed]

314. Chiche, J.; Reverso-Meinietti, J.; Mouchotte, A.; Rubio-Patino, C.; Mhaidly, R.; Villa, E.; Bossowski, J.P.; Proics, E.; Grima-Reyes, M.; Paquet, A.; et al. GAPDH Expression Predicts the Response to R-CHOP, the Tumor Metabolic Status, and the Response of DLBCL Patients to Metabolic Inhibitors. *Cell Metab.* **2019**, *29*, 1243–1257. [CrossRef] [PubMed]

315. Yang, X.; Niu, B.; Wang, L.; Chen, M.; Kang, X.; Wang, L.; Ji, Y.; Zhong, J. Autophagy inhibition enhances colorectal cancer apoptosis induced by dual phosphatidylinositol 3-kinase/mammalian target of rapamycin inhibitor NVP-BEZ235. *Oncol. Lett.* **2016**, *12*, 102–106. [CrossRef]

316. Rangwala, R.; Chang, Y.C.; Hu, J.; Algazy, K.M.; Evans, T.L.; Fecher, L.A.; Schuchter, L.M.; Torigian, D.A.; Panosian, J.T.; Troxel, A.B.; et al. Combined MTOR and autophagy inhibition: Phase I trial of hydroxychloroquine and temsirolimus in patients with advanced solid tumors and melanoma. *Autophagy* **2014**, *10*, 1391–1402. [CrossRef] [PubMed]

317. Haas, N.B.; Appleman, L.J.; Stein, M.; Redlinger, M.; Wilks, M.; Xu, X.; Onorati, A.; Kalavacharla, A.; Kim, T.; Zhen, C.J.; et al. Autophagy Inhibition to Augment mTOR Inhibition: A Phase I/II Trial of Everolimus and Hydroxychloroquine in Patients with Previously Treated Renal Cell Carcinoma. *Clin. Cancer Res.* **2019**, *25*, 2080–2087. [CrossRef] [PubMed]

318. Mahoney, S.J.; Narayan, S.; Molz, L.; Berstler, L.A.; Kang, S.A.; Vlasuk, G.P.; Saiah, E. A small molecule inhibitor of Rheb selectively targets mTORC1 signaling. *Nat. Commun.* **2018**, *9*, 548. [CrossRef]

319. Niessner, H.; Beck, D.; Sinnberg, T.; Lasithiotakis, K.; Maczey, E.; Gogel, J.; Venturelli, S.; Berger, A.; Mauthe, M.; Toulany, M.; et al. The farnesyl transferase inhibitor lonafarnib inhibits mTOR signaling and enforces sorafenib-induced apoptosis in melanoma cells. *J. Investig. Derm.* **2011**, *131*, 468–479. [CrossRef]

320. Becher, O.J.; Gilheeney, S.W.; Khakoo, Y.; Lyden, D.C.; Haque, S.; De Braganca, K.C.; Kolesar, J.M.; Huse, J.T.; Modak, S.; Wexler, L.H.; et al. A phase I study of perifosine with temsirolimus for recurrent pediatric solid tumors. *Pediatr. Blood Cancer* **2017**, *64*. [CrossRef]

321. Gupta, S.; Argiles, G.; Munster, P.N.; Hollebecque, A.; Dajani, O.; Cheng, J.D.; Wang, R.; Swift, A.; Tosolini, A.; Piha-Paul, S.A. A Phase I Trial of Combined Ridaforolimus and MK-2206 in Patients with Advanced Malignancies. *Clin. Cancer Res.* **2015**, *21*, 5235–5244. [CrossRef]

322. Xing, Y.; Lin, N.U.; Maurer, M.A.; Chen, H.; Mahvash, A.; Sahin, A.; Akcakanat, A.; Li, Y.; Abramson, V.; Litton, J.; et al. Phase II trial of AKT inhibitor MK-2206 in patients with advanced breast cancer who have tumors with PIK3CA or AKT mutations, and/or PTEN loss/PTEN mutation. *Breast Cancer Res.* **2019**, *21*, 78. [CrossRef] [PubMed]

323. Jonasch, E.; Hasanov, E.; Corn, P.G.; Moss, T.; Shaw, K.R.; Stovall, S.; Marcott, V.; Gan, B.; Bird, S.; Wang, X.; et al. A randomized phase 2 study of MK-2206 versus everolimus in refractory renal cell carcinoma. *Ann. Oncol.* **2017**, *28*, 804–808. [CrossRef] [PubMed]

324. Oki, Y.; Fanale, M.; Romaguera, J.; Fayad, L.; Fowler, N.; Copeland, A.; Samaniego, F.; Kwak, L.W.; Neelapu, S.; Wang, M.; et al. Phase II study of an AKT inhibitor MK2206 in patients with relapsed or refractory lymphoma. *Br. J. Haematol.* **2015**, *171*, 463–470. [CrossRef] [PubMed]

325. Lynch, J.T.; McEwen, R.; Crafter, C.; McDermott, U.; Garnett, M.J.; Barry, S.T.; Davies, B.R. Identification of differential PI3K pathway target dependencies in T-cell acute lymphoblastic leukemia through a large cancer cell panel screen. *Oncotarget* **2016**, *7*, 22128–22139. [CrossRef] [PubMed]

326. Liang, X.; Lan, C.; Jiao, G.; Fu, W.; Long, X.; An, Y.; Wang, K.; Zhou, J.; Chen, T.; Li, Y.; et al. Therapeutic inhibition of SGK1 suppresses colorectal cancer. *Exp. Mol. Med.* **2017**, *49*, e399. [CrossRef] [PubMed]

327. Matschke, J.; Wiebeck, E.; Hurst, S.; Rudner, J.; Jendrossek, V. Role of SGK1 for fatty acid uptake, cell survival and radioresistance of NCI-H460 lung cancer cells exposed to acute or chronic cycling severe hypoxia. *Radiat. Oncol.* **2016**, *11*, 75. [CrossRef]

328. Talarico, C.; D'Antona, L.; Scumaci, D.; Barone, A.; Gigliotti, F.; Fiumara, C.V.; Dattilo, V.; Gallo, E.; Visca, P.; Ortuso, F.; et al. Preclinical model in HCC: The SGK1 kinase inhibitor SI113 blocks tumor progression in vitro and in vivo and synergizes with radiotherapy. *Oncotarget* **2015**, *6*, 37511–37525. [CrossRef]

329. Yang, C.; Huang, X.; Liu, H.; Xiao, F.; Wei, J.; You, L.; Qian, W. PDK1 inhibitor GSK2334470 exerts antitumor activity in multiple myeloma and forms a novel multitargeted combination with dual mTORC1/C2 inhibitor PP242. *Oncotarget* **2017**, *8*, 39185–39197. [CrossRef]

330. Hopkins, B.D.; Pauli, C.; Du, X.; Wang, D.G.; Li, X.; Wu, D.; Amadiume, S.C.; Goncalves, M.D.; Hodakoski, C.; Lundquist, M.R.; et al. Suppression of insulin feedback enhances the efficacy of PI3K inhibitors. *Nature* **2018**, *560*, 499–503. [CrossRef]

331. Liang, X.; Lan, C.; Zhou, J.; Fu, W.; Long, X.; An, Y.; Jiao, G.; Wang, K.; Li, Y.; Xu, J.; et al. Development of a new analog of SGK1 inhibitor and its evaluation as a therapeutic molecule of colorectal cancer. *J. Cancer* **2017**, *8*, 2256–2262. [CrossRef]

332. Earwaker, P.; Anderson, C.; Willenbrock, F.; Harris, A.L.; Protheroe, A.S.; Macaulay, V.M. RAPTOR up-regulation contributes to resistance of renal cancer cells to PI3K-mTOR inhibition. *PLoS ONE* **2018**, *13*, e0191890. [CrossRef] [PubMed]

333. Lamhamedi-Cherradi, S.E.; Menegaz, B.A.; Ramamoorthy, V.; Vishwamitra, D.; Wang, Y.; Maywald, R.L.; Buford, A.S.; Fokt, I.; Skora, S.; Wang, J.; et al. IGF-1R and mTOR Blockade: Novel Resistance Mechanisms and Synergistic Drug Combinations for Ewing Sarcoma. *J. Natl. Cancer Inst.* **2016**, *108*. [CrossRef] [PubMed]

334. Holmes, B.; Benavides-Serrato, A.; Saunders, J.T.; Landon, K.A.; Schreck, A.J.; Nishimura, R.N.; Gera, J. The protein arginine methyltransferase PRMT5 confers therapeutic resistance to mTOR inhibition in glioblastoma. *J. Neurooncol.* **2019**. [CrossRef] [PubMed]

335. Makinoshima, H.; Umemura, S.; Suzuki, A.; Nakanishi, H.; Maruyama, A.; Udagawa, H.; Mimaki, S.; Matsumoto, S.; Niho, S.; Ishii, G.; et al. Metabolic Determinants of Sensitivity to Phosphatidylinositol 3-Kinase Pathway Inhibitor in Small-Cell Lung Carcinoma. *Cancer Res.* **2018**, *78*, 2179–2190. [CrossRef] [PubMed]

336. Wei, W.; Shin, Y.S.; Xue, M.; Matsutani, T.; Masui, K.; Yang, H.; Ikegami, S.; Gu, Y.; Herrmann, K.; Johnson, D.; et al. Single-Cell Phosphoproteomics Resolves Adaptive Signaling Dynamics and Informs Targeted Combination Therapy in Glioblastoma. *Cancer Cell* **2016**, *29*, 563–573. [CrossRef]

337. Jacinto, E.; Werlen, G. mTOR: The mTOR complexes in T cell development and immunity. In *Encyclopedia of Inflammatory Diseases*; Springer: Berlin, Germany, 2015.

338. Thomson, A.W.; Turnquist, H.R.; Raimondi, G. Immunoregulatory functions of mTOR inhibition. *Nat. Rev. Immunol.* **2009**, *9*, 324–337. [CrossRef]

339. Patel, C.H.; Leone, R.D.; Horton, M.R.; Powell, J.D. Targeting metabolism to regulate immune responses in autoimmunity and cancer. *Nat. Rev. Drug Discov.* **2019**, *18*, 669–688. [CrossRef]

340. Frauwirth, K.A.; Riley, J.L.; Harris, M.H.; Parry, R.V.; Rathmell, J.C.; Plas, D.R.; Elstrom, R.L.; June, C.H.; Thompson, C.B. The CD28 signaling pathway regulates glucose metabolism. *Immunity* **2002**, *16*, 769–777. [CrossRef]

341. Michalek, R.D.; Gerriets, V.A.; Jacobs, S.R.; Macintyre, A.N.; MacIver, N.J.; Mason, E.F.; Sullivan, S.A.; Nichols, A.G.; Rathmell, J.C. Cutting edge: Distinct glycolytic and lipid oxidative metabolic programs are essential for effector and regulatory CD4+ T cell subsets. *J. Immunol.* **2011**, *186*, 3299–3303. [CrossRef]

342. Geltink, R.I.K.; Kyle, R.L.; Pearce, E.L. Unraveling the Complex Interplay Between T Cell Metabolism and Function. *Annu. Rev. Immunol.* **2018**, *36*, 461–488. [CrossRef]

343. Huijts, C.M.; Lougheed, S.M.; Bodalal, Z.; van Herpen, C.M.; Hamberg, P.; Tascilar, M.; Haanen, J.B.; Verheul, H.M.; de Gruijl, T.D.; van der Vliet, H.J.; et al. The effect of everolimus and low-dose cyclophosphamide on immune cell subsets in patients with metastatic renal cell carcinoma: Results from a phase I clinical trial. *Cancer Immunol. Immunother.* **2019**, *68*, 503–515. [CrossRef] [PubMed]

344. Battaglia, M.; Stabilini, A.; Migliavacca, B.; Horejs-Hoeck, J.; Kaupper, T.; Roncarolo, M.G. Rapamycin promotes expansion of functional CD4+CD25+FOXP3+ regulatory T cells of both healthy subjects and type 1 diabetic patients. *J. Immunol.* **2006**, *177*, 8338–8347. [CrossRef] [PubMed]

345. Hirayama, Y.; Gi, M.; Yamano, S.; Tachibana, H.; Okuno, T.; Tamada, S.; Nakatani, T.; Wanibuchi, H. Anti-PD-L1 treatment enhances antitumor effect of everolimus in a mouse model of renal cell carcinoma. *Cancer Sci.* **2016**, *107*, 1736–1744. [CrossRef]

346. Cao, G.; Wang, Q.; Li, G.; Meng, Z.; Liu, H.; Tong, J.; Huang, W.; Liu, Z.; Jia, Y.; Wei, J.; et al. mTOR inhibition potentiates cytotoxicity of Vgamma4 gammadelta T cells via up-regulating NKG2D and TNF-alpha. *J. Leukoc. Biol.* **2016**, *100*, 1181–1189. [CrossRef] [PubMed]

347. Crompton, J.G.; Sukumar, M.; Roychoudhuri, R.; Clever, D.; Gros, A.; Eil, R.L.; Tran, E.; Hanada, K.; Yu, Z.; Palmer, D.C.; et al. Akt inhibition enhances expansion of potent tumor-specific lymphocytes with memory cell characteristics. *Cancer Res.* **2015**, *75*, 296–305. [CrossRef] [PubMed]

348. Alizadeh, D.; Wong, R.A.; Yang, X.; Wang, D.; Pecoraro, J.R.; Kuo, C.F.; Aguilar, B.; Qi, Y.; Ann, D.K.; Starr, R.; et al. IL15 Enhances CAR-T Cell Antitumor Activity by Reducing mTORC1 Activity and Preserving Their Stem Cell Memory Phenotype. *Cancer Immunol. Res.* **2019**, *7*, 759–772. [CrossRef]

349. Pearce, E.L.; Walsh, M.C.; Cejas, P.J.; Harms, G.M.; Shen, H.; Wang, L.S.; Jones, R.G.; Choi, Y. Enhancing CD8 T-cell memory by modulating fatty acid metabolism. *Nature* **2009**, *460*, 103–107. [CrossRef]

350. Araki, K.; Turner, A.P.; Shaffer, V.O.; Gangappa, S.; Keller, S.A.; Bachmann, M.F.; Larsen, C.P.; Ahmed, R. mTOR regulates memory CD8 T-cell differentiation. *Nature* **2009**, *460*, 108–112. [CrossRef]

351. Pollizzi, K.N.; Patel, C.H.; Sun, I.H.; Oh, M.H.; Waickman, A.T.; Wen, J.; Delgoffe, G.M.; Powell, J.D. mTORC1 and mTORC2 selectively regulate CD8(+) T cell differentiation. *J. Clin. Investig.* **2015**, *125*, 2090–2108. [CrossRef]

352. Turner, A.P.; Shaffer, V.O.; Araki, K.; Martens, C.; Turner, P.L.; Gangappa, S.; Ford, M.L.; Ahmed, R.; Kirk, A.D.; Larsen, C.P. Sirolimus enhances the magnitude and quality of viral-specific CD8+ T-cell responses to vaccinia virus vaccination in rhesus macaques. *Am. J. Transpl.* **2011**, *11*, 613–618. [CrossRef]

353. Mannick, J.B.; Morris, M.; Hockey, H.P.; Roma, G.; Beibel, M.; Kulmatycki, K.; Watkins, M.; Shavlakadze, T.; Zhou, W.; Quinn, D.; et al. TORC1 inhibition enhances immune function and reduces infections in the elderly. *Sci. Transl. Med.* **2018**, *10*. [CrossRef] [PubMed]

354. Gammon, J.M.; Gosselin, E.A.; Tostanoski, L.H.; Chiu, Y.C.; Zeng, X.; Zeng, Q.; Jewell, C.M. Low-dose controlled release of mTOR inhibitors maintains T cell plasticity and promotes central memory T cells. *J. Control. Release: Off. J. Control. Release Soc.* **2017**, *263*, 151–161. [CrossRef] [PubMed]

355. Chen, Y.L.; Lin, H.W.; Sun, N.Y.; Yie, J.C.; Hung, H.C.; Chen, C.A.; Sun, W.Z.; Cheng, W.F. mTOR Inhibitors Can Enhance the Anti-Tumor Effects of DNA Vaccines through Modulating Dendritic Cell Function in the Tumor Microenvironment. *Cancers* **2019**, *11*, 617. [CrossRef] [PubMed]

356. Amiel, E.; Everts, B.; Freitas, T.C.; King, I.L.; Curtis, J.D.; Pearce, E.L.; Pearce, E.J. Inhibition of mechanistic target of rapamycin promotes dendritic cell activation and enhances therapeutic autologous vaccination in mice. *J. Immunol.* **2012**, *189*, 2151–2158. [CrossRef]

357. Figlin, R.A.; de Souza, P.; McDermott, D.; Dutcher, J.P.; Berkenblit, A.; Thiele, A.; Krygowski, M.; Strahs, A.; Feingold, J.; Boni, J.; et al. Analysis of PTEN and HIF-1alpha and correlation with efficacy in patients with advanced renal cell carcinoma treated with temsirolimus versus interferon-alpha. *Cancer* **2009**, *115*, 3651–3660. [CrossRef]

358. Cho, D.; Signoretti, S.; Dabora, S.; Regan, M.; Seeley, A.; Mariotti, M.; Youmans, A.; Polivy, A.; Mandato, L.; McDermott, D.; et al. Potential histologic and molecular predictors of response to temsirolimus in patients with advanced renal cell carcinoma. *Clin. Genitourin. Cancer* **2007**, *5*, 379–385. [CrossRef]

359. Janku, F.; Wheler, J.J.; Naing, A.; Stepanek, V.M.; Falchook, G.S.; Fu, S.; Garrido-Laguna, I.; Tsimberidou, A.M.; Piha-Paul, S.A.; Moulder, S.L.; et al. PIK3CA mutations in advanced cancers: Characteristics and outcomes. *Oncotarget* **2012**, *3*, 1566–1575. [CrossRef]

360. Batsios, G.; Viswanath, P.; Subramani, E.; Najac, C.; Gillespie, A.M.; Santos, R.D.; Molloy, A.R.; Pieper, R.O.; Ronen, S.M. PI3K/mTOR inhibition of IDH1 mutant glioma leads to reduced 2HG production that is associated with increased survival. *Sci. Rep.* **2019**, *9*, 10521. [CrossRef]

361. Zhang, Y.; Kwok-Shing Ng, P.; Kucherlapati, M.; Chen, F.; Liu, Y.; Tsang, Y.H.; de Velasco, G.; Jeong, K.J.; Akbani, R.; Hadjipanayis, A.; et al. A Pan-Cancer Proteogenomic Atlas of PI3K/AKT/mTOR Pathway Alterations. *Cancer Cell* **2017**, *31*, 820–832. [CrossRef]

362. Dienstmann, R.; Serpico, D.; Rodon, J.; Saura, C.; Macarulla, T.; Elez, E.; Alsina, M.; Capdevila, J.; Perez-Garcia, J.; Sanchez-Olle, G.; et al. Molecular profiling of patients with colorectal cancer and matched targeted therapy in phase I clinical trials. *Mol. Cancer* **2012**, *11*, 2062–2071. [CrossRef]

363. Martinez-Carreres, L.; Puyal, J.; Leal-Esteban, L.C.; Orpinell, M.; Castillo-Armengol, J.; Giralt, A.; Dergai, O.; Moret, C.; Barquissau, V.; Nasrallah, A.; et al. CDK4 regulates lysosomal function and mTORC1 activation to promote cancer cell survival. *Cancer Res.* **2019**. [CrossRef] [PubMed]

364. Zibelman, M.; Wong, Y.N.; Devarajan, K.; Malizzia, L.; Corrigan, A.; Olszanski, A.J.; Denlinger, C.S.; Roethke, S.K.; Tetzlaff, C.H.; Plimack, E.R. Phase I study of the mTOR inhibitor ridaforolimus and the HDAC inhibitor vorinostat in advanced renal cell carcinoma and other solid tumors. *Investig. New Drugs* **2015**, *33*, 1040–1047. [CrossRef] [PubMed]

365. Murugan, A.K.; Liu, R.; Xing, M. Identification and characterization of two novel oncogenic mTOR mutations. *Oncogene* **2019**, *38*, 5211–5226. [CrossRef] [PubMed]

366. Kong, Y.; Si, L.; Li, Y.; Wu, X.; Xu, X.; Dai, J.; Tang, H.; Ma, M.; Chi, Z.; Sheng, X.; et al. Analysis of mTOR Gene Aberrations in Melanoma Patients and Evaluation of Their Sensitivity to PI3K-AKT-mTOR Pathway Inhibitors. *Clin. Cancer Res.* **2016**, *22*, 1018–1027. [CrossRef] [PubMed]

367. Xu, J.; Pham, C.G.; Albanese, S.K.; Dong, Y.; Oyama, T.; Lee, C.H.; Rodrik-Outmezguine, V.; Yao, Z.; Han, S.; Chen, D.; et al. Mechanistically distinct cancer-associated mTOR activation clusters predict sensitivity to rapamycin. *J. Clin. Investig.* **2016**, *126*, 3526–3540. [CrossRef] [PubMed]

368. Yamaguchi, H.; Kawazu, M.; Yasuda, T.; Soda, M.; Ueno, T.; Kojima, S.; Yashiro, M.; Yoshino, I.; Ishikawa, Y.; Sai, E.; et al. Transforming somatic mutations of mammalian target of rapamycin kinase in human cancer. *Cancer Sci.* **2015**, *106*, 1687–1692. [CrossRef] [PubMed]

369. Sato, T.; Nakashima, A.; Guo, L.; Coffman, K.; Tamanoi, F. Single amino-acid changes that confer constitutive activation of mTOR are discovered in human cancer. *Oncogene* **2010**, *29*, 2746–2752. [CrossRef]

370. Verheijen, R.B.; Atrafi, F.; Schellens, J.H.M.; Beijnen, J.H.; Huitema, A.D.R.; Mathijssen, R.H.J.; Steeghs, N. Pharmacokinetic Optimization of Everolimus Dosing in Oncology: A Randomized Crossover Trial. *Clin. Pharm.* **2018**, *57*, 637–644. [CrossRef]

371. Lunova, M.; Smolkova, B.; Lynnyk, A.; Uzhytchak, M.; Jirsa, M.; Kubinova, S.; Dejneka, A.; Lunov, O. Targeting the mTOR Signaling Pathway Utilizing Nanoparticles: A Critical Overview. *Cancers* **2019**, *11*, 82. [CrossRef]

Review

Mutations That Confer Drug-Resistance, Oncogenicity and Intrinsic Activity on the ERK MAP Kinases—Current State of the Art

Karina Smorodinsky-Atias [1],[†], Nadine Soudah [1],[†] and David Engelberg [1],[2],[3],*

[1] Department of Biological Chemistry, The Institute of Life Science, The Hebrew University of Jerusalem, Jerusalem 91904, Israel; karinasun@gmail.com (K.S.-A.); nadine.soudah@mail.huji.ac.il (N.S.)

[2] CREATE-NUS-HUJ, Molecular Mechanisms Underlying Inflammatory Diseases (MMID), National University of Singapore, 1 CREATE WAY, Innovation Wing, Singapore 138602, Singapore

[3] Department of Microbiology, Yong Loo Lin School of Medicine, National University of Singapore, Singapore 117456, Singapore

* Correspondence: engelber@mail.huji.ac.il or MICDE@nus.edu.sg; Tel.: +972-2-6584718 or +972-54-2066378

† These authors contributed equally to this work.

Received: 10 December 2019; Accepted: 2 January 2020; Published: 6 January 2020

Abstract: Unique characteristics distinguish extracellular signal-regulated kinases (Erks) from other eukaryotic protein kinases (ePKs). Unlike most ePKs, Erks do not autoactivate and they manifest no basal activity; they become catalysts only when dually phosphorylated on neighboring Thr and Tyr residues and they possess unique structural motifs. Erks function as the sole targets of the receptor tyrosine kinases (RTKs)-Ras-Raf-MEK signaling cascade, which controls numerous physiological processes and is mutated in most cancers. Erks are therefore the executers of the pathway's biology and pathology. As oncogenic mutations have not been identified in Erks themselves, combined with the tight regulation of their activity, Erks have been considered immune against mutations that would render them intrinsically active. Nevertheless, several such mutations have been generated on the basis of structure-function analysis, understanding of ePK evolution and, mostly, via genetic screens in lower eukaryotes. One of the mutations conferred oncogenic properties on Erk1. The number of interesting mutations in Erks has dramatically increased following the development of Erk-specific pharmacological inhibitors and identification of mutations that cause resistance to these compounds. Several mutations have been recently identified in cancer patients. Here we summarize the mutations identified in Erks so far, describe their properties and discuss their possible mechanism of action.

Keywords: MAPK kinase; ERK1; ERK2; CD domain; Rolled; SCH772984; VRT-11E; *sevenmaker*

1. Introduction

The unusual biochemical properties of the extracellular signal-regulated Kinases (Erks), their numerous biological functions and their critical roles in essentially all types of cancer, make these enzymes important subjects for research, and attractive targets for therapeutic purposes. Indeed, more than 50,000 studies have addressed aspects of the biochemistry, biology and pathology of Erks. Nevertheless, serious obstacles, which seem to be related to the unusual characteristics of the Erk enzymes, have been hindering the research. One of the hurdles has been the lack of key reagents, such as intrinsically/constitutively active mutants of Erks, and another is the absence of specific pharmacological inhibitors, not to mention clinically relevant inhibitors. The unavailability of these tools was unexpected, because useful inhibitors and a variety of active mutants were readily developed for most other protein kinases, including those that function upstream and downstream of Erk and those that are similar to Erks, such as p38s and JNKs [1–16].

This situation has been changing dramatically in the last decade. An arsenal of over a dozen useful inhibitors was finally developed, and, soon after, numerous mutations that render Erks resistant to these drugs were identified. Other mutations in ERKs were found in screens for cells in culture that acquired resistance to inhibitors of Raf and MEK. The mutations that cause drug resistance joined a small number of mutations that had been generated on the basis of gain-of-function mutations in lower organisms, or via structural studies. Finally, sequencing of genomes of tens of thousands of cancer patients led to the discovery of a few more mutations in ERKs.

Thus, a large number of interesting mutations in ERKs has been finally gathered (Table 1). The effects of many of these mutations on the structure, biochemistry, biology, or pathology of Erks have not yet been fully characterized, but some notions are emerging. This review summarizes our current knowledge of ERK mutations and describes their effect on the catalytic, physiological, pharmacological and pathological properties of Erks.

1.1. The Erk MAP Kinases

1.1.1. The Erk MAP Kinases Are Conserved in All Eukaryotes and Carry Out a Plethora of Functions

Erk proteins form a small subgroup within the family of MAP kinases. In mammals this group is encoded by two genes, ERK1 and ERK2, and by several splicing variants thereof. Erk1 and Erk2 are expressed in all cells of the organism and are critical for the functionality of all tissues and body systems. An indication of the remarkable competency of Erks is the large number of substrates they phosphorylate—497 have been identified so far [17]. For comprehensive reviews on the Erks, see [18].

Erks are highly conserved in evolution structurally and functionally, so that many discoveries with Erks' orthologs of *S. cerevisiae* (Fus3, Kss1 and Slt2/Mpk1; [19–22]) or of *D. melanogaster* (Rolled; [23,24]) are directly relevant to the mammalian molecules.

Mammalian Erk1 and Erk2 share 83% sequence identity and 88% similarity (alignment of the human proteins) and seem to be equally activated in response to relevant signals, suggesting that many of their activities are redundant. Observations that raised the possibility of distinct functions for each isoform were made primarily with knockout mice [25–34]. The most significant finding in this regard was that knocking out *ERK2* resulted in embryonic lethality, whereas knocking out *ERK1* had only mild effects [35]. Yet, overexpression of Erk1 in mice knocked-out for *ERK2*, restores viability and the mice are normal and fertile [36]. It seems, therefore, that the physiological functions of Erk1 and Erk2 are almost fully redundant and the dramatic difference in the phenotype between ERK1$^{-/-}$ and ERK2$^{-/-}$ mice stems solely from the fact that in most tissues Erk2 is expressed at much higher levels than Erk1 [36]. Also pointing to similarity in structure-function relationships are the observations that the majority (but not all) of the mutations identified recently and discussed in this review confer similar effects on the Erk1 and Erk2 proteins. Finally, the newly developed pharmacological inhibitors manifest similar (but not identical) efficacy towards the two isoforms, although it should be noted that some of those inhibitors have not yet been tested against both isoforms. These observations combined suggest minor differences in functionality between Erk1 and Erk2 native proteins.

1.1.2. Erks Are Targets of the Proto-Oncogenic RTK-Ras-Raf-MEK Pathway

Erks function as the downstream targets of the receptor tyrosine kinase (RTK)-Ras-Raf-MEK pathway, which regulates a large number of biological processes in all cell types and in all developmental stages (for reviews on the RTK-Ras-Raf-MEK pathways see [37–41]). Although in particular cell-types and under some conditions Raf and MEK may phosphorylate various substrates [17], in response to most signals Erks seem to be the only targets of this cascade and therefore mediate most, if not all, of the effects of the pathway [42]. Erks are activated by all 20 subfamilies of RTKs [43], including the clinically important epidermal growth factor receptors (EGFRs), nerve growth factor receptors (NGFRs), vascular endothelial growth factor receptors (VEGFRs), platelet-derived growth factor receptors (PDGFRs), fibroblast growth factor receptors (FGFRs) and insulin receptors (InsRs) [43].

A series of consecutive reactions leads from ligand-bound receptor to Erk activation. Briefly, upon association with its ligand, the RTK dimerizes and trans-autophosphorylates on several tyrosine residues at its cellular domain [44]. The phosphorylated tyrosines serve as scaffolds for SH2- and PTB-containing cytoplasmic enzymes [38]. One of the protein complexes that bind to a phosphotyrosine on the RTK is Grb2-Sos, which in turn activates the small GTPase Ras. Active, GTP-bound, Ras recruits Raf proteins (A-Raf; B-Raf and c-Raf/Raf1) [45], the MAP3Ks of the Erk pathway. Raf kinases phosphorylate the MAP2Ks Mek1 and Mek2. Phosphorylated Meks dually phosphorylate Erk1/2 on neighboring Thr and Tyr residues, part of a TEY motif located at the activation loop. Several additional MAP3Ks may activate Mek, depending on the context of the cell and the type of stimulus (i.e., MOS [46], TPL2/Cot [47] and MLTK [48]). With only a few exceptions, Meks are the only known activators of Erk1/2. Without MEK-mediated dual phosphorylation, Erks are catalytically inactive. Erks can also be activated by GPCRs involving different subunits of G-proteins or β-arrestin, in a ligand-independent mechanism [49–52]. Thus, a variety of ligands, which activate either RTKs or GPCRs, as well as various environmental changes, lead to Erk activation. In addition to the interaction with the direct upstream activators, Erks interact with scaffold proteins such as KSR [53] and Mek partner-1 (MP-1) [54], which facilitate the association of the various cascade components thereby increasing the efficiency of their activation [55]. Activated Erks phosphorylate their substrates on Ser or Thr residues, in all cellular compartments. Cytoplasmic substrates include protein kinases, such as Rsk1/2, Mnk1/2 and Msk1 [56–58]. Nuclear targets include transcription factors of the Fos, Myc and Ets families [59]. Erks also phosphorylate upstream pathway components, such as Raf-1, B-Raf and Mek, as part of positive and negative feedback mechanisms [60–64]. Another manner of negative feedback is the Erk-induced expression of its own deactivating phosphatases [65]. For review on Erk substrates and downstream targets see [17].

1.1.3. Erk Activation Is Achieved by Dual Phosphorylation of a TEY Motif within the Activation Loop

The difficulty in obtaining mutations that render Erks intrinsically active may stem in part from its tight regulation, supported by unique structure-function properties that distinguish them from most other eukaryotic protein kinases (ePKs). Almost all ePKs share a common kinase domain, which includes a highly conserved ATP binding site, a catalytic site and an activation loop. ePKs reside in equilibrium between active and inactive conformations, so that these catalytically-relevant sites are functional only in the active conformation. The kinase domain of all ePKs, including Erks, consists of a small N-lobe and a larger C-lobe (Figure 1A). The N-lobe contains 5 β-strands and a single helix (αC-helix), which is dynamic and occupies the space between the lobes. The αC-helix contains a conserved Glu, which, in the active conformation, forms a salt bridge with a Lys residue located within the AXK motif in β3 strand. This bridge is important and conserved in all ePKs and ensures anchoring and proper orientation of the ATP molecule. The C-lobe, which is mainly α-helical, binds and brings substrates adjacent to the ATP. A short (20–30 amino acids long) fragment located between the N- and C-lobes, known as the activation segment, contains some important elements, such as the DFG and APE motifs, the P+1 site, the catalytic and the activation loops (Figure 1A; [66]). The DFG motif is important for proper positioning of the ATP for phosphate transfer. While the Asp in the DFG is critical for recognizing the Mg^{+2} ions, the Phe forms hydrophobic interactions with the αC-helix and also with the catalytic Asp of the Y/HRD motif. The Y/HRD motif belongs to the catalytic loop responsible for catalysis. The conserved catalytic Asp of the Y/HRD motif functions by orienting the phosphate accepting hydroxyl as well as a proton-transfer acceptor. The Tyr/His residue of this motif, which is also conserved, is part of the R-spine and forms hydrophobic interactions with the DFG. An important conserved moiety is a phosphoacceptor (commonly a threonine) within the activation loop. In most ePKs, phosphorylation of this Thr is a pre-requisite for activity, as it is essential for shifting the equilibrium towards the active conformation. This phosphorylation induces several structural changes, including a conformational change of the DFG motif (to DFG 'in'), rotation of the αC-helix, which enables the formation of a Glu–Lys salt bridge, and a domain closure between the N- and C-lobes, which ultimately stabilize the regulatory and the catalytic spines of the enzyme [67,68].

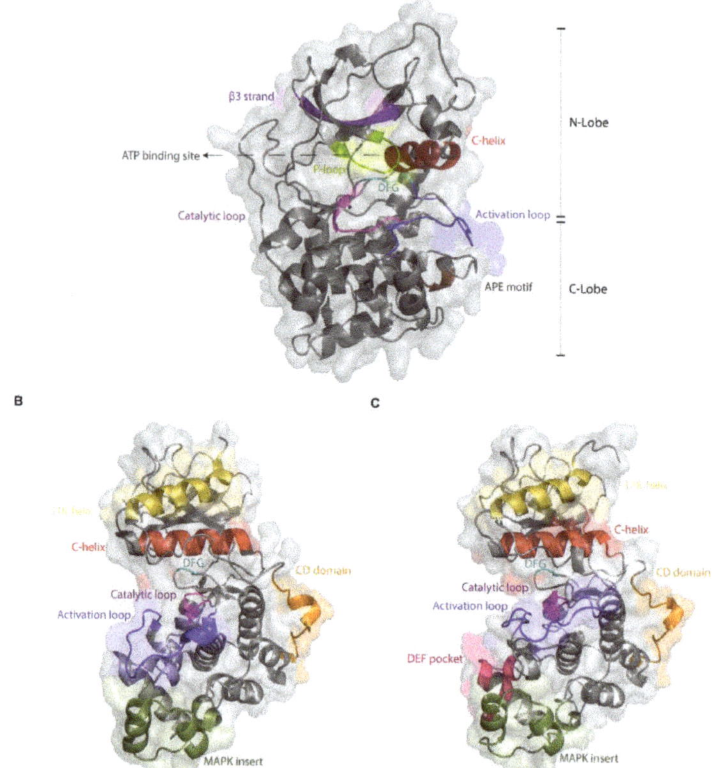

Figure 1. The kinase fold of Erks is highly similar to that of other ePKs, but they possess additional, specific domains. Shown are the crystal structures of (**A**) PKA (PDB 1FMO), (**B**) unphosphorylated Erk2 (PDB 4S31) and (**C**) dually phosphorylated Erk2 (PDB 2ERK). All panels show a cartoon representation covered with a transparent molecular surface with important regions presented and colored accordingly. Note the L16 helix and MAPK insert, not present in PKA, and the DEF pocket that forms only in phosphorylated Erk2.

As this dramatic shift from the non-active to active conformation is a result of the single phosphorylation event, the activation-loop phosphoacceptor Thr is an obvious target for regulation and for mutagenesis aimed at generating activating variants. In several ePKs, converting this Thr to Glu resulted in a constitutive activation of the kinase. However, for most ePKs, genetic manipulations are not required to achieve constant activity, because these enzymes are capable of autoactivating in a mechanism that probably involves dimerization, which enforces a 'prone-to-autophosphorylate' conformation [69–71]. The rate of this spontaneous autophosphorylation is different in each ePK, but in the majority of cases it is sufficient to give rise to a significant activity [69]. In Erks, autoactivation is extremely inefficient and almost non-measurable [72,73]. Erks are fully dependent, therefore, on MAP2Ks for activation loop phosphorylation and induction of catalysis. The lack of autophosphorylation/autoactivation capability in Erk molecules makes overexpression a non-useful experimental approach for studying their biological and pathological effects. As the overall structure of the kinase domain of Erk is very similar to that of the other eEPKs (Figure 1), the explanation for the lack of autophosphorylation and basal catalytic activity of Erks is not trivial [69]. Not only that Erks are incapable of autophosphorylation as opposed to most ePKs, several other structural features also distinguish them from common ePKs. For example, although, like most ePKs, Erks seem to reside in equilibrium between two conformations (termed L and R; [74]) these conformations differ from the classical active and inactive conformations of ePKs. For example, no conformational change in the DFG motif (from 'out' to 'in') is apparent in the structure.

The differences in the biochemical properties between MAPKs and other ePKs may be associated with structural motifs, not part of the kinase domain, which are not present in other ePKs. Two such prominent motifs are the MAPK insert and the C-terminal extension, which includes a domain termed the L16 helix (Figure 1B). However, a bioinformatics-based evolutionary study suggests that the inability to autophosphorylate and the dependence on MEK may stem from minor structural differences, and not necessarily involving the MAPK insert, or the C-terminal extension [75]. This study reconstituted an inferred common ancestor of Erk1, Erk2 and Erk5 that is able to autophosphorylate and an ancestor of Erk1 and Erk2 that cannot. Analysis of the two ancestors suggested that a single amino acid deletion in the linker loop connecting the αC-helix and the β3 strand (position 74 in modern Erk1) and a mutation in the gatekeeper residue (Gln122 in modern Erk1) account for the loss of autophosphorylation and dependence of modern Erk on its upstream activator. Indeed, inserting these two modifications into modern Erk1 was sufficient to generate Erk1 molecules that, when tested in kinase assays in vitro, showed high autophosphorylation ability and consequently catalytic capabilities similar to those of Mek-phosphorylated Erk1 [75]. This study clearly points at residues and domains that could be manipulated in an effort to generate intrinsically active, Mek-independent, Erks. These same residues were identified, in fact, as candidates for mutagenesis by other approaches as well [76].

Other unique Erk domains are the substrate binding motifs. Erks possess two distinct sites through which substrates, activators and deactivators can bind. The first is the common docking (CD) site, which is located about 10Å from the active site of Erk2 (Figure 1B) and is composed of amino acids such as Asp316 and Asp319. The second docking site of Erk2 is the hydrophobic DEF pocket, which is composed of residues Met197, Leu198, Tyr231, Leu232, Leu235 and Tyr261, and exists only in dually phosphorylated Erk adjacent to the catalytic site (Figure 1C). Important mutations, discussed here, occurred in these two domains [73,77,78].

1.1.4. ERK Molecules Are Highly Active in Most Cancers, but Oncogenic Mutations in ERK Themselves Are Very Rare

All upstream components of the Erk signaling cascade are frequently mutated in cancer [79], and it is believed, therefore, that Erks are abnormally overactive in essentially all cancer cases [80]. Accordingly, the Ras-Raf-MEK-Erk cascade has become a major target for anti-cancer therapy [81]. Oncogenic mutations or other genetic alterations (e.g., gene amplification) have been found in RTKs, Ras, Raf and MEKs, but no activating mutations or genetic alterations in Erk molecules themselves have been reported as oncogenic in tumor viruses or in patients. However, as the only known substrates of Rafs are the MEKs, and the only known substrates of MEKs are the Erks, it implies that the biological and pathological/oncogenic effects of the pathway are mediated exclusively via the Erk proteins (note some reports on deviations from the linearity of the RTK-Ras-Raf-MEK-Erk tier: [41,49–51], reviewed in [17]). It is not clear, therefore, why mutations in Erks are rarely found in cancer patients. This situation could be taken as another indication for the unusual tight regulation bestowed on Erks by the specific structural motifs, immunizing them against mutations that render them spontaneously active and oncogenic. Indeed, although some mutations that render Erks intrinsically active and even oncogenic (one mutation) have been discovered in the laboratory, they do not activate Erk to the maximal levels possible (that of Mek-activated Erk), and their oncogenic effect is markedly weaker than that of active, oncogenic, Ras or Raf [82,83].

Nevertheless, mutations in ERKs have been identified in a small number of patients (Table 1) and at least one of those, E320K in ERK2, seems to appear in a few dozen patients suffering from cervical and head and neck carcinoma (Table 1B). Perhaps activating mutations in ERKs are not fully oncogenic and do not have a causative effect, but may promote the disease.

1.1.5. Erk Inhibitors Have Been Recently Developed

Not only that Erks are obvious targets for anti-cancer therapy because they are the downstream components of the RTK-Ras-Raf-MEK pathway, in the majority of cases of tumors resistant to EGFR, B-Raf and MEK inhibitors, re-activation of Erk is observed [84]. These findings reinforce the need for direct Erk inhibitors. Specific inhibition of Erks should also be a powerful tool for research. For unknown reasons, developing pharmacological Erk inhibitors has lagged behind the development of inhibitors against the other MAP kinases, JNK and p38 and has required unusual efforts. Morris et al., for example, screened approximately five million compounds and performed multiple improvement steps in order to discover SCH772984, a small molecule that inhibits both Erk isoforms with an IC_{50} at the nano-molar range [85]. Further development of this inhibitor provided an orally administered analog, MK-8353 [86], which is being tested in phase-I clinical trials. Yet another potent Erk inhibitor is BVD-523 (Ulixertinib), a selective and reversible Erk1 and Erk2 ATP competitive molecule [87], which is currently in phase II clinical trials. GDC-0994 (Ravoxertinib) [88], a pyrazolylpyrrole-based inhibitor that was optimized specifically towards Erk by using structure-guided methods [89], is also undergoing clinical testing. Additional compounds that exhibit selectivity towards Erk are LY3214996 [90], FR180204 [91], VRT-11E [92] and the Erk dimerization inhibitor DEL22379 [93]. For a comprehensive description of Erk inhibitors, see [94].

In parallel to the biochemical and pharmacological characterization of the inhibitors, a large number of mutations that render Erk proteins resistant to them have been reported [95–97] (Table 1).

2. Identification of Various Mutations in ERKs and Study of Their Properties

2.1. Almost All Known Mutations in ERKs Have Been Identified Experimentally

Unlike the many mutations known in RTKs, Ras, Raf and MEK, mostly identified in tumors, almost all mutations known in ERKs have been identified in laboratory setups. Only a few have been identified in cancer patients, and even for those, it is not clear whether they are associated with the disease. Experimental systems were initially designed for identification of intrinsically active Erks. Later, following the development of Erk inhibitors, genetic screens were developed for identification of mutations that would cause drug resistance. The mutations identified in patients, in screens for drug-resistant molecules and for intrinsically active Erks are summarized in Table 1. It should be noted that numeration of the mutations in ERK1 (in text and in Table 1A) refer to the sequence of the human protein and numeration of mutations in ERK2 (in text and in Table 1B) refer to the sequence of rat protein. Notably, mutations that had been discovered (until 2006) in ERK orthologs in lower organisms were summarized in [98] and will not be discussed here.

2.2. Mutations Produced on the Basis of Structure-Function Studies

Original attempts to develop intrinsically active Erk molecules took the conventional approach of trying to mimic the activatory phosphorylation of the activation loop. As no phosphomimetic residue is available for tyrosine, this approach was limited to modifying the Thr of the TEY motif and turned out to be ineffective [99,100]. In fact, not only was changing this Thr to Glu unsuccessful, but the resulting $Erk2^{T183E}$ enzyme showed lower activity than $Erk2^{WT}$, even when phosphorylated by MEK [98,101,102]. Furthermore, even when mutations that render Erk2 intrinsically active were discovered, combining them with the T183E mutation did not create a more active molecule [103].

Other mutations devised on the basis of structure-function understanding, however, did lead to the development of interesting mutants. For example, mutating the gatekeeper residue of Erk2 resulted in an intrinsically active variant (Q103G and Q103A) [76]. Furthermore, mutating residues that interact with Gln103 provided even more active variants (when tested as purified recombinant proteins), primarily Erk2^{I82A} and Erk2^{I84A} [76]. As described above, the gatekeeper residue was also discovered as a site that distinguishes between an inferred ancestor kinase, which is capable of autophosphorylation, and the modern Erk1 and Erk2. Mutating the gatekeeper residue in Erk1 on the basis of comparison to the inferred ancestor (inserting the Q122M mutation) rendered the mutant intrinsically active [75].

The mechanism that renders all these mutants intrinsically active was shown to be the acquisition of an autophosphorylation capability. Namely, the mutations did not impose adoption of the native conformation, but rather unleashed an obstructed autophosphorylation capability and allowed autoactivation. These observations suggest that Erks are similar to most other ePKs that possess the autophosphorylation machinery, but this activity in Erks is not spontaneous. Structural blockers of autophosphorylation in Erks are not known implying that activating mutations could reveal them. An interesting mechanism of action for how substitutions at Gln103 or Ile84 unblock autophosphorylation was suggested by Emrick et al. It was proposed that the mutations induce a pathway of intramolecular interactions leading to flexibility in the activation lip, thereby enabling the phosphoacceptors to reach the catalytic Asp (D147 in Erk2). In the inactive form of Erk2, the intramolecular pathway includes hydrophobic interaction between Leu73, Gln103 or Ile84 and phe166 of the DFG motif, which in turn is linked to L168 through the backbone. The latter forms side-chain interactions with Val186 from the activation lip. Emrick et al. further suggested that T188 forms hydrogen bond with Asp147 and Lys149. These interactions together impede autoactivation by holding the activation lip in a stable conformation. It is therefore not surprising that mutating Leu73, Gln103 or Ile84 would lead to the movement of Phe166 and thus affect the hydrophobic interaction between Leu168 and Val186 and the hydrogen bond between Thr188 and Asp147/Lys149. This would cause an increase in flexibility of the activation loop and eventually autophosphorylation.

Although, when tested in vitro as recombinant proteins, these mutations render the Erk molecules intrinsically active, the significance of the mutants in the gatekeeper and nearby residues in living cells is not clear. While Erk2^{I84A} seems to be spontaneously active when expressed in HEK293 cells, it is not spontaneously active in NIH3T3 cells. The equivalent Erk1 mutant, Erk1^{I103A} is not spontaneously active in either cell line [82]. None of the mutants can oncogenically transform NIH3T3 cells [82], but, intriguingly, several mutations in the gatekeeper of Erk2 (Q103) were identified in screens for drug-resistant Erk2 molecules (Table 1B) and a mutation in I82 (I82T) was found in one cancer patient (Table 1B). Perhaps conversion of these residues to particular amino acids (e.g., Thr), other than those tested so far (i.e., Ala) would render them more active in cells and possibly even oncogenic.

Another mutation that was generated in Erk2 on the basis of structural studies is S151D. This mutation was designed following an alignment of the conserved sequence DLKPSN in MKK1 with the sequence DLKPEN in cAMP-dependent protein kinase. This mutant resulted in a 15-fold enhancement of MKK1 activity [99] and was therefore attempted in Erk2 [103]. Erk2^{S151D} manifested MEK-independent activity that was about 15-fold higher than that of Erk2^{WT}, but was just 1.5% of that of MEK-phosphorylated Erk2^{WT} [73,103].

2.3. Only a Few of the Mutations Identified via Genetic Screens in ERK Orthologs of Evolutionarily Low Organisms Are Relevant to Mammalian Erks

High throughput screens in *S. cerevisiae* and *D. melanogaster*, provided a variety of gain-of-function mutants of Erk orthologs in these organisms [73,77,104,105] (Reviewed in [98]). These mutations allowed important insights into the modes of regulation of the given Erk ortholog in each case and to interesting cross talks between yeast MAPK pathways. The relevance of most of those to mammalian Erks is, however, unclear, because some of the mutations occurred in residues that are not conserved in the mammalian enzymes, and of the conserved residues only a handful were tested [73,103,106]. Overall, very few of the mutations turned out to be relevant to mammalian Erks, but these are of significant importance.

For example, insertion of the L73P mutation to Erk2 (equivalent to the L63P mutation in Fus3 [107]) rendered Erk2 intrinsically active, as tested by an in vitro kinase assay with purified recombinant proteins, but to an activity level of approximately 1% of the activity displayed by Mek-activated Erk2 [103]. Combining L73P with other mutations, such as S151D and D319N, created a more active enzyme [103]. It required a combination of three mutations, L73P+S151D+D319N to create an Erk2 protein with a MEK-independent activity that was 100-fold higher than that of wild type Erk2 in vitro. Notably however, this activity is just about 6% of the MEK-phosphorylated Erk2 activity [103]. Thus, these mutants are *bona fide* intrinsically active, but their activity is not very high. Interestingly Leu73 is part of the hydrophobic cluster affected by the "gatekeeper mutations". It seems that mutations in Ser151 also interfere with the contacts of the catalytic base Asp147 with Thr188, resulting in increased activation lip flexibility and activation of the phosphoacceptors Tyr185 and Thr183.

The only Erk mutant that has been shown so far to oncogenically transform cells in cultures, Erk1^{R84S}, was generated on the basis of a mutation in the yeast ortholog Mpk1/Slt2. The mutation in Mpk1/Slt2, R68S, was identified in a screen that looked for Mpk1 mutants that rescue the phenotype of cells lacking the relevant MEKs [73]. Six Mpk1 mutants were isolated, but only the R68S mutation was found relevant to Erks of higher eukaryotes, including of *Drosophila* and mammals [73]. *Drosophila* and mammalian Erks carrying the equivalent mutation (R80S in *Drosophila*'s ERK/Rolled; R84S in mammalian Erk1; R65S in mammalian Erk2) displayed high spontaneous intrinsic catalytic activity (>30% of the activity of Mek-activated Erk), independent of Mek activation in in vitro assays [73,82], in cell cultures [82] and in vivo, in transgenic mice and flies [83,108]. Furthermore, Erk1^{R84S} and Rolled^{R80S} were shown to function as oncogenes, capable of transforming NIH3T3 cells and to give rise to tumors in the fly, respectively [82,83]. Erk1^{R84S} was also shown to cause mild cardiac hypertrophy, when expressed as a transgene in the heart of mice [108]. The basic mechanism of action of the R80S/R84S/R65S mutation is similar to that of the other intrinsically active variants described above, namely, it bestowed upon the Rolled and Erk proteins an efficient autophosphorylation capability [73,82,83]. The structural basis for this capability is not clear, in part due to the extreme flexibility of Arg65 within the Erk2 structure. In many Erk2 structures it accommodates a different conformation (Figure 2).

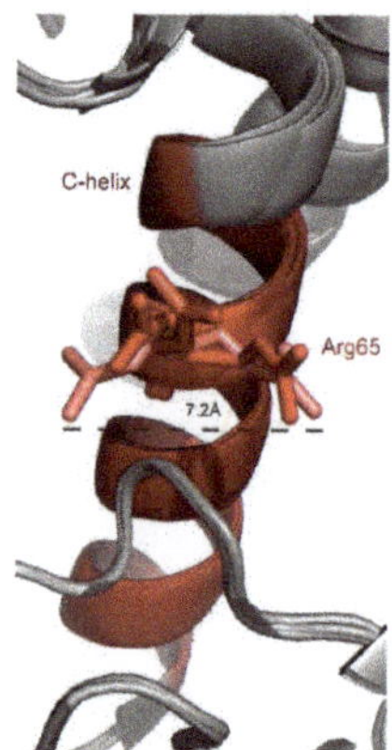

Figure 2. Unusual flexibility of the Arg65 residue at the αC-helix of different Erk2 crystal structures. 5 different crystal structures of Erk2 (PDB 1ERK, 4ERK, 4S31, 4GT3 and 5UMO) were superimposed and a zoom in into the αC-helix (colored in red) is presented. Arg65 is shown in sticks and the distance between the two extreme orientations is calculated.

Arg65 is located at a pivotal position in the αC-helix, the conserved helix within the N lobe, and interacts with the L16 domain, which is flexible and unstructured in the inactive form. Interestingly, in spite of the different conformation adopted by Arg65 in the various crystal structures of Erk2 (Figure 2), in many of them Arg65 is in association with amino acids of the conserved DFG motif

(D165, F166, G167). In the structure of Erk2^{WT}, PDB 1ERK, or 5UMO, Arg65 seems to have two possible conformations so that it forms a hydrogen bond with either the side chain of Asp165 or the backbone of Gly167. In the structure of Erk2^{R65S} (determined at a high resolution of 1.48Å (PDB 4SZ2)) the substituted serine is smaller than arginine and is not in association with the DFG motif. Instead, a new hydrogen bond is formed between Ser65 and Tyr34 of the P-loop (see Figure 8 in reference [82]). In addition, Ser65 stabilizes Thr183 from the activation loop and it is hypothesized that the Asp165 of the DFG motif can interact with ATP. It is noteworthy that in the crystal structure of dually phosphorylated Erk2 (PDB 2ERK), Thr183 interacts with the αC-helix, particularly with Arg65 via a water molecule [109].

Interestingly, similar to the case of Erk2^{R65S}, in the crystal structure of the intrinsically active Erk2^{I84A} (PDB code 4S30), a mutant that also autophosphorylates efficiently, the interaction of Arg65 with the DFG is also abolished, although the mutated residue is located in a distance from the αC-helix. As discussed above, the I84A mutation affects a hydrophobic cluster, involving, amongst other residues, Leu73 of the αC-helix, which may in turn divert Arg65. Also, in the Erk2^{I84A} structure in complex with AMP-PNP (PDB 4S34), a shift of Tyr34, causes it to form a hydrogen bond with Thr66 in addition to the Pi-Pi interactions with Tyr62 of the αC-helix observed in Erk2^{WT}. Tyr34 in Erk2 plays a pivotal role in catalysis and the DFG, especially Asp165, is involved in ATP binding by interacting with the gamma phosphate of ATP. Erk2 bearing mutations in Tyr34 (Y34H/N, in ERK1: Y53H) or Tyr62 (Y62N, in ERK1: Y81C) were shown to acquire resistance to the Erk inhibitors SCH77984 and VRT-11E in both ERK1 and ERK2 in two different screens (see below [95,96]). Finally, the association between Arg65 of Erk2 and the DFG is also abolished in other intrinsically active mutants, including mutations found in the CD site (PDB 6OT6) [110].

It is thus conceivable that the association of Arg65 with the DFG motif is crucial for blocking autophosphorylation activity by acting as a barrier between the ATP binding pocket, DFG motif and the activation loop. Tyr34 and Tyr62 of the P loop, which slightly change their conformation in several active mutants, may also play some role in suppressing spontaneous activity. Notably, mutations in Arg84 of Erk1, equivalent to Arg65 in Erk2, were identified in two cancer patients (Table 1A).

Intriguingly, another gain-of-function mutation, Y268C [73], was identified in the same genetic screen in yeast that provided the R68S mutation in Mpk1/Slt2. It was later shown that Y268A is also a gain-of-function mutation in Mpk1 [78]. Tyr268 is located at the heart of the DEF pocket, and in mammalian Erk2 a mutation in the equivalent Tyr, Y261A, is a partial loss-of-function mutation because Erk2^{Y261A} cannot bind and phosphorylate some of its substrates and cannot execute some of its biological functions [29,111,112]. As the mechanism through which Mpk1$^{Y268A/C}$ function as gain-of-function mutants is unknown, it is also unexplained why the equivalent mutations in Erk2 cause a loss of function [78].

Similar to the mammalian Erks, Rolled, the Erk ortholog in *Drosophila melanogaster*, is involved in numerous processes in the fly, including the development of the eye [23,24]. A screen aimed at isolation of mutants that facilitate proper eye development even in the absence of the ligand that activates the Erk pathway in the eye, identified a mutant fly that was termed *sevenmaker*. The mutation it carried was found to be D334N in the *Drosophila's* ERK/Rolled [77]. An equivalent mutation was isolated in another screen in yeast designed for identifying Fus3 molecules that are not inhibited by Hog1 [107], in a large-scale screen performed in mammalian cells for gain-of-function and inhibitor-resistant Erk2 mutants (D319 in Erk2) [95], and in cancer patients. As the *sevenmaker* mutation occurs at the CD site, its proposed mechanism of action is an elevated resistance to MAPK phosphatases resulting in increased sensitivity to low levels of MEK's activity [106]. This notion was based on the observation that the Erk2^{D319N} protein was less susceptible to inactivation by the MAPK phosphatase CL100 than the Erk2^{WT} recombinant protein, while both undergo inactivation in a dose-dependent manner [106]. Inserting this mutation into mammalian Erk2 showed minimal effect on catalytic activity in in vitro experiments and in cell culture [73,103,106]. It is not clear how the *sevenmaker* mutation affects Erk's association with phosphatases, but not with activators and substrates that also utilize the CD site.

Misiura et al. have shown, quantitatively, that the *sevenmaker* mutation increases the catalytic activity of Erk by changing the interaction energies. It specifically modifies the enzyme's susceptibility to deactivation by phosphatases, while not affecting the activation process by the MAPK kinase [113]. Nonetheless, the *sevenmaker* mutation may also affect catalysis *per se*. Indeed, recombinant Erk2^{D319N} does not manifest unusual catalytic properties, but when combining the D319N mutation with an activating mutation there is a dramatic elevation of catalysis [73,83,103]. Since Erk phosphatases do not exist in in vitro assays, the effect of the *sevenmaker* mutation should be explained by other mechanisms. Probably, in addition to reducing the affinity to phosphatases, the *sevenmaker* mutation also confers a conformational change that further stabilizes the "prone-to-autophosphorylate" conformation [69] induced by another mutation on the same protein. Supporting the role of the *sevenmaker* residue not only in substrate binding, but also in activation of catalysis, Molecular Dynamic simulations suggest that stabilization of the active conformation of Erk2, following phosphorylation of Thr183, is associated with disruption of several hydrogen bonding involving Asp334 [114]. Furthermore, in the structure of Erk2^{D319N} the interaction of Arg65 with the DFG is lost, a property of several of the autophosphorylating Erk mutants. Aside from this disruption, the crystal structure of Erk2^{D319N} is, essentially, indistinguishable from that of Erk2^{WT} [110].

The *sevenmaker* site seems to be a hot spot for mutations as it is being re-discovered in different screens. Mutations in Asp319 of Erk2 (D319N, D319V) or mutations in the neighboring residue (E320K and E320V), as well as in Glu79, which associates with D319, (E79K), were identified in a comprehensive screen that searched for gain-of-function and drug-resistant mutations, using A375 cells (see below; [95]). The *sevenmaker* mutation, itself, D319N, was also reported in four cases of carcinoma (COSMIC ID: D319N in ERK2—COSM98175) and three other patients carried another substitution in the 319 position (Table 1B). Interestingly, the ERK1 *sevenmaker* site was not found to be mutated in any screen, or in cancer patients.

2.4. Mutations in ERKs Are Very Rare in Cancer Patients, but Some Are Similar to Those Identified in Laboratory Models

Although several dozens of mutations in ERK1 and ERK2 have been identified in cancer patients (Table 1) the rate of mutations in ERK in patients is very low and most of the mutations appeared in only one of the samples tested. Mutations in ERK2 were observed in 179 out of 60,712 unique samples, approximately 0.3%. The COSMIC database lists 148 reported ERK1 mutations, out of 47,784 tumor samples tested (also approximately 0.3%). It is not clear whether these mutations have any causative effect on the malady. An exception is the E320K mutation, which was observed in 27 patients of squamous cell carcinoma (COSMIC ID: E320K in ERK2—COSM461148). Mutations in Glu320 were also identified in a screen for drug resistant Erk molecules. Notably Glu320 is neighboring Asp319 (the location of the *sevenmaker* mutation) within the CD site. Just like the *sevenmaker* mutation, the E320K does not affect the enzyme's intrinsic catalytic properties as tested in vitro with recombinant Erk2^{E320K}, and in transient transfections of HEK293 and NIH3T3 cells ([110]; Smorodinsky-Atias and Engelberg, unpublished observation). It may function, therefore, similar to the mutations in Asp319 by reducing the protein association with phosphatases (also supported by Brenan et al. [95]). Yet, unlike D319N, the E320K mutation enforces significant structural changes on the crystal structure of Erk2 and on its biophysical properties. Also, when equivalent mutations were inserted to the *Drosophila ERK/Rolled* they conferred different properties on the protein, suggesting that D319N and E320K function via different mechanisms [110,115]. More mutations of note identified in patients occurred in Glu79 (COSMIC ID: in ERK2—COSM444794), Ser140 (COSMIC ID: in ERK2—COSM3552430) and Pro56 (COSMIC ID: in ERK2—COSM4471756) residues. These substitutions were subsequently discovered as gain-of-function mutations in the screen that tested all possible missense mutations of ERK2 [95] (see below).

Nearly all of the ERK1 mutations recorded in patients, appeared only once. Only 18 mutations occurred in two or three independent samples. Notably, two patients carried the R84H mutation in

ERK1. As discussed above, another mutation at the same location, R84S, was shown to render Erk1 capable to oncogenically transform cells in culture [82]. This finding calls for testing the oncogenic potential of Erk1^{R84H} and perhaps of more Erk1 molecules in which Arg84 was substituted.

2.5. A Large Number of Mutations Can Render Erks Resistant to Pharmacological Inhibitors

The development of specific inhibitors towards Erks was the impetus for a series of studies that searched for mutations that cause drug resistance. Goetz et al., constructed a library of randomly mutagenized ERK1 and ERK2 cDNAs and induced its expression in A375 melanoma cells (harboring the BRAF^{V660E} oncogene) in the presence of either the Erk inhibitor VRT-11E, the MEK inhibitor trametinib, or with a combination of trametinib and the Raf inhibitor dabrafenib [96]. Overall, sequencing the ERK1 or ERK2 molecules in cell populations that survived the treatment identified 33 mutations in ERK1 (in 28 amino acids) and 24 in ERK2 (in 20 amino acids). In a separate screen for A375 colonies resistant to VRT-11E, another five mutations in ERK2 were discovered. All mutations are presented in Table 1. Only five of the mutations that caused resistance to VRT-11E were identified in both isoforms. These were (in ERK1/ERK2 order) Y53H/Y34H/N, G54A/G35S, P75L/P56L, Y81C/Y62N and C82Y/C63Y. A mutation at only one of the residues that caused resistance to the MEK inhibitor was also common in the two isoforms, Y148H/Y129N/H/F/C/S. Importantly, some of the residues that were found mutated in this screen were also reported to be mutated in patients, including Arg84 and Gly186 in Erk1 and Asp319 and Glu320 in Erk2. Another important finding of this study, with significant implications to therapeutic strategies, is that Erk variants that are resistant to RAF/MEK inhibitors are sensitive to Erk inhibition and vice versa [96].

As the mutations were not tested on purified Erk proteins it is not known how they affect the intrinsic biochemical and structural properties of the enzymes leaving the detailed mechanism of the acquired drug resistance open for future studies. It is possible however that mutations identified in the Erk-inhibitor screen interfere with drug binding as they cluster in proximity to the ATP/drug binding pocket. These mutations are located in the glycine-rich loop and the loop between β3 and αC-helix (in ERK1: I48N, Y53H, G54A, S74G, P75L. In ERK2: Y34H, G71S, P56L). Mutations residing in the αC-helix itself (ERK1: Y81C, C82Y, ERK2: Y62N, C63Y) may function similarly. ERK1 mutants that were identified in the Raf/MEK inhibitors screen probably function via a different mechanism. They are distributed along the molecule, but seem to cluster in domains important for catalysis and may render the kinase catalytically active. A206V/A187V and S219P/S200P (of ERK1/ERK2), for example, reside in the activation lip, R84H, C82Y and Q90R of Erk1 map to the αC-helix and Y148H of Erk1 and Y129F, D319G and E320K of Erk2 are located in the CD site. Mutations in the *sevenmaker* residue were already discussed above and probably function by reducing affinity to phosphatases in combination with some effect on catalysis. Another mutation, R84H in Erk1, occurred at the same residue in which the only oncogenic mutation identified so far in Erks, R84S, also occurred [82]. As Erk1^{R84S} was shown to be intrinsically catalytically active and spontaneously active when expressed in culture cells and in transgenic mice [82,108], it could be that Erk1^{R84H} also acquired similar properties and is independent of upstream activation, making it resistant to Raf and MEK inhibitors. Indeed, kinase assays using the various mutants, expressed in and immunoprecipitated from A375 cells, showed that the mutants maintained activity in the presence of either VRT-11E or SCH772984, another Erk inhibitor [96]. On the basis of the immunoprecipitation kinase assay it could be suggested that the mutants arising from the screen in the presence of RAF/MEK inhibitors acquired the capability to maintain sufficient kinase activity, even under these conditions, thereby rescuing the transformed cells from the inhibitors. As some of these mutations are found in patients, this conclusion is therapeutically relevant and may suggest that RAF and MEK inhibitors should be contraindicated for patients that harbor these mutants in the tumor.

Intriguingly, the mechanisms of action of the various mutants in vivo may be different for Erk1 and Erk2 as some of the mutations seem to be isoform-specific. Mutations in αC-helix were found only in ERK1 (C82Y, R84H and Q90R), while mutations in the CD site, such as D319G and E320K were only found in Erk2. Although the differences in the occurrence of the mutations could be a result of an unsaturated screen, the notion that the mutations are isoform-specific is supported by the situations observed in patients. The E320K mutation of Erk2 was found in 27 cancer patients while the equivalent mutation in Erk1 was not reported. Similarly, the *sevenmaker* site of Erk2, D319, which was mutated in seven patients, was not found so far to be mutated in ERK1 in patients. In line with these observations Goetz et al. inserted the equivalents of the D319G and E320K mutations into ERK1, and observed that the resulting proteins, Erk1^{D338N} and Erk1^{E339K} were not resistant to inhibitors, at least in the cell assay [96]. It is reasonable to conclude that different mutations may render Erk1 and Erk2 resistant to drugs. Given that the biological functions of the two isoforms is almost fully redundant [36] and that the inhibitors affect both isoforms similarly *in vitro*, it is currently difficult to explain the dichotomy in the mutations that cause resistance of each isoform.

A mutation that renders Erks resistant to SCH772984 was identified when cells of the colorectal cancer cell line HCT-116 (harboring a mutated KRAS) were serially passaged in the presence of increasing concentration of the inhibitor for 4 months [97]. Sequencing the ERK1/2 genes isolated from resistant clones, revealed a reoccurring point mutation, G186D, in ERK1 [97]. Gly186 resides in the DFG motif of the activation segment. The mutant displayed a several-fold reduction in binding affinity to the inhibitor, compared to the wild-type protein. Crystal structure of SCH772984-associated Erk2 suggests that this reduction is a direct result of a steric clash imposed by the aspartic acid in the active site, which destabilized the binding of the inhibitor. The same Erk1^{G186D} mutant did not provide resistance to another ATP-competitive Erk inhibitor (VRT-11E), an observation that is explained by the structural difference between the two inhibitors, significantly altering the interactions with the binding pocket, predominantly the distance of the molecule from the new aspartic acid [97]. Notably, the orthologous residue in Erk2, Gly167 was found mutated to Asp in a screen for Erk2 mutants that are resistant to SCH772984 and VRT-11E in A375 cells. In this screen Gly167 was mutated to many other residues, but only the Erk2^{G167D} was selected as rendering the kinase drug-resistance [95]. The interference of Asp at position 169 for drug binding seems very particular.

Erk2^{G167D} was in fact isolated in a large-scale screen that led to the discovery of many more mutants. In this effort, Brenan et al. employed saturation mutagenesis and were able to screen a library of 6810 variants of ERK2, out of 6821 possible, each carrying a point mutation. They searched for gain-of-function, loss-of-function, as well as for drug-resistant mutants. The ERK2 mutants library was introduced into A375 cells and inducibly expressed, under the premise that cells expressing gain-of-function mutants will proliferate slower, while cells expressing loss-of-function mutants will proliferate faster, than cells expressing Erk2^{WT} [95]. The relative abundance of Erk2 molecules that were expressed in the cells after 96 h was determined by parallel sequencing.

Indeed, mutants considered to be carrying GOF mutations on the basis of previous screens or appearance in tumors, such as Glu320 and Asp319, were depleted from the proliferative culture. Mutations in an additional 19 residues caused the Erk2 molecules to be depleted to a degree equivalent to or greater than that of Erk2^{Glu320} and Erk2^{Asp319}. The 19 residues were Glu79, Gly83, Gly8, Leu333, Pro56, Val47, Ser358, Val16, Pro354, Arg13, Glu320, Glu58, Ala3, Ala350, Ala7, Ser140, Thr24, Gln15, Asp319, Gly20 and Phe346 (from the most depleted to the least). Many of the mutants considered to have acquired gain-of-function mutations were shown to be catalytically active when immunoprecipitated from cells treated with the RAF inhibitor trametinib. As the mutants were not tested as recombinant purified proteins it is not clear whether they possess any unusual catalytic properties, and, specifically, if they are intrinsically active.

The mutants library was also inducibly expressed in A375 cells exposed to sublethal doses of VRT-11E, or SCH772984 in order to discover Erk2 mutants that are resistant to these inhibitors. The relative enrichment of the Erk2 mutants was quantified after 12 days of exposure to inhibitors. The rationale in this experiment was that cells harboring an Erk2 mutant insensitive to an inhibitor would allow proliferation in its presence. Mutations in 12 residues rendered Erk2 resistant to VRT-11E (Arg13, Ile29, Gly30, Ala33, Met36, Val37, Arg65, Gln95, Met96, Asp98, Thr108, Leu154), mutations in

9 residues confer resistance to SCH772984 (Glu31, Tyr41, Val47, Lys53, Glu68, Leu67, Ile101, Asp122, Asp125) and mutations in 18 residues render Erk2 resistant to both inhibitors (Tyr34, Gly35, Cys38, Asp42, Val49, Ile54, Ser55, Pro56, Phe57, Gln60, Tyr62, Cys63, Thr66, Glu69, Leu73, Gln103, Gly167, Ile345). Curiously, half of the mutations that cause resistant to VRT-11E occurred in residues that make direct contacts with the inhibitor, but none of the mutants that confer specific resistance to SCH772984 are mutated in residues that contact the inhibitor. Projection of the mutations identified in this study into cancer associated mutations showed that within the top 20 residues harboring GOF mutations, five were reported to be mutated in patients (E320K/V, D319N/V, E79K, P56L and S140L). Erk2 mutants carrying any of these GOF mutations were found to rescue A375 cells from the anti-proliferative effect of Raf/MEK inhibitors, but only one, Erk2$^{P56L/G}$ was able to rescue the cells from SCH772984. All GOF mutations clustered within the CD site, whereas LOF mutations altered the DEF pocket. The LOF mutations in the DEF pocket seem dominant since when combined, on a single Erk2 molecule, with a GOF mutation in the CD site the resulting protein was not able to rescue cells and promote the downstream signaling. It seems that the mechanism of action of the GOF mutants residing in the CD site is similar to that of the *sevenmaker* mutation. This hypothesis was validated by co-expressing Erk2 mutants with BRAF^{V600E} and dual specificity phosphatase (DUSP) in 293T cells and monitoring Erk2 phosphorylation. Erk2 molecules mutated at the CD site (E320K/V; D319N/V E79K) exhibited sustained phosphorylation levels relative to Erk2^{WT} [95]. It is worth noting that the GOF mutant P56L showed a reduced level of phosphorylation, suggesting that constant phosphorylation in the presence of DUSP is not a general property of all GOF mutations. Most mutants identified in this comprehensive study await further biochemical, structural and pharmacological analysis.

3. Discussion

Table 1 lists an impressive number of mutations in Erk1 and Erk2, many of them discovered in screens for drug-resistant mutants. The mutations that confer drug resistance outscore the very few mutations that render Erks intrinsically active and the single mutant that was shown so far to possess oncogenic properties. The mutations causing drug-resistant occur throughout the protein so that various mechanisms are involved. It is not clear whether the list in Table 1 reflects all relevant mutations possible in Erks. Indeed, most of the screens that provided the currently known mutations were high throughput, but they were performed in a particular context of a given cell line, or against specific pharmacological inhibitors. Also, it is not clear if these screens themselves were saturated. It is a plausible assumption therefore that further screening, particularly in novel experimental setups, would result in yet unidentified mutations. Also, once Erk inhibitors reach the clinic, patients may develop resistance by acquiring mutations other than those discovered in the laboratory. As the research of Erk mutations is relatively young many properties of the known mutation are still elusive. For most of the mutations that cause resistant to inhibitors the mechanism of action is not known and for many of them even the effects on Erks' conformation and catalytic properties are yet to be revealed. Finally, it may take a long time to reveal the role (if any) of the rare mutations identified in cancer patients, in disease etiology. Most mutations that were identified in a single patient, may be bystanders with no role in the disease, while others may contribute to disease development or even disease onset. Prime suspects for a causative role are the R84H mutation in Erk1, because an equivalent mutation, R84S, was shown to be oncogenic in cells in culture, and E320K in Erk2 that is found in patients in a higher rate than all other mutations.

3.1. Mutations Identified in Erks Could Not Have Been Predicted by Structural or Mechanistic Analysis

It is not currently possible to predict the location and type of more Erk mutations. Although Erk proteins have been the subject of comprehensive structural studies, including via X-ray crystallography, NMR and HX-MS approaches, and although critical features of their structure-function properties have been revealed, it is not clear how to translate this knowledge into predicting specific mutations that will modify Erks' biochemical or pharmacological properties or biological functions. Also, modifying

Erks' biological effects requires understanding of additional mechanisms, responsible for sub-cellular localization and interaction with partners and scaffold proteins. Some of these mechanisms could be isoform-specific and, consequently, mutations that affect these processes may be different in Erk1 and Erk2, similar to the case of the CD site mutations that seem relevant only to Erk2 and mutation in the αC-helix, which are more relevant to Erk1. As a result of our inability to translate structural and mechanistic knowledge into mutation design, most mutations identified so far were discovered via unbiased screens, and even following their isolation, their mechanism of action is not understood.

3.2. Mutations Identified in Erks Disclose 3 Hotspots for Mutagenesis, Perhaps Reflecting Some Prevailing Mechanisms

Although mutations discussed in this review are spread along the Erk molecules, some hotspots re-appear in several laboratory screens as well as in cancer patients. The *sevenmaker* residue and its neighbor at the CD site, Glu320, are prominent examples, relevant for Erk2. αC-helix, mainly Arg84/Arg65 (in ERK1/ERK2), is another hotspot, which seems relevant primarily to Erk1. Yet another important residue is the gatekeeper that was discovered as a target for mutagenesis by studying the evolution of Erks and by structural approaches. Mutations that occur in the CD site affect protein-protein interactions, primarily with phosphatases, while mutations in the gatekeeper, in the residues that are in proximity to it and in residues of the αC-helix, significantly increase the commonly negligible intrinsic catalytic activity of the kinases. All mutations that caused elevation of the basal catalytic activity did so via identical mechanism, increasing the autocatalytic capability of the proteins. So far, no mutation was discovered that allows Erks to bypass the requirement of activation loop phosphorylation and enforces by itself adoption of the active conformation. Rather, the mutations discovered so far caused the Erk molecule to acquire a 'prone to autophosphorylate' conformation. Erks may be immune against mutations that induce the active conformation by themselves.

4. Conclusions

Clinical and Biochemical Lessons from the Erk Mutant

Most of the mutations described in this review seem to be directly relevant to understanding cancer etiology and patients response to drugs. An important lesson is that Erks, at least Erk1, could become oncogenic. This finding strongly suggests that the oncogenicity of the RTK-Ras-Raf-MEK pathway is mediated primarily via Erks, reinforcing the effort to inhibit Erk as a powerful anti-cancer approach. The analysis of the mutants may further suggest that isoform-specific inhibitors should be developed with higher priority to Erk1-specific inhibitors. This would be a difficult challenge for drug developers. The mutations that cause drug resistance could be already taken into consideration when a therapeutic strategy is planned. Namely, drugs should be applied according to the mutation that appears in the tumor. This requires deep understanding of the effect of each of the many mutations discovered so far (Table 1) on Erks' biochemistry, biology and pathology.

Table 1. (**A**) Mutations identified in ERK1 (MAPK3; mutations numeration is according to the sequence of the human ERK1). (**B**) Mutations identified in ERK2 (MAPK1) (numeration of the mutants is according to the sequence of rat ERK2).

Mutation	Mode of Identification	Reference
	(A)	
A6_Q7insA	In cancer patients	COSMIC, cBioPortal
A6dup	In cancer patients	COSMIC
Q7R	In cancer patients	COSMIC
Q7H	In cancer patients	COSMIC, cBioPortal
G10R	In cancer patients	COSMIC
G11V	In cancer patients	COSMIC
G11_Splice	In cancer patients	COSMIC, cBioPortal
G12E	In cancer patients	COSMIC
E13K	In cancer patients	COSMIC, cBioPortal
E13*	In cancer patients	COSMIC, cBioPortal
R15G	In cancer patients	COSMIC, cBioPortal
R16I	In cancer patients	COSMIC, cBioPortal
E18Q	In cancer patients	COSMIC, cBioPortal, TumorPortal
V20V	In cancer patients	COSMIC
G23G	In cancer patients, and in a screen for mutants resistant to Erk inhibitors	COSMIC, [96]
V24F	In cancer patients	COSMIC, cBioPortal
V24S	In cancer patients	cBioPortal
V24fs*8	In cancer patients	cBioPortal
P25S	In cancer patients	COSMIC, cBioPortal
E27fs*35	In cancer patients	COSMIC, cBioPortal
E27G	In cancer patients	COSMIC
M30I	In cancer patients	COSMIC, cBioPortal

Table 1. *Cont.*

Mutation	Mode of Identification	Reference
G33W	In cancer patients	cBioPortal
P35S	In cancer patients	COSMIC
D37D	In cancer patients	COSMIC, TumorPortal
Q46H	In cancer patients	cBioPortal
I48N	In a screen for mutants resistant to VRT-11E	[96]
G51S	In a screen for mutants resistant to RAF/MEK inhibitors	[96]
A52P	In cancer patients	COSMIC, cBioPortal
Y53Y	In cancer patients	COSMIC
Y53H	In a screen for mutants resistant to VRT-11E and SCH772984	[96]
Y53C	In a screen for mutants resistant to Erk inhibitors	[96]
G54S	In a screen for mutants resistant to VRT-11E and SCH772984	[96]
G54C	In cancer patients	cBioPortal
G54A	In a screen for mutants resistant to Erk inhibitors	[96]
S57G	In a screen for mutants resistant to RAF/MEK inhibitors	[96]
X57_splice	In cancer patients	cBioPortal
S58L	In cancer patients	cBioPortal, TumorPortal
Y60C	In a screen for mutants resistant to Erk inhibitors	[96]
D61E	In cancer patients	COSMIC, cBioPortal
H62_S74>R	In cancer patients	COSMIC, cBioPortal
V63M	In cancer patients	COSMIC
R64C	In cancer patients	COSMIC, cBioPortal
R64L	In cancer patients	cBioPortal
K65R	In a screen for mutants resistant to Erk and RAF/MEK inhibitors	[96]
R67C	In cancer patients	COSMIC

Table 1. *Cont.*

Mutation	Mode of Identification	Reference
V68L	In cancer patients	cBioPortal
K72N	In cancer patients	COSMIC, cBioPortal
K72R	In a screen for mutants resistant to Erk inhibitors	[96]
I73S	In a screen for mutants resistant to Erk inhibitors	[96]
I73M	In Cancer patients	COSMIC, cBioPortal
S74G	In a screen for mutants resistant to VRT-11E	[96]
Insertion 74N	An activating mutation. On the basis of inferred ancestor	[75]
P75P	In Cancer patients	TumorPortal
P75L	In a screen for mutants resistant to VRT-11E and SCH772984	[96]
P75S	In a screen for mutants resistant to VRT-11E and SCH772984	[96]
E77E	In cancer patients, and in a screen for mutants resistant to Erk inhibitors	COSMIC, cBioPortal, TumorPortal, [96]
Q79H	In cancer patients	cBioPortal
Q79*	In cancer patients	COSMIC, cBioPortal, TumorPortal
Y81C	In a screen for mutants resistant to VRT-11E	[96]
C82Y	In cancer patients	cBioPortal, [96]
R84H	In cancer patients, and in a screen for mutants resistant to trametinib and dabrafenib	COSMIC, cBioPortal,[96]
R84S	In a screen for mutants resistant to RAF/MEK inhibitors, and in a screen for MEK-independent mutants	[73,96]
T85T	In cancer patients	COSMIC
L86L	In cancer patients	COSMIC
L86R	In a screen for mutants resistant to RAF/MEK inhibitors	[96]
L86P	In a screen for mutants resistant to RAF/MEK inhibitors	[96]
R87W	In cancer patients	COSMIC, cBioPortal

Table 1. *Cont.*

Mutation	Mode of Identification	Reference
Q90R	In a screen for mutants resistant to trametinib and dabrafenib	[96]
R94C	In cancer patients	COSMIC, cBioPortal, TumorPortal
R96H	In cancer patients	COSMIC, cBioPortal
R96C	In cancer patients	cBioPortal
R96R	In cancer patients	COSMIC
H97H	In cancer patients	COSMIC
E98K	In cancer patients	COSMIC, cBioPortal
E98D	In cancer patients	COSMIC, cBioPortal
I101I	In cancer patients	COSMIC
G102D	In cancer patients	COSMIC, cBioPortal
I103A	On the basis of structural considerations	[76]
R104Q	In cancer patients	cBioPortal
L107L	In cancer patients	TumorPortal
R108W	In cancer patients	COSMIC, cBioPortal
A109P	In cancer patients	COSMIC
Q122M	An activating mutation. On the basis of inferred ancestor	[75]
L124L	In cancer patients	COSMIC
E126*	In cancer patients	COSMIC, cBioPortal, TumorPortal
S135N	In a screen for mutants resistant to RAF/MEK inhibitors	[96]
Q136fs*4	In cancer patients	cBioPortal
Q136*	In cancer patients	cBioPortal
Q136H	In cancer patients	cBioPortal
L138L	In cancer patients	COSMIC

Table 1. *Cont.*

Mutation	Mode of Identification	Reference
H142H	In cancer patients	COSMIC
Y145*	In cancer patients	COSMIC
Y148H	In a screen for mutants resistant to trametinib and dabrafenib	[96]
Y148C	In cancer patients	COSMIC, cBioPortal
R152L	In cancer patients	TumorPortal
R152W	In cancer patients	COSMIC, cBioPortal, TumorPortal
G153S	In cancer patients	COSMIC, cBioPortal
G153G	In cancer patients	COSMIC
H158L	In cancer patients	COSMIC
S159S	In cancer patients	COSMIC, TumorPortal
A160T	In cancer patients	COSMIC, cBioPortal
A160A	In cancer patients	COSMIC
L163L	In cancer patients	COSMIC
R165L	In cancer patients	COSMIC
P169L	In cancer patients	cBioPortal
N171D	In cancer patients	COSMIC, cBioPortal
T177I	In cancer patients	COSMIC, cBioPortal
C178R	In cancer patients	COSMIC
C178Y	In cancer patients	COSMIC
C178C	In cancer patients	COSMIC
D179N	In cancer patients	COSMIC
I182N	In cancer patients	cBioPortal
I182-splice	In cancer patients	TumorPortal
F185F	In cancer patients	COSMIC

Table 1. *Cont.*

Mutation	Mode of Identification	Reference
F185I	In cancer patients	cBioPortal
G186R	In cancer patients	COSMIC, cBioPortal, TumorPortal
G186D	In a screen for mutants resistant to VRT-1E and SCH772984	[96,97]
R189W	In cancer patients	COSMIC, cBioPortal
R189Q	In cancer patients	COSMIC
R189R	In cancer patients	COSMIC
I190T	In cancer patients	COSMIC, cBioPortal
P193T	In cancer patients	COSMIC, cBioPortal
P193H	In cancer patients	COSMIC, cBioPortal
P193S	In cancer patients	cBioPortal, TumorPortal
E194Q	In cancer patients	COSMIC, cBioPortal
T198T	In cancer patients	COSMIC
G199D	In cancer patients	COSMIC, cBioPortal
T202M	In cancer patients	COSMIC, cBioPortal
E203K	In cancer patients	cBioPortal
A206V	In a screen for mutants resistant to trametinib and dabrafenib	[96]
T207T	In cancer patients	COSMIC, TumorPortal
R211Q	In cancer patients	COSMIC, cBioPortal
R211P	In cancer patients	COSMIC, cBioPortal
R211W	In cancer patients	cBioPortal
E214D	In cancer patients	cBioPortal
M216I	In cancer patients, and in a screen for mutants resistant to RAF/MEK inhibitors	COSMIC, cBioPortal,[96]
N218N	In cancer patients	COSMIC

Table 1. *Cont.*

Mutation	Mode of Identification	Reference
S219F	In cancer patients	COSMIC, cBioPortal
S219P	In a screen for mutants resistant to trametinib and dabrafenib	[96]
X221_splice	In cancer patients	cBioPortal
D227N	In cancer patients	cBioPortal
V231L	In cancer patients	COSMIC
A236T	In cancer patients	COSMIC
S240C	In cancer patients	COSMIC
R242R	In cancer patients	TumorPortal
L251L	In cancer patients	COSMIC
Q253P	In cancer patients	COSMIC, cBioPortal
I257V	In cancer patients	COSMIC, cBioPortal
I260H	In cancer patients	COSMIC, cBioPortal
Q266*	In cancer patients	COSMIC, cBioPortal
L269P	In cancer patients	COSMIC, cBioPortal, TumorPortal
I273M	In Cancer patients	cBioPortal
R278*	In cancer patients, and in a screen for mutants resistant to RAF/MEK inhibitors	cBioPortal, [96]
L281I	In cancer patients	cBioPortal
V290A	In cancer patients	COSMIC, cBioPortal, TumorPortal
F296F	In cancer patients	COSMIC
D300E	In cancer patients	COSMIC, cBioPortal
A303V	In cancer patients, and in a screen for mutants resistant to RAF/MEK inhibitors	COSMIC, cBioPortal, [96]
L304P	In Cancer patients	cBioPortal
L306L	In Cancer patients, and in a screen for mutants resistant to Erk inhibitors	cBioPortal, [96]
T312S	In cancer patients	cBioPortal

Table 1. *Cont.*

Mutation	Mode of Identification	Reference
N314I	In cancer patients	COSMIC, cBioPortal, TumorPortal
N314N	In cancer patients	COSMIC
P315H	In cancer patients	COSMIC
R318W	In cancer patients	COSMIC, cBioPortal
R318R	In cancer patients	COSMIC
V321fs*40	In cancer patients	cBioPortal
V321W	In cancer patients	cBioPortal
P328L	In cancer patients	COSMIC, cBioPortal
E331E	In cancer patients	COSMIC
Q332H	In cancer patients	COSMIC
D335N	In cancer patients	COSMIC, cBioPortal
P336Q	In cancer patients	COSMIC
T337T	In cancer patients	COSMIC
E339V	In cancer patients	COSMIC, cBioPortal
E343K	In cancer patients	cBioPortal
P345T	In cancer patients	COSMIC
F346I	In cancer patients	COSMIC, cBioPortal
F346F	In cancer patients	COSMIC
F348F	In cancer patients	COSMIC
A349T	In cancer patients	cBioPortal
R359W	In cancer patients	COSMIC, cBioPortal, TumorPortal
R359Q	In cancer patients	COSMIC
R359L	In cancer patients	COSMIC, cBioPortal

Table 1. *Cont.*

Mutation	Mode of Identification	Reference
E362K	In cancer patients	COSMIC, cBioPortal
E362*	In cancer patients	COSMIC, TumorPortal
F365C	In cancer patients	cBioPortal
Q366H	In cancer patients	cBioPortal
E367D	In cancer patients	COSMIC
P373P	In cancer patients	COSMIC, TumorPortal
G374*	In cancer patients	COSMIC
G374K	In cancer patients	COSMIC, cBioPortal
L376R	In cancer patients	COSMIC, cBioPortal
A378G	In cancer patients	COSMIC, cBioPortal
(B)		
A2V	In cancer patients	COSMIC, cBioPortal
A6_A7delAA	In cancer patients	COSMIC
A5delA	In cancer patients	COSMIC
A7T	In cancer patients	COSMIC, cBioPortal
G8S	In cancer patients	COSMIC, cBioPortal
R13P	In a screen for mutants that are resistant to VRT-11E and SCH772984	[95]
Q15Q	In cancer patients	COSMIC
P21S	In cancer patients	COSMIC, cBioPortal
I29M	In Cancer patients, and in a screen for mutants that are resistant to VRT-11E	COSMIC, cBioPortal,[95]
I29Q	In a screen for mutants that are resistant to VRT-11E	[95]
I29R	In a screen for mutants that are resistant to VRT-11E	[95]
I29L	In a screen for mutants that are resistant to VRT-11E	[95]
I29E	In a screen for mutants that are resistant to VRT-11E	[95]

Table 1. *Cont.*

Mutation	Mode of Identification	Reference
I29K	In a screen for mutants that are resistant to VRT-11E	[95]
I29H	In a screen for mutants that are resistant to VRT-11E	[95]
I29Y	In a screen for mutants that are resistant to VRT-11E	[95]
I29D	In a screen for mutants that are resistant to VRT-11E	[95]
I29C	In a screen for mutants that are resistant to VRT-11E	[95]
I29W	In a screen for mutants that are resistant too VRT-11E	[95]
I29N	In a screen for mutants that are resistant to VRT-11E	[95]
G30P	In a screen for mutants that are resistant to VRT-11E	[95]
E31P	In a screen for mutants that are resistant to SCH772984	[95]
E31Q	In cancer patients	COSMIC, cBioPortal
G32D	In cancer patients	COSMIC, cBioPortal
G32C	In cancer patients	cBioPortal
A33N	In a screen for mutants that are resistant to VRT-11E	[95]
Y34Y	In cancer patients	COSMIC
Y34H	In a screen for mutants that are resistant to VRT-11E and SCH772984	[95,96]
Y34V	In a screen for mutants that are resistant to VRT-11E and SCH772984	[95]
Y34T	In a screen for mutants that are resistant to VRT-11E and SCH772984	[95]
Y34Q	In a screen for mutants that are resistant to VRT-11E and SCH772984	[95]
Y34G	In a screen for mutants that are resistant to VRT-11E and SCH772984	[95]
Y34S	In a screen for mutants that are resistant to VRT-11E and SCH772984	[95]
Y34C	In a screen for mutants that are resistant to VRT-11E and SCH772984	[95]
Y34I	In a screen for mutants that are resistant to VRT-11E and SCH772984	[95]
Y34D	In a screen for mutants that are resistant to VRT-11E and SCH772984	[95]
Y34R	In a screen for mutants that are resistant to VRT-11E and SCH772984	[95]

Table 1. *Cont.*

Mutation	Mode of Identification	Reference
Y34N	In a screen for mutants that are resistant to VRT-11E and SCH772984	[95,96]
Y34L	In a screen for mutants that are resistant to VRT-11E and SCH772984	[95]
Y34M	In a screen for mutants that are resistant to VRT-11E and SCH772984	[95]
G35D	In a screen for mutants that are resistant to VRT-11E and SCH772984	[95]
G35T	In a screen for mutants that are resistant to VRT-11E and SCH772984	[95]
G35K	In a screen for mutants that are resistant to VRT-11E and SCH772984	[95]
G35S	In a screen for mutants that are resistant to VRT-11E and SCH772984	[95,96]
G35A	In a screen for mutants that are resistant to VRT-11E and SCH772984	[95]
G35N	In a screen for mutants that are resistant to VRT-11E	[95]
G35P	In a screen for mutants that are resistant to VRT-11E	[95]
G35C	In a screen for mutants that are resistant to VRT-11E and SCH772984	[95]
M36P	In a screen for mutants that are resistant to VRT-11E	[95]
V37A	In a screen for mutants that are resistant to VRT-11E	[95]
C38P	In a screen for mutants that are resistant to VRT-11E and SCH772984	[95]
C38R	In a screen for mutants that are resistant to SCH772984	[95]
Y41W	In a screen for mutants that are resistant to SCH772984	[95]
Y41E	In a screen for mutants that are resistant to SCH772984	[95]
D42H	In a screen for mutants that are resistant to VRT-11E and SCH772984	[95]
V44F	In cancer patients	cBioPortal
V47A	In a screen for mutants that are resistant to SCH772984	[95]
R48*	In cancer patients	COSMIC, cBioPortal
V49K	Causes resistance to SCH772984	[95]
V49H	In a screen for mutants that are resistant to VRT-11E and SCH772984	[95]
A50S	In Cancer patients	COSMIC, TumorPortal

Table 1. *Cont.*

Mutation	Mode of Identification	Reference
K53G	In a screen for mutants that are resistant to SCH772984	[95]
I54H	In a screen for mutants that are resistant to SCH772984	[95]
I54D	In a screen for mutants that are resistant to SCH772984	[95]
54W	In a screen for mutants that are resistant to SCH772984	[95]
I54K	In a screen for mutants that are resistant to SCH772984	[95]
I54Y	In a screen for mutants that are resistant to SCH772984	[95]
I54E	In a screen for mutants that are resistant to SCH772984	[95]
I54Q	In a screen for mutants that are resistant to VRT-11E and SCH772984	[95]
I54S	In a screen for mutants that are resistant to VRT-11E and SCH772984	[95]
I54G	In a screen for mutants that are resistant to VRT-11E and SCH772984	[95]
I54P	In a screen for mutants that are resistant to VRT-11E	[95]
S55P	In a screen for mutants that are resistant to VRT-11E	[95]
S55G	In a screen for mutants that are resistant to VRT-11E	[95]
S55F	In a screen for mutants that are resistant to VRT-11E and SCH772984	[95]
P56L	In cancer patients, In a screen for mutants that are resistant to VRT-11E and SCH772984	COSMIC, cBioPortal, [95,96]
P56S	In a screen for mutants that are resistant to VRT-11E and SCH772984	[95,96]
P56W	In a screen for mutants that are resistant to VRT-11E and SCH772984	[95]
P56R	In a screen for mutants that are resistant to VRT-11E and SCH772984	[95]
P56K	In a screen for mutants that are resistant to VRT-11E and SCH772984	[95]
P56A	In a screen for mutants that are resistant to VRT-11E and SCH772984	[95]
P56M	In a screen for mutants that are resistant to VRT-11E and SCH772984	[95]
P56N	In a screen for mutants that are resistant to VRT-11E and SCH772984	[95]

Table 1. *Cont.*

Mutation	Mode of Identification	Reference
P56G	In a screen for mutants that are resistant to VRT-11E and SCH772984	[95]
P56Y	In a screen for mutants that are resistant to VRT-11E and SCH772984	[95]
P56F	In a screen for mutants that are resistant to VRT-11E and SCH772984	[95]
P56Q	In a screen for mutants that are resistant to VRT-11E	[95]
P56V	In a screen for mutants that are resistant to VRT-11E	[95]
P56T	In a screen for mutants that are resistant to VRT-11E	[95,96]
P56I	In a screen for mutants that are resistant to VRT-11E	[95]
F57G	In a screen for mutants that are resistant to VRT-11E	[95]
F57P	In a screen for mutants that are resistant to VRT-11E	[95]
F57S	In a screen for mutants that are resistant to VRT-11E and SCH772984	[95]
F57R	In a screen for mutants that are resistant to VRT-11E and SCH772984	[95]
E58Q	In a screen for mutants that are resistant to SCH772984	[95]
E58S	In a screen for mutants that are resistant to SCH772984	[95]
Q60P	In a screen for mutants that are resistant to VRT-11E and SCH772984	[95]
T61T	In cancer patients	COSMIC
T61I	In cancer patients	COSMIC
Y62G	In a screen for mutants that are resistant to VRT-11E	[95]
Y62E	In a screen for mutants that are resistant to VRT-11E	[95]
Y62D	In a screen for mutants that are resistant to VRT-11E	[95]
Y62S	In a screen for mutants that are resistant to VRT-11E	[95]
Y62C	In a screen for mutants that are resistant to VRT-11E	[95]
Y62T	In a screen for mutants that are resistant to VRT-11E	[95]
Y62Q	In a screen for mutants that are resistant to VRT-11E	[95]
Y62A	In a screen for mutants that are resistant to VRT-11E	[95]

Table 1. *Cont.*

Mutation	Mode of Identification	Reference
Y62P	In a screen for mutants that are resistant to VRT-11E	[95]
Y62V	In a screen for mutants that are resistant to VRT-11E and SCH772984	[95]
Y62M	In a screen for mutants that are resistant to VRT-11E and SCH772984	[95]
Y62K	In a screen for mutants that are resistant to VRT-11E and SCH772984	[95]
Y62I	In a screen for mutants that are resistance to VRT-11E and SCH772984	[95]
Y62L	In a screen for mutants that are resistance to VRT-11E and SCH772984	[95]
Y62R	In a screen for mutants that are resistant to VRT-11E and SCH772984	[95]
Y62N	In a screen for mutants that are resistant to VRT-11E	[95,96]
C63F	In a screen for mutants that are resistant to VRT-11E	[95]
C63W	In a screen for mutants that are resistant to VRT-11E	[95]
C63fs*3	In Cancer patients	COSMIC, cBioPortal
C63Y	In a screen for mutants that are resistant to VRT-11E and SCH772984	[95,96]
R65I	In cancer patients, and in a screen for mutants that are resistant to VRT-11E	COSMIC, cBioPortal,[95]
R65K	In a screen for mutants that are resistant to Erk inhibitors and RAF/MEK inhibitors	[96]
R65S	Genetic screen for Mpk1 intrinsically active mutants	[73]
T66T	In Cancer patients	COSMIC
T66M	In a screen for mutants that are resistant to VRT-11E	[95,96]
T66Q	In a screen for mutants that are resistant to VRT-11E	[95,96]
T66F	In a screen for mutants that are resistant to VRT-11E	[95,96]
T66I	In a screen for mutants that are resistant to VRT-11E	[95,96]
T66L	In a screen for mutants that are resistant to VRT-11E	[95,96]
T66P	In a screen for mutants that are resistant to VRT-11E	[95,96]

Table 1. *Cont.*

Mutation	Mode of Identification	Reference
T66D	In a screen for mutants that are resistant to VRT-11E and SCH772984	[95,96]
T66Y	In a screen for mutants that are resistant to VRT-11E and SCH772984	[95,96]
T66H	In a screen for mutants that are resistant to VRT-11E	[95,96]
T66N	In a screen for mutants that are resistant to VRT-11E and SCH772984	[95,96]
L67L	In cancer patients	COSMIC
R68R	In cancer patients	COSMIC
E69P	In a screen for mutants that are resistant to VRT-11E	[95,96]
E69C	In a screen for mutants that are resistant to VRT-11E	[95,96]
E69G	In a screen for mutants that are resistant to VRT-11E	[95,96]
E69K	In a screen for mutants that are resistant to VRT-11E and SCH772984	[95,96]
E69A	In a screen for mutants that are resistant to VRT-11E and SCH772984	[95,96]
I72fs*8	In cancer patients	COSMIC
L73E	In a screen for mutants that are resistant to VRT-11E	[95,96]
L73H	In a screen for mutants that are resistant to VRT-11E	[95,96]
L73R	In a screen for mutants that are resistant to VRT-11E	[95,96]
L73W	In a screen for mutants that are resistant to VRT-11E and SCH772984	[95]
L73P	In a screen for mutants that are resistant to VRT-11E, and on the basis of genetic screen for intrinsically active FUS3	[95,103]
R75C	In cancer patients	COSMIC, cBioPortal
R77S	In cancer patients	cBioPortal
R77K	In cancer patients	COSMIC, cBioPortal
E79K	In cancer patients	COSMIC, cBioPortal, TumorPortal
N80fs*18	In cancer patients	COSMIC, cBioPortal
I82T	In cancer patients	COSMIC, cBioPortal
I82A	Activating mutation. On the basis of structural considerations	[76]

Table 1. *Cont.*

Mutation	Mode of Identification	Reference
I84A	Activating mutation. On the basis of structural considerations	[76]
D86-del	In cancer patients	cBioPortal
I88F	In cancer patients	COSMIC
I93I	In cancer patients	COSMIC, TumorPortal
Q95R	In a screen for mutants that are resistant to VRT-11E	[95]
M96W	In a screen for mutants that are resistant to VRT-11E	[95]
M96I	In cancer patients	cBioPortal
D98N	In cancer patients	cBioPortal
D98M	In a screen for mutants that are resistant to VRT-11E	[95]
I101Q	In a screen for mutants that are resistant to SCH772984	[95]
I101W	In a screen for mutants that are resistant to SCH772984	[95]
I101Y	In a screen for mutants that are resistant to SCH772984	[95]
I101R	In cancer patients	COSMIC
V102V	In a screen for mutants that are resistant to Erk inhibitors and RAF/MEK inhibitor	[96]
Q103A	Generated in order to study the biological effect the gatekeeper residue	[76]
Q103G	Generated in order to study the biological effect the gatekeeper residue	[76]
Q103I	In a screen for mutants that are resistant to SCH772984	[95]
Q103F	In a screen for mutants that are resistant to SCH772984	[95]
Q103T	In a screen for mutants that are resistant to SCH772984	[95]
Q103W	In a screen for mutants that are resistant to VRT-11E and SCH772984	[95]
Q103V	In a screen for mutants that are resistant to VRT-11E and SCH772984	[95]
Q103N	In a screen for mutants that are resistant to VRT-11E	[95]
Q103Y	In a screen for mutants that are resistant to VRT-11E and SCH772984	[95]
D104D	In a screen for mutants that are resistant to Erk inhibitors	[96]

Table 1. *Cont.*

Mutation	Mode of Identification	Reference
D104H	In cancer patients	cBioPortal
T108P	In a screen for mutants that are resistant to VRT-11E	[95]
L110R	In cancer patients	cBioPortal
L114S	In cancer patients	cBioPortal
L119I	In cancer patients	COSMIC, cBioPortal, TumorPortal
D122T	In a screen for mutants that are resistant to VRT-11E and SCH772984	[95]
I124F	In cancer patients	COSMIC
C125I	In a screen for mutants that are resistant to VRT-11E and SCH772984	[95]
Y129N	In a screen for mutants that are resistant to RAF/MEK inhibitors	[96]
Y129H	In a screen for mutants that are resistant to RAF/MEK inhibitors	[96]
Y129F	In a screen for mutants that are resistant to trametinib and dabrafenib	[96]
Y129C	In Cancer patients, and In a screen for mutants that are resistant to trametinib and dabrafenib	COSMIC, cBioPortal, [96]
Y129S	In a screen for mutants that are resistant totrametinib and dabrafenib	[96]
Q130E	In cancer patients	COSMIC, cBioPortal
L132P	In cancer patients	COSMIC, cBioPortal
R133K	In cancer patients	COSMIC, cBioPortal
G134E	In cancer patients	COSMIC, cBioPortal
I138I	In cancer patients	COSMIC
H139Y	In cancer patients	COSMIC, cBioPortal, TumorPortal
H139R	In cancer patients	cBioPortal
S140L	In cancer patients	COSMIC, cBioPortal
A141A	In cancer patients	COSMIC, TumorPortal
N142K	In cancer patients	COSMIC, cBioPortal
N142N	In cancer patients	COSMIC

Table 1. *Cont.*

Mutation	Mode of Identification	Reference
H145Y	In cancer patients	COSMIC, cBioPortal
H145R	In cancer patients	cBioPortal
R146S	In cancer patients	COSMIC, cBioPortal
R146L	In cancer patients	cBioPortal, TumorPortal
R146C	In cancer patients	cBioPortal
R146H	In cancer patients	COSMIC, cBioPortal
D147Y	In cancer patients	cBioPortal
S151D	On the basis of alignment with MKK1	[103]
L154N	In a screen for mutants that are resistant to VRT-11E	[95]
L154G	In a screen for mutants that are resistant to VRT-11E	[95]
L155L	In cancer patients	COSMIC, TumorPortal
D160N	In cancer patients	COSMIC, cBioPortal, TumorPortal
D160G	In cancer patients	COSMIC, cBioPortal, TumorPortal
I163I	In cancer patients	cBioPortal
C164R	In cancer patients	COSMIC, cBioPortal
D165G	In cancer patients	COSMIC, cBioPortal
G167D	In a screen for mutants that are resistant to VRT-11E and SCH772984	[95,97]
L168L	In a screen for mutants that are resistant to Erk inhibitors	[96]
R170H	In cancer patients	COSMIC, cBioPortal, TumorPortal
V171I	In cancer patients	COSMIC
P174T	In cancer patients	COSMIC, cBioPortal
P174S	In cancer patients	COSMIC, cBioPortal
D177G	In cancer patients	cBioPortal
T179fs*29	In cancer patients	COSMIC, cBioPortal
F181S	In cancer patients	COSMIC

Table 1. *Cont.*

Mutation	Mode of Identification	Reference
E184*	In cancer patients	cBioPortal
A187V	In a screen for mutants that are resistant to trametinib and dabrafenib	[96]
R189C	In cancer patients	COSMIC, cBioPortal
R189H	In cancer patients	COSMIC, cBioPortal
W190L	In cancer patients	cBioPortal
E195*	In cancer patients	cBioPortal
L198F	In cancer patients	COSMIC
S200P	In a screen for mutants that are resistant to trametinib and dabrafenib	[96]
G202C	In cancer patients	COSMIC, cBioPortal,[96]
G202S	In cancer patients	cBioPortal,[96]
G202G	In cancer patients	COSMIC, [96]
Y203N	In a screen for mutants that are resistant to RAF/MEK inhibitors	[96]
T204I	In cancer patients	COSMIC, cBioPortal
I209V	In cancer patients	COSMIC, cBioPortal
V212A	In a screen for mutants that are resistant to RAF/MEK inhibitors	[96]
E218K	In cancer patients	COSMIC
S220Y	In cancer patients	COSMIC, cBioPortal
I225F	In cancer patients	COSMIC, cBioPortal
P227S	In cancer patients	COSMIC
P227L	In cancer patients	COSMIC, cBioPortal
G228E	In cancer patients	COSMIC, cBioPortal
D233E	In cancer patients	COSMIC, cBioPortal
D233G	In cancer patients	cBioPortal

Table 1. *Cont.*

Mutation	Mode of Identification	Reference
D233V	In cancer patients	COSMIC, cBioPortal
D233*	In cancer patients	TumorPortal
G240_splice	In cancer patients	cBioPortal
L242I	In cancer patients	COSMIC, cBioPortal
L242F	In cancer patients	COSMIC, cBioPortal, TumorPortal
S244F	In cancer patients	COSMIC, cBioPortal
S244S	In cancer patients	COSMIC
L250L	In a screen for mutants that are resistant to RAF/MEK inhibitors	[96]
N255S	In cancer patients	COSMIC, cBioPortal
R259G	In cancer patients	COSMIC, cBioPortal
N260T	In cancer patients	COSMIC, cBioPortal, TumorPortal
Y261C	In cancer patients	COSMIC, cBioPortal
L263F	In a screen for mutants that are resistant to Erk inhibitors	[96]
S264F	In cancer patients	COSMIC
L265P	In cancer patients	COSMIC
P266L	In cancer patients	COSMIC, cBioPortal
P272S	In cancer patients	COSMIC
L276L	In cancer patients	COSMIC, TumorPortal
L276M	In cancer patients	cBioPortal
F277F	In cancer patients	COSMIC
K283T	In cancer patients	cBioPortal
L285M	In a screen for mutants that are resistant to RAF/MEK inhibitors	[96]
L288L	In cancer patients	COSMIC
D289G	In cancer patients	COSMIC, cBioPortal, TumorPortal
D289H	In cancer patients	cBioPortal

Table 1. *Cont.*

Mutation	Mode of Identification	Reference
P296T	In cancer patients	COSMIC
E301K	In a screen for mutants that are resistant to RAF/MEK inhibitors	[96]
A305S	In cancer patients	cBioPortal
L311P	In cancer patients	COSMIC
Y314F	In cancer patients	COSMIC, cBioPortal, TumorPortal
Y314C	In cancer patients	cBioPortal
D316N	In cancer patients	COSMIC, cBioPortal
P317S	In cancer patients	COSMIC, cBioPortal
P317P	In cancer patients	COSMIC
S318C	In cancer patients	COSMIC
D319N	In cancer patients, and in a genetic screen in *Drosophila*	COSMIC, cBioPortal, [77]
D319A	In cancer patients	COSMIC
D319G	In a screen for mutants that are resistant to trametinib and dabrafenib	cBioPortal, [96]
D319V	In cancer patients	COSMIC, cBioPortal
D319E	In cancer patients	COSMIC, cBioPortal
E320K	In a screen for mutants that are resistant to trametinib and dabrafenib	CISMIC, cBioPortal, TumorPortal,[96]
E320*	In cancer patients	cBioPortal, TumorPortal
E320N	In cancer patients	cBioPortal
E320A	In cancer patients	COSMIC, cBioPortal
E320V	In cancer patients	COSMIC
P321P	In a screen for mutants that are resistant to Erk inhibitors	[96]
P321S	In a screen for mutants that are resistant to RAF/MEK inhibitors	[96]
P321L	In a screen for mutants that are resistant to RAF/MEK inhibitors	[96]
I322V	In cancer patients	cBioPortal

Table 1. *Cont.*

Mutation	Mode of Identification	Reference
A323T	In cancer patients	cBioPortal
A323S	In cancer patients	cBioPortal
A323V	In cancer patients	cBioPortal
E324*	In cancer patients	COSMIC, cBioPortal
F329F	In cancer patients	COSMIC
F329Y	In cancer patients	COSMIC
D330N	In cancer patients	COSMIC, cBioPortal
M331I	In cancer patients	COSMIC, cBioPortal
D335N	In a screen for mutants that are resistant to RAF/MEK inhibitors	[96]
E343*	In cancer patients	COSMIC
I345H	In a screen for mutants that are resistant to to VRT-11E	[95]
I345M	In a screen for mutants that are resistant to VRT-11E	[95]
I345F	In a screen for mutants that are resistant to VRT-11E	[95]
I345L	In a screen for mutants that are resistant to VRT-11E	COSMIC
I345Y	In a screen for mutants that are resistant to VRT-11E and SCH772984	[95]
I345W	In a screen for mutants that are resistant to VRT-11E and SCH772984	[95]
E347*	In cancer patients	COSMIC
E347K	In cancer patients, and in a screen for mutants that are resistant to RAF/MEK inhibitors	COSMIC, cBioPortal,[96]
T349T	In cancer patients	COSMIC
A350S	In cancer patients	COSMIC
A350V	In cancer patients	COSMIC
R351K	In a screen for mutants that are resistant to RAF/MEK inhibitors	[96]
Y356D	In cancer patients	COSMIC, cBioPortal
Y356Y	In cancer patients	COSMIC
R357T	In cancer patients	cBioPortal

Funding: Research in David Engelberg laboratory was funded by the Israel Science Foundation grant number (1463/18) and by the Singapore National Research Foundation under its HUJ-NUS partnership program at the Campus for Research Excellence and Technology Enterprise (CREATE). David Engelberg holds a Wolfson Family Chair in Biochemistry.

Conflicts of Interest: The authors declare no conflict of interest.

References

1. Askari, N.; Diskin, R.; Avitzour, M.; Capone, R.; Livnah, O.; Engelberg, D. Hyperactive variants of p38α induce, whereas hyperactive variants of p38γ suppress, activating protein 1-mediated transcription. *J. Biol. Chem.* **2007**, *282*, 91–99. [CrossRef] [PubMed]

2. Avitzour, M.; Diskin, R.; Raboy, B.; Askari, N.; Engelberg, D.; Livnah, O. Intrinsically active variants of all human p38 isoforms. *FEBS J.* **2007**, *274*, 963–975. [CrossRef] [PubMed]

3. Fujimura, T. Current status and future perspective of robot-assisted radical cystectomy for invasive bladder cancer. *Int. J. Urol.* **2019**, *26*, 1033–1042. [CrossRef] [PubMed]

4. Roskoski, R., Jr. Targeting oncogenic Raf protein-serine/threonine kinases in human cancers. *Pharmacol. Res.* **2018**, *135*, 239–258. [CrossRef]

5. Keith, W.M.; Kenichi, N.; Sara, W. Recent advances of MEK inhibitors and their clinical progress. *Curr. Top. Med. Chem.* **2007**, *7*, 1364–1378.

6. Zhao, Y.; Adjei, A.A. The clinical development of MEK inhibitors. *Nat. Rev. Clin. Oncol.* **2014**, *11*, 385–400. [CrossRef]

7. Roskoski, R., Jr. Allosteric MEK1/2 inhibitors including cobimetanib and trametinib in the treatment of cutaneous melanomas. *Pharmacol. Res.* **2017**, *117*, 20–31. [CrossRef]

8. Dominguez, C.; Powers, D.A.; Tamayo, N. p38 MAP kinase inhibitors: Many are made, but few are chosen. *Curr. Opin. Drug Discov. Dev.* **2005**, *8*, 421–430.

9. Yong, H.Y.; Koh, M.S.; Moon, A. The p38 MAPK inhibitors for the treatment of inflammatory diseases and cancer. *Expert Opin. Investig. Drugs* **2009**, *18*, 1893–1905. [CrossRef]

10. Pettus, L.H.; Wurz, R.P. Small molecule p38 MAP kinase inhibitors for the treatment of inflammatory diseases: Novel structures and developments during 2006–2008. *Curr. Top. Med. Chem.* **2008**, *8*, 1452–1467. [CrossRef]

11. Messoussi, A.; Feneyrolles, C.; Bros, A.; Deroide, A.; Daydé-Cazals, B.; Chevé, G.; Van Hijfte, N.; Fauvel, B.; Bougrin, K.; Yasri, A. Recent progress in the design, study, and development of c-Jun N-terminal kinase inhibitors as anticancer agents. *Chem. Biol.* **2014**, *21*, 1433–1443. [CrossRef] [PubMed]

12. Mansour, S.J.; Matten, W.T.; Hermann, A.S.; Candia, J.M.; Rong, S.; Fukasawa, K.; Woude, G.V.; Ahn, N.G. Transformation of mammalian cells by constitutively active MAP kinase kinase. *Science* **1994**, *265*, 966–970. [CrossRef] [PubMed]

13. Kolibaba, K.S.; Druker, B.J. Protein tyrosine kinases and cancer. *Biochim. Biophys. Acta* **1997**, *1333*, F217–F248. [CrossRef]

14. Garnett, M.J.; Marais, R. Guilty as charged: B-RAF is a human oncogene. *Cancer Cell* **2004**, *6*, 313–319. [CrossRef] [PubMed]

15. Raingeaud, J.; Whitmarsh, A.J.; Barrett, T.; Derijard, B.; Davis, R.J. MKK3-and MKK6-regulated gene expression is mediated by the p38 mitogen-activated protein kinase signal transduction pathway. *Mol. Cell. Biol.* **1996**, *16*, 1247–1255. [CrossRef] [PubMed]

16. Wick, M.J.; Wick, K.R.; Chen, H.; He, H.; Dong, L.Q.; Quon, M.J.; Liu, F. Substitution of the autophosphorylation site Thr516 with a negatively charged residue confers constitutive activity to mouse 3-phosphoinositide-dependent protein kinase-1 in cells. *J. Biol. Chem.* **2002**, *277*, 16632–16638. [CrossRef]

17. Yang, L.; Zheng, L.; Chng, W.J.; Ding, J.L. Comprehensive Analysis of ERK1/2 Substrates for Potential Combination Immunotherapies. *Trends Pharmacol. Sci.* **2019**, *40*, 897–910. [CrossRef]

18. Kyriakis, J.M.; Avruch, J. Mammalian MAPK signal transduction pathways activated by stress and inflammation: A 10-year update. *Physiol. Rev.* **2012**, *92*, 689–737. [CrossRef]

19. Brewster, J.L.; Gustin, M.C. Hog1: 20 years of discovery and impact. *Sci. Signal* **2014**, *7*, re7. [CrossRef]

20. Engelberg, D.; Perlman, R.; Levitzki, A. Transmembrane signaling in Saccharomyces cerevisiae as a model for signaling in metazoans: State of the art after 25 years. *Cell. Signal.* **2014**, *26*, 2865–2878. [CrossRef]

21. Saito, H. Regulation of cross-talk in yeast MAPK signaling pathways. *Curr. Opin. Microbiol.* **2010**, *13*, 677–683. [CrossRef] [PubMed]

22. Chen, R.E.; Thorner, J. Function and regulation in MAPK signaling pathways: Lessons learned from the yeast Saccharomyces cerevisiae. *Biochim. Biophys. Acta* **2007**, *1773*, 1311–1340. [CrossRef] [PubMed]

23. Shilo, B.Z. The regulation and functions of MAPK pathways in Drosophila. *Methods* **2014**, *68*, 151–159. [CrossRef] [PubMed]

24. Ashton-Beaucage, D.; Therrien, M. *How Genetics Has Helped Piece Together the MAPK Signaling Pathway;* Humana Press: New York, NY, USA, 2017; Volume 1487, pp. 1–21.

25. Bost, F.; Aouadi, M.; Caron, L.; Even, P.; Belmonte, N.; Prot, M.; Dani, C.; Hofman, P.; Pagès, G.; Pouysségur, J.; et al. The extracellular signal-regulated kinase isoform ERK1 is specifically required for in vitro and in vivo adipogenesis. *Diabetes* **2005**, *54*, 402–411. [CrossRef] [PubMed]

26. Bourcier, C.; Jacquel, A.; Hess, J.; Peyrottes, I.; Angel, P.; Hofman, P.; Auberger, P.; Pouysségur, J.; Pagès, G. p44 mitogen-activated protein kinase (extracellular signal-regulated kinase 1)-dependent signaling contributes to epithelial skin carcinogenesis. *Cancer Res.* **2006**, *66*, 2700–2707. [CrossRef] [PubMed]

27. Guihard, S.; Clay, D.; Cocault, L.; Saulnier, N.; Opolon, P.; Souyri, M.; Pagès, G.; Pouysségur, J.; Porteu, F.; Gaudry, M. The MAPK ERK1 is a negative regulator of the adult steady-state splenic erythropoiesis. *Blood* **2010**, *115*, 3686–3694. [CrossRef]

28. Lefloch, R.; Pouysségur, J.; Lenormand, P. Single and combined silencing of ERK1 and ERK2 reveals their positive contribution to growth signaling depending on their expression levels. *Mol. Cell. Biol.* **2008**, *28*, 511–527. [CrossRef]

29. Shin, S.; Dimitri, C.A.; Yoon, S.O.; Dowdle, W.; Blenis, J. ERK2 but not ERK1 induces epithelial-to-mesenchymal transformation via DEF motif-dependent signaling events. *Mol. Cell* **2010**, *38*, 114–127. [CrossRef]

30. Voisin, L.; Saba-El-Leil, M.K.; Julien, C.; Frémin, C.; Meloche, S. Genetic demonstration of a redundant role of extracellular signal-regulated kinase 1 (ERK1) and ERK2 mitogen-activated protein kinases in promoting fibroblast proliferation. *Mol. Cell. Biol.* **2010**, *30*, 2918–2932. [CrossRef]

31. Frémin, C.; Ezan, F.; Boisselier, P.; Bessard, A.; Pagès, G.; Pouysségur, J.; Baffet, G. ERK2 but not ERK1 plays a key role in hepatocyte replication: An RNAi-mediated ERK2 knockdown approach in wild-type and ERK1 null hepatocytes. *Hepatology* **2007**, *45*, 1035–1045. [CrossRef]

32. Radtke, S.; Milanovic, M.; Rossé, C.; De Rycker, M.; Lachmann, S.; Hibbert, A.; Kermorgant, S.; Parker, P.J. ERK2 but not ERK1 mediates HGF-induced motility in non-small cell lung carcinoma cell lines. *J. Cell Sci.* **2013**, *126 Pt 11*, 2381–2391. [CrossRef]

33. Chang, S.F.; Lin, S.S.; Yang, H.C.; Chou, Y.Y.; Gao, J.I.; Lu, S.C. LPS-Induced G-CSF Expression in Macrophages Is Mediated by ERK2, but Not ERK1. *PLoS ONE* **2015**, *10*, e0129685. [CrossRef] [PubMed]

34. Samuels, I.S.; Karlo, J.C.; Faruzzi, A.N.; Pickering, K.; Herrup, K.; Sweatt, J.D.; Saitta, S.C.; Landreth, G.E. Deletion of ERK2 mitogen-activated protein kinase identifies its key roles in cortical neurogenesis and cognitive function. *J. Neurosci.* **2008**, *28*, 6983–6995. [CrossRef] [PubMed]

35. Hatano, N.; Mori, Y.; Oh-hora, M.; Kosugi, A.; Fujikawa, T.; Nakai, N.; Niwa, H.; Miyazaki, J.I.; Hamaoka, T.; Ogata, M. Essential role for ERK2 mitogen-activated protein kinase in placental development. *Genes Cells* **2003**, *8*, 847–856. [CrossRef] [PubMed]

36. Frémin, C.; Saba-El-Leil, M.K.; Lévesque, K.; Ang, S.L.; Meloche, S. Functional Redundancy of ERK1 and ERK2 MAP Kinases during Development. *Cell Rep.* **2015**, *12*, 913–921. [CrossRef]

37. Marshall, C.J. Specificity of receptor tyrosine kinase signaling: Transient versus sustained extracellular signal-regulated kinase activation. *Cell* **1995**, *80*, 179–185. [CrossRef]

38. McKay, M.M.; Morrison, D.K. Integrating signals from RTKs to ERK/MAPK. *Oncogene* **2007**, *26*, 3113–3121. [CrossRef]

39. Wee, P.; Wang, Z. Epidermal Growth Factor Receptor Cell Proliferation Signaling Pathways. *Cancers* **2017**, *9*, 52.

40. Katz, M.; Amit, I.; Yarden, Y. Regulation of MAPKs by growth factors and receptor tyrosine kinases. *Biochim. Biophys. Acta* **2007**, *1773*, 1161–1176. [CrossRef]

41. Schlessinger, J. Receptor tyrosine kinases: Legacy of the first two decades. *Cold Spring Harb. Perspect. Biol.* **2014**, *6*, a008912. [CrossRef]

42. Liu, F.; Yang, X.; Geng, M.; Huang, M. Targeting ERK, an Achilles' Heel of the MAPK pathway, in cancer therapy. *Acta Pharm. Sin. B* **2018**, *8*, 552–562. [CrossRef] [PubMed]

43. Lemmon, M.A.; Schlessinger, J. Cell signaling by receptor tyrosine kinases. *Cell* **2010**, *141*, 1117–1134. [CrossRef] [PubMed]

44. van der Geer, P.; Hunter, T.; Lindberg, R.A. Receptor Protein-Tyrosine Kinases and Their Signal-Transduction Pathways. *Annu. Rev. Cell Biol.* **1994**, *10*, 251–337. [CrossRef] [PubMed]

45. Wellbrock, C.; Karasarides, M.; Marais, R. The RAF proteins take centre stage. *Nat. Rev. Mol. Cell Biol.* **2004**, *5*, 875–885. [CrossRef] [PubMed]

46. Okazaki, K.; Sagata, N. The Mos/MAP kinase pathway stabilizes c-Fos by phosphorylation and augments its transforming activity in NIH 3T3 cells. *EMBO J.* **1995**, *14*, 5048–5059. [CrossRef]

47. Das, S.; Cho, J.; Lambertz, I.; Kelliher, M.A.; Eliopoulos, A.G.; Du, K.; Tsichlis, P.N. Tpl2/cot signals activate ERK, JNK, and NF-kappaB in a cell-type and stimulus-specific manner. *J. Biol. Chem.* **2005**, *280*, 23748–23757. [CrossRef]

48. Gotoh, I.; Adachi, M.; Nishida, E. Identification and characterization of a novel MAP kinase kinase kinase, MLTK. *J. Biol. Chem.* **2001**, *276*, 4276–4286. [CrossRef]

49. Shenoy, S.K.; Drake, M.T.; Nelson, C.D.; Houtz, D.A.; Xiao, K.; Madabushi, S.; Reiter, E.; Premont, R.T.; Lichtarge, O.; Lefkowitz, R.J. beta-arrestin-dependent, G protein-independent ERK1/2 activation by the beta2 adrenergic receptor. *J. Biol. Chem.* **2006**, *281*, 1261–1273. [CrossRef]

50. Choi, M.; Staus, D.P.; Wingler, L.M.; Ahn, S.; Pani, B.; Capel, W.D.; Lefkowitz, R.J. G protein-coupled receptor kinases (GRKs) orchestrate biased agonism at the beta2-adrenergic receptor. *Sci. Signal* **2018**, *11*, eaar7084. [CrossRef]

51. Jain, R.; Watson, U.; Vasudevan, L.; Saini, D.K. ERK Activation Pathways Downstream of GPCRs. *Int. Rev. Cell Mol. Biol.* **2018**, *338*, 79–109.

52. Watson, U.; Jain, R.; Asthana, S.; Saini, D.K. Spatiotemporal Modulation of ERK Activation by GPCRs. *Int. Rev. Cell Mol. Biol.* **2018**, *338*, 111–140. [PubMed]

53. Morrison, D.K. KSR: A MAPK scaffold of the Ras pathway? *J. Cell Sci.* **2001**, *114*, 1609–1612. [PubMed]

54. Sharma, C.; Vomastek, T.; Tarcsafalvi, A.; Catling, A.D.; Schaeffer, H.J.; Eblen, S.T.; Weber, M.J. MEK partner 1 (MP1): Regulation of oligomerization in MAP kinase signaling. *J. Cell. Biochem.* **2005**, *94*, 708–719. [CrossRef] [PubMed]

55. Morrison, D.K.; Davis, R.J. Regulation of MAP kinase signaling modules by scaffold proteins in mammals. *Annu. Rev. Cell Dev. Biol.* **2003**, *19*, 91–118. [CrossRef] [PubMed]

56. Frodin, M.; Gammeltoft, S. Role and regulation of 90 kDa ribosomal S6 kinase (RSK) in signal transduction. *Mol. Cell. Endocrinol.* **1999**, *151*, 65–77. [CrossRef]

57. Reyskens, K.M.; Arthur, J.S.C. Emerging Roles of the Mitogen and Stress Activated Kinases MSK1 and MSK2. *Front. Cell Dev. Biol.* **2016**, *4*, 56. [CrossRef]

58. Gaestel, M. Specificity of signaling from MAPKs to MAPKAPKs: kinases' tango nuevo. *Front. Biosci.* **2008**, *13*, 6050–6059. [CrossRef]

59. Cruzalegui, F.H.; Cano, E.; Treisman, R. ERK activation induces phosphorylation of Elk-1 at multiple S/T-P motifs to high stoichiometry. *Oncogene* **1999**, *18*, 7948–7957. [CrossRef]

60. Dougherty, M.K.; Müller, J.; Ritt, D.A.; Zhou, M.; Zhou, X.Z.; Copeland, T.D.; Conrads, T.P.; Veenstra, T.D.; Lu, K.P.; Morrison, D.K. Regulation of Raf-1 by direct feedback phosphorylation. *Mol. Cell* **2005**, *17*, 215–224. [CrossRef]

61. Fritsche-Guenther, R.; Witzel, F.; Sieber, A.; Herr, R.; Schmidt, N.; Braun, S.; Brummer, T.; Sers, C.; Blüthgen, N. Strong negative feedback from Erk to Raf confers robustness to MAPK signalling. *Mol. Syst. Biol.* **2011**, *7*, 489. [CrossRef]

62. Shin, S.Y.; Rath, O.; Choo, S.M.; Fee, F.; McFerran, B.; Kolch, W.; Cho, K.H. Positive- and negative-feedback regulations coordinate the dynamic behavior of the Ras-Raf-MEK-ERK signal transduction pathway. *J. Cell Sci.* **2009**, *122 Pt 3*, 425–435. [CrossRef]

63. Dong, C.; Waters, S.B.; Holt, K.H.; Pessin, J.E. SOS phosphorylation and disassociation of the Grb2-SOS complex by the ERK and JNK signaling pathways. *J. Biol. Chem.* **1996**, *271*, 6328–6332.

64. Xu, B.E.; Wilsbacher, J.L.; Collisson, T.; Cobb, M.H. The N-terminal ERK-binding site of MEK1 is required for efficient feedback phosphorylation by ERK2 in vitro and ERK activation in vivo. *J. Biol. Chem.* **1999**, *274*, 34029–34035. [CrossRef] [PubMed]

65. Ekerot, M.; Stavridis, M.P.; Delavaine, L.; Mitchell, M.P.; Staples, C.; Owens, D.M.; Keenan, I.D.; Dickinson, R.J.; Storey, K.G.; Keyse, S.M. Negative-feedback regulation of FGF signalling by DUSP6/MKP-3 is driven by ERK1/2 and mediated by Ets factor binding to a conserved site within the DUSP6/MKP-3 gene promoter. *Biochem. J.* **2008**, *412*, 287–298. [CrossRef] [PubMed]

66. Nolen, B.; Taylor, S.; Ghosh, G. Regulation of protein kinases; controlling activity through activation segment conformation. *Mol. Cell* **2004**, *15*, 661–675. [CrossRef] [PubMed]

67. Kornev, A.P.; Haste, N.M.; Taylor, S.S.; Eyck, L.F. Surface comparison of active and inactive protein kinases identifies a conserved activation mechanism. *Proc. Natl. Acad. Sci. USA* **2006**, *103*, 17783–17788. [CrossRef]

68. Kornev, A.P.; Taylor, S.S. Dynamics-Driven Allostery in Protein Kinases. *Trends Biochem. Sci.* **2015**, *40*, 628–647. [CrossRef]

69. Beenstock, J.; Mooshayef, N.; Engelberg, D. How Do Protein Kinases Take a Selfie (Autophosphorylate)? *Trends Biochem. Sci.* **2016**, *41*, 938–953. [CrossRef]

70. Shi, F.; Telesco, S.E.; Liu, Y.; Radhakrishnan, R.; Lemmon, M.A. ErbB3/HER3 intracellular domain is competent to bind ATP and catalyze autophosphorylation. *Proc. Natl. Acad. Sci. USA* **2010**, *107*, 7692–7697. [CrossRef]

71. Pike, A.C.; Rellos, P.; Niesen, F.H.; Turnbull, A.; Oliver, A.W.; Parker, S.A.; Turk, B.E.; Pearl, L.H.; Knapp, S. Activation segment dimerization: A mechanism for kinase autophosphorylation of non-consensus sites. *EMBO J.* **2008**, *27*, 704–714. [CrossRef]

72. Seger, R.; Ahn, N.G.; Boulton, T.G.; Yancopoulos, G.D.; Panayotatos, N.; Radziejewska, E.; Ericsson, L.; Bratlien, R.L.; Cobb, M.H.; Krebs, E.G. Microtubule-associated protein 2 kinases, ERK1 and ERK2, undergo autophosphorylation on both tyrosine and threonine residues: Implications for their mechanism of activation. *Proc. Natl. Acad. Sci. USA* **1991**, *88*, 6142–6146. [CrossRef] [PubMed]

73. Levin-Salomon, V.; Kogan, K.; Ahn, N.G.; Livnah, O.; Engelberg, D. Isolation of intrinsically active (MEK-independent) variants of the ERK family of mitogen-activated protein (MAP) kinases. *J. Biol. Chem.* **2008**, *283*, 34500–34510. [CrossRef] [PubMed]

74. Pegram, L.M.; Liddle, J.C.; Xiao, Y.; Hoh, M.; Rudolph, J.; Iverson, D.B.; Vigers, G.P.; Smith, D.; Zhang, H.; Wang, W.; et al. Activation loop dynamics are controlled by conformation-selective inhibitors of ERK2. *Proc. Natl. Acad. Sci. USA* **2019**, *116*, 15463–15468. [CrossRef] [PubMed]

75. Sang, D.; Pinglay, S.; Wiewiora, R.P.; Selvan, M.E.; Lou, H.J.; Chodera, J.D.; Turk, B.E.; Gumus, Z.H.; Holt, L.J. Ancestral reconstruction reveals mechanisms of ERK regulatory evolution. *eLife* **2019**, *8*, e38805. [CrossRef] [PubMed]

76. Emrick, M.A.; Lee, T.; Starkey, P.J.; Mumby, M.C.; Resing, K.A.; Ahn, N.G. The gatekeeper residue controls autoactivation of ERK2 via a pathway of intramolecular connectivity. *Proc. Natl. Acad. Sci. USA* **2006**, *103*, 18101–18106. [CrossRef] [PubMed]

77. Brunner, D.; Oellers, N.; Szabad, J.; Biggs, W.H.; Zipursky, S.L.; Hafen, E. A gain-of-function mutation in Drosophila MAP kinase activates multiple receptor tyrosine kinase signaling pathways. *Cell* **1994**, *76*, 875–888. [CrossRef]

78. Goshen-Lago, T.; Goldberg-Carp, A.; Melamed, D.; Darlyuk-Saadon, I.; Bai, C.; Ahn, N.G.; Admon, A.; Engelberg, D. Variants of the yeast MAPK Mpk1 are fully functional independently of activation loop phosphorylation. *Mol. Biol. Cell* **2016**, *27*, 2771–2783. [CrossRef]

79. Dhillon, A.S.; Hagan, S.; Rath, O.; Kolch, W. MAP kinase signalling pathways in cancer. *Oncogene* **2007**, *26*, 3279–3290. [CrossRef]

80. Hanahan, D.; Weinberg, R.A. Hallmarks of cancer: The next generation. *Cell* **2011**, *144*, 646–674. [CrossRef]

81. Samatar, A.A.; Poulikakos, P.I. Targeting RAS-ERK signalling in cancer: Promises and challenges. *Nat. Rev. Drug Discov.* **2014**, *13*, 928–942. [CrossRef]

82. Smorodinsky-Atias, K.; Goshen-Lago, T.; Goldberg-Carp, A.; Melamed, D.; Shir, A.; Mooshayef, N.; Beenstock, J.; Karamansha, Y.; DArlyuk-Saadon, I.; Livnah, O.; et al. Intrinsically active variants of Erk oncogenically transform cells and disclose unexpected autophosphorylation capability that is independent of TEY phosphorylation. *Mol. Biol. Cell* **2016**, *27*, 1026–1039. [CrossRef] [PubMed]

83. Kushnir, T.; Bar-Cohen, S.; Mooshayef, N.; Lange, R.; Bar-Sinai, A.; Rozen, H.; Salzberg, A.; Engelberg, D.; Paroush, Z. An Activating Mutation in ERK Causes Hyperplastic Tumors in a scribble Mutant Tissue in Drosophila. *Genetics* **2019**. [CrossRef] [PubMed]

84. Groenendijk, F.H.; Bernards, R. Drug resistance to targeted therapies: Deja vu all over again. *Mol. Oncol.* **2014**, *8*, 1067–1083. [CrossRef] [PubMed]

85. Morris, E.J.; Jha, S.; Restaino, C.R.; Dayananth, P.; Zhu, H.; Cooper, A.; Carr, D.; Deng, Y.; Jin, W.; Black, S.; et al. Discovery of a novel ERK inhibitor with activity in models of acquired resistance to BRAF and MEK inhibitors. *Cancer Discov.* **2013**, *3*, 742–750. [CrossRef]

86. Moschos, S.J.; Sullivan, R.J.; Hwu, W.J.; Ramanathan, R.K.; Adjei, A.A.; Fong, P.C.; Shapira-Frommer, R.; Tawbi, H.A.; Rubino, J.; Rush, T.S.; et al. Development of MK-8353, an orally administered ERK1/2 inhibitor, in patients with advanced solid tumors. *JCI Insight* **2018**, *3*, e92352. [CrossRef]

87. Germann, U.A.; Furey, B.F.; Markland, W.; Hoover, R.R.; Aronov, A.M.; Roix, J.J.; Hale, M.; Boucher, D.M.; Sorrell, D.A.; Martinez-Botella, G.; et al. Targeting the MAPK Signaling Pathway in Cancer: Promising Preclinical Activity with the Novel Selective ERK1/2 Inhibitor BVD-523 (Ulixertinib). *Mol. Cancer Ther.* **2017**, *16*, 2351–2363. [CrossRef]

88. Blake, J.F.; Burkard, M.; Chan, J.; Chen, H.; Chou, K.J.; Diaz, D.; Dudley, D.A.; Gaudino, J.J.; Gould, S.E.; Grina, J.; et al. Discovery of (S)-1-(1-(4-Chloro-3-fluorophenyl)-2-hydroxyethyl)-4-(2-((1-methyl-1H-pyrazol-5-yl)amino)pyrimidin-4-yl)pyridin-2(1H)-one (GDC-0994), an Extracellular Signal-Regulated Kinase 1/2 (ERK1/2) Inhibitor in Early Clinical Development. *J. Med. Chem.* **2016**, *59*, 5650–5660. [CrossRef]

89. Aronov, A.M.; Baker, C.; Bemis, G.W.; Cao, J.; Chen, G.; Ford, P.J.; Germann, U.A.; Green, J.; Hale, M.R.; Jacobs, M.; et al. Flipped out: Structure-guided design of selective pyrazolylpyrrole ERK inhibitors. *J. Med. Chem.* **2007**, *50*, 1280–1287. [CrossRef]

90. Bhagwat, S.V.; McMillen, W.T.; Cai, S.; Zhao, B.; Whitesell, M.; Kindler, L.; Flack, R.S.; Wu, W.; Huss, K.; Anderson, B.; et al. Discovery of LY3214996, a selective and novel ERK1/2 inhibitor with potent antitumor activities in cancer models with MAPK pathway alterations. *Cancer Res.* **2017**, *77*. [CrossRef]

91. Ohori, M.; Kinoshita, T.; Okubo, M.; Sato, K.; Yamazaki, A.; Arakawa, H.; Nishimura, S.; Inamura, N.; Nakajima, H.; Neya, M.; et al. Identification of a selective ERK inhibitor and structural determination of the inhibitor-ERK2 complex. *Biochem. Biophys. Res. Commun.* **2005**, *336*, 357–363. [CrossRef]

92. Aronov, A.M.; Tang, Q.; Martinez-Botella, G.; Bemis, G.W.; Cao, J.; Chen, G.; Ewing, N.P.; Ford, P.J.; Germann, U.A.; Green, J.; et al. Structure-guided design of potent and selective pyrimidylpyrrole inhibitors of extracellular signal-regulated kinase (ERK) using conformational control. *J. Med. Chem.* **2009**, *52*, 6362–6368. [CrossRef] [PubMed]

93. Herrero, A.; Pinto, A.; Colon-Bolea, P.; Casar, B.; Jones, M.; Agudo-Ibanez, L.; Vidal, R.; Tenbaum, S.P.; Nuciforo, P.; Valdizan, E.M.; et al. Small Molecule Inhibition of ERK Dimerization Prevents Tumorigenesis by RAS-ERK Pathway Oncogenes. *Cancer Cell* **2015**, *28*, 170–182. [CrossRef] [PubMed]

94. Ryan, M.B.; Corcoran, R.B. Therapeutic strategies to target RAS-mutant cancers. *Nat. Rev. Clin. Oncol.* **2018**, *15*, 709–720. [CrossRef] [PubMed]

95. Brenan, L.; Andreev, A.; Cohen, O.; Pantel, S.; Kamburov, A.; Cacchiarelli, D.; Persky, N.S.; Zhu, C.; Bagul, M.; Goettz, E.M.; et al. Phenotypic Characterization of a Comprehensive Set of MAPK1/ERK2 Missense Mutants. *Cell Rep.* **2016**, *17*, 1171–1183. [CrossRef]

96. Goetz, E.M.; Ghandi, M.; Treacy, D.J.; Wagle, N.; Garraway, L.A. ERK mutations confer resistance to mitogen-activated protein kinase pathway inhibitors. *Cancer Res.* **2014**, *74*, 7079–7089. [CrossRef]

97. Jha, S.; Morris, E.J.; Hruza, A.; Mansueto, M.S.; Schroeder, G.K.; Arbanas, J.; McMasters, D.; Restaino, C.R.; Dayanath, P.; Black, S.; et al. *Dissecting* Therapeutic Resistance to ERK Inhibition. *Mol. Cancer Ther.* **2016**, *15*, 548–559. [CrossRef]

98. Askari, N.; Diskin, R.; Avitzour, M.; Yaakov, G.; Livnah, O.; Engelberg, D. MAP-quest: Could we produce constitutively active variants of MAP kinases? *Mol. Cell. Endocrinol.* **2006**, *252*, 231–240. [CrossRef]

99. Cowley, S.; Paterson, H.; Kemp, P.; Marshall, C.J. Activation of Map Kinase Kinase Is Necessary and Sufficient for Pc12 Differentiation and for Transformation of Nih 3t3 Cells. *Cell* **1994**, *77*, 841–852. [CrossRef]

100. Huang, S.; Jiang, Y.; Li, Z.; Nishida, E.; Mathias, P.; Lin, S.; Ulevitch, R.J.; Nemerow, G.R.; Hanj, J. Apoptosis signaling pathway in T cells is composed of ICE/Ced-3 family proteases and MAP kinase kinase 6b. *Immunity* **1997**, *6*, 739–749. [CrossRef]

101. Prowse, C.N.; Deal, M.S.; Lew, J. The complete pathway for catalytic activation of the mitogen-activated protein kinase, ERK2. *J. Biol. Chem.* **2001**, *276*, 40817–40823. [CrossRef]

102. Robbins, D.J.; Zhen, E.; Owaki, H.; Vanderblit, C.A.; Ebert, D.; Geppert, T.D.; Cobb, M.H. Regulation and properties of extracellular signal-regulated protein kinases 1 and 2 in vitro. *J. Biol. Chem.* **1993**, *268*, 5097–5106. [PubMed]

103. Emrick, M.A.; Hoofnagle, A.N.; Miller, A.S.; Ten Eyck, L.F.; Ahn, N.G. Constitutive activation of extracellular signal-regulated kinase 2 by synergistic point mutations. *J. Biol. Chem.* **2001**, *276*, 46469–46479. [CrossRef] [PubMed]

104. Brill, J.A.; Elion, E.A.; Fink, G.R. A role for autophosphorylation revealed by activated alleles of FUS3, the yeast MAP kinase homolog. *Mol. Biol. Cell* **1994**, *5*, 297–312. [CrossRef] [PubMed]

105. Madhani, H.D.; Styles, C.A.; Fink, G.R. MAP kinases with distinct inhibitory functions impart signaling specificity during yeast differentiation. *Cell* **1997**, *91*, 673–684. [CrossRef]

106. Bott, C.M.; Thorneycroft, S.G.; Marshall, C.J. The sevenmaker gain-of-function mutation in p42 MAP kinase leads to enhanced signalling and reduced sensitivity to dual specificity phosphatase action. *FEBS Lett.* **1994**, *352*, 201–205. [CrossRef]

107. Hall, J.P.; Cherkasova, V.; Elion, E.; Gustin, M.C.; Winter, E. The osmoregulatory pathway represses mating pathway activity in Saccharomyces cerevisiae: Isolation of a FUS3 mutant that is insensitive to the repression mechanism. *Mol. Cell. Biol.* **1996**, *16*, 6715–6723. [CrossRef]

108. Mutlak, M.; Schlesinger-Laufer, M.; Haas, T.; Shofti, R.; Ballan, N.; Lewis, Y.E.; Zuler, M.; Zohar, Y.; Caspi, L.H.; Kehat, I. Extracellular signal-regulated kinase (ERK) activation preserves cardiac function in pressure overload induced hypertrophy. *Int. J. Cardiol.* **2018**, *270*, 204–213. [CrossRef]

109. Canagarajah, B.J.; Khokhlatchev, A.; Cobb, M.H.; Golsdmisth, E.J. Activation mechanism of the MAP kinase ERK2 by dual phosphorylation. *Cell* **1997**, *90*, 859–869. [CrossRef]

110. Taylor, C.A.; Cormier, K.W.; Keenan, S.E.; Earnest, S.; Stippec, S.; Wichaidit, C.; Juang, Y.C.; Wang, J.; Shvartsman, S.Y.; Goldsmith, E.J.; et al. Functional divergence caused by mutations in an energetic hotspot in ERK2. *Proc. Natl. Acad. Sci. USA* **2019**, *116*, 15514–15523. [CrossRef]

111. Lee, T.; Hoofnagle, A.N.; Kabuyama, Y.; Stoud, J.; Min, X.; Goldsmith, E.J.; Chen, L.; Resing, K.A.; Ahn, N.G. Docking motif interactions in MAP kinases revealed by hydrogen exchange mass spectrometry. *Mol. Cell* **2004**, *14*, 43–55. [CrossRef]

112. Dimitri, C.A.; Dowdle, W.; MacKeiga, J.P.; Blenis, J.; Murphy, L.O. Spatially separate docking sites on ERK2 regulate distinct signaling events in vivo. *Curr. Biol.* **2005**, *15*, 1319–1324. [CrossRef]

113. Misiura, M.; Kolomeisky, A.B. Kinetic network model to explain gain-of-function mutations in ERK2 enzyme. *J. Chem. Phys.* **2019**, *150*, 155101. [CrossRef] [PubMed]

114. Barr, D.; Oashi, T.; Burkhard, K.; Lucius, S.; Samadani, R.; Zhang, J.; Shapiro, P.; MacKerell, A.D.; van der Vaart, A. Importance of domain closure for the autoactivation of ERK2. *Biochemistry* **2011**, *50*, 8038–8048. [CrossRef] [PubMed]

115. Mahalingam, M.; Arvind, R.; Ida, H.; Murugan, A.K.; Yamaguchi, M.; Tsuchida, N. ERK2 CD domain mutation from a human cancer cell line enhanced anchorage-independent cell growth and abnormality in Drosophila. *Oncol. Rep.* **2008**, *20*, 957–962. [PubMed]

Article

The Pyrazolo[3,4-d]pyrimidine Derivative, SCO-201, Reverses Multidrug Resistance Mediated by ABCG2/BCRP

Sophie E. B. Ambjørner [1], Michael Wiese [2], Sebastian Christoph Köhler [2], Joen Svindt [3], Xamuel Loft Lund [1], Michael Gajhede [1], Lasse Saaby [4], Birger Brodin [4], Steffen Rump [5], Henning Weigt [6], Nils Brünner [1,7] and Jan Stenvang [1,7,*]

[1] Department of Drug Design and Pharmacology, University of Copenhagen, 1353 Copenhagen, Denmark; sophie.ambjoerner@gmail.com (S.E.B.A.); lxs184@alumni.ku.dk (X.L.L.); mig@sund.ku.dk (M.G.); nb@scandiononcology.com (N.B.)

[2] Pharmaceutical Institute, University of Bonn, 53012 Bonn, Germany; mwiese@uni-bonn.de (M.W.); skoehler@uni-bonn.de (S.C.K.)

[3] Department of Biology, University of Copenhagen, 1353 Copenhagen, Denmark; joen.svindt@gmail.com

[4] Department of Pharmacy, University of Copenhagen, 1353 Copenhagen, Denmark; lasse.saaby@sund.ku.dk (L.S.); birger.brodin@sund.ku.dk (B.B.)

[5] SRConsulting, 31319 Sehnde, Germany; rump@s-r-consulting.com

[6] Division of Chemical Safety and Toxicity, Fraunhofer Institute of Toxicology and Experimental Medicine, 30625 Hannover, Germany; henning.weigt@item.fraunhofer.de

[7] Scandion Oncology A/S, Symbion, 1353 Copenhagen, Denmark

* Correspondence: js@scandiononcology.com; Tel.: +45-60-777-678

Received: 4 February 2020; Accepted: 1 March 2020; Published: 4 March 2020

Abstract: ATP-binding cassette (ABC) transporters, such as breast cancer resistance protein (BCRP), are key players in resistance to multiple anti-cancer drugs, leading to cancer treatment failure and cancer-related death. Currently, there are no clinically approved drugs for reversal of cancer drug resistance caused by ABC transporters. This study investigated if a novel drug candidate, SCO-201, could inhibit BCRP and reverse BCRP-mediated drug resistance. We applied in vitro cell viability assays in SN-38 (7-Ethyl-10-hydroxycamptothecin)-resistant colon cancer cells and in non-cancer cells with ectopic expression of BCRP. SCO-201 reversed resistance to SN-38 (active metabolite of irinotecan) in both model systems. Dye efflux assays, bidirectional transport assays, and ATPase assays demonstrated that SCO-201 inhibits BCRP. In silico interaction analyses supported the ATPase assay data and suggest that SCO-201 competes with SN-38 for the BCRP drug-binding site. To analyze for inhibition of other transporters or cytochrome P450 (CYP) enzymes, we performed enzyme and transporter assays by in vitro drug metabolism and pharmacokinetics studies, which demonstrated that SCO-201 selectively inhibited BCRP and neither inhibited nor induced CYPs. We conclude that SCO-201 is a specific, potent, and potentially non-toxic drug candidate for the reversal of BCRP-mediated resistance in cancer cells.

Keywords: multidrug resistance in cancer; drug efflux pumps; ATP-binding cassette transporter; breast cancer resistance protein (BCRP); ABCG2; pyrazolo-pyrimidine derivative; SCO-201

---

1. Introduction

Chemotherapy resistance is considered the single most important obstacle to greater success with chemotherapy for cancer patients [1–3]. Although many cancer patients initially benefit from chemotherapy treatment, a large proportion of treatments fail due to acquisition of resistance to multiple anti-cancer drugs. This phenomenon is known as multidrug resistance (MDR) and refers to the concurrent development of cross-resistance to many chemically diverse anti-cancer agents [4]. MDR results in poor prognosis and decreased survival rate of cancer patients, and strategies to circumvent MDR are therefore highly needed [3,5]. The mechanisms underlying MDR are complex and include many different tumor survival mechanisms [6]. Overexpression of drug expelling ATP-binding cassette (ABC) transporters seems to be an important mechanism of MDR in cancer cells [7]. Today, the most extensively studied and characterized ABC transporters, found to be involved in cancer MDR, are (a) multidrug resistance protein 1/P-glycoprotein (MDR1/P-gp), encoded by the *ABCB1* gene; (b) multidrug resistance-associated protein 1 (MRP1), encoded by the *ABCC1* gene; and (c) breast cancer resistance protein (BCRP), encoded by the *ABCG2* gene [8]. ABC transporters are normally expressed in tissues such as the intestines, brain, liver, and placenta, where they prevent xenobiotic substrates from accumulating [7]. The ABC transporters are transmembrane proteins that utilize ATP hydrolysis to drive the active transport of substrates from the cytoplasmic site to the extracellular space [9]. The transporters consist of two transmembrane domains (TMDs), able to undergo a conformational change that triggers the removal of the substrate, and two cytoplasmic nucleotide-binding domains (NBDs) that bind and hydrolyze ATP [10]. Due to a broad drug specificity, ABC transporters can efflux many different anticancer agents, thus resulting in MDR [7,9]. BCRP (ABCG2) is a 72 kDa half-transporter that acts as a homomeric dimer, and so far, BCRP is known to mediate resistance to a variety of anti-cancer agents, among these the chemotherapeutic agents SN-38, topotecan, mitoxantrone, doxorubicin, and daunorubicin [11–16]. SN-38 (Figure 1) is the active metabolite of irinotecan (Camptosar) and is especially important in the treatment of gastrointestinal cancers such as colorectal cancer [17] and pancreatic cancer (European Society for Medical Oncology (ESMO) guidelines for pancreatic cancer). Several studies have indicated that high cancer cell levels of BCRP is the key player in SN-38 resistance, and BCRP thus hinders successful treatment of metastatic gastrointestinal cancer patients [11–16]. Mitoxantrone was the first chemotherapy to be identified as a substrate of BCRP, and BCRP was found to be involved in mitoxantrone-resistant breast cancer, thus giving BCRP its name [13].

During the last 40 years, researchers have tried to develop non-toxic, highly potent, and efficacious drugs that are able to reverse ABC-transporter-mediated MDR [7,9,17–19]. These MDR-reversing agents, also known as re-sensitizing agents or chemo-sensitizers, act by either inhibiting the expression of ABC transporters or by directly inhibiting the transport function, and thereby restore the sensitivity of the cancer cells to anti-cancer agents [9,10]. The compound fumitremorgin C was the first BCRP inhibitor to be identified, and although it was found to have a high inhibitory potency, neurotoxic side effects prevented the clinical use of this compound [20,21]. To prevent these side effects, researchers synthesized new different fumitremorgin C analogues, for instance, the potent BCRP inhibitor Ko143 [22,23]. Nonetheless, these analogues, including Ko143, were not stable in plasma, still caused the side effects, and could not be used in the clinic [23]. Other known ABC transporter inhibitors include verapamil, tariquidar, and valspodar (PSC833), which all inhibit MDR1/P-gp [9]. However, despite a long list of different potent inhibitors, none of these have been approved for clinical use. The lack of ABC transporter inhibitors in clinical use can be attributed to several issues: (1) the inhibitors specifically only inhibit one transporter, (2) the inhibitors exhibit a significant degree of toxicity, (3) clinical studies were poorly designed—inhibitors were not combined with the drug that the patients had proved to be resistant to—and the studies lacked randomization, and (4) lack of companion diagnostic tests to optimize patients' selection and treatment [1,7,9]. Thus, new strategies are greatly needed to improve the treatment success and survival rate of cancer patients with MDR.

To identify potential new compounds that interfere with common drug resistance mechanisms, such as the overexpression of BCRP, we previously established the DEN-50R screening platform. This platform consists of isogenic pairs of drug-sensitive and drug-resistant patient-derived cancer cell lines, for instance, colorectal, breast, prostate, and pancreatic cancer [24]. These resistant cell lines were established by exposing chemotherapy-sensitive cells to gradually increasing concentrations of chemotherapy over a period of 8–10 months [25]. We thoroughly characterized these drug-resistant cell lines to identify important drug resistance mechanisms [25–30]. In accordance with several other studies with in vitro model systems [31], we found that BCRP overexpression was a key player of resistance to SN-38 [25]. Using our DEN-50R screening platform, we found that pyrazolo-pyrimidine derivatives might serve as potential inhibitors of drug resistance in these cell lines. One of the hits from the drug screening was the pyrazolo-pyrimidine derivative SCO-201 (previously OBR-5-340) (Figure 1). These preliminary data indicated that SCO-201, which is previously known as a potent viral capsid inhibitor, might serve as a potential inhibitor of drug resistance [32,33]. Interestingly, in a study by Burkhart et al. (2009) [34], it was demonstrated that pyrazolo-pyrimidine derivatives comprise a prominent structural class of selective and potent ABC transporter inhibitors, with low toxicity and low risk of increased chemotherapy-mediated toxicity [19,34].

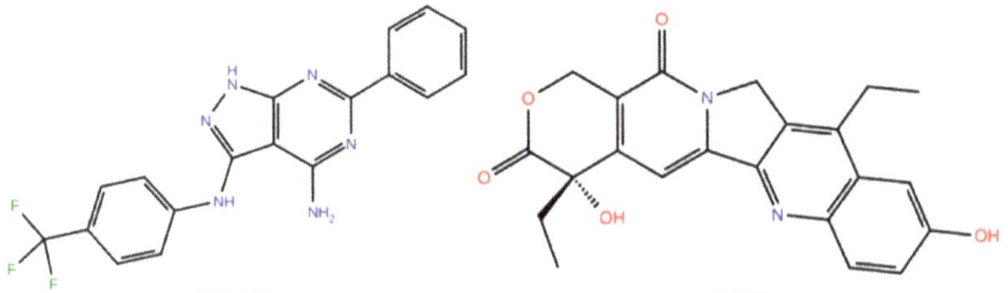

Figure 1. Chemical structures of the pyrazolo[3,4-d]pyrimidine derivative SCO-201 and the active metabolite of irinotecan, SN-38. Graphics produced using Maestro, Schrödinger 2019-3, limited liability company (LLC), New York, NY, 2019. SN-38 structure obtained from PubChem Database [35,36].

On the basis of these findings, the aim of this study was to clarify if SCO-201 re-sensitizes cancer cells to chemotherapy substrates of BCRP. An additional aim was to investigate any potential risks of pharmacokinetic interactions of SCO-201, which was investigated by testing the effect of SCO-201 on cytochrome P450 enzymes and efflux/uptake transporters. Our study is the first to show that SCO-201 competitively inhibits the transport activity of BCRP, triggers the accumulation of BCRP dye substrate, and re-sensitizes cancer cells to chemotherapy in two different in vitro models of BCRP-mediated resistance. Moreover, in vitro drug metabolism and pharmacokinetics (DMPK) data suggest that SCO-201 will not influence metabolism of other drugs by the cytochrome P450 (CYP) system. This indicates that SCO-201 is not likely to increase the risk of pharmacokinetic interactions with chemotherapy that are metabolized by the CYP system. The present data warrant further studies including regulatory toxicity and ADME (absorption, distribution, metabolism, and excretion) studies according to good laboratory practice (GLP) in order to initiate clinical development [35,36].

2. Materials and Methods

2.1. Reagents and Antibodies

DMSO, SN-38, Ko143, topotecan, mitoxantrone, Hoechst 33342, MTT reagent (3-(4,5-dimethylthiazol-2-yl)-2,5-diphenyltetrazolium bromide), primary antibody for β-actin (mAb A5441), PREDEASY SB-MDR1/P-gp Hi5, and SB-BCRP-M ATPase assays kits (Solvo Biotechnology) were all purchased from Sigma-Aldrich/Merck (Schnelldorf, Germany). PSC833 was purchased from Tocris/Bio-techne (Abingdon, UK). The primary antibody for BCRP (mAb BXP-21) was purchased from Abcam (Cambridge, UK).

Plasticware, such as T-75 culture flasks and Transwell permeable supports (1.12 cm^2, 0.4 μm pores), were purchased from Corning, Fisher Scientific (Slangerup, Denmark). Cell culture reagents such as Hank's balanced salt solution (HBSS) was purchased from Life Technologies (Taastrup, Denmark); fetal bovine serum (FBS) from Gibco, Fischer Scientific (Slangerup, Denmark); penicillin and streptomycin were acquired from Bio Whittaker Cambrex (Vallensbaek, Denmark); Dulbecco's modified Eagle's medium, bovine serum albumin (BSA), and Minimum Essential Media (MEM) nonessential amino acids were purchased from Sigma-Aldrich (Brøndby, Denmark); whereas 2-[4-(2-hydroxyethyl)piperazin-1-yl] ethanesulfonic acid (HEPES) was purchased from AppliChem GmbH (Darmstadt, Germany). [3H]-Estrone-3-sulfate (51.8 Ci/mmol), [14C]-mannitol (0.06 Ci/mmol), and Ultima Gold scintillation fluid was purchased from PerkinElmer (Boston, MA, USA).

All drugs were dissolved in DMSO, except mitoxantrone, which was dissolved in ethanol.

2.2. Cell Lines and Culture Conditions

All cell lines were maintained at 37 °C in a humidified 5% CO_2 atmosphere. The Madin-Darby Canine Kidney (MDCK)-II-BCRP cell line was a kind gift from Dr. A. Schinkel (The Netherlands Cancer Institute, Amsterdam, The Netherlands [29,30]) and cultured in Dulbecco's modified Eagle's medium (DMEM) supplemented with 10% fetal calf serum (FCS), 50 μg/mL streptomycin, 50 U/mL penicillin G, and 2 mM L-glutamine. The parental drug-sensitive ($HT29_{PAR}$), obtained from the National Cancer Institute (NCI)/Development Therapeutics Program, and SN-38-resistant ($HT29_{SN-38-RES}$) cell lines were cultured in Gibco Roswell Park Memorial Institute (RPMI) 1640–GlutaMAX medium supplemented with 10% FCS [25].

2.3. Western Blot Analysis

The proteins of HT29 cell lysates were separated with SDS-PAGE and transferred to nitrocellulose membranes. After blocking the membrane with 5% skimmed milk in 1× Tris-Buffered Saline, 0.1% Tween® 20 Detergent (TBS-T), the membrane were incubated with primary monoclonal antibodies: anti-BCRP (BXP-21; 1:1000) and β-actin (1:500,000) overnight at 4 °C, and thereafter with HRP-conjugated secondary antibody (1:4000) for 1 h. Bands were detected using Enhanced Chemiluminescence (ECL) peroxide solution and luminol/enhancer solution (Clarity Western ECL Substrate, Bio-Rad Laboratories (Copenhagen, Denmark)) for 5 min. Images were obtained with the UVP Biospectrum imaging system (VisionWorks software, version LS 7.0.1).

2.4. MDR Reversal Analysis with MTT Assay

Cells were seeded in Nunc 96-well plates (2000 or 8000 cells/well for the MDCK-II and HT29 cells, respectively). Following cell attachment (12–24 h), drugs were added to a total volume of 200 μL. Control conditions consisted of full growth medium. All treatments were performed in triplicate. Following 72 h of drug exposure, MTT reagent was added for either 1 h (MDCK-II) or 3 h (HT29). For cell lysis and solubilization of the formazan crystals, DMSO was added to the MDCK-II cells, whereas acidified (0.02 M HCl) sodium dodecyl sulphate was added to the HT29 cells. Optical densities were measured with a microplate spectrophotometer at either 544 nm and 710 nm (background) for the MDCK-II cells or 570 nm and 670 nm (background) for the HT29 cells. Background optical density values were subtracted, and the average optical densities were calculated. Cell viability was calculated as percentage of untreated control cells. The mean IC_{50} values were determined using GraphPad Prism (version 6.0, San Diego, CA, USA). The drug sensitivity analysis was carried out at least three independent times and representative data is shown.

2.5. Bidirectional Transport Assay

Bidirectional transport experiments with [3H]-estrone-3-sulfate were completed with monolayers of Caco-2 cells from the American Type Culture Collection (ATCC) cultured on Transwell permeable supports for 27 days in Dulbecco's modified Eagle's medium, supplemented with 10% FBS, 10 μL·mL^{-1} nonessential amino acids (×100), and 100 U·mL^{-1} to 100 μg·mL^{-1} penicillin-streptomycin solution. The transepithelial electrical resistance was measured across Caco-2 cell monolayers with an Endohm 12-cup electrode chamber (World Precision Instruments Inc., Sarasota, FL, USA) connected volt meter (EVOM, World Precision Instruments Inc., Sarasota, FL, USA) to ensure that cell monolayers were electrically tight before initiating transport experiments. Cell monolayers were allowed to equilibrate to room temperature before the resistance was measured. Prior to initiating the transport experiments, the cell monolayers were pre-incubated in transport buffer (HBSS supplemented with 10 mM HEPES, pH 7.4, and 0.05 % BSA) for 30 min. Transport experiments were started by replacing the blank transport buffer in the donor compartment with transport buffer containing [3H]-estrone-3-sulfate (1 µCi/mL) and [14C]-mannitol (0.8 µCi/mL) with or without 10 µM SCO-201. For transport experiments in the apical to basolateral direction, samples of 100 µL were taken from the basoteral compartment (volume 1 mL) at t = 15, 30, 45, 60, 90, and 120 min. From transport experiments in the basolateral to apical direction, samples of 50 µL were taken from the apical compartment (volume 0.5 mL) at the same time points. The withdrawn sample volume from the acceptor compartments were immediately replaced with an equal volume of blank transport buffer. The withdrawn samples were pipetted into scintillation vials and mixed with 2 mL of scintillation fluid. The radioactivity of the samples was determined by means of liquid scintillation (Packard Tri-Carb 2910 TR, PerkinElmer, Waltham, MA, USA). Transport of mannitol was measured to validate the barrier integrity of the Caco-2 cell monolayers. The overall average apparent permeability of mannitol across Caco-2 cell monolayers was $2.1 \pm 0.4 \times 10^{-7}$ cm·s^{-1} ($n = 3$, total $N = 9$), which is within the expected range for mannitol permeability across intact monolayers of Caco-2 cells.

Data treatment: The accumulated amount of compound (Qt, nmol) appearing in the donor compartment was plotted against time. The steady-state flux of compound was calculated as the slope of the linear part of this plot, thus correcting for lag-time effects. The apparent permeability was subsequently calculated with Equation (1):

$$P = \frac{J}{C\_0} = \frac{Q\_t}{(C_0 * A\_t)} \tag{1}$$

where J represents the steady state flux (nmol·cm^{-2}·min^{-1}), C_0 represents the initial concentration in the donor compartment, A_t denotes the area of the permeable support (1.12 cm^2), and Q_t is the accumulated amount of compound (nmol) in the receiver compartment at time t (min).

The ratio between apparent permeability in the basolateral to apical direction and the apparent permeability in the opposite direction (efflux ratio $= \frac{PB-A}{PA-B}$) was used as a measure of active efflux transport.

2.6. Cellular Dye Efflux Assay

2.6.1. HT29 Cells

HT29 cells were seeded either into a 96-well Nunclon plate (Thermo Fisher Scientific, Roskilde, Denmark) at a density of 6000 cells/well (for Celigo Imaging Cytometry) or into a Nunc 6-well plate at a density of 150,000 cells/well (for fluorescence microscopy). After 24 h incubation for cell attachment, the cells were incubated with either drug, DMSO or medium for 1 h, then stained with 5 µg/mL Hoechst 33,342 and incubated for 1 h at 37 °C. Then, the cells were washed with ice-cold PBS to remove excess Hoechst dye. Drugs were added again and the cells were incubated for 1h at 37 °C. The plates were analyzed with either fluorescence microscopy (6-well plates) or imaging cytometry (96-well plates). For the Celigo Imaging Cytometry (Lawrence, MA, USA), the application "Target 1 + 2 (merge)" was used, and the mean fluorescence intensities were measured and data presented as percentage of parental control.

2.6.2. MDCK-II-BCRP Cells

The inhibitory effect of SCO-201 on BCRP was determined in the Hoechst 33,342 accumulation assay as described earlier [37]. Briefly, cells were pre-incubated with SCO-201 for 30 min and then Hoechst 33,342 was added to a final concentration of 1 mM. Fluorescence was measured immediately in constant intervals (60 s) for a period of 120 min with an excitation of 355 nm and an emission wavelength of 460 nm at 37 °C using microplate readers (POLARstar and FLUOstar optima by BMG Labtech, Offenburg, Germany). Background fluorescence was subtracted and the average fluorescence between 100 and 109 min obtained in the steady state was calculated and plotted against the logarithm of the compound concentration. Dose–response curves were fitted by nonlinear regression using the four-parameter or three-parameter logistic equation, whichever was statistically preferred (GraphPad Prism, version 6.0, San Diego, CA, USA).

2.7. ATPase Assay

The effect of SCO-201 on the ATPase activity of human BCRP was measured using the PREDEASY ATPase assay system (Solvo Biotechnology, (Sigma-Aldrich/Merck, Schnelldorf, Germany). The assay is a modification of the method of Müller and Sarkardi et al. [38], and the procedure was carried out according to the instructions provided by the manufacturer. Briefly, recombinant BCRP membranes (provided by Solvo Biotechnology) were incubated in the presence or absence of vanadate and different concentrations of either SCO-201 or SN-38, and incubated at 37 °C for 10 min. To test the effect of SCO-201 on the sulfasalazine-stimulated ATPase activity of BCRP, the membranes were prepared with sulfasalazine, prior to the incubation with SCO-201. To test the ability of SCO-201 to hinder the stimulation of BCRP by SN-38, the membranes were prepared with either 0.5 or 1.5 µM SCO-201 and then incubated with different concentrations of SN-38. After incubation with test compounds, MgATP was added to each well and incubated at 37 °C for 10 min. The ATPase reaction was stopped by the addition of 1x Developer solution at room temperature (RT). Two minutes after, 100 µL Blocker solution was added and the plate was incubated for 30 min at 37 °C for 30 min. The optical densities were measured at 620 nm using a PowerWave X Microplate spectrophotometer (BioTek, Bad Friedrichshall, Germany).

2.8. Molecular Interaction Modelling

Docking of SCO-201 and SN-38 in human BCRP transporter were performed using Glide, Schrödinger Release, 2019-3, limited liability company (LLC) [39–41]. The 3D structure of BCRP was obtained from the Research Collaboratory for Structural Bioinformatics (RCSB) Protein Data Bank (PDB ID: 6ETI) [16] and prepared using Protein Preparation Wizard, Schrödinger 2019-3, LLC [42]. SCO-201 and SN-38 were prepared using Ligprep, Schrödinger 2019-3, LLC, and docked using flexible XP docking with sampling of both nitrogen inversions and ring conformations. Further characterization of the binding between BCRP and the ligands were performed using the Desmond Molecular Dynamics System, D.E. Shaw Research, Schrödinger, 2019-2, LLC [43]. A standard membrane was fitted to the transmembrane domain of the transporter, and the system was saturated with ions and water molecules. Simulation was run for 10ns and analyzed visually and using the Simulations Interactions Diagram, Desmond, Schrödinger, 2019-2, LLC [43]. 2D docking graphics were produced from Desmond, Schrödinger, 2019-2, LLC, and 3D molecular graphic images of docking were established using the PyMOL Molecular Graphics System, Version 2.0, Schrödinger, LLC.

2.9. In Vitro DMPK Analysis: Transporter Inhibition Analysis

Cells were seeded in a 96-well plate (20,000 cells/well) and used on days 2 or 3 post-seeding. SCO-201 was prepared in assay buffer (HBSS-HEPES, pH 7.4), added to the cell plate and pre-incubated at 37 °C for 15 min. SCO-201 was tested at either a single concentration (10 µM by default) or multiple concentrations (0.03, 0.1, 0.3, 1, 3, 10, 30, and 100 µM by default), with a final DMSO concentration of 1%. Subsequently, substrate was added to the plate followed by 20 min incubation at 37 °C. The plate was then washed with cold assay buffer followed by fluorescence reading on a plate reader. The tested transporters, cell lines, substrates, and reference inhibitors are shown in the Supplementary Materials (Tables S1 and S2).

2.10. In Vitro DMPK Analysis: CYP Inhibition

The following procedure was used to asses if SCO-201 inhibits the activity of common CYP enzymes in pooled human liver microsomes in 96-well plate format. SCO-201 was pre-incubated with substrate and human liver microsomes (mixed gender, pool of 50 donors, 0.1 mg/mL) in phosphate buffer (pH 7.4) for 5 min in a 37 °C shaking waterbath. SCO-201 was tested at either a single concentration (10 µM by default) with 0.1% DMSO or multiple concentrations (0.03, 0.1, 0.3, 1, 3, 10, 30, and 100 µM by default) with up to 1% DMSO for IC_{50} determination. The reaction was initiated by adding a Nicotinamide adenine dinucleotide phosphate (NADPH)-generating system. The reaction was allowed for 10 min and stopped by transferring the reaction mixture to acetonitrile/methanol. Samples were mixed and centrifuged. Supernatants were used for HPLC-MS/MS of the respective metabolite. Tested CYP enzymes, substrates, metabolites, and reference inhibitors are shown in the Supplementary Materials (Tables S2 and S3).

Data analysis: Peak areas corresponding to the metabolite were recorded. The percent of control activity was calculated by comparing the peak area in the presence of the test compound to the control samples containing the same solvent. Subsequently, the percent inhibition was calculated by subtracting the percent control activity from 100. The IC_{50} value (concentration causing a half-maximal inhibition of the control value) was determined by non-linear regression analysis of the concentration–response curve using the Hill equation.

2.11. In Vitro DMPK Analysis: CYP Induction

The following procedure was carried out to test whether SCO-201 induces CYP1A, CYP2B6, and CYP3A activities in human hepatocytes. The procedure was designed in accordance with the FDA Guidance for Industry on Drug Interaction Studies (2006). Male and female human hepatocytes were thawed and plated into collagen-coated 96-well plates in serum-containing medium (plating medium) at a density of 0.7×10^6 viable cells/mL. The hepatocytes were cultured in a humidified incubator at 37 °C and 5% CO_2. At 4 h post plating, human hepatocytes were washed once with fresh plating medium, followed by overnight incubation. At 24 h after plating, the plating medium was removed, and the hepatocytes were overlaid with extracellular matrix (ECM) (Sigma) or Matrigel (BD) in the serum-free medium (incubation medium), and then incubated for another 24 h. Incubation medium with 0.1% DMSO was used as the negative control. After the 2-day recovery period, the hepatocytes were treated with SCO-201 or a known inducer (Table S3) in the incubation medium on day 3 and day 4. The known inducer was tested as the positive control. On day 5, the medium was removed, and the cells were incubated with the respective CYP substrate in Krebs–Henseleit buffer (pH 7.4) containing 3 mM salicylamide for 30 min. The reaction was terminated by transferring the incubation mixture to an equal volume of acetonitrile/methanol mixture (1/1, *v/v*). Samples were mixed and centrifuged. Supernatants are used for HPLC-MS/MS analysis of the corresponding metabolite. Peak areas corresponding to the metabolite were recorded. The assay was rendered valid if enzyme activity with the positive control was at least twofold greater than negative control.

2.12. Statistical Analyses

Statistical analyses were performed using Microsoft Excel. Means and standard deviations were calculated for all quantitative data. For data represented in percentage (i.e., cell viability), the standard deviations, determined from triplicate experiments, were calculated and displayed on the graphs as standard deviation percentages: $Stdv\% = Stdv * \left(\frac{\%of\,control}{OD average} \right)$. Two-tailed, type 3 Student's *t*-tests were applied on datasets where relevant, in order to determine any significant statistical differences. Statistical analysis of the data in the bidirectional transport assay was performed by comparing group means with a Student's *t*-test (two-tailed) or ANOVA, followed by Bonferroni's multiple comparisons test. The significance level was set to 5%, and thus *p*-values less than 0.05 were considered significant.

3. Results

3.1. SCO-201 Reversed BCRP-mediated Drug Resistance in Chemotherapy Resistant Cells

To assess the potential BCRP-dependent re-sensitizing effects of SCO-201, we used the BCRP-transduced canine kidney subline, MDCK-II-BCRP, which in several studies has been shown to express high levels of BCRP [37,44]. The MDCK-II-BCRP cells are known to be less sensitive to chemotherapy substrates of BCRP, such as SN-38, topotecan, and mitoxantrone, compared to their parental counterpart, MDCK-II-WT [37,44]. We therefore tested the potential of SCO-201 to restore the drug sensitivity of the MDCK-II-BCRP cells to SN-38 and mitoxantrone by treating the cells with either chemotherapy alone or in combination with SCO-201, or the BCRP inhibitor Ko143 for comparison. As seen in Figure 2A-C, SCO-201 was able to completely restore the response to SN-38 and mitoxantrone in the MDCK-II-BCRP cells similar to the BCRP-inhibitor Ko143. The IC_{50} values are shown in Table 1. Treatment with SCO-201 and SN-38 or mitoxantrone resulted in decreased IC_{50} values for the MDCK-II-BCRP cells that were comparable to the IC_{50} values found for the MDCK-II-WT cells. Altogether, these results show that SCO-201 significantly re-sensitized MDCK-II-BCRP cells to both SN-38 and mitoxantrone. This is proof-of-concept that SCO-201 can affect BCRP-mediated resistance in a model system where the resistance is engineered by ectopic overexpression of BCRP.

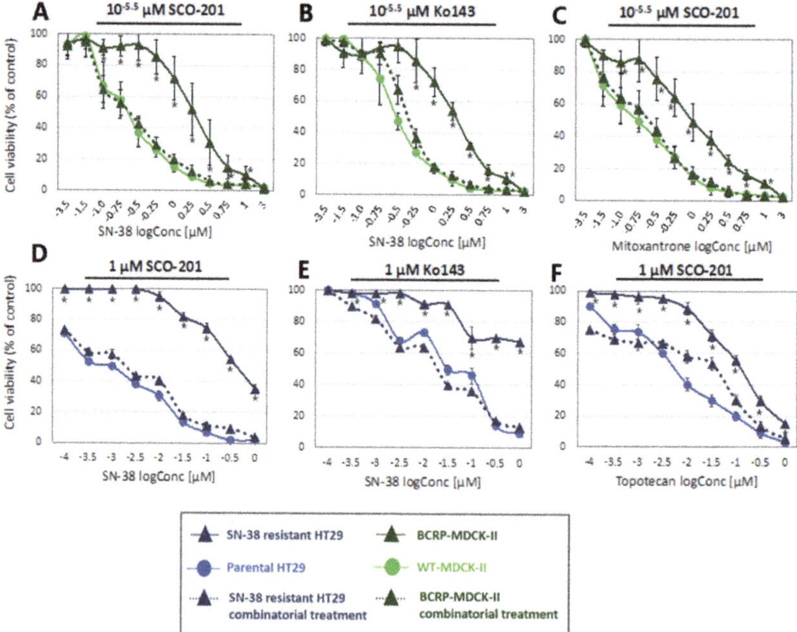

Figure 2. SCO-201 reversed drug resistance in multidrug resistance (MDR) cells. Cell viability of wild-type or breast cancer resistance protein (BCRP)-overexpressing MDCK-II cells and SN-38-sensitive or -resistant HT29 cells. Cell viability is indicated as percentage of untreated control, and error bars indicate percentage SD determined on the basis of $n = 3$–4. Note that some error bars might be invisible due to the size of the data labels. Control conditions consisted of full growth medium. (**A–C**) Wild-type and BCRP-overexpressing MDCK-II cells treated with either chemotherapy alone or in combination with $10^{-5.5}$ µM SCO-201 or Ko143. (**D–F**) Parental and SN-38-resistant HT29 cells treated with either chemotherapy alone or in combination with 1 µM SCO-201 or Ko143. Statistical difference ($p < 0.05$) between mono-treatment and combination treatment of the resistant cell lines is marked by * (Student's *t*-test (two-tailed, type 3)).

Table 1. IC_{50} values of anti-cancer drugs in the presence/absence of SCO-201 or Ko143.

	IC_{50} Value (μM) for Each Cell Line			
Drug/Drug Combination	**MDCK-II-WT**	**MDCK-II-BCRP**	**HT29$_{PAR}$**	**HT29$_{SN-38-RES}$**
SN-38	0.214 ± 0.023	1.956 ± 0.080	0.016 ± 0.010	0.463 ± 0.363
SN-38 + SCO-201		0.238 ± 0.029		0.007 ± 0.005
SN-38 + Ko143		0.437 ± 0.020		0.014 ± 0.021
Mitoxantrone	0.157 ± 0.023	1.060 ± 0.095		
Mitoxantrone + SCO-201		0.214 ± 0.030		
Topotecan			0.008 ± 0.003	0.121 ± 0.024
Topotecan + SCO-201				0.088 ± 0.030

Taken together, our results show that SCO-201 could successfully reverse resistance to the anti-cancer BCRP substrates, SN-38, topotecan, and mitoxantrone, similarly to the BCRP inhibitor Ko143 in BCRP over-expressing MDCK-II-BCRP and HT29$_{SN-38-RES}$ cells. This indicates that SCO-201 could be a modulator of BCRP activity. Supplementary studies showed that SCO-201 has a dose-dependent effect with SN-38 in both MDCK-II-BCRP and HT29$_{SN-38-RES}$ cells (Figures S3 and S4).

To further investigate the re-sensitizing effects of SCO-201 and to apply a more complex model system of resistance, we tested the effects of SCO-201 in our DEN-50R in vitro model system of acquired SN-38 resistance in colorectal cancer. To generate this system, the colorectal adenocarcinoma cell line, HT29, was subjected to gradually increasing SN-38 concentrations for a period of $\approx$10 months, resulting in an SN-38-resistant cell line (HT29$_{SN-38-RES}$). Genome-wide expression mRNA profiling revealed that BCRP was highly upregulated (25-fold) in the HT29$_{SN-38-RES}$ cells, compared to their parental counterpart (HT29$_{PAR}$) (GEO—Gene Expression Omnibus, NCBI, accession number GSE42387) [25]. We confirmed the BCRP overexpression with Western blot analysis (Figure S1). As seen in the blot, two bands could be observed for BCRP in the HT29$_{SN-38-RES}$ cells, most likely due to the glycosylation states of BCRP [45].

We tested the potential re-sensitizing effects of SCO-201 in the HT29$_{SN-38-RES}$ cells by treating the cells with either chemotherapy alone or in combination with SCO-201 or Ko143. As seen in Figure 2D–F, SCO-201 was able to significantly restore the response to both SN-38 and topotecan in the HT29$_{SN-38-RES}$ cells, similar to the response observed for combinatorial treatment with Ko143. The IC_{50} values are shown in Table 1. The IC_{50} values of SN-38 and topotecan decreased several folds for the HT29$_{SN-38-RES}$ cells, following treatment with SCO-201, and they almost completely reached the IC_{50} values for the HT29$_{PAR}$ cells (Table 1).

To investigate if these observations were due to general damaging effects on the cells that could lead to general increase in sensitivity to chemotherapy, we applied oxaliplatin, which is not a substrate for BCRP. When oxaliplatin was combined with either SCO-201 or Ko143 in the HT29$_{SN-38-RES}$ cells, no added effects were observed (Figure S2). This suggests that the re-sensitizing effects were not due to general cellular effects of either SCO-201 or Ko143.

3.2. SCO-201 Inhibited the BCRP-Mediated Flux Across Cell Membranes

To more directly investigate whether SCO-201 inhibits BCRP-mediated efflux transport, a series of bidirectional transport experiments with the prototypical BCRP substrate [³H]-estrone-3-sulfate were completed across monolayers of Caco-2 cells (Figure 3). In the absence of SCO-201, the apparent permeability of [³H]-estrone-3-sulfate in the efflux direction (basolateral to apical) was $2.7 \pm 0.2 \times 10^{-5}$ cm/second, whereas it was considerably lower in the opposite direction with an apparent permeability of $2.2 \pm 0.01 \times 10^{-6}$ cm/second. The resulting efflux ratio in the absence of SCO-201 was 12.2, which indicated a marked polarized transport in the efflux direction for [³H]-estrone-3-sulfate across monolayers of Caco-2 cells. In the presence of 10 μM SCO-201, the apparent B-A (basolateral to apical) permeability was significantly reduced to $9.6 \pm 0.7 \times 10^{-6}$ cm/second ($p < 0.0001$), whereas the apparent permeability in the A-B (apical to basolateral) direction was significantly increased to $3.2 \pm 0.3 \times 10^{-6}$ cm/second ($p = 0.0061$). Correspondingly, the calculated efflux ratio was markedly reduced to 2.9, which together with the reduction in efflux transport of [^{3}H]-estrone-3-sulfate were clear indications that SCO-201 had an inhibitory effect on BCRP-mediated efflux transport of [^{3}H]-estrone-3-sulfate across Caco-2 cell monolayers (Figure 3).

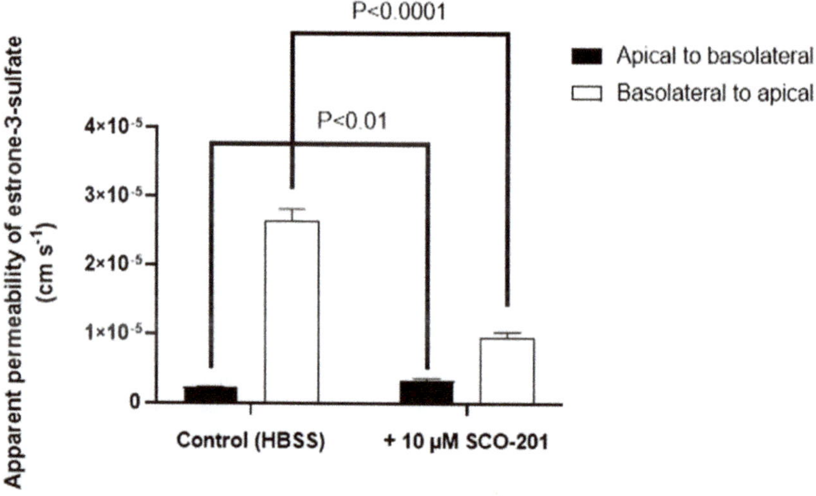

Figure 3. Bidirectional transport of estrone-3-sulfate across Caco-2 cell monolayers in the presence or absence of 10 µM SCO-201. Permeability (Papp) values were calculated from steady-state fluxes as described the in the Methods section (Section 2.5). Filled bars show PA-B (apical to basolateral) values, open bars show PB-A (basolateral to apical) values. Values are means ± SD of three individual passages, with three individual permeable supports for each transport direction per passage (n = 3–5, total N = 9). The p-values are indicated in the figure.

To further elucidate whether SCO-201 modulates BCRP and in this way triggers intracellular accumulation of chemotherapy, we conducted dye efflux studies on wild-type and SN-38-resistant HT29 cells, the latter of which overexpresses BCRP. The cells were stained with Hoechst in the presence or absence of SCO-201, Ko143, or the MDR1/P-gp inhibitor PSC833 as a negative control. Figure 4 presents results from fluorescence microscopy and imaging cytometry analysis. Accumulation of Hoechst could clearly be detected in the parental cells, whereas only low levels of Hoechst accumulation could be detected in the BCRP-overexpressing chemotherapy-resistant cells (Figure 4A,B). When the resistant cells were treated with either SCO-201 or Ko143, Hoechst accumulated to the level of the parental control cells, whereas treatment with PSC833 did not have any effects (Figure 4A,B). In a similar experiment, we quantified the dose-dependent effects of SCO-201 compared with Ko143 and evaluated the outcome with imaging cytometry. Figure 4C shows the dose-dependent effect of SCO-201 on intracellular accumulation of dye in Hoechst-stained cells compared to Ko143. As seen from the IC$_{50}$ values, the potency of Ko143 (IC$_{50}$ = 0.37 µM) and SCO-201 (IC$_{50}$ = 0.45 µM) was almost identical. Altogether, these results indicate that SCO-201, like Ko143, modulates the efflux of dye via BCRP, resulting in an accumulation of Hoechst dye in the HT29$_{SN-38-RES}$ cells. The same tendency could be observed when we quantified the effect of SCO-201 on the accumulation of Hoechst or Pheophorbic A in the MDCK-II-BCRP cells (Figure S5 and S6). These Supplementary studies also indicated that SCO-201 is likely not an inhibitor of MDR1/P-gp.

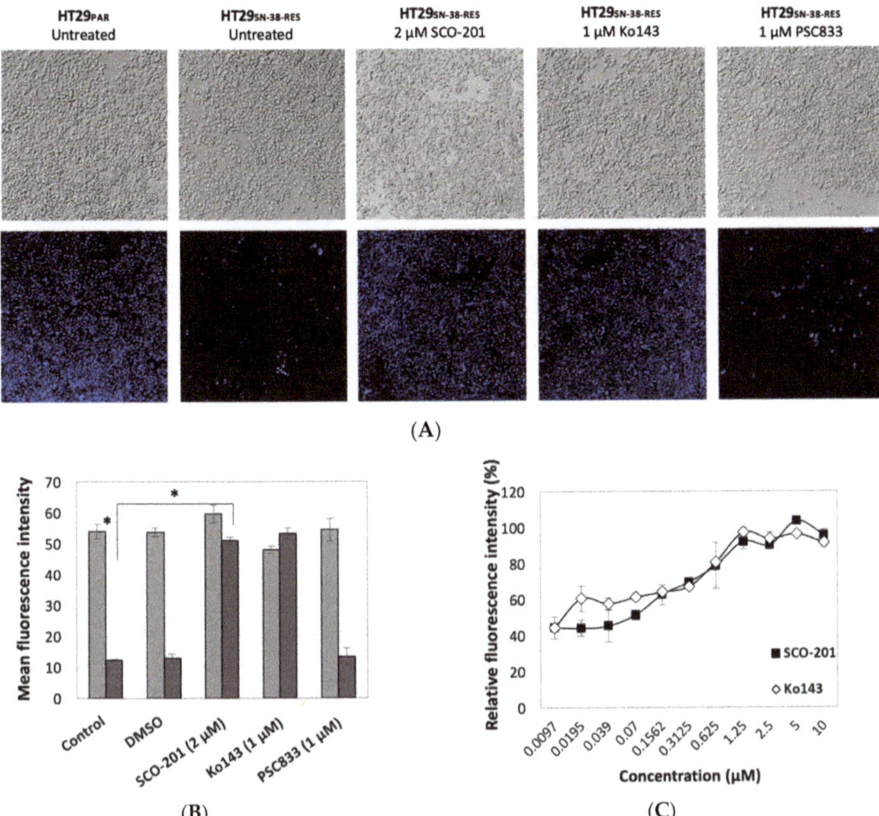

Figure 4. SCO-201 inhibited the efflux of Hoechst 33,342 from HT29$_{SN-38-RES}$ cells. (**A**) Fluorescence micrographs of Hoechst-stained parental HT29 cells (HT29$_{PAR}$) and HT29$_{SN-38-RES}$ cells that were incubated with either SCO-201, the BCRP-inhibitor Ko143, or the MDR1-inhibitor valspodar (PSC833). Full growth medium was used for the control condition. Untreated parental HT29 cells were included as a positive control, indicating the maximal accumulation of Hoechst dye, as these cells do not overexpress BCRP. (**B**) Mean fluorescence intensity of Hoechst-stained SN-38-sensitive (light grey) and SN-38 resistant (dark grey) HT29 cells treated with either DMSO, SCO-201, Ko143, or PSC833. Full growth medium was used for the control condition. The asterisks (*) indicate statistical significance ($p < 0.05$). (**C**) The dose-dependent effects of SCO-201 and Ko143 on the accumulation of Hoechst, indicated by the increase in relative fluorescence intensity of Hoechst-stained HT29$_{SN-38-RES}$ cells. Error bars in (**B**,**C**) indicate SD determined from triplicate experiments.

3.3. The Drug-Stimulated ATPase Activity of BCRP Was Competitively Inhibited by SCO-201

Following our results from the flux studies (Figures 3 and 4), we further evaluated whether SCO-201 indeed is a modulator of BCRP transport activity. To evaluate the modulatory effects of SCO-201, we conducted ATPase studies to test the effect of SCO-201 on the ATPase activity of BCRP in the presence or absence of another activating BCRP substrate. Figure 5A shows the relative ATPase activities of BCRP incubated with either SCO-201 alone, or in the presence of the strong BCRP activator, sulfasalazine. Sulfasalazine alone resulted in a maximal ATPase activity of 104.8 nmol Pi/mg protein/min compared to the basal activity of 50.68 nmol Pi/mg protein/min (data not shown). As seen on Figure 5A, SCO-201 was not in itself a stimulator of BCRP ATPase activity, and in the absence of sulfasalazine, the ATPase activity of BCRP decreased dose-dependently upon incubation with SCO-201.

When SCO-201 was added to sulfasalazine-stimulated BCRP, the ATPase activity of BCRP decreased in a dose-dependent manner (Figure 5A). Then, we tested the effect of two different concentrations of SCO-201 on SN-38-stimulated BCRP. Figure 5B,C shows the relative SN-38-stimulated ATPase activity of BCRP in the presence or absence of either 0.5 µM or 1.5 µM SCO-201. In the presence of SCO-201, the SN-38-stimulated ATPase activity of BCRP was shifted downwards, and the higher the SCO-201 concentration, the higher the SN-38 concentration was needed to reach the maximal activity of SN-38-stimulated BCRP. This revealed a typical competitive inhibition mechanism, showing that SCO-201 competed for the drug binding to the active site of BCRP. Altogether our data indicate that SCO-201 competitively inhibits the drug-stimulated ATPase activity of BCRP, suggesting that SCO-201 is a direct modulator of BCRP.

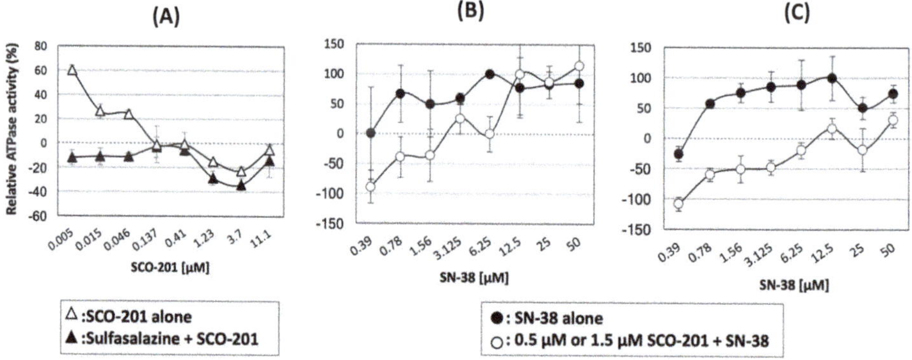

Figure 5. SCO-201 inhibited drug-stimulated ATPase activity of BCRP. (**A**) The effect of increasing concentrations of SCO-201 on basal and sulfasalazine-stimulated ATPase activity of BCRP. The data are indicated as relative ATPase activity, normalized with respect to basal (untreated) and maximal (sulfasalazine-stimulated) ATPase activity of BCRP. Sulfasalazine alone resulted in a maximal ATPase activity of 104.8 nmol Pi/mg protein/min compared to the baseline activity of 50.68 nmol Pi/mg protein/min. (**B,C**) The effect of 0.5 µM (**B**) and 1.5 µM (**C**) SCO-201 on SN-38-stimulated ATPase activity of BCRP. Error bars on all graphs represent SD determined from duplicates.

3.4. Molecular Binding Model Further Supported a Competitive Action of SCO-201

Thus far, our results have indicated that SCO-201 competitively inhibits the transport of BCRP substrates, such as SN-38, and that SCO-201 directly interacts with BCRP. To further support these results, we performed in silico molecular docking simulations to identify the binding sites of SCO-201 or SN-38 in BCRP. SN-38 and SCO-201 were successfully docked into the 3D structure of the BCRP and the resulting models showed that both SN-38 and SCO-201 were predicted to bind in the same binding cavity (Figure 6). Docking scores were -12.24 for SN-38 and -8.66 for SCO-201, indicating that both ligands could bind in the ligand binding site of the transporter.

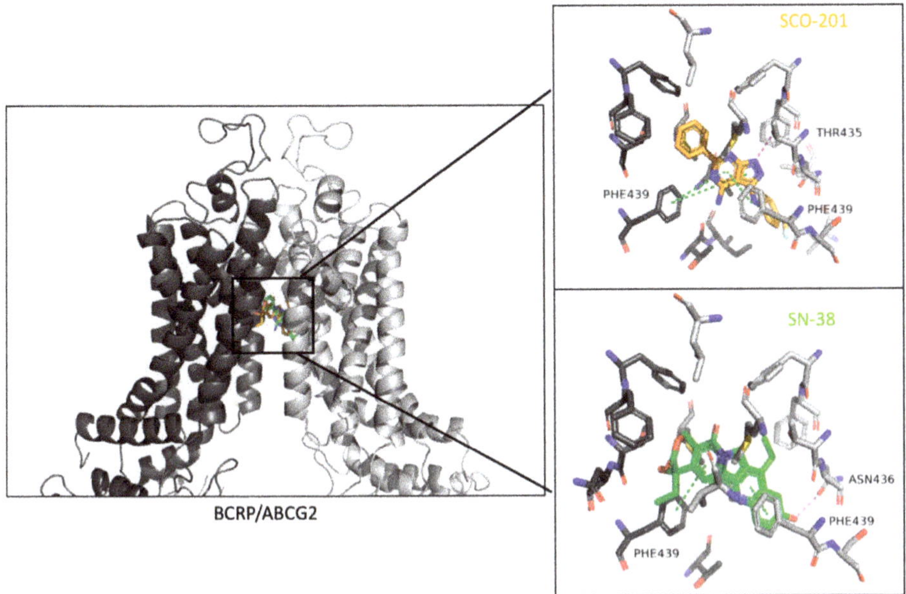

Figure 6. Docking of SN-38 and SCO-201 in the human BCRP transporter. The transporter is a homomeric dimer with chain A (dark grey) and chain B (light grey). SN-38 (green) and SCO-201 (orange) bind in the same binding cavity when docked using Glide, Schrödinger, 2019-3, LLC [39–41]. Both the substrate SN-38 and the proposed inhibitor SCO-201 interacted with Phenylalanine (PHE)-439 in both chains of the protein through hydrophobic Pi-stacking interactions. SCO-201 formed a hydrogen bond to Threonine (THR)-435, whereas SN-38 formed a hydrogen bond to Aspargine (ASN)-436. Pi-stacking interactions are colored green and hydrogen bonds are colored violet.

The molecular dynamic simulations demonstrated that both ligands remained in the binding sites throughout the sampled time period (Videos S1 and S2). The simulation interaction diagram (Figure 7) showed that SN-38 interacted through pi-stacking interactions with Phenylalanine (PHE)-439 of both protein chains, whereas SCO-201 predominantly interacted with the PHE-439 residue of the B chain and only to a lesser extent with the residue of the A protein chain (Figure 7; Figures S7 and S8). SN-38 further formed a hydrogen bond to Aspargine (ASN)-436 residue of the B chain of the transporter, whereas SCO-201 formed a hydrogen bond to both the Threonine (THR)-435 and ASN-436 residues of the B chain (Figure 7). As SCO-201 binds in the same binding cavity and interacts with some of the residues that the substrate SN-38 interacts with, it is likely that SCO-201 is a competitive inhibitor of the BCRP transporter by blocking substrate access.

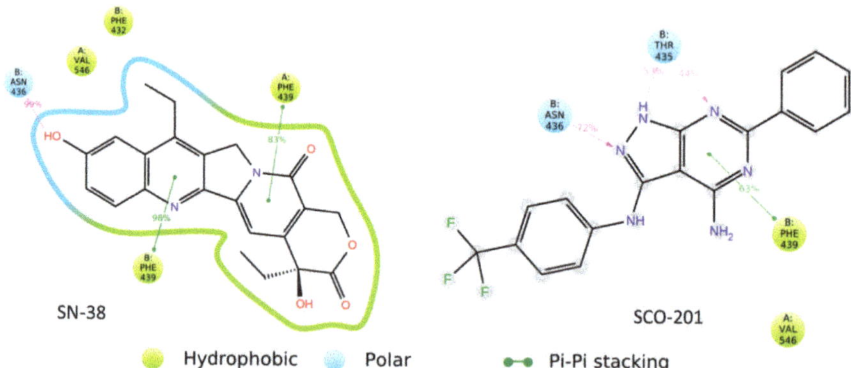

Figure 7. Binding interactions shown between SN-38 (left) and SCO-201 (right) to the transporter BCRP. Both molecules interacted with the PHE-439 residue of the B chain of BCRP. The interaction percentages indicate accounting for number of frames over the 10 ns simulation where the binding was present. Timetable of ligand–protein interactions are available in Figures S7 and S8. Interactions shown were present in over 30% of the simulation frames. Graphical representation was produced using Desmond, Schrödinger, 2019-3, LLC [43].

3.5. In Vitro DMPK Data Suggest That SCO-201 Is Not Likely to Increase The Risk of Pharmacokinetic Interactions

In clinical trials with ABC transporter inhibitors, pharmacokinetic interactions gave rise to increased serum levels of chemotherapy, thus enhancing the toxic effects of the chemotherapy, and dose reductions were therefore needed [34–38]. These dose reductions resulted in patients being under- or even overdosed, as the pharmacokinetic profile of each individual patient was difficult to predict [46–50]. On the basis of this, we aimed to test the potential inhibitory effects of SCO-201 on common transporters, playing a key role in pharmacokinetics, in order to predict the possibility that SCO-201 would negatively influence the pharmacokinetic profile of co-administered drugs. Specifically, we tested the inhibitory effects of SCO-201 on MDR1/P-gp as well as on several members of the solute carrier (SLC) family involved in drug pharmacokinetics, in accordance with the European Medicines Agency (EMA) Guidance of regulatory requirements for toxicological assessment of small molecules.

MDR1/P-gp and members of the SLC family including the organic anion transporting polypeptides (OATP) OATP1B1, OATP1B3, organic anion transporter (OAT) OAT1 and organic cation transporters (OCT) OCT2 act as major determinants of the absorption, distribution, excretion, and toxicity (ADME-tox) properties of drugs [51,52]. To investigate if SCO-201 inhibits any of the aforementioned transporters, we conducted a cell-based fluorometric drug transporter inhibition assay. We included BCRP as a positive control. The results are presented in Table 2 and show that SCO-201 inhibited BCRP as expected but is not likely an inhibitor of either MDR1/P-gp or any of the tested SLC family members as none of the fluorometric substrates accumulated. In addition, it is seen that the IC$_{50}$ value for BCRP in CHO cells was 1.7 µM, showing that SCO-201 is a potent inhibitor of BCRP (Table 2 and Figure S7).

Table 2. Summary of drug transporter inhibition.

Transport Protein	Cell Line	Substrate	IC$_{50}$ (M)
OCT2	OCT2-CHO	ASP+	NC
BCRP	BCRP-CHO	Hoechst 33342	1.7×10^{-6}
OAT1	OAT-CHO	CF	NC
OAT3	OAT3-CHO	CF	NC
OATP1B1	OATP1B1-CHO	FMTX	NC
OATP1B3	OATP1B3-CHO	FMTX	NC
P-gp (MDR1)	MDR1-MDCK-II	Calcein-AM	NC

NC: not calculable.

To further predict any potential risk of drug–drug interactions, we tested the effect of SCO-201 on key oxidative metabolic drug enzymes of the cytochrome P450 (CYP) family. Specifically, we tested the potential inhibition or induction by SCO-201 on a range of common CYPs playing a key role in determining the pharmacokinetic profile of drugs [53]. We used a standard CYP inhibition assay based on human liver microsomes, and the IC_{50} values of SCO-201 towards CYP1A, CYP2B6, CYP2C8, CYP2C9, CYP2C19, CYP2D6, and CYP3A (with two substrates) were determined in a range of concentrations (from 0.03 to 100 µM). The results are shown in Table 3 and found that no IC_{50} value was less than 100 µM, suggesting that SCO-201 is not likely an inhibitor of these CYP isoforms and may not cause CYP inhibition when its plasma concentration is below 100 µM.

Table 3. Cytochrome P450 (CYP) inhibition.

CYP	Test	Substrate	IC_{50} (M)
CYP1A	Human hepatocytes	Phenacetin substrate	$>1 \times 10^{-4}$
CYP2B6	Human hepatocytes	Bupropion substrate	NC
CYP2C8	Human liver microsomes	Paclitaxel	$>1 \times 10^{-4}$
CYP2C9	Human liver microsomes	Diclofenac	$>1 \times 10^{-4}$
CYP2C19	Human liver microsomes	Omeprazole	NC
CYP2D6	Human liver microsomes	Dextromethorphan	NC
CYP3A	Human liver microsomes	Midazolam	NC
		Testosterone	$>1 \times 10^{-4}$

NC: not calculable.

Inductions of CYP1A2, CYP2B6, and CYP3A4 by SCO-201 were tested at 1, 10, and 100 µM using both enzyme activity and mRNA level changes as the end-points. The results are presented in Table 4. Using enzyme activity as the end-point, the results were all below the cutoff value (40% of positive control), suggesting that SCO-201 is not an inducer for these CYP isoforms. Using mRNA level as the end-point, the results were also below the cutoff values, similarly suggesting that SCO-201 does not induce the CYP isoforms. However, there was a trend in both enzyme and mRNA assays that fold induction decreased with the increase in test concentrations. This trend may have resulted from cytotoxicity toward the hepatocytes.

Table 4. CYP induction.

CYP	Mean Fold Induction of mRNA at 1×10^{-4} M			Mean Fold Enzyme Activity Induction at 1×10^{-4} M		
	Donor 1	Donor 2	Donor 3	Donor 1	Donor 2	Donor 3
CYP1A				1	0.7	0.7
CYP1A2	0.24	0.39	0.26			
CYP2B6	0.17	0.52	0.45	0.8	0.9	0.6
CYP3A				1.3	2.0	0.4
CYP3A	0.25	0.25	0.32			

Fold induction = (activity of test compound treated cells)/(activity of negative control). mRNA fold induction = (mRNA level in test compound treated cells)/(mRNA levels in vehicle treated controls).

In conclusion, our results from these in vitro analyses might imply that SCO-201 does not significantly negatively influence the pharmacokinetic profile of co-administered drugs. This suggest that SCO-201 may be of a new generation of ABC transporter modulators with low risk of increased chemotherapy-mediated toxicity.

4. Discussion

ABC transporter-mediated resistance to multiple anti-cancer drugs is one of the major reasons for cancer treatment failure [1–5]. Overexpression of the transporter BCRP prevents chemotherapeutic agents such as SN-38, mitoxantrone, and fluoruracil from remaining inside cancer cells, and in this way, protects the cancer cells from being killed by these drugs. BCRP expression in cancer cells confers drug resistance in leukemia, and higher levels are reported in solid tumors from the digestive tract, endometrium, lung, and melanoma, although, contrarily, expression is generally low in breast cancer tumours [54]. There is significant association between BCRP expression and tumor response to chemotherapy and progression-free survival [1–4].

In this study, we showed that the pyrazolo-pyrimidine derivative SCO-201 can reverse MDR in vitro by competitively inhibiting the transport function of BCRP. Firstly, we tested the ability of SCO-201 to re-sensitize drug-resistant MDCK-II-BCRP and HT29$_{SN-38-RES}$ cells to chemotherapy, and these data demonstrated that SCO-201 can successfully reverse resistance in these cells (Figure 2). To investigate the potential mechanism of action of SCO-201, we conducted dye efflux assay and examined the intracellular accumulation of Hoechst 33,342 in BCRP-overexpressing cells, when treated with SCO-201, by fluorescence microscopy and imaging cytometry (Figure 4). Our results indicated that SCO-201 triggers the accumulation of Hoechst 33,342 in the BCRP-expressing cells, similarly to Ko143. We subsequently performed ATPase assay in order to examine if the effect of SCO-201 was caused by a direct inhibition of the transport function of BCRP. These results showed that SCO-201 competitively inhibits the drug-stimulated activity of BCRP (Figure 5).

To further support these results, we performed molecular docking and molecular dynamics simulations, and found that SCO-201 and SN-38 were predicted to bind in the same binding pocket of BCRP. Both SN-38 and SCO-201 interacted with PHE-439 of the B protein chain through Pi-stacking hydrophobic interactions, and ASN-436 through hydrogen bonds (Figures 6 and 7). From the cryo-EM structure of BCRP (PDB ID: 6ETI), stacking interaction was also seen between the inhibitor and PHE-439, which emphasized the importance of these residues both in transport and in inhibition of BCRP. It further supports SCO-201 being a competitive inhibitor of the BCRP transporter [16].

There are challenges of bringing ABC transporter inhibitors to clinical use, reflected by the fact that after 40 years of research there are still no approved ABC transporter inhibitors for use in the clinical setting [1]. To date, three generations of different MDR1/P-gp inhibitors have been tested and developed pre-clinically and clinically [1,7,9]. The first generation of inhibitors tested in clinical trials were not specifically developed to modulate ABC transporters but were drugs already in clinical use (e.g., verapamil and cyclosporine A). These were weak inhibitors and needed high doses. Such high doses in combination with anti-cancer drugs (e.g., mitoxantrone, daunorubicin, and etoposide) caused toxic side effects and had low therapeutic response [7–10]. Therefore, these were quickly replaced by more potent and specific second-generation inhibitors (e.g., R-verapamil and PSC-833 (Valspodar)) in order to reduce possible primary toxicities. In acute myeloid leukemia (AML) patients, the combination of PSC-833 with anti-cancer drugs seemed to be beneficial for some patients. However, the second-generation inhibitors were also shown to be inhibitors of CYPs and displayed pharmacokinetic interactions leading to increased toxicity [7–10]. Several phase III clinical studies with PSC-833 revealed that the combination with chemotherapeutic agents did not prolong the survival of cancer patients [47,48,50,55]. Third-generation inhibitors (e.g., laniquidar (R101933), ONT-093 (OC14–093), zosuqiodar (LY335979), elacridar (GF120918), and tariquidar (XR9576)) were up to 200-fold more potent and had low pharmacokinetic interaction due to a limited CYP3A inhibition [56]. The third-generation inhibitors are well tolerated in humans, safe to combine with chemotherapy due to less systemic pharmacokinetic interactions than previous MDR1 inhibitors, and were found to cause potent MDR1/P-gp inhibition in humans [57–63]. Furthermore, scanning imaging of tumors for contents of (99m)Tc-sestamibi before and after dosing of third generation inhibitors could possibly be applied to identify subgroups of anti-cancer-resistant cancer patients who may benefit from a combination of inhibitor and anti-cancer drug.

Thus, the development from first to third generation inhibitors has to a large degree abolished the pharmacokinetic interactions and related toxicities with the tested inhibitors and anti-cancer drugs. Even if toxicities should be observed with novel inhibitors and combinations with anti-cancer drugs, this can be taken care of by starting a patient with an ABC transporter inhibitor and reduced dose of chemotherapy. If no severe side effects are noted, the dose of chemotherapy can be increased at the next cycle. It is important to remember that, even with a need for lowering the dose of chemotherapy, a significant anti-tumor effect might be obtained due to the simultaneous inhibition of drug efflux pumps in the cancer cells. Another problem with the clinical studies, which tested the efficacy of ABC transporter inhibitors in combination with chemotherapy, was the general lack of randomization, and in most studies the ABC transporter inhibitor was not combined with the chemotherapy that the patient had developed resistance against. Finally, the clinical studies also lacked the inclusion of predictive biomarkers. In some of the studies, a few of the patients had ABC cassette proteins measured in their tumor tissue, but in none of the studies was this performed on a fresh tumor biopsy, and was instead performed on the primary biopsy obtained at the time of diagnosis and prior to any chemotherapy, which does not necessarily reflect the expression of ABC transporter proteins in the resistant tumor cells.

To our knowledge, no second or third generation BCRP inhibitors have been developed and tested in clinical studies. The BCRP inhibitors tested so far are first generation inhibitors, which are developed to inhibit other targets, and have pharmacokinetics interactions [64]. Some of the mechanisms by which ABC transporter inhibitors could alter the pharmacokinetics of the anti-cancer agent include competition for CYPs, intestinal or liver metabolism, inhibition of ABC transporter-mediated biliary excretion or intestinal transport, or inhibition of renal excretion and elimination [65]. This means that for an inhibitor to succeed, it needs to be non-toxic itself, and have no or low risk of interaction with important pharmacokinetic proteins, such as CYPs.

In this study, in vitro DMPK analyses of SCO-201 demonstrated no inhibition or induction of CYPs or SLCs whatsoever, suggesting a reduced risk of drug–drug interactions with other drugs, such as chemotherapy (Tables 2–4). As mentioned, an obstacle for inhibitors to succeed in clinical development is the fact that these also inhibit ABC transporters found in healthy tissues, which may lead to increased toxic effects of chemotherapy. This is especially a problem with broad-spectrum inhibitors that interact with many different efflux and uptake transporters, such as other ABC transporters or SLCs. However, by applying a specific inhibitor, this will allow the other ABC transporters in healthy tissues to compensate for the inhibition of the specific transporter, thereby protecting the healthy tissue from the toxic effects of the chemotherapy. In contrast, the anti-cancer drug has induced an up-regulation of the specific transporter in the resistant cancer cells, which will therefore be re-sensitized to the toxic effects of the anti-cancer drug. Therefore, it is highly likely that the more specific the inhibitor, the lower is the risk of increased toxicity. The specificity of SCO-201 to BCRP could provide benefits to the safety and tolerability profile in co-medication in cancer treatment compared to application of broad-spectrum inhibitors. Thereby, a BCRP-specific inhibition reduces the risk of potential drug–drug interactions in co-medication, thus decreasing the risk for increased chemotherapy-mediated toxicity.

To our best knowledge, a specific BCRP inhibitor has never been in clinical testing. Future in vivo studies of SCO-201 in combination with BCRP chemotherapy substrates should be conducted to further examine the pharmacology and test for potential increased chemotherapy-mediated toxicity. In such studies, we should utilize all the prior knowledge obtained with ABC transporter inhibitors. This means that we should carefully select patients with acquired drug resistance (a prior benefit to the chemotherapy in question), test for biomarkers such as cancer cell ABCG2 expression, start with a reduced dose of chemotherapy (the chemotherapy that the patient had acquired resistance against, and it should be an ABCG2 substrate drug) in combination with the inhibitor, perform randomization of patients in order to include time-dependent end-points such as progression free survival and overall survival, and perform a post-treatment association study between patient outcome and biomarkers. In vivo pharmacokinetic studies of other pyrazolo-pyrimidine derivatives, such as Reversan [19,34], have indicated that these do not cause increased toxicity of chemotherapy, and therefore it is possible that SCO-201 also will not cause these unwanted toxic effects in vivo.

5. Conclusions

Altogether, our data suggest that SCO-201 is a potential new drug candidate for the reversal of BCRP-mediated resistance in cancer. SCO-201 appears to be a specific and potent inhibitor of BCRP without affecting the CYP450 levels. Additionally, SCO-201 is stable in serum, has a favorable pharmacokinetic, toxicological, and pharmacodynamic profile in mice, and is orally active [33]. We conclude that SCO-201 is a highly promising drug candidate for drug-resistant cancer where overexpression of BCRP is the key mechanism of drug resistance.

Supplementary Materials: The following are available online at http://www.mdpi.com/2073-4409/9/3/613/s1. Figure S1: Western blot analysis of BCRP expression in SN-38-sensitive and SN-38-resistant HT29 cells. Figure S2: Cell viability assay with HT29 SN38-resistant colon cancer cells. Figure S3: Cell viability assay with mitoxantrone-resistant MDCK-II ATP-binding cassette (ABC)G2 cells. Figure S4: Cell viability assay with HT29 SN38-resistant colon cancer cells. Figure S5: Comparison of ABCG2 (BCRP) inhibition by SCO-201 and Ko143 as determined in the Hoechst 33,342 accumulation assay in MDCK cells. Figure S6: Comparison of SCO-201 and Ko143 as determined in the pheophorbide A accumulation assay in MDCK cells. Figure S7: SCO-201 (OBR-5-340)-mediated inhibition of BCRP. Figure S8: Timeline representation of interaction and contacts between residues in the binding cavity of the BCRP transporter and SCO-201. Figure S9: Timeline representation of interaction and contacts between residues in the binding cavity of the BCRP transporter and SN-38. Table S1: In vitro DMPK analysis of transporter inhibition. Table S2: In vitro DMPK analysis: CYP inhibition. Table S3: In vitro DMPK analysis: CYP induction.

Author Contributions: Conceptualization, J.S. (Jan Stenvang), N.B., S.R., H.W., and S.E.B.A.; methodology, J.S. (Jan Stenvang) and N.B.; validation, S.E.B.A., M.W., S.C.K., L.S., and B.B.; formal analysis, S.E.B.A., J.S. (Joen Svindt), M.W., S.K., X.L.L., M.G., L.S., and B.B.; investigation, S.E.B.A. X.L.L., M.G., L.S., and B.B.; data curation, S.A., S.R., H.W., X.L.L., M.G., L.S., B.B., and J.S. (Joen Svindt); writing—original draft preparation, S.E.B.A.; writing—review and editing, S.R., H.W., N.B., and J.S. (Jan Stenvang); visualization, S.E.B.A., X.L.L., M.G., and L.S.; supervision, J.S. (Jan Stenvang), N.B., S.R., and H.W.; project administration, J.S. (Jan Stenvang); funding acquisition, J.S. (Jan Stenvang) and N.B. All authors have read and agreed to the published version of the manuscript.

Funding: This research was funded by Savvaerksejer Jeppe Juhl og Hustru Ovita Juhls Mindelegat, grant file number 102-5212/18-3000, The Lundbeck Foundation Research Initiative on Brain Barriers and Drug Delivery and the APC was funded by Scandion Oncology.

Acknowledgments: We would like to thank Signe Lykke Nielsen for her technical support.

Conflicts of Interest: Steffen Rump (S.R.) and Henning Weigt (H.W.) hold patent application PCT/EP2016/053843 on pyrazolo-pyrimidine derivatives as BCRP inhibitors. Jan Stenvang (J.S.) and Nils Brünner (N.B.) are co-founders of Scandion Oncology A/S who own all rights to SCO-201. The funders had no role in the design of the study; in the collection, analyses, or interpretation of data; in the writing of the manuscript; or in the decision to publish the results.

References

1. Robey, R.W.; Pluchino, K.M.; Hall, M.D.; Fojo, A.T.; Bates, S.E.; Gottesman, M.M. Revisiting the role of ABC transporters in multidrug-resistant cancer. *Nat. Rev. Cancer* **2018**, *18*, 452–464. [CrossRef]

2. Hammond, W.A.; Swaika, A.; Mody, K. Pharmacologic resistance in colorectal cancer: A review. *Adv. Med. Oncol.* **2016**, *8*, 57–84. [CrossRef]

3. Lage, H. An overview of cancer multidrug resistance: A still unsolved problem. *Cell. Mol. Life Sci.* **2008**, *65*, 3145. [CrossRef] [PubMed]

4. Ren, F.; Shen, J.; Shi, H.; Hornicek, F.J.; Kan, Q.; Duan, Z. Novel mechanisms and approaches to overcome multidrug resistance in the treatment of ovarian cancer. *Biochim. Biophys. Acta Rev. Cancer* **2016**, *1866*, 266–275. [CrossRef] [PubMed]

5. Leonard, G.D. The Role of ABC Transporters in Clinical Practice. *Oncologist* **2003**, *8*, 411–424. [CrossRef] [PubMed]

6. Gottesman, M.M. Mechanisms of Cancer Drug Resistance. *Annu. Rev. Med.* **2002**, *53*, 615–627. [CrossRef] [PubMed]

7. Gottesman, M.M.; Fojo, T.; Bates, S.E. Multidrug resistance in cancer: Role of ATP–dependent transporters. *Nat. Rev. Cancer* **2002**, *2*, 48–58. [CrossRef] [PubMed]

8. Fletcher, J.I.; Haber, M.; Henderson, M.J.; Norris, M.D. ABC transporters in cancer: More than just drug efflux pumps. *Nat. Rev. Cancer* **2010**, *10*, 147–156. [CrossRef]

9. Ambudkar, S.V.; Dey, S.; Hrycyna, C.A.; Ramachandra, M.; Pastan, I.; Gottesman, M.M. Biochemical, Cellular, And Pharmacological Aspects of The Multidrug Transporter. *Annu. Rev. Pharm. Toxicol.* **1999**, *39*, 361–398. [CrossRef]

10. Li, W.; Zhang, H.; Assaraf, Y.G.; Zhao, K.; Xu, X.; Xie, J.; Yang, D.H.; Chen, Z.S. Overcoming ABC transporter-mediated multidrug resistance: Molecular mechanisms and novel therapeutic drug strategies. *Drug Resist. Updates* **2016**, *27*, 14–29. [CrossRef]

11. Sarkadi, B.; Özvegy-Laczka, C.; Német, K.; Váradi, A. ABCG2—A transporter for all seasons. *FEBS Lett.* **2004**, *567*, 116–120. [CrossRef] [PubMed]

12. Doyle, L.; Ross, D.D. Multidrug resistance mediated by the breast cancer resistance protein BCRP (ABCG2). *Oncogene* **2003**, *22*, 7340–7358. [CrossRef] [PubMed]

13. Litman, T.; Brangi, M.; Hudson, E.; Fetsch, P.; Abati, A.; Ross, D.D.; Miyake, K.; Resau, J.H.; Bates, S.E. The multidrug-resistant phenotype associated with overexpression of the new ABC half-transporter, MXR (ABCG2). *J. Cell Sci.* **2000**, *113*, 2011–2021. [PubMed]

14. Robey, R.W.; Polgar, O.; Deeken, J.; To, K.W.; Bates, S.E. ABCG2: Determining its relevance in clinical drug resistance. *Cancer Metastasis Rev.* **2007**, *26*, 39–57. [CrossRef]

15. Özvegy, C.; Litman, T.; Szakács, G.; Nagy, Z.; Bates, S.; Váradi, A.; Sarkadi, B. Functional Characterization of the Human Multidrug Transporter, ABCG2, Expressed in Insect Cells. *Biochem. Biophys. Res. Commun.* **2001**, *285*, 111–117. [CrossRef]

16. Jackson, S.M.; Manolaridis, I.; Kowal, J.; Zechner, M.; Taylor, N.M.I.; Bause, M.; Bauer, S.; Bartholomaeus, R.; Bernhardt, G.; Koenig, B.; et al. Structural basis of small-molecule inhibition of human multidrug transporter ABCG2. *Nat. Struct. Mol. Biol.* **2018**, *25*, 333–340. [CrossRef]

17. Fojo, T.; Bates, S. Strategies for reversing drug resistance. *Oncogene* **2003**, *22*, 7512–7523. [CrossRef]

18. Shukla, S.; Wu, C.-P.; Ambudkar, S.V. Development of inhibitors of ATP-binding cassette drug transporters-present status and challenges. *Expert Opin. Drug Metab. Toxicol.* **2008**, *4*, 205–223. [CrossRef]

19. Falasca, M.; Linton, K.J. Investigational ABC transporter inhibitors. *Expert Opin. Investig. Drugs* **2012**, *21*, 657–666. [CrossRef]

20. Ahmed-Belkacem, A.; Pozza, A.; Macalou, S.; Pérez-Victoria, J.M.; Boumendjel, A.; Di Pietro, A. Inhibitors of cancer cell multidrug resistance mediated by breast cancer resistance protein (BCRP/ABCG2). *Anti-Cancer Drugs* **2006**, *17*, 239–243. [CrossRef]

21. Rabindran, S.K.; He, H.; Singh, M.; Brown, E.; Collins, K.I.; Annable, T.; Greenberger, L.M. Reversal of a Novel Multidrug Resistance Mechanism in Human Colon Carcinoma Cells by Fumitremorgin C. *Cancer Res.* **1998**, *58*, 5850–5858. [PubMed]

22. Allen, J.D.; van Loevezijn, A.; Lakhai, J.M.; van der Valk, M.; van Tellingen, O.; Reid, G.; Schellens, J.H.; Koomen, G.J.; Schinkel, A.H. Potent and specific inhibition of the breast cancer resistance protein multidrug transporter in vitro and in mouse intestine by a novel analogue of fumitremorgin C. *Mol. Cancer* **2002**, *1*, 417–425.

23. Weidner, L.D.; Zoghbi, S.S.; Lu, S.; Shukla, S.; Ambudkar, S.V.; Pike, V.W.; Mulder, J.; Gottesman, M.M.; Innis, R.B.; Hall, M.D. The Inhibitor Ko143 Is Not Specific for ABCG2. *J. Pharm. Exp. Ther.* **2015**, *354*, 384–393. [CrossRef] [PubMed]

24. Stenvang, J.M.; Moreira, J.M.A.; Jensen, N.F.; Nielsen, S.L.; Orntoft, T.; Lassen, U.; Hansen, S.N.; Jandu, H.; Andreasen, M.; Noer, J.B.; et al. DEN-50R-establishment of a novel and unique cell line based drug screening platform for cancer treatment. In Proceedings of the AACR-NCI-EORTC International Conference on Molecular Targets and Cancer Therapeutics, Boston, MA, USA, 5–9 November 2015.

25. Jensen, N.F.; Stenvang, J.; Beck, M.K.; Hanáková, B.; Belling, K.C.; Do, K.N.; Viuff, B.; Nygård, S.B.; Gupta, R.; Rasmussen, M.H.; et al. Establishment and characterization of models of chemotherapy resistance in colorectal cancer: Towards a predictive signature of chemoresistance. *Mol. Oncol.* **2015**, *9*, 1169–1185. [CrossRef] [PubMed]

26. Lin, X.; Stenvang, J.; Rasmussen, M.H.; Zhu, S.; Jensen, N.F.; Tarpgaard, L.S.; Yang, G.; Belling, K.; Andersen, C.L.; Li, J.; et al. The potential role of Alu Y in the development of resistance to SN38 (Irinotecan) or oxaliplatin in colorectal cancer. *BMC Genom.* **2015**, *16*, 404. [CrossRef]

27. Guo, J.; Xu, S.; Huang, X.; Li, L.; Zhang, C.; Pan, Q.; Ren, Z.; Zhou, R.; Ren, Y.; Zi, J.; et al. Drug Resistance in Colorectal Cancer Cell Lines is Partially Associated with Aneuploidy Status in Light of Profiling Gene Expression. *J. Proteome Res.* **2016**, *15*, 4047–4059. [CrossRef]

28. Hansen, S.N.; Westergaard, D.; Thomsen, M.B.; Vistesen, M.; Do, K.N.; Fogh, L.; Belling, K.C.; Wang, J.; Yang, H.; Gupta, R.; et al. Acquisition of docetaxel resistance in breast cancer cells reveals upregulation of ABCB1 expression as a key mediator of resistance accompanied by discrete upregulation of other specific genes and pathways. *Tumour. Biol.* **2015**, *36*, 4327–4338. [CrossRef]

29. Jandu, H.; Aluzaite, K.; Fogh, L.; Thrane, S.W.; Noer, J.B.; Proszek, J.; Do, K.N.; Hansen, S.N.; Damsgaard, B.; Nielsen, S.L.; et al. Molecular characterization of irinotecan (SN-38) resistant human breast cancer cell lines. *BMC Cancer* **2016**, *16*, 34. [CrossRef]

30. Hansen, S.N.; Ehlers, N.S.; Zhu, S.; Thomsen, M.B.; Nielsen, R.L.; Liu, D.; Wang, G.; Hou, Y.; Zhang, X.; Xu, X.; et al. The stepwise evolution of the exome during acquisition of docetaxel resistance in breast cancer cells. *BMC Genom.* **2016**, *17*, 442. [CrossRef]

31. Bates, S.E.; Medina-Pérez, W.Y.; Kohlhagen, G.; Antony, S.; Nadjem, T.; Robey, R.W.; Pommier, Y. ABCG2 Mediates Differential Resistance to SN-38 (7-Ethyl-10-hydroxycamptothecin) and Homocamptothecins. *J. Pharm. Exp.* **2004**, *310*, 836–842. [CrossRef]

32. Braun, H.; Makarov, V.A.; Riabova, O.B.; Komarova, E.S.; Richter, M.; Wutzler, P.; Schmidtke, M. OBR-5-340—A Novel Pyrazolo-Pyrimidine Derivative with Strong Antiviral Activity Against Coxsackievirus B3 In Vitro and In Vivo. *Antivir. Res.* **2011**, *90*, A29. [CrossRef]

33. Makarov, V.A.; Braun, H.; Richter, M.; Riabova, O.B.; Kirchmair, J.; Kazakova, E.S.; Seidel, N.; Wutzler, P.; Schmidtke, M. Pyrazolopyrimidines: Potent Inhibitors Targeting the Capsid of Rhino- and Enteroviruses. *ChemMedChem* **2015**, *10*, 1629–1634. [CrossRef] [PubMed]

34. Burkhart, C.A.; Watt, F.; Murray, J.; Pajic, M.; Prokvolit, A.; Xue, C.; Flemming, C.; Smith, J.; Purmal, A.; Isachenko, N.; et al. Small-molecule multidrug resistance-associated protein 1 inhibitor reversan increases the therapeutic index of chemotherapy in mouse models of neuroblastoma. *Cancer Res.* **2009**, *69*, 6573–6580. [CrossRef] [PubMed]

35. Kim, S.; Chen, J.; Cheng, T.; Gindulyte, A.; He, J.; He, S.; Li, Q.; Shoemaker, B.A.; Thiessen, P.A.; Yu, B.; et al. PubChem 2019 update: Improved access to chemical data. *Nucleic Acids Res.* **2019**, *47*, D1102–D1109. [CrossRef]

36. National Center for Biotechnology Information. PubChem Database. 7-Ethyl-10-hydroxycamptothecin, CID= 104842. Available online: https://pubchem.ncbi.nlm.nih.gov/compound/7-Ethyl-10-hydroxycamptothecin (accessed on 28 January 2020).

37. Pick, A.; Müller, H.; Wiese, M. Structure-activity relationships of new inhibitors of breast cancer resistance protein (ABCG2). *Bioorg. Med. Chem.* **2008**, *16*, 8224–8236. [CrossRef] [PubMed]

38. Müller, M.; Bakos, E.; Welker, E.; Váradi, A.; Germann, U.A.; Gottesman, M.M.; Morse, B.S.; Roninson, I.B.; Sarkadi, B. Altered drug-stimulated ATPase activity in mutants of the human multidrug resistance protein. *J. Biol. Chem.* **1996**, *271*, 1877–1883. [CrossRef]

39. Friesner, R.A.; Murphy, R.B.; Repasky, M.P.; Frye, L.L.; Greenwood, J.R.; Halgren, T.A.; Sanschagrin, P.C.; Mainz, D.T. Extra Precision Glide: Docking and Scoring Incorporating a Model of Hydrophobic Enclosure for Protein–Ligand Complexes. *J. Med. Chem.* **2006**, *49*, 6177–6196. [CrossRef]

40. Halgren, T.A.; Murphy, R.B.; Friesner, R.A.; Beard, H.S.; Frye, L.L.; Pollard, W.T.; Banks, J.L. Glide: A New Approach for Rapid, Accurate Docking and Scoring. 2. Enrichment Factors in Database Screening. *J. Med. Chem.* **2004**, *47*, 1750–1759. [CrossRef]

41. Friesner, R.A.; Banks, J.L.; Murphy, R.B.; Halgren, T.A.; Klicic, J.J.; Mainz, D.T.; Repasky, M.P.; Knoll, E.H.; Shelley, M.; Perry, J.K.; et al. Glide: A New Approach for Rapid, Accurate Docking and Scoring. 1. Method and Assessment of Docking Accuracy. *J. Med. Chem.* **2004**, *47*, 1739–1749. [CrossRef]

42. Sastry, G.M.; Adzhigirey, M.; Day, T.; Annabhimoju, R.; Sherman, W. Protein and ligand preparation: Parameters, protocols, and influence on virtual screening enrichments. *J. Comput. Aided Mol. Des. Des.* **2013**, *27*, 221–234. [CrossRef]

43. Bowers, A.K.J.; Chow, E.; Xu, H.; Dror, R.O.; Eastwood, M.P.; Gregersen, B.A.; Klepeis, J.L.; Kolossvary, I.; Moraes, M.A.; Sacerdoti, F.D.; et al. Scalable algorithms for molecular dynamics simulations on commodity clusters. In Proceedings of the 2006 ACM/IEEE Conference on Supercomputing, Tampa, FL, USA, 11–17 November 2006; p. 84-es.

44. Krapf, M.K.; Gallus, J.; Vahdati, S.; Wiese, M. New Inhibitors of Breast Cancer Resistance Protein (ABCG2) Containing a 2,4-Disubstituted Pyridopyrimidine Scaffold. *J. Med. Chem.* **2018**, *61*, 3389–3408. [CrossRef] [PubMed]

45. Diop, N.K.; Hrycyna, C.A. N-Linked Glycosylation of the Human ABC Transporter ABCG2 on Asparagine 596 Is Not Essential for Expression, Transport Activity, or Trafficking to the Plasma Membrane. *Biochemistry* **2005**, *44*, 5420–5429. [CrossRef] [PubMed]

46. Szakács, G.; Paterson, J.K.; Ludwig, J.A.; Booth-Genthe, C.; Gottesman, M.M. Targeting multidrug resistance in cancer. *Nat. Rev. Drug Discov.* **2006**, *5*, 219–234. [CrossRef] [PubMed]

47. Baer, M.R.; George, S.L.; Dodge, R.K.; O'Loughlin, K.L.; Minderman, H.; Caligiuri, M.A.; Anastasi, J.; Powell, B.L.; Kolitz, J.E.; Schiffer, C.A.; et al. Phase 3 study of the multidrug resistance modulator PSC-833 in previously untreated patients 60 years of age and older with acute myeloid leukemia: Cancer and Leukemia Group B Study 9720. *Blood* **2002**, *100*, 1224–1232. [CrossRef]

48. Lhommé, C.; Joly, F.; Walker, J.L.; Lissoni, A.A.; Nicoletto, M.O.; Manikhas, G.M.; Baekelandt, M.M.O.; Gordon, A.N.; Fracasso, P.M.; Mietlowski, W.L.; et al. Phase III Study of Valspodar (PSC 833) Combined With Paclitaxel and Carboplatin Compared With Paclitaxel and Carboplatin Alone in Patients With Stage IV or Suboptimally Debulked Stage III Epithelial Ovarian Cancer or Primary Peritoneal Cancer. *J. Clin. Oncol.* **2008**, *26*, 2674–2682. [CrossRef]

49. Leonard, G.D.; Polgar, O.; Bates, S.E. ABC transporters and inhibitors: New targets, new agents. *Curr. Opin. Investig. Drugs* **2002**, *3*, 1652–1659.

50. Greenberg, P.L.; Lee, S.J.; Advani, R.; Tallman, M.S.; Sikic, B.I.; Letendre, L.; Dugan, K.; Lum, B.; Chin, D.L.; Dewald, G.; et al. Mitoxantrone, etoposide, and cytarabine with or without valspodar in patients with relapsed or refractory acute myeloid leukemia and high-risk myelodysplastic syndrome: A phase III trial (E2995). *J. Clin. Oncol.* **2004**, *22*, 1078–1086. [CrossRef]

51. Kovacsics, D.; Patik, I.; Özvegy-Laczka, C. The role of organic anion transporting polypeptides in drug absorption, distribution, excretion and drug-drug interactions. *Expert Opin. Drug Metab. Toxicol.* **2017**, *13*, 409–424. [CrossRef]

52. Szakács, G.; Váradi, A.; Özvegy-Laczka, C.; Sarkadi, B. The role of ABC transporters in drug absorption, distribution, metabolism, excretion and toxicity (ADME–Tox). *Drug Discov. Today* **2008**, *13*, 379–393. [CrossRef]

53. Zanger, U.M.; Schwab, M. Cytochrome P450 enzymes in drug metabolism: Regulation of gene expression, enzyme activities, and impact of genetic variation. *Pharmacol. Ther.* **2013**, *138*, 103–141. [CrossRef]

54. Noguchi, K.; Katayama, K.; Sugimoto, Y. Human ABC transporter ABCG2/BCRP expression in chemoresistance: Basic and clinical perspectives for molecular cancer therapeutics. *Pharmgenom. Pers. Med.* **2014**, *7*, 53–64. [CrossRef] [PubMed]

55. Friedenberg, W.R.; Rue, M.; Blood, E.A.; Dalton, W.S.; Shustik, C.; Larson, R.A.; Sonneveld, P.; Greipp, P.R. Phase III study of PSC-833 (valspodar) in combination with vincristine, doxorubicin, and dexamethasone (valspodar/VAD) versus VAD alone in patients with recurring or refractory multiple myeloma (E1A95): A trial of the Eastern Cooperative Oncology Group. *Cancer* **2006**, *106*, 830–838. [CrossRef]

56. Romanov, R.A.; Bystrova, M.F.; Rogachevskaya, O.A.; Sadovnikov, V.B.; Shestopalov, V.I.; Kolesnikov, S.S. The ATP permeability of pannexin 1 channels in a heterologous system and in mammalian taste cells is dispensable. *J. Cell Sci.* **2012**, *125*, 5514. [CrossRef]

57. Kelly, R.J.; Draper, D.; Chen, C.C.; Robey, R.W.; Figg, W.D.; Piekarz, R.L.; Chen, X.; Gardner, E.R.; Balis, F.M.; Venkatesan, A.M.; et al. A pharmacodynamic study of docetaxel in combination with the P-glycoprotein antagonist tariquidar (XR9576) in patients with lung, ovarian, and cervical cancer. *Clin. Cancer Res.* **2011**, *17*, 569–580. [CrossRef] [PubMed]

58. Abraham, J.; Edgerly, M.; Wilson, R.; Chen, C.; Rutt, A.; Bakke, S.; Robey, R.; Dwyer, A.; Goldspiel, B.; Balis, F.; et al. A phase I study of the P-glycoprotein antagonist tariquidar in combination with vinorelbine. *Clin. Cancer Res. Off. J. Am. Assoc. Cancer Res.* **2009**, *15*, 3574–3582. [CrossRef] [PubMed]

59. Pusztai, L.; Wagner, P.; Ibrahim, N.; Rivera, E.; Theriault, R.; Booser, D.; Symmans, F.W.; Wong, F.; Blumenschein, G.; Fleming, D.R.; et al. Phase II study of tariquidar, a selective P-glycoprotein inhibitor, in patients with chemotherapy-resistant, advanced breast carcinoma. *Cancer* **2005**, *104*, 682–691. [CrossRef] [PubMed]

60. Fox, E.; Widemann, B.C.; Pastakia, D.; Chen, C.C.; Yang, S.X.; Cole, D.; Balis, F.M. Pharmacokinetic and pharmacodynamic study of tariquidar (XR9576), a P-glycoprotein inhibitor, in combination with doxorubicin, vinorelbine, or docetaxel in children and adolescents with refractory solid tumors. *Cancer Chemother. Pharm.* **2015**, *76*, 1273–1283. [CrossRef]

61. Ruff, P.; Vorobiof, D.A.; Jordaan, J.P.; Demetriou, G.S.; Moodley, S.D.; Nosworthy, A.L.; Werner, I.D.; Raats, J.; Burgess, L.J. A randomized, placebo-controlled, double-blind phase 2 study of docetaxel compared to docetaxel plus zosuquidar (LY335979) in women with metastatic or locally recurrent breast cancer who have received one prior chemotherapy regimen. *Cancer Chemother. Pharmacol.* **2009**, *64*, 763–768. [CrossRef]

62. Lê, L.H.; Moore, M.J.; Siu, L.L.; Oza, A.M.; MacLean, M.; Fisher, B.; Chaudhary, A.; de Alwis, D.P.; Slapak, C.; Seymour, L.; et al. Phase I study of the multidrug resistance inhibitor zosuquidar administered in combination with vinorelbine in patients with advanced solid tumours. *Cancer Chemother. Pharm.* **2005**, *56*, 154–160. [CrossRef]

63. Fracasso, P.M.; Goldstein, L.J.; de Alwis, D.P.; Rader, J.S.; Arquette, M.A.; Goodner, S.A.; Wright, L.P.; Fears, C.L.; Gazak, R.J.; Andre, V.A.M.; et al. Phase I Study of Docetaxel in Combination with the P-Glycoprotein Inhibitor, Zosuquidar, in Resistant Malignancies. *Clin. Cancer Res.* **2004**, *10*, 7220. [CrossRef]

64. Brackman, D.J.; Giacomini, K.M. Reverse Translational Research of ABCG2 (BCRP) in Human Disease and Drug Response. *Clin. Pharm.* **2018**, *103*, 233–242. [CrossRef] [PubMed]

65. Relling, M.V. Are the Major Effects of P-Glycoprotein Modulators Due to Altered Pharmacokinetics of Anticancer Drugs? *Ther. Drug Monit.* **1996**, *18*, 350–356. [CrossRef] [PubMed]

Article

The Vacuolar H$^+$ ATPase α3 Subunit Negatively Regulates Migration and Invasion of Human Pancreatic Ductal Adenocarcinoma Cells

Mette Flinck [1], **Sofie Hagelund** [1], **Andrej Gorbatenko** [2], **Marc Severin** [1], **Elena Pedraz-Cuesta** [1], **Ivana Novak** [1], **Christian Stock** [3] and **Stine Falsig Pedersen** [1,*]

[1] Section for Cell Biology and Physiology, Department of Biology, Faculty of Science, University of Copenhagen, DK-2100 Copenhagen, Denmark; mette.flinck@bio.ku.dk (M.F.); knr881@alumni.ku.dk (S.H.); trv586@alumni.ku.dk (M.S.); elenapedraz@gmail.com (E.P.-C.); inovak@bio.ku.dk (I.N.)

[2] Department of Pathology, Icahn School of Medicine at Mount Sinai, New York, NY 10029, USA; andrej.gorbatenko@yahoo.com

[3] Department of Gastroentero-, Hepato- and Endocrinology, Hannover Medical School, D-30625 Hannover, Germany; Stock.Christian@mh-hannover.de

* Correspondence: sfpedersen@bio.ku.dk

Received: 12 December 2019; Accepted: 13 February 2020; Published: 18 February 2020

Abstract: Increased metabolic acid production and upregulation of net acid extrusion render pH homeostasis profoundly dysregulated in many cancers. Plasma membrane activity of vacuolar H$^+$ ATPases (V-ATPases) has been implicated in acid extrusion and invasiveness of some cancers, yet often on the basis of unspecific inhibitors. Serving as a membrane anchor directing V-ATPase localization, the a subunit of the V0 domain of the V-ATPase (ATP6V0a1-4) is particularly interesting in this regard. Here, we map the regulation and roles of ATP6V0a3 in migration, invasion, and growth in pancreatic ductal adenocarcinoma (PDAC) cells. a3 mRNA and protein levels were upregulated in PDAC cell lines compared to non-cancer pancreatic epithelial cells. Under control conditions, a3 localization was mainly endo-/lysosomal, and its knockdown had no detectable effect on pH$_i$ regulation after acid loading. V-ATPase inhibition, but not a3 knockdown, increased HIF-1α expression and decreased proliferation and autophagic flux under both starved and non-starved conditions, and spheroid growth of PDAC cells was also unaffected by a3 knockdown. Strikingly, a3 knockdown increased migration and transwell invasion of Panc-1 and BxPC-3 PDAC cells, and increased gelatin degradation in BxPC-3 cells yet decreased it in Panc-1 cells. We conclude that in these PDAC cells, a3 is upregulated and negatively regulates migration and invasion, likely in part via effects on extracellular matrix degradation.

Keywords: PDAC; TCIRG1; ATP6V0a3; invasion; migration; matrix degradation; proliferation; pH-regulation; autophagy

---

1. Introduction

Pancreatic cancer, of which pancreatic ductal adenocarcinoma (PDAC) comprises about 90% of cases, is one of the deadliest cancers globally, with a 5-year survival rate of less than 10% and predicted to be the second leading cause of cancer-related death in the USA by 2030 [1,2]. Reliable biomarkers are lacking, and most cases are diagnosed so late that surgical treatment is unfeasible. The standard of care is chemotherapy in the form of gemcitabine plus nab-paclitaxel, or FOLFIRINOX (leucovorin, 5-fluorouracil, irinotecan, and oxaliplatin). However, response rates are low and treatments prolong life only for weeks to a few months [1]. Hence, novel diagnostic and treatment options are urgently needed.

Solid tumors exhibit metabolic changes and high proliferative rates, causing increased acid generation and making cancer cells dependent on the upregulation of net acid extrusion [3,4]. In conjunction with restricted diffusion, this renders the tumor microenvironment highly acidic, while intracellular pH (pH_i) is normal or alkaline [4,5]. Collectively this favors cancer cell proliferation and survival and promotes motility and invasiveness [3–5]. Well-studied net acid extruding ion transporters that are often highly expressed in cancer cells include Na^+/H^+ exchanger 1 (NHE1), Na^+,HCO_3^- cotransporters (NBCs), and monocarboxylate transporters (MCTs) [4,5]. V-type H^+-ATPases (V-ATPases) have also been studied in the context of cancer [6,7]. In contrast to NHE1, NBCs and MCTs, V-ATPases predominantly localize to endosomes and lysosomes, the Golgi apparatus, and other intracellular compartments [8–13] and are only found in the plasma membrane in specialized cell types and some cancer cells [6,14,15]. V-ATPases play a pivotal role in controlling the luminal pH of endosomes, lysosomes, and the Golgi apparatus [6,7,14,16]. Through this, as well as through acidification-independent scaffolding functions, they regulate endocytic trafficking, autophagy, macropinocytosis, lysosomal degradation, metabolism, protein glycosylation, and signaling pathways including notch-, Wnt-, and epidermal growth factor receptor (EGFR) signaling [6,7,14–16].

V-ATPases consist of a peripheral V_1 section responsible for ATP hydrolysis and a membrane-integral V_0 section, responsible for H^+ translocation. V_1 comprises subunits A–H, and V_0 subunits a, d, e, c, c″ and accessory subunit Ac45. The a subunit (ATP6V0a), of which there are four isoforms, a1–4, forms a hemichannel mediating H^+ transport and a transmembrane anchor important for V-ATPase localization [6,7,14]. The isoform expression pattern of ATP6V0a is cell type specific. ATP6V0a3 (a3) is the predominant a isoform in osteoclasts, where it localizes the V-ATPase to the plasma membrane, and a3 mutations are responsible for inherited forms of osteopetrosis [17].

Several highly invasive breast cancer cells exhibit elevated plasma membrane V-ATPase expression, and their in vitro invasion was shown to be reduced by V-ATPase inhibitors bafilomycin and concanamycin A (ConA) in a manner proposed to involve plasma membrane-localized V-ATPases [18,19]. Similar findings were reported in melanoma cells [20], whereas in prostate cancer cells, invasiveness was inhibited by bafilomycin in the absence of detectable V-ATPase expression in the plasma membrane [9,10]. In breast cancer cells, a3 knockdown (KD) reduced invasiveness, yet only by at most ~25% [21,22], and a role for a4 rather than a3 in invasiveness was proposed in 4T1-12B breast cancer cells [23].

V-ATPase function is of particular interest in PDAC, given the reliance of this exceptionally aggressive cancer on nutrient scavenging, and increased lysosomal catabolism, processes critically dependent on V-ATPase activity [24–26]. In PDAC patient tissue, expression of V-ATPase subunits V1E [27] and V0c [28] was reported to correlate with cancer stage. Furthermore, a3 was detected in the plasma membrane of invasive PDAC cells [27,29]. However, V-ATPase inhibition did not consistently reduce invasion in the PDAC cell lines [27,29], and conversely to what would be expected for a stimulatory role in invasion, V-ATPase inhibition upregulated the activity of matrix metalloprotease-2 (MMP-2) [27], a major matrix-degrading MMP in pancreatic cancers [30]. In contrast, V-ATPase activity is important for the degradation of MT1-MMP (also known as MMP14), which is highly expressed in PDAC cells [31,32] and is a key regulator of invasion [33]. Thus, a clear link between a3 and PDAC growth and invasiveness is lacking.

The aim of this study is to characterize the regulation and roles of a3 in PDAC cells. We report that a3 was upregulated in PDAC cell lines compared to non-cancerous pancreatic epithelial cells. In PDAC cells, a3 localized mainly to an endosomal/lysosomal compartment and its knockdown had no detectable effect on pH_i regulation after an acid load. V-ATPase inhibition, but not a3 knockdown, decreased PDAC cell proliferation and autophagic flux. Notably, a3 KD increased migration and invasion of Panc-1 and BxPC-3 PDAC cells, and increased gelatin matrix degradation by BxPC-3 cells but decreased it in Panc-1 cells. Thus, in these cancer cells, a3 is upregulated and negatively regulates migration and invasiveness. No major roles of a3 in favoring PDAC development could be detected, although their existence in settings not studied here remains possible.

2. Methods

2.1. Reagents

Unless otherwise mentioned reagents were from Sigma-Aldrich (St. Louis, MO, USA) and were of the highest analytical grade. Antibodies against AMPK, phospho-Thr172-AMPK, LC3B, GAPDH, Golgin 97, Rab7, PARP, phospho-Ser780 pRb, and phospho-Ser15 p53 were obtained from Cell Signaling Technology (Danvers, MA, USA). Antibodies against V-ATPase B2 subunit, E-cadherin, p21, and LAMP1 were obtained from Santa Cruz Biotechnology (Santa Cruz, CA, USA), and antibody against β-actin was from Sigma-Aldrich. Antibodies against alpha-adaptin 2, MT1-MMP (MMP14) and p62 were from Abcam (Cambridge, UK). HIF-1α and p150^{Glued} antibodies were from BD Biosciences (San Jose, CA, USA). Antibodies against Giantin and HSP47 were from Enzo Lifesciences (Farmingdale, NY, USA). HRP-conjugated secondary antibodies for Western blotting (goat-anti-mouse (GAM) and goat-anti-rabbit (GAR)) were from DAKO (Glostrup, Denmark). Rhodamin phalloidin, Alexa Fluor 568 conjugated GAM, and Alexa Fluor 488 conjugated GAR secondary antibodies for immunofluorescence were from Invitrogen (Carlsbad, CA, USA). Antibody against Cortactin was from Merck (Darmstadt, Germany). Concanamycin A, Forskolin, and antibody against TCIRG1 (V-ATPase a3 subunit) were from Sigma-Aldrich (St. Louis, MO, USA).

2.2. Cell Culture and Treatments

BxPC-3 cells (American Type Culture Collection, ATCC) were cultured in RPMI-1640 medium (Gibco, #61870-010, 11 mM glucose) and Panc-1 cells (ATCC) in Dulbecco's modified Eagle medium (DMEM, Gibco, #32430-027, 25 mM glucose), high glucose, all supplemented with Glutamax (Life Technologies, Camarillo, CA, USA), 10% fetal bovine serum (Gibco, #10 106-177), and 100 U/mL penicillin, 100 µg/mL streptomycin (Pen/Strep, Invitrogen, Carlsbad, CA #15140-148). For starvation conditions, cells were grown in DMEM (Gibco, #11966-027) or RPMI-1640 (Gibco, #11879020) media without glucose, 1% fetal bovine serum and 100 U/mL penicillin, 100 µg/mL streptomycin. Immortalized normal human pancreatic ductal epithelial (HPDE H6c7) cells were kindly provided by Dr M.-S. Tsao, Ontario Cancer Institute, Toronto, Canada [34,35] and cultured in kerantinocyte basal medium supplemented with epidermal growth factor and bovine pituitary extract. All cell cultures were maintained at 37 °C, 95% humidity and 5% CO2.

2.3. siRNA and Transfection

The pre-designed siRNAs against a3 and a negative control siRNA were from Origene (Rockville, MD, USA, #SR306986). Plasmids used were LAMP1-mCherry in pEGFP-N1 (GFP removed), a gift from Bin Liu, Danish Cancer Society Research Center, Denmark, and human TCIRG1 (a3) in pMA, a gift from Johan Richter, Lund University, Sweden. Both were verified by sequencing. Cells were transfected with the relevant siRNA (10 nM final concentration) or plasmid using Lipofectamine 2000 or -3000 (Life Technologies, Camarillo, CA, USA, #11668019), 48 h before start of the experiment. Then, 4 h after transfection, the medium was changed to either growth or starvation medium.

2.4. SDS-PAGE and Western Blotting

After treatments as indicated, cells were lysed in 95 °C SDS lysis buffer (1% SDS, 0.1 M Tris-HCl, 0.1 M NaVO$_3$, pH 7.5) supplemented with Complete™ protease inhibitor mix (Roche Diagnostics GmbH, Germany, #11836153001). Lysates were homogenized by sonication (PowerMED, Portland, Maine), centrifuged (Micromax RF, Thermo) for 5 min at 20,000× g at 4 °C, and protein concentrations determined using DC Protein assay kit (BioRad, Hercules, CA, USA, #500-0113, #500-0114, #500-0115). Samples were equalized with ddH$_2$O and NuPAGE LDS 4x sample buffer (5 mM Tris-Cl pH 6.8, 10% SDS, 1% bromophenol blue, 10% glycerol; Life Technologies, Carlsbad, CA, USA, #NP0007) and dithiothreitol added. Equal amounts of protein per lane were separated by SDS-PAGE, using Tris/glycine/SDS running buffer (BioRad, Hercules, CA, USA, #161-0732), precast Criterion 10% TGX gels (BioRad, Hercules, CA, USA, #567-1034 (18-wells) or #567-1035 (26-wells)), and BenchMark protein ladder (Life Technologies, Carlsbad, CA, USA, #10747-012). Proteins were transferred to Trans-Blot Turbo 0.2 μm nitrocellulose membranes (BioRad, Hercules, CA, USA, #170-4159). Membranes were Ponceau S stained (Sigma-Aldrich, St. Louis, MO, USA, #P7170-1L), blocked for 1 h at 37 °C in 5% nonfat dry milk in TBST (0.01 M Tris/HCl, 0.15 M NaCl, 0.1% Tween 20, pH 7.4), incubated with primary and secondary antibodies, and developed using ECL (Pierce™ ECL Western Blotting Substrate (Bio-Rad, Hercules, CA, USA, Cat. #1705061) or SignalFire (Cell Signaling, Danvers, MA, USA, #6883) and the Fusion Fx system (Vilber Lourmat, Marne-la-Vallé, France) for HRP-conjugated secondary antibodies. Densitometric analyses were carried out using UN-SCAN-IT 6.1 (Silk Scientific, Orem, Utah), or ImageJ software v1.52s.

2.5. Quantitative Real-Time PCR (qPCR)

Isolation of total RNA was performed using *NucleoSpin®RNA II* (Macherey-Nagel, Germany) according to the manufacturer's instructions. RNA was reverse-transcribed using Superscript III Reverse Transcriptase (Invitrogen, Carlsbad, CA, USA, #18080044) and cDNA amplified by qPCR using SYBR Green (Roche, Basel, Switzerland, #04913914001) in an ABI7900 qPCR machine, in triplicate and using the steps: 95 °C for 10 min, 40 cycles of [95 °C for 30 s, 55 °C for 1 min, 72 °C for 30 s], 95 °C for 1 min. Primers were designed using NCBI/ Primer-BLAST (www.ncbi.nlm.nih.gov) and synthesized by Eurofins Genomics, Ebersberg, Germany (ATP6V0a1 and ATP6V0a2 and β-actin) or Invitrogen, Carlsbad, Ca, USA (ATP6V0a3, ATP6V1B2). Primer sequences: ATP6V0a1, sense 5′-GAGGAGGCAGACGAGTTTGA-3′; antisense 5′-CCGGTCCCGCTGTACAATTT-3′, ATP6V0a2, sense 5′-GGTTATCGCGCTCTTTGCAG-3′; antisense 5′-TTCTACCCAGTGGAGGCGTA-3′, ATP6V0a3, sense 5′-GTGAATGGCTGGAGCTCCGATGA-3′; antisense 5′-AGGCCTATGCGCATCACCATGG-3′ and ATP6V1B2, sense 5′- AGTCAGTCGGAACTA CCTCTC-3′; antisense 5′-CATCCGGTAAGGTCAAATGGAC-3′; β-actin sense 5′-AGCGAGCATCC CCCAAAGTT-3′, antisense 5′-GGGCACGAAGGCTCATCATT-3′. mRNA levels were determined using the comparative threshold cycle (Ct) method, normalized to β-actin, and were expressed relative to that in HPDE cells or relative mock ctrl.

2.6. Immunofluorescence Imaging

Cells grown on glass coverslips were washed in ice-cold phosphate-buffered saline (PBS), fixed in 2% paraformaldehyde (Sigma, St. Louis, MO, USA, #47608) for 15 min at room temperature, washed in TBST (2 × 5 min), permeabilized for 5 min in 0.1% Triton x-100 (Sigma-Aldrich, St. Louis, MO, USA, #T8787) in TBST, blocked for 30 min in 5% BSA in TBST, and incubated at room temperature (RT) for 1.5 h or overnight at 4 °C with primary antibodies diluted in TBST + 1% BSA. The next day, preparations were washed in TBST + 1% BSA (3 × 5 min), and incubated for 1 h at room temperature with the relevant fluorophore-conjugated secondary antibodies diluted in TBST + 1% BSA. Finally, preparations were washed in TBST + 1% BSA for 3 × 5 min, of which the second wash contained 4′,6-diamidino-2-phenylindole (DAPI; Invitrogen, Carlsbad, CA, USA, #C10595) for nuclear staining. Coverslips were mounted in N-propyl-gallate antifade mounting media (Sigma, St. Louis, MO, USA #P-3130) on glass slides and sealed with nail polish. Cells were visualized using the 60X/1.35 Oil or 40X/1.0 NA objective of an Olympus BX63 or IX83 epifluorescence microscope. Z-stacks were deconvoluted in Olympus cellSens software using a constrained iterative algorithm. No or negligible labeling was seen in the absence of primary antibody. Overlays and brightness/contrast adjustment was carried out using ImageJ software. No other image adjustment was performed.

2.7. Gelatin Degradation Assay

Coverslips were coated with 60 °C preheated Oregon-green conjugated gelatin (Invitrogen, Carlsbad, CA, USA, #G13186) at 0.5 mg/mL in PBS + 2% sucrose, then gently aspirated with a soft vacuum source to form a thin uniform coat. Each coverslip was placed in a light-protected 12-well plate and left to air dry until the coating became whitish. Then, 1 mL of pre-chilled 0.5% glutaraldehyde solution (Sigma, St. Louis, MO, USA, #G6257) was added to each well and incubated for 15 min on ice. Coverslips were washed 3 times with PBS at RT, followed by addition of 1 mL freshly prepared sodium borohydride (5 mg/mL) (Sigma, St. Louis, MO, USA, #452882) solution to each well, incubation for 3 min, and wash in PBS. The desired number of coated coverslips was transferred into a new sterile 12-well plate, followed by seeding of cells to reach a confluence of 60% on the day of fixation. Cells were prepared for immunofluorescence analysis as above, using DAPI to visualize nuclei and Alexa Fluor 568–phalloidin to visualize F-actin (Invitrogen, Carlsbad, CA, USA, # R415).

2.8. Measurements of Intracellular pH (pH$_i$)

Cells seeded on 15 mm glass coverslips were transfected with mock or a3 siRNA as above. After 48 h, cells were loaded for 30 min with 2 µM BCECF-AM (ThermoFischer Scientific, Waltham, MA, USA #B1150) in Ringer (in mM: NaCl 125, KCl 5, CaCl$_2$ 1, MgCl$_2$ 0.5, Na$_2$HPO$_4$ 1, glucose 11, HEPES 25), followed by two washes in this solution. Coverslips were mounted in the perfused chamber of an Imic2000 microscope equipped with a gravity flow perfusion system, and images acquired using a 40× oil immersion objective (Olympus, Tokyo, Japan), a PolychromeV monochromator light source (Till Photonics, Gräfelfing, Germany), appropriate Chroma filter set (Chroma Technology, Bellows Falls, VT, USA), and an Ixon 885 camera (Andor, Belfast, N. Ireland). Every 2 s, fluorescence signals were collected at 470–550 nm after excitation at 440 and 485 nm, using Till Photonics Live Acquisition software. Experiments were done at 37 °C, in the absence of CO$_2$/HCO$_3^-$ to facilitate isolation of the contribution of V-ATPases. After achieving a stable baseline pH$_i$, cells were exposed to 20 mM NH$_4$Cl in Ringer for 10 min, followed by a ~3 min exposure to Na$^+$-free saline (replaced by N-methyl-D-glucammonium (NMDG$^+$)). Calibration was performed using the high-K/nigericin technique, essentially as in [36]. Intrinsic intracellular buffering capacity (β_i) was calculated from the change in pH$_i$ upon washout of NH$_4$Cl under Na$^+$-free conditions, using the equation $\beta_i = \Delta[NH_4^+]_i/\Delta pH_i$. $[NH_4^+]_i$ was calculated using the Henderson–Hasselbalch equation and assuming a pKa of 8.9. Net Na$^+$-and CO$_2$/HCO$_3^-$-independent acid efflux (J) was calculated as the change in pH$_i$ during the last min of the Na$^+$-free phase, multiplied by β_i, as: $J. = dpH_i/dt \times \beta_i$.

2.9. Single Cell Migration Analysis

T-12.5 flasks were coated with a substratum composed of extracellular matrix-mimicking components including laminins (4.46%), fibronectin (4.46%), collagen IV (0.60%), collagen III (1.34%), collagen I (89.15%). The coating was let solidify overnight in an incubator at 37 °C. Transfected cells were thinly seeded at a density of 2×10^4 cells per flask and allowed to adhere for 3–4 h at 37 °C, 5% CO$_2$ prior to placing flasks in a heated chamber (37 °C) on the stage of an inverted microscope (Axio Vert.A1; 20× objective; Zeiss, Oberkochen, Germany) for recording. A representative field displaying at least eight cells was chosen for evaluation. Migration was monitored for 10 h with an ORCA C8484-05G02 video camera (Hamamatsu, Herrsching, Germany) controlled by HC-IMAGE-Live software (version 4.2.5; Hamamatsu). Images were taken at 10 min intervals and stored as stacks of TIF-files. Single cell migration was manually tracked and data extracted using ImageJ (http://rsb.info.nih.gov/ij/). Migration is represented by the movement of the nucleus during a defined time interval. Migration velocity (µm min^{-1}) was calculated as the mean of the velocities for each time (10 min) interval, and is a measure of the average speed of movement during the experiment. Translocation was calculated as the mean distance between the position of the nucleus at the beginning and at the end of the experiment. The total path length was determined as the sum of the single distances covered within each time interval. A directionality index as measure of sustained, directed migration was calculated dividing the translocation (start-end distance) by the total path length. Thus, a directionality index of "1" would represent a straight linear forward movement.

2.10. Cell Invasion Analyses

Invasion was analyzed using 24-well 8 μm transwell BioCoat Chambers (BD Biosciences, Franklin Lakes, NJ, USA). Briefly, cells were starved in low serum media (1% FBS) supplemented with 0.1% BSA. After 24 h, cells were harvested and resuspended in serum-free medium. Then, 2.5×10^4 cells/mL cells were seeded onto the growth factor reduced Matrigel invasion chambers (BD Biosciences #354483) to analyze invasion and, in parallel wells, onto cell culture inserts (BD Biosciences, #354578) to analyze migration. Medium containing 10% FBS was added to the lower chambers and chambers were incubated at 37 °C and 5% CO_2 for 24 h. After washing with Gurr buffer, non-invading cells were gently removed using a cotton swab. Invasive cells located on the lower side of the chamber were fixed in ice cold absolute methanol for 30 min and stained with 30% Giemsa staining solution for 30 min. Membranes were washed 3× in Gurr buffer and air-dried. Stained membranes were cut out and placed on a glass slide. Ten random fields of view per membrane were imaged using the 40× objective of an Olympus Bx63 microscope and an Olympus D73 camera. Data are shown as the ratio of invaded/migrated cells over these 10 fields of view.

2.11. Proliferation Assay

Cells were seeded in 96-well plates in triplicates and at 0.3×10^3 and 10^4 cells per well for each condition. After 24 h, BrdU labeling reagent was added (Roche, Basel, Schwitzerland, #11647229001). After 4 h, cells were fixed and stained with BrdU antibody according to the manufacturer's protocol (Roche, Basel, Schwitzerland, #11647229001). BrdU incorporation was measured on FLUOstar OPTIMA microplate reader (BMG Labtech, Ortenberg, Germany).

2.12. 3D spheroid Growth and CellTiter Glo Assay

First, 1000 Panc-1 or BxPC-3 cells were seeded per well in round bottomed, ultra-low attachment 96-well plates (Corning, NY, USA, # 7007) in 200 μL growth medium. Cells were subsequently spun down at 750 RCF for 15 min, and were grown for 9 days at 37 °C with 95% humidity and 5% CO_2. Media for Panc-1 spheroids was supplemented with GelTrex LDEV-Free reduced growth factor basement membrane matrix (ThermoFisher Scientific, Waltham, MA, USA, #A14132029). Then, 100 μL medium was exchanged every second day. Light microscopic (Leica MZ16, Germany) images of the spheroids at 10× magnification were acquired on days 2, 4, 7, and 9. On day 9, 100 μL medium was replaced with 50 μL CellTiter-Glo®3D Reagent (Promega, Madison, WI, USA, #G9683). The plates were shaken for 5 min, incubated for 25 min at room temperature and luminescence recorded using a FLUOStar Optima microplate reader (BMG Labtech, Ortenberg, Germany).

2.13. Statistics

Unless otherwise specified, data are presented as mean ± S.E.M. of at least three independent experiments. Data analysis was performed using SPSS software and Graphpad Prism 6.0. Multiple groups were analyzed by one-way ANOVA with Dunnett's post-hoc test as relevant; Student's *t*-test was used for comparisons of datasets with only two groups; *p*-values of <0.05 were considered statistically significant.

3. Results

3.1. a3 is Upregulated in PDAC Cells and Mainly Localizes to Late Endosomes and Lysosomes

To study the potential roles of a3 in PDAC cells we first determined mRNA- and protein levels of this subunit compared to B2 (the only B subunit isoform in somatic cells and thus a proxy for total V-ATPase expression) in non-cancerous, immortalized human pancreatic epithelial cells (HPDE cells) and three human PDAC cell lines: Panc-1, BxPC-3, and AsPC-1. Both mRNA and protein levels of B2 were unaltered or slightly reduced in PDAC cells compared to HPDE cells (Figure 1A,B). In contrast, a3 mRNA and protein levels were increased in all PDAC cells, with 6- to 8-fold upregulation in AsPC-1 compared to that in HPDE cells (Figure 1C,D). BxPC-3 and Panc-1 cells were selected for further analysis as examples of PDAC cells expressing wild type (BxPC-3) or constitutively active (Panc-1) KRAS [37], which was recently shown to be important for V-ATPase function in PDAC cells [15].

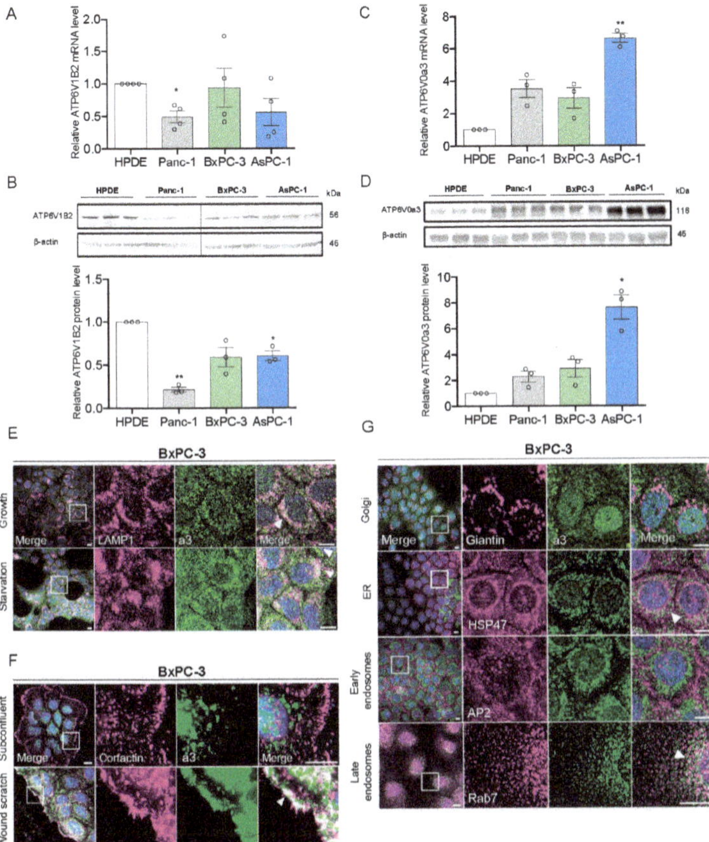

Figure 1. a3 is upregulated in PDAC cells and mainly localizes to late endosomes and lysosomes. (**A,C**) Relative mRNA levels of ATP6V1B2 and ATP6V0a3 in PDAC cells assessed by qRT-PCR. Data is normalized to β-actin, expressed relative to that in HPDE cells, and shown as mean with S.E.M. error bars, of 3 (ATP6V0a3) to 4 (ATP6V1B2) independent experiments per cell line. * $p < 0.05$ and ** $p < 0.01$: Significantly different from the level in HPDE cells, using one-way ANOVA and Dunnett's post hoc test. (**B,D**) Protein levels of ATP6V1B2 and ATP6V0a3 in PDAC cells, assayed by Western blotting. Upper panels show representative blots, and lower panels show quantification normalized to β-actin (loading ctrl.) and the level in HPDE cells. Data is shown as mean with S.E.M. error bars, of 3 independent experiments per cell line. * $p < 0.05$ and ** $p < 0.01$: Significantly different from the level in HPDE cells, using one-way ANOVA and Dunnett's *post hoc* test. (**E,F**) BxPC-3 cells cultured under growth, starvation, subconfluent or wound scratch conditions as indicated were subjected to immunofluorescence analysis for a3 (green) and LAMP1 or cortactin (magenta). Nuclei were stained with DAPI. Scalebar = 10 μm. White arrowheads indicate colocalization. Images represent 2–5 independent experiments. (**G**) Immunofluorescence analysis of a3 (green) and either Giantin (Golgi), HSP47 (ER), AP2 (early endosomes), or Rab7 (late endosomes) (all magenta) in BxPC-3 cells. Nuclei were stained with DAPI. Scalebar = 10 μm. White arrowheads indicate colocalization. Images are representative of 2–3 independent experiments.

In confluent BxPC-3 cells (Figure 1E, top panel), the majority of a3 labeling was intracellular and punctate, and overlapped substantially with staining for LAMP1 to visualize lysosomes (white arrowheads). A similar pattern was seen in Panc-1 cells (Figure 1A, top panel) cells and this colocalization was also observed when a3 and LAMP1 were overexpressed (Figure S1B,C). In BxPC-3 cells, there was partial colocalization of a3 with Rab7 (late endosomes), and to a lesser extent with heat shock protein (HSP)47 (endoplasmic reticulum), but negligible colocalization with Giantin (cis- and medial Golgi) and AP2 (early endosomes) (Figure 1G). To address if a3 membrane localization was dependent on confluence or nutrient status, we determined a3 localization (i) following glucose- and serum starvation (Figure 1E, Figure S1A, lower panels), (ii) in subconfluent cultures (Figure 1F, Figure S1D–F), and (iii) after introduction of a wound scratch (Figure 1F; Figure S1D). Colocalization between a3 and LAMP1 was not altered by starvation (Figure 1E, Figure S1A, lower panels). There was no detectable colocalization of a3 with cortactin (plasma membrane) under subconfluent conditions (Figure 1F), yet after a wound scratch, a distinct relocalization of a3 to the plasma membrane was seen in BxPC-3 cells (Figure 1F). In Panc-1 cells, no a3 membrane localization was detected under these conditions (Figure S1D). In contrast, in subconfluent Panc-1 cells, but not in subconfluent BxPC-3 cells, a3 localized in a highly perinuclear pattern surrounding the Golgi apparatus (Figure S1E,F). The same pattern was seen for LAMP1 (Figure S1E,F), consistent with translocation of lysosomes to the perinuclear region (see Discussion).

These results show that a3 expression is increased in PDAC cells compared to immortalized pancreatic epithelial cells, and that a3 localizes predominantly to late endosomes and lysosomes in Panc-1 and BxPC-3 cells.

3.2. a3 Is Not. Upregulated by Starvation and Does Not. Contribute to pH$_i$ Recovery After Acidification

In mammalian cells, V-ATPase assembly and function are increased by glucose starvation [38]. We therefore tested the effect of glucose and serum starvation (0 mM glucose, 1% serum) on a3 and V-ATPase expression. Protein (Figure 2A–D) and mRNA (Figure S2E,I) expression of B2 tended to be increased by starvation and were largely unaltered by a3 KD, which also had no detectable effect on a1 and a2 mRNA levels (Figure S2B–I), as previously reported in osteoclasts [39]. In complete medium, 50–60% knockdown of a3 was obtained in both Panc-1 and BxPC3 cells (Figure 2A–D). Under starvation conditions, only about 25% a3 KD could be obtained at the protein level (Figure 2A–D) while it was about 50% at the mRNA level (Figure S2D,H), precluding further analysis of the effect of a3 KD under starvation conditions. To stimulate V-ATPase membrane localization, we used forskolin to increase cAMP, which promotes V-ATPase membrane insertion in some cell types [40]. B2 localization was not detectably affected by a3 KD, neither under growth (Figure 2E) nor starvation conditions (Figure S2A). Forskolin treatment did indeed elicit redistribution of B2 positive vesicles from mostly perinuclear to more widely distributed throughout the cells, but failed to induce detectable plasma membrane insertion (Figure 2E).

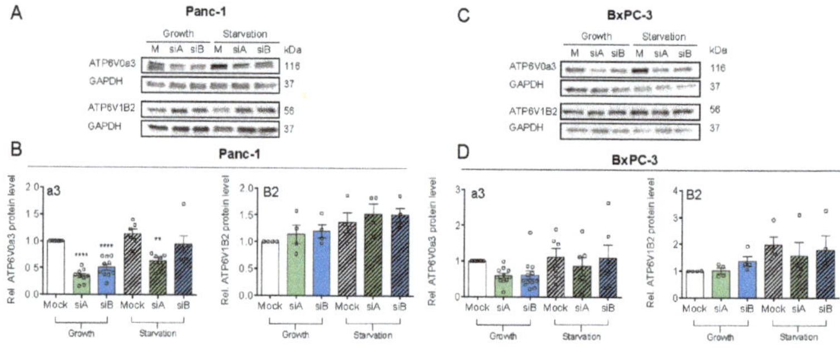

Figure 2. *Cont.*

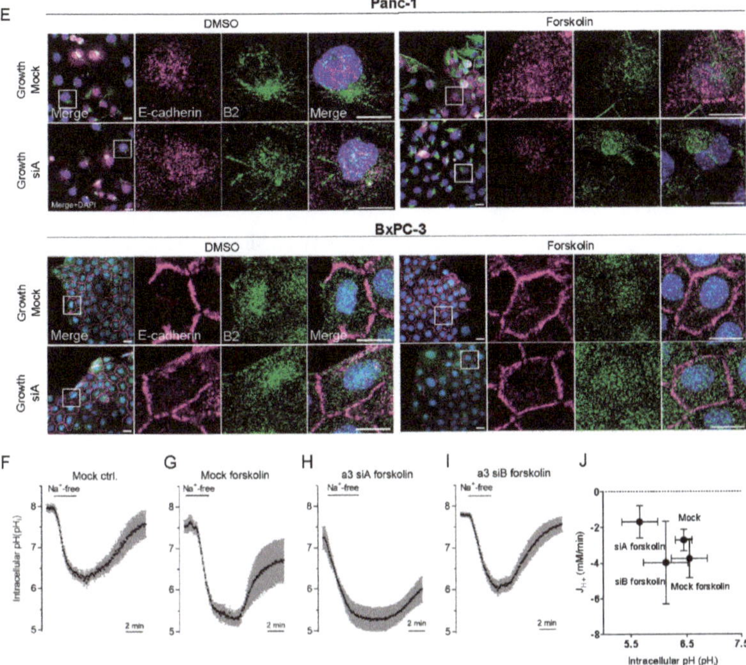

Figure 2. a3 is not upregulated in starvation and its KD does not inhibit pH_i recovery. Panc-1 and BxPC-3 cells were transfected with mock siRNA or 2 different siRNAs targeting a3 and grown under growth or starvation conditions for 48 h, lysed and subjected to Western blotting for a3 and B2. Upper panels (**A,C**) show representative blots and lower panels (**B,D**) show densitometric quantification normalized to GAPDH (loading ctrl.) and the level in respective mock controls. Data is shown as mean with S.E.M. error bars, of 3 to 12 independent experiments per cell line and siRNA. ** $p < 0.01$ and **** $p < 0.0001$: Significantly different from the level in control conditions (mock siRNA), using one-way ANOVA and Dunnett's post hoc test. Rel. = relative. (**E**) Panc-1 and BxPC-3 were transfected with Mock siRNA or 2 different siRNAs targeting a3, and were further treated with 10 μM forskolin or vehicle ctrl. (DMSO). Cells were grown for 48 h, and subjected to immunofluorescence analysis for B2 (green) and E-cadherin (magenta). Nuclei were stained with DAPI. Scalebar = 20 μm. (**F–J**) BxPC-3 cells were transfected with mock siRNA (**F,G**) or siRNA-A or -B against a3 (**H,I**) as indicated, and 48 h later, were subjected to live imaging of pH_i using BCECF-AM under CO_2/HCO_3^- free conditions. Where indicated, forskolin (10 μM) was present during the 30 min loading of with BCECF. The horizontal bar indicates the switch from standard HEPES-buffered Na^+-containing Ringer with 20 mM NH_4Cl to induce the acid load, to NMDG-Cl Ringer to observe Na^+-independent pH_i recovery. (**J**) The Na^+-and CO_2/HCO_3^- independent net acid efflux over time was calculated as the change in pH_i during the last minute of the Na^+-free phase, multiplied by β_i to obtain the flux, J_{H+}, and plotted against the mean pH_i during the minute of the measurement. BCECF ratios were calibrated to pH_i using the high K^+/nigericin technique. Data shown are representative of (**F–I**) or summarized (**J**) from 3 independent experiments. Error bars show SD of measurements on multiple cells in the same experiment (**F–I**) or S.E.M. of 3 independent experiments.

We next asked if the V-ATPase, and specifically a3, is important for pH_i regulation in PDAC cells. BxPC3 cells were treated with forskolin, and recovery of pH_i after an NH_4Cl prepulse-induced intracellular acid load determined. To isolate the V-ATPase contribution from that of HCO_3^-- and Na^+-dependent transporters, recovery was quantified in absence of these ions. Under these conditions, pH_i recovery was absent, both in mock- and a3 KD cells (Figure 2F–I). Figure 2J summarizes these data as the net H$^+$ efflux (J_{H+}) in the absence of Na$^+$ and as a function of pH$_i$. Comparable data were obtained in Panc-1 cells, using two different siRNAs (n = 2 per condition, not shown). When Na$^+$ was re-introduced, BxPC-3 cells recovered pH$_i$ efficiently after the acid load (Figure 2F–I), consistent with the very prominent Na$^+$/H$^+$ activity in these cells [41].

These results show that a3 KD does not detectably alter V-ATPase localization in PDAC cells, nor does the V-ATPase contribute detectably to pH$_i$ recovery after acidification.

3.3. V-ATPase Inhibition, but not a3 KD, Reduces Proliferation and Increases Cell Death

To investigate the roles of the V-ATPase and specifically a3 in proliferation and stress signaling in PDAC cells, we determined the effect of the V-ATPase inhibitor ConA on levels of hyperphosphorylated retinoblastoma protein (p-pRb, proliferation marker) and p21 (Figure 3A–C). The p-pRb level was dose-dependently reduced by ConA, yet was conversely slightly increased by a3 KD (Figure 3D,E,G,H). Accordingly, BrdU incorporation was reduced by ConA treatment yet increased by a3 KD (Figure 3J–K). Both V-ATPase inhibition and a3 KD increased p21 expression (Figure 3A,C,D–I), with no effect on p-p53 (Figure S3F–H), and ConA treatment, yet not a3 KD, increased PARP cleavage (Figure S3A–E). To determine the possible PDAC cell dependence on a3 in a 3D setting more similar to the tumor microenvironment, 3D spheroids were generated from Panc-1 cells 48 h after a3 KD and spheroid growth monitored for 9 days, followed by viability quantification. Also under these conditions, a3 KD did not reduce growth or viability (Figure 3L–N). Similar data were obtained in BxPC-3 cells (Figure S3I–K).

These results show that V-ATPase inhibition strongly reduces proliferation and induces cell death in PDAC cells, while a3 KD slightly increases proliferation and does not elicit cell death.

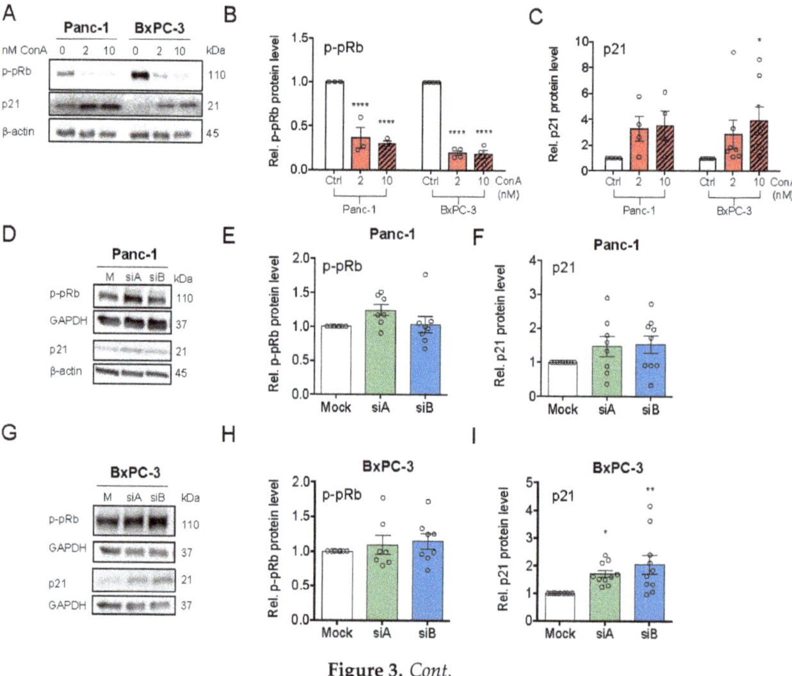

Figure 3. *Cont.*

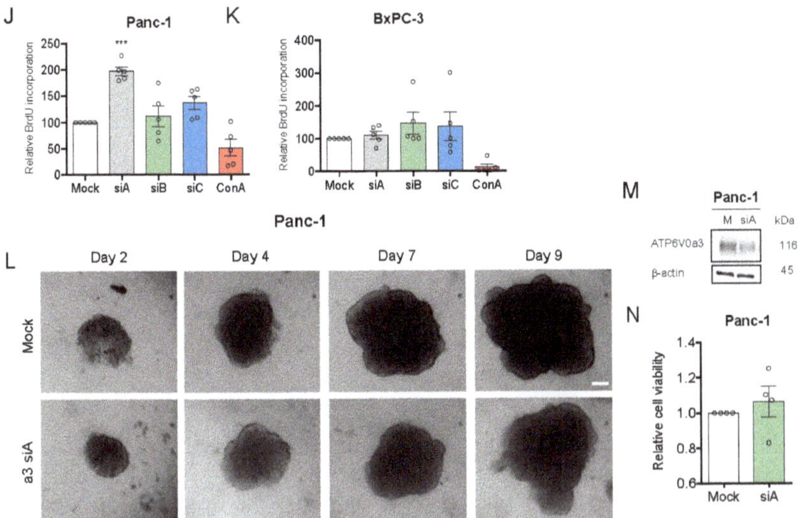

Figure 3. V-ATPase inhibition, but not a3 KD, reduces proliferation and increases cell death. (**A–C**) Panc-1 and BxPC-3 cells were treated with 2 or 10 nM ConA or vehicle ctrl. (DMSO) for 48 h and subjected to Western blotting for p-pRb and p21. (**A**) Representative blots. (**B,C**) Blot quantification. Data are normalized to β-actin (loading ctrl.) and the level in ctrl. conditions. Data is shown as mean with S.E.M. error bars, of 3 to 5 independent experiments per cell line. * $p < 0.05$ and **** $p < 0.0001$: Significantly different from the level in HPDE cells, using one-way ANOVA and Dunnett's post hoc test. Rel. = relative. (**D–I**) Panc-1 (**D–F**) and BxPC-3 cells (**G–I**) were transfected with mock siRNA or 2 different siRNAs targeting a3, and grown for 48 h, lysed and subjected to Western blotting for p-pRb and p21. (**D,G**) Representative blots. (**E,F,H,I**) Quantification, normalized to GAPDH (loading ctrl.) and the level in the respective mock ctrl. Data is shown as mean with S.E.M. error bars, of 3 to 11 independent experiments per cell line and siRNA. * $p < 0.05$ and ** $p < 0.01$: Significantly different from the level in control conditions (mock siRNA), using one-way ANOVA and Dunnett's post hoc test. Rel. = relative. (**J,K**) Panc-1 and BxPC-3 cells were transfected with mock siRNA or 3 different siRNAs targeting a3 or treated with 10 nM ConA, followed by BrdU analysis. Data are shown as mean with S.E.M. error bars, of 5 independent experiments per cell line. *** $p < 0.001$: Significantly different from the level in mock, using one-way ANOVA and Dunnett's post hoc test. Rel. = relative. (**L**) Panc-1 cells were transfected with 2 different a3 siRNAs followed by seeding as 3D spheroids 48 h after transfection. (**L**) Representative images of Panc-1 spheroids after 9 days of growth. Scalebar = 100 µm. After 9 days, Panc-1 spheroids were subjected to Western blot analysis of a3 protein level to evaluate KD of a3 as compared to mock ctrl. (n = 3) (**M**) or to CellTiter Glo assay to measure cell viability as normalized to mock ctrl. (n = 4) (**N**).

3.4. V-ATPase Inhibition Increases HIF-1α Protein Level, AMPK Activity and Alters Autophagic Flux

Consistent with previous work [42], ConA treatment strongly increased HIF-1α protein in both Panc-1 and BxPC3 cells (Figure 4A–C), while only a slight trend in the same direction was seen upon a3 KD (Figure 4D–F). Given the major role of HIF-1α as a metabolic regulator, we reasoned that V-ATPase inhibition might elicit metabolic stress. In congruence with this notion, treatment with the V-ATPase inhibitor ConA, but not a3 KD, increased the level of phosphorylated, active AMP kinase (AMPK) (Figure 4G–K).

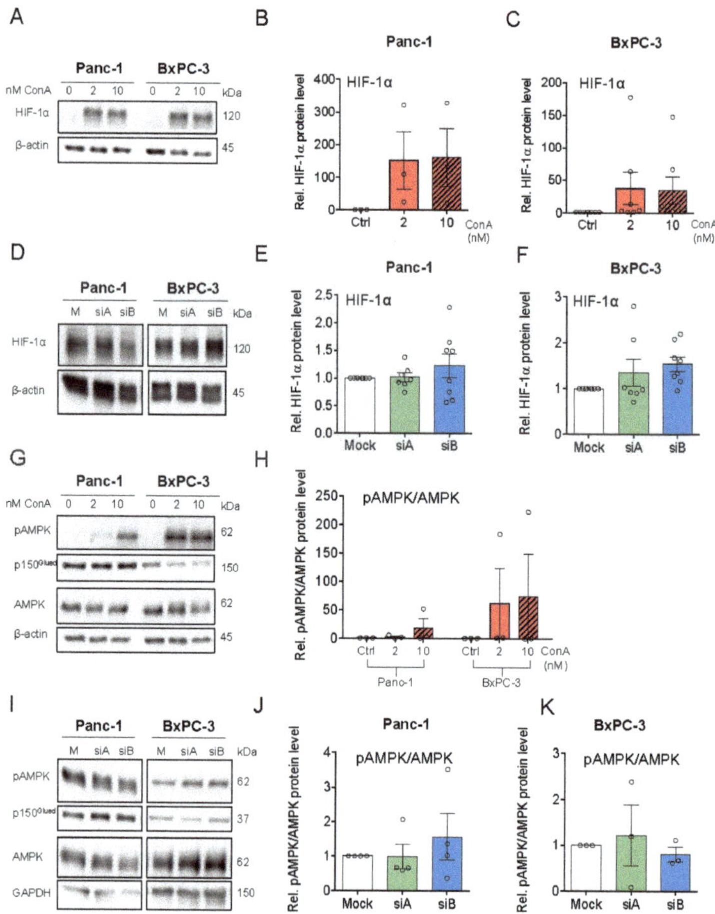

Figure 4. V-ATPase inhibition increases HIF-1α stabilization and AMPK activity and a similar trend is seen after a3 KD. Panc-1 and BxPC-3 cells were treated with either DMSO ctrl., 2 or 10 nM ConA for 48 h (**A–C,G,H**), or were transfected with Mock siRNA or 2 different siRNAs targeting a3, and grown for 48h (**D–F,I–K**), followed by Western blot analysis. (**A,G** and **D,I**) shows representative blots of HIF-1α, AMPK, and pAMPK in cells treated with ConA (n = 3–6) or in cells transfected with a3 siRNAs (n = 3–8). **B,E,F,H,J,K**) show quantification of the protein levels of HIF-1α (**B,C,E,F**) and pAMPK/AMPK (**G–K**) in cells treated with ConA or siRNA targeting a3. Data was normalized to beta-actin or GAPDH (loading ctrl.) and the level in respective DMSO or Mock ctrl. Data is shown as mean with S.E.M. error bars, of 3–6 independent experiments per cell line and siRNA. Rel. = relative.

An increase in AMPK activity signals energy stress and induces upregulation of autophagy [43]. We therefore assessed the impact of ConA and a3 KD on autophagic markers and compared it to that of starvation, a known inducer of autophagy. A shift of LC3B-I to its PE-conjugated form, LC3B-II, and a shift from diffuse cytosolic to punctate staining of LC3B and the cargo receptor p62/SQSTM1 (p62) are indicative of increased autophagy, whereas increased accumulation of both LC3 and p62 indicates stalling of autophagy [44–46]. As seen from the representative blots, and consistent with the known ability of ConA and other V-ATPase inhibitors to stall autophagic flux [47], ConA treatment increased total LC3B, mainly as LC3B-II in Panc-1 cells and LC3B-I in BxPC-3 cells, and increased p62 in both cell lines (Figure 5A–C). A requirement for V-ATPase activity for LC3 lipidation, as seen in BxPC-3 cells, is a hallmark of unconventional autophagy [48], and this finding seems consistent with a previous report of differential regulation of LC3 lipidation in BxPC-3 and Panc-1 cells [49]. a3 KD modestly increased LC3B-II/LC3B-I in both cell types (Figure 5D,E), with little if any change in p62 (Figure 5D,F). In Panc-1 cells, glucose starvation, as well as ConA treatment, elicited p62 aggregation, and a3 KD markedly increased the colocalization of p62 and LC3B (Figure 5G). In BxPC-3 cells, ConA-induced p62 puncta were much smaller than those seen in Panc-1 cells, and starvation and a3 KD had no detectable effect on p62 and LC3B localization (Figure 5H).

Collectively, these results confirm the major impact of V-ATPase inhibition on the autophagic machinery of PDAC cells and indicate the role of a3 in this process in Panc-1 cells, albeit in absence of significant changes in the LC3B-II/-I ratio and p62 level.

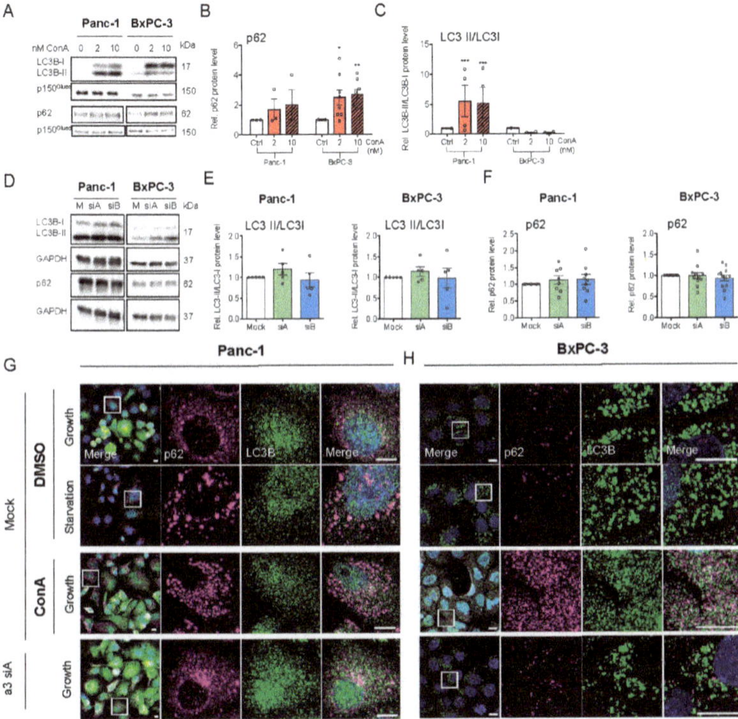

Figure 5. V-ATPase inhibition alters autophagic flux and a similar trend is seen after a3 KD. Panc-1 and BxPC-3 cells were treated with either DMSO ctrl., 2 or 10 nM ConA for 48 h (**A–C**), or were transfected with Mock siRNA or 2 different siRNAs targeting a3, and grown for 48 h (**D–F**), followed by lysing and Western blot analysis. (**A,D**) shows representative blots of LC3B and p62 in cells either treated with ConA (n = 3–7) (**A**) or transfected with siRNA against a3 (n = 3–12) (**D**). (**B,C,E,F**) show densitometric quantification of the protein levels of p62 (**B,F**) and LC3II/LCI (**C,E**). Data were normalized to respective loading ctrl. (p150^{Glued} or GAPDH) and the level in respective DMSO or mock ctrl. Data are shown as mean with S.E.M. error bars, of 3 to 12 independent experiments per cell line. * $p < 0.05$, ** $p < 0.01$, and *** $p < 0.001$: Significantly different from the level in control conditions (mock siRNA), using one-way ANOVA and Dunnett's post hoc test. Rel. = relative. (**G,H**) Panc-1 and BxPC-3 cells were transfected with mock siRNA or siRNA targeting a3 and were simultaneously treated with either DMSO or 10 nM ConA, and were subjected to either growth or starvation conditions for 48 h. The cells were subsequently subjected to immunofluorescence analysis of LC3B (green) and p62 (magenta). Nuclei were stained with DAPI. Scalebar = 10 μm. Images represent 4 (Panc-1) and 5 (BxPC-3) independent experiments.

3.5. a3 Negatively Regulates Migration and Invasion of PDAC Cells

To our knowledge, there are no published reports of the effect of a3 KD on PDAC cell migration. Panc-1 and BxPC-3 cell migration were assessed on a matrix simulating the composition of PDAC ECM (collagen I, laminins, fibronectin, collagen III, and collagen IV [50]). Mock- and a3 KD cells were seeded sparsely and single cell migration analyzed as total distance migrated, directionality index, migration velocity, and translocation (see Materials and Methods). Figure 6A shows trajectories of single migrating cells, normalized to a common starting point (see Videos S1–S4), and Figure 6B–I show the summarized data. Strikingly, a3 KD significantly increased all four parameters in BxPC-3 cells, and all except directionality in Panc-1 cells. Similar results were obtained when cell migration was measured on matrigel (Figure S4A–H). Thus, a3 KD significantly increased migration in both cell lines and on two different matrices.

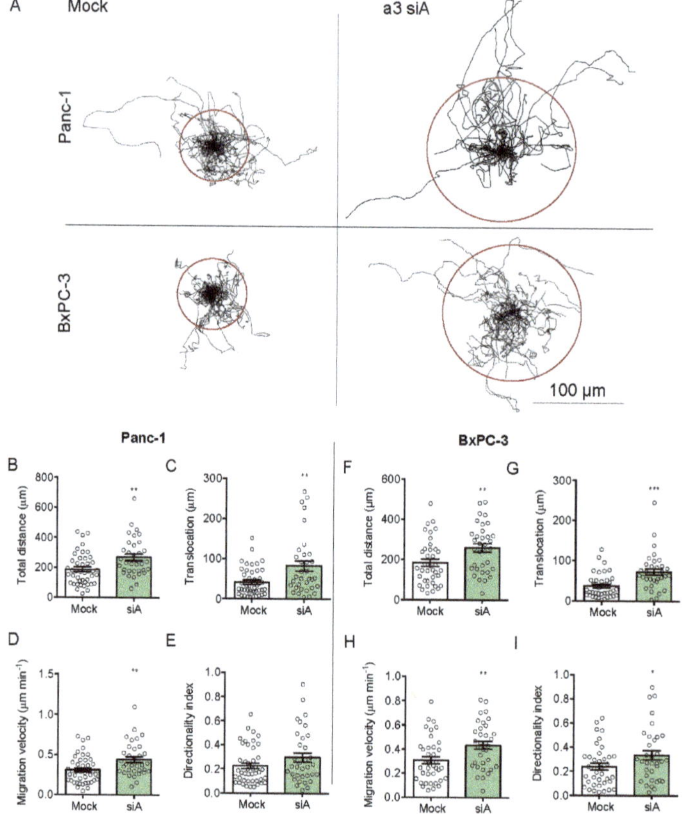

Figure 6. a3 negatively regulates the 2D migration of PDAC cells. (**A**) Trajectories of migrating mock- and a3 siRNA-transfected Panc-1 and BxPC-3 cells. The trajectories obtained were standardized to the same starting post represented by the center of a circle. The radii of the circles represent the mean distances (=translocation) covered within 10 h. Scalebar = 100 µm. (**B–I**) Total distance covered, translocation, velocity, and directionality index, calculated based on all data as in (**A**). Data are representative of 4 independent experiments (44 mock and 34 a3 siRNA cells) for Panc-1 cells and 3 experiments (38 mock and 34 a3 siRNA cells) for BxPC-3 cells. Data are shown as mean with SEM error bars. * $p < 0.05$, ** $p < 0.01$, and *** $p < 0.001$: Significantly different from the level in control conditions (mock siRNA), using Student's *t*-test.

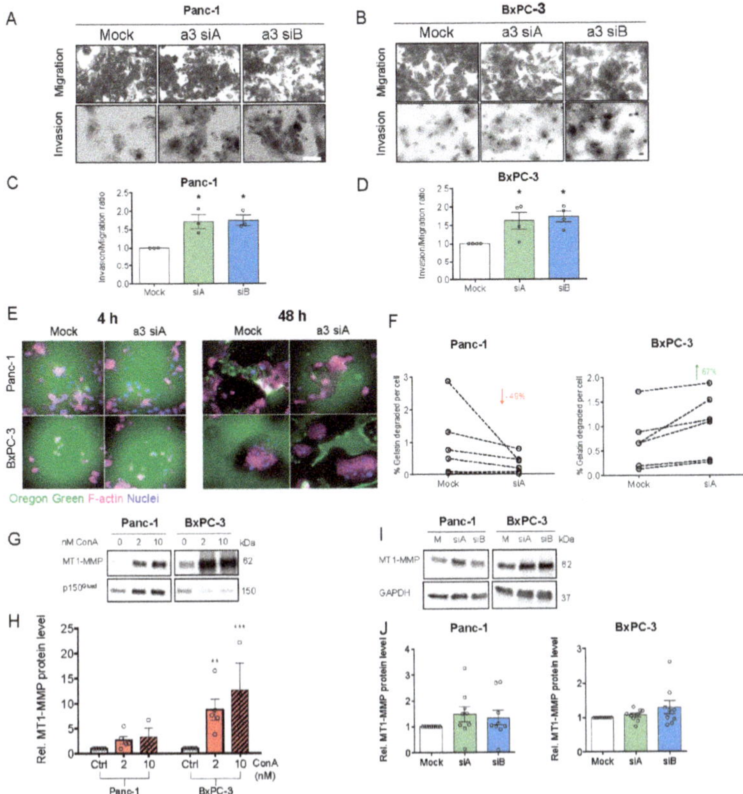

Figure 7. a3 negatively regulates the invasion of PDAC cells through matrigel. (**A,B**) Panc-1 or BxPC-3 cells were transfected with mock siRNA or 2 different siRNAs targeting a3 prior to seeding onto matrigel invasion chambers. Experiments were terminated after another 22 to 24 h, as described in Materials and Methods. Upper panels (**A,B**) show representative images of cells adhering to the lower chamber stained with Giemsa solution after the invasive process. Scalebars = 40 μm. (**C,D**) show relative invasion, as determined by the number of cells invaded through matrigel-coated inserts. Data are shown as mean ± S.E.M. of 3 to 4 independent experiments per condition. * $p < 0.05$: Significantly different from the level in control conditions (mock siRNA), using one-way ANOVA and Dunnett's post hoc test. (**E**) Panc-1 and BxPC-3 cells were transfected with mock siRNA or siRNA targeting a3 and grown for 48 h, followed by seeding onto Oregon-green conjugated gelatin-coated coverslips and growth for 4 or 48 h. The cells were subsequently subjected to immunofluorescence analysis of F-actin (Alexa-568-phalloidin (magenta), and Oregon-green gelatin). Nuclei were stained with DAPI. Scalebar = 40 μm. (**F**) Shows the quantification of gelatin degradation (performed in ImageJ) in Panc-1 and BxPC-3 cells after 48 h, normalized to the number of cells. Data represents 6 independent experiments per condition. A Student's *t*-test was used to test for significant differences between mock and siA conditions. (**G–J**) Panc-1 and BxPC-3 cells were treated with DMSO (ctrl.), 2 or 10 nM ConA for 48 h (**G–H**) or were transfected with Mock siRNA or 2 different siRNAs targeting a3 and grown for 48 h (**I,J**), followed by Western blot analysis. (**G,I**) Representative blots of MT1-MMP and loading controls. (**H–J**) Quantification of the MT1-MMP protein level. Data was normalized to GAPDH or p150^{Glued} (loading ctrl.) and the level in respective mock growth or DMSO ctrl. cells. Data is shown as mean with S.E.M. error bars, of 3 to 5 independent experiments per cell line and treatment. * $p < 0.05$, ** $p < 0.01$, and *** $p < 0.005$: Significantly different from the level in control conditions (mock siRNA or DMSO ctrl.), using one-way ANOVA and Dunnett's post hoc test. Rel. = relative.

In congruence with this, a3 KD significantly increased the invasion of both cell types through matrigel (Figure 7A–D). To understand the mechanisms through which a3 KD increased invasion, we assessed the roles of the V-ATPase and specifically a3 in the regulation of matrix degradation. After a3 KD, PDAC cells were seeded on Oregon-green gelatin-covered coverslips and the gelatin was evaluated after 4 and 48 h (Figure 7E,F). Interestingly, a3 KD decreased matrix degradation by Panc-1 cells by about 50%, yet increased it in BxPC-3 cells by about 67% (Figure 7F). V-ATPase activity was reported to be important for the degradation of MT1-MMP (also known as MMP14), which plays a key role in PDAC cell invasion [33] and is an activator of MMP2 and, via MMP2, also MMP9 [31,51]. In congruence with this, MT1-MMP expression was increased by ConA treatment (Figure 7G,H), yet not or only marginally by a3 KD (Figure 7I,J).

These results show that a3 negatively regulates migration and invasion of Panc-1 and BxPC-3 cells, yet the mechanisms may differ between the two cell lines and do not seem to involve MT1-MMP upregulation.

4. Discussion

PDAC is an exceptionally aggressive cancer type relying heavily on nutrient scavenging and autophagy, processes dependent on endo-/lysosomal V-ATPase activity [24–26]. It was previously reported that in some but not all PDAC cells, V-ATPases localize to the plasma membrane and their inhibition attenuates invasion [27]. However, in addition to these cell-type differences, including no net effect on Panc-1 cell invasion, conclusions regarding invasion were drawn based on scratch wound assays and migration into an agar disc, i.e., very far from the matrix conditions in PDAC tumors [27]. Here, we focused on the ATP6V0a3 (a3) subunit, which was proposed to target the V-ATPase to the plasma membrane in some invasive breast cancer and melanoma cells [20–22] and was previously detected in the plasma membrane of PDAC cells [27]. We found mRNA and protein levels of a3 but not of B2 to be increased in PDAC cell lines compared to immortalized human pancreatic duct epithelial cells. This suggests that a3 expression is increased in the PDAC cell lines studied, in the absence of total V-ATPase upregulation, in line with recent work in breast cancer cells [52].

Under control conditions, a3 mainly localized to LAMP1- and Rab7-positive vesicles in agreement with previous work [15,39], i.e., it was predominantly late endosomal/lysosomal. Although previous work proposed a3 localization to the plasma membrane in PDAC [27,29], the overall localization pattern appears similar to that found in the present study, with a marginal fraction of total a3 staining found in the plasma membrane. In agreement with our findings under control conditions in PDAC cells, little if any V-ATPase expression was detectable in the plasma membrane of melanocytes [12]. In contrast, the introduction of a wound scratch elicited significant redistribution of a3 to the plasma membrane in BxPC-3 cells, consistent with previous findings [23]. This agrees well with previous work showing a role of a3-Rab7 interactions in secretory lysosome trafficking in osteoclasts [39]. Interestingly, a3 membrane translocation did not happen in subconfluently seeded cells with no wound scratch, suggesting that it may be caused by the wounding of the cells, which releases motility-stimulating molecules such as ATP.

An increase in cAMP elicits V-ATPase membrane insertion in some cell types [40], yet in our hands, only a redistribution of a3 to more peripheral intracellular regions was detectable upon forskolin treatment of PDAC cells; a3 knockdown had no effect on pH_i recovery after acidification, in absence or presence of forskolin. Thus, even if a minor fraction of V-ATPases not detectable in our immunofluorescence analysis was present in the plasma membrane, this seems to be of little if any functional relevance for pH_i regulation under the conditions studied. While a small contribution of V-ATPase activity to pH_i regulation has been reported in intact pancreatic ducts [53], our data are consistent with pH_i recovery after an acid load being largely or fully accounted for by the activity of Na^+/H^+ exchangers and $Na^+-HCO_3^-$ cotransporters in most cancer cells [54,55] and with recent work showing that depletion of V-ATPases did not alter submembranous pH in Panc-1 cells [15].

V-ATPases play important roles in nutrient response and proliferation [56–58]. ATPase inhibition decreased proliferation and increased p21 expression in PDAC cells, and increased HIF-1α protein level and AMPK activity, in congruence with previous work [59,60]. While the upregulation of HIF-1α by V-ATPase inhibition was previously reported, the proposed mechanisms differ between studies, likely reflecting both cell-type-, inhibitor- and dosage-dependent effects [59,60]. HIF-1α upregulation by bafilomycin A in PC3 prostate cancer cells was pH_i independent [60], in congruence with the lack of effect of a3 KD on pH_i in our study. In PC3 cells, HIF-1α upregulation by bafilomycin was found to reflect a bafilomycin-induced interaction of ATP6V0C with HIF-1α, resulting in reduced HIF-1α degradation [60]. HIF-1α undergoes not only proteasomal but also lysosomal degradation (specifically chaperone-mediated autophagy) [61]. Thus, HIF-1α upregulation by interfering with V-ATPase function may at least in part reflect its reduced lysosomal degradation. Finally, it was recently reported that the HIF-1α upregulation upon V-ATPase inhibition was downstream from disturbed cellular iron metabolism due to interference with transferrin receptor trafficking [62]. Similar to our findings, p21 was also upregulated in a HIF-1α-dependent but a p53-increase-independent manner by bafilomycin A in SiHa cervical cancer cells [60]. These two events may be linked, as in PC3 cells, HIF-1α was only marginally transcriptionally active towards its normal targets after bafilomycin-induced stabilization, yet elicited the upregulation of p21 by dissociating c-myc from the p21 repressor [60].

V-ATPase activity is essential for the maintenance of autophagic flux, and consistent with this, V-ATPase inhibition caused p62- and LC3B accumulation and a change in autophagic vesicle morphology in both cell types. a3 KD had no detectable effect on p62, and only caused a slight increase in the LC3B-II/LC3B-I ratio. However, under starved conditions, a marked increase in p62/LC3 colocalization was seen after a3 KD in Panc-1 cells, indicative of reduced autophagic flux. This is in congruence with the notion that the V0 domain is important for the docking and fusion between secretory vesicles, although the role of a3 in this process is subject to controversy [39,63]. Whether the role of a3/V-ATPase in cell death/proliferation signaling is dependent on altered H^+ pumping by the V-ATPase remains unresolved. Clearly, however, not all functions of neither a3 nor the V-ATPase per se are dependent on H^+-transport [6,64].

Contrary to our expectation based on a series of papers on this topic in breast cancer [18,19,21], a3 KD significantly increased both 2D migration and invasion. This finding was consistent between Panc-1 and BxPC-3 cells and on two different migration matrices. In conjunction with the lack of detectable a3 plasma membrane localization under control conditions in these cell lines [21], this strongly suggests that the role of a3 in PDAC cell invasiveness differs from that proposed in melanoma and breast cancer cells [18,20–22]. A major difference between this and previous work is that in Panc-1 and BxPC-3 cells, a4 was undetectable using qPCR analysis, consistent with in silico analyses suggesting minimal if any a4 expression. It is notable that also in MDA-MB-231 cells, KD of a4, but not of a3, reduced plasma membrane V-ATPase expression and a4 KD inhibited invasion more than did a3 KD [21]. Furthermore, in 4T1-12B breast cancer cells, CRISPR/Cas9 knockout of a3 had no effect on migration, invasion, and plasma membrane V-ATPase localization, whereas all of these parameters were inhibited by a4 knockout [23]. We therefore conclude that the evidence for a role of a3 in plasma membrane targeting and invasion of cancer cells is limited to a few cell types, and at least in some cases seems linked to the expression of a4.

Consistent with previous reports [33,65], MT1-MMP activity was upregulated by V-ATPase inhibition. However, MT1-MMP activity was at most marginally affected by a3 KD and thus seems unlikely to account for the stimulation of migration and invasion by a3 KD. Conversely, MT1-MMP knockdown reverted a proposed stimulatory effect of ConA on autophagy in glioblastoma cells [66]. This possibility was not further addressed here, given that in our hands, ConA blocks autophagic flux, consistent with the generally reported effect of this and other V-ATPase inhibitors [47]. The stimulatory effect of a3 KD on migration was similar on matrigel and a complex matrix corresponding closely to the PDAC tumor microenvironment [50]. Consistent with this, ECM degradation was increased upon a3 KD in BxPC-3 cells. However, conversely, ECM degradation was inhibited by a3 KD in Panc-1 cells. BxPC-3 cells are KRAS wild type and harbor a homozygous deletion of SMAD4, while Panc-1 cells have a constitutively active KRAS mutation and are SMAD4 wild type [37]. Given the known roles of both mutations in regulating epithelial-to-mesenchymal transition and invasive behavior in PDAC [67,68], this may play a role in the observed difference between the cell types; however, this remains to be addressed.

Thus, further studies are required to determine the mechanisms involved in the unexpected stimulation of migration and invasion by a3 KD. a3 has been assigned important roles in both plasma membrane cholesterol localization in PDAC cells [15] and secretory lysosome fusion in osteoclasts [39], and involvement of both of these processes is possible and should be addressed in future work. We speculate that the stimulation of migration and invasion by a3 knockdown demonstrated in this work is likely related to roles in cell motility of V-ATPase dependent processes such as many cellular signaling pathways and endo-lysosomal trafficking [6,7,14,16]. Importantly, our work confirmed that a3 is upregulated in PDAC cells, and given the complexity of the in vivo microenvironment of PDAC tumors, it remains possible that a3 is a relevant target in anticancer therapy in some PDAC types or conditions. The involvement of the V-ATPases per se, and the a subunits, in particular, in cancer, are still incompletely understood. This is exemplified by the striking yet unexplained differences between the roles of different a subunits in invasion/migration in different cancers [18,20–23], but the roles of a3-containing V-ATPases in other cancer hallmarks, such as chemotherapy resistance [7] and macropinocytosis [15], should be explored in future studies.

In conclusion, the a3 subunit of the V-ATPase was upregulated in PDAC cells compared to non-cancer pancreatic epithelial cells. Under unstimulated conditions, a3 localization was mainly endo-/lysosomal, and its knockdown had no detectable effect on pH_i regulation after acid loading. V-ATPase inhibition increased HIF-1α expression and decreased proliferation and autophagic flux under both starved and non-starved conditions, whereas a3 KD had little or no effect on these parameters. Also, spheroid growth of PDAC cells was unaffected by a3 KD. Remarkably, a3 KD increased migration and invasion of Panc-1 and BxPC-3 PDAC cells, and increased gelatin degradation in BxPC-3 cells, yet decreased it in Panc-1 cells. We conclude that in these PDAC cell lines, a3 negatively regulates migration and invasion and that a3-dependence of extracellular matrix degradation likely contributes to, but cannot fully account for, this effect.

Supplementary Materials: The following are available online at https://zenodo.org/record/3696682#.XmDDrfSYOpp, Figure S1: (A,D) Panc-1 cells subjected to growth, starvation, subconfluent or wound scratch conditions were subjected to immunofluorescence analysis for a3 (green) and LAMP1 or Cortactin (magenta). Nuclei were stained with DAPI. Scalebar = 10 µm. (B,C) Panc-1 or BxPC-3 cells subjected to growth conditions were transfected with empty vector (ctrl.) or a3-GFP and LAMP1-mCherry vector and subjected to immunofluorescence analysis for LAMP1 (magenta). Nuclei were stained with DAPI. Scalebar =10 µm. (E,F) Subconfluent Panc-1 or BxPC-3 cells were subjected to immunofluorescence analysis for a3 (green) and Golgin-97, Rab7 or LAMP1 (all magenta). Nuclei were stained with DAPI. Scalebar = 10 µm. Images are representative of 2–5 independent experiments, Figure S2: (A) Panc-1 or BxPC-3 cells were transfected with Mock siRNA or siRNA targeting a3, and were grown under starvation conditions for 48 h, in presence of either 10 M DMSO or Forskolin, prior to immunofluorescence analysis, using anti-B2 antibody (green) in conjunction with anti-E-cadherin (magenta). Nuclei were stained with DAPI. Scalebar = 20 µm B-I) Relative mRNA levels of ATP6V0a1, ATP6V0a2, ATP6V0a3 and ATP6V1B2 in siRNA-transfected Panc-1 and BxPC-3 cells were assessed by qRT-PCR. Data was normalized to the mRNA level of beta-actin and the level in respective Mock siRNA ctrls. Data is shown as mean with S.E.M. error bars, of three independent experiments per cell line. One-Way ANOVA and Dunnett's post hoc test was used to test for significant differences between Mock siRNA and a3 siRNA. Rel. = relative, Figure S3: Panc-1 and BxPC-3 cells were treated with either DMSO ctrl, 2 or 10 nM ConA for 48 h (A,B), or were transfected with Mock siRNA or 2 different siRNAs targeting a3, and grown for 48 h (C–H), followed by lysing and Western blot analysis of cPARP/PARP (A–E) and p-p53 (F–H). (A,C,F) show representative blots of cPARP/PARP (A,C) and p-p53 (F) and (B,D,E,G,H) show densitometric quantification normalized to respective loading ctrls (β-actin, GAPDH or p150Glued) and the level in either Mock ctrl or DMSO ctrl cells. Data is shown as mean with S.E.M. error bars, of 3–5 independent experiments per cell line. Rel. = relative. (I) BxPC-3 cells were transfected with Mock siRNA or siRNAs targeting a3 followed by seeding as 3D spheroids 48 h after transfection. (I) shows representative images of Panc-1 spheroids after 9 days of growth. Scalebar = 100 µm. After 9 days, BxPC-3 spheroids were subjected to Western blot analysis of a3 protein level to evaluate KD of a3 as compared to mock ctrl (n = 3) (J). BxPC-3 spheroids were subjected to CellTiter Glo assay to measure cell viability as normalized to mock ctrl (n = 5) (K). A Student's *t*-test was used to test for significant differences between Mock siRNA and a3 siRNA, Figure S4: Migration of mock- and a3 siRNA-transfected Panc-1 BxPC-3 cells on Matrigel. (A–G) Total distance covered, translocation, velocity, and directionality index, calculated based on all trajectory data (not shown). Data are representative of 3 independent experiments (27 mock and 34 a3 siRNA cells) for Panc-1 cells and 3 experiments (42 mock and 33 a3 siRNA cells) for BxPC-3 cells. Data given as means with S.E.M. error bars. ** ($p < 0.01$), *** ($p < 0.001$), and **** ($p < 0.0001$): Significantly different from the level in control conditions (mock siRNA), Student's *t*-test, Video S1: Time-lapse imaging of migrating Panc-1 cells transfected with mock siRNA, Video S2: Time-lapse imaging of migrating Panc-1 cells transfected with a3 siRNA, Video S3: Time-lapse imaging of migrating BxPC-3 cells transfected with mock siRNA, Video S4: Time-lapse imaging of migrating BxPC-3 cells transfected with a3 siRNA.

Author Contributions: Conceptualization of this project was by A.G., M.F., C.S., I.N. and S.F.P., with inputs from all co-authors. Experiments and most of the data analysis were carried out by S.H., M.F., A.G., M.S. and E.P.-C. Data was analyzed and figures prepared by S.H., M.F., A.G., M.S., E.P.-C. and S.F.P. Supervision was by S.F.P. and M.F. The original manuscript draft was prepared by A.G. and S.F.P., and the final manuscript was written by S.F.P., with comments and inputs from all co-authors. All authors have read and agreed to the published version of the manuscript.

Funding: This work was supported by the Marie Curie Initial Training Network IonTraC (Grant Agreement No. 289648), the Marie Curie Initial Training Network pHIoniC (H2020-MSCA-ITN-2018, grant 8138), the Hartmann Foundation (SFP), and University of Copenhagen (EU bonus grant to IN).

Acknowledgments: We are grateful to Katrine F. Mark and Pernille Roshof for expert technical assistance, to S.C. Kong for initial experiments for Figure 1A–D and contributions to Figure 6, and to Daniel Sauter for assistance in setting up the live imaging experiments. The HPDE cell line was a gift from M. S. Tsao from University Health Network, Toronto. LAMP1-mCherry was a gift from Bin Liu, Danish Cancer Society Research Center, Denmark and human TCIRG1 (a3) a gift from Johan Richter, Lund University, Sweden.

Conflicts of Interest: The authors declare no conflict of interest.

References

1. Gong, J.; Tuli, R.; Shinde, A.; Hendifar, A.E. Meta-analyses of treatment standards for pancreatic cancer. *Mol. Clin. Oncol.* **2016**, *4*, 315–325. [CrossRef]

2. Rahib, L.; Smith, B.D.; Aizenberg, R.; Rosenzweig, A.B.; Fleshman, J.M.; Matrisian, L.M. Projecting cancer incidence and deaths to 2030: The unexpected burden of thyroid, liver, and pancreas cancers in the United States. *Cancer Res.* **2014**, *74*, 2913–2921. [CrossRef]

3. White, K.A.; Grillo-Hill, B.K.; Barber, D.L. Cancer cell behaviors mediated by dysregulated pH dynamics at a glance. *J. Cell Sci.* **2017**, *130*, 663–669. [CrossRef]

4. Stock, C.; Pedersen, S.F. Roles of pH and the Na$^+$/H$^+$ exchanger NHE1 in cancer: From cell biology and animal models to an emerging translational perspective? *Semin.Cancer Biol.* **2017**, *43*, 5–16. [CrossRef]

5. Boedtkjer, E.P.; Pedersen, S.F. The acidic tumor microenvironment as a driver of cancer. *Ann. Rev. Physiol.* **2020**, *82*, 103–126. [CrossRef] [PubMed]

6. Forgac, M. Vacuolar ATPases: Rotary proton pumps in physiology and pathophysiology. *Nat. Rev. Mol. Cell Biol.* **2007**, *8*, 917–929. [CrossRef] [PubMed]

7. Whitton, B.; Okamoto, H.; Packham, G.; Crabb, S.J. Vacuolar ATPase as a potential therapeutic target and mediator of treatment resistance in cancer. *Cancer Med.* **2018**, *7*, 3800–3811. [CrossRef] [PubMed]

8. Toyomura, T.; Murata, Y.; Yamamoto, A.; Oka, T.; Sun-Wada, G.H.; Wada, Y.; Futai, M. From lysosomes to the plasma membrane: Localization of vacuolar-type H$^+$-ATPase with the a3 isoform during osteoclast differentiation. *J. Biol. Chem.* **2003**, *278*, 22023–22030. [CrossRef]

9. Michel, V.; Licon-Munoz, Y.; Trujillo, K.; Bisoffi, M.; Parra, K.J. Inhibitors of vacuolar ATPase proton pumps inhibit human prostate cancer cell invasion and prostate-specific antigen expression and secretion. *Int. J. Cancer* **2013**, *132*, E1–E10. [CrossRef]

10. Smith, G.A.; Howell, G.J.; Phillips, C.; Muench, S.P.; Ponnambalam, S.; Harrison, M.A. Extracellular and Luminal pH Regulation by Vacuolar H$^+$-ATPase Isoform Expression and Targeting to the Plasma Membrane and Endosomes. *J. Biol. Chem.* **2016**, *291*, 8500–8515. [CrossRef]

11. Sun-Wada, G.H.; Tabata, H.; Kawamura, N.; Aoyama, M.; Wada, Y. Direct recruitment of H+-ATPase from lysosomes for phagosomal acidification. *J. Cell Sci.* **2009**, *122*, 2504–2513. [CrossRef] [PubMed]

12. Tabata, H.; Kawamura, N.; Sun-Wada, G.H.; Wada, Y. Vacuolar-type H$^+$-ATPase with the a3 isoform is the proton pump on premature melanosomes. *Cell Tissue Res.* **2008**, *332*, 447–460. [CrossRef] [PubMed]

13. Sun-Wada, G.H.; Toyomura, T.; Murata, Y.; Yamamoto, A.; Futai, M.; Wada, Y. The a3 isoform of V-ATPase regulates insulin secretion from pancreatic beta-cells. *J. Cell Sci.* **2006**, *119*, 4531–4540. [CrossRef] [PubMed]

14. Ramirez, C.; Hauser, A.D.; Vucic, E.A.; Bar-Sagi, D. Plasma membrane V-ATPase controls oncogenic RAS-induced macropinocytosis. *Nature* **2019**, *576*, 477–481. [CrossRef] [PubMed]

15. Sun-Wada, G.H.; Wada, Y. Role of vacuolar-type proton ATPase in signal transduction. *Biochim. Biophys. Acta* **2015**, *1847*, 1166–1172. [CrossRef]

16. Marshansky, V.; Futai, M. The V-type H$^+$-ATPase in vesicular trafficking: Targeting, regulation and function. *Curr. Opin. Cell Biol.* **2008**, *20*, 415–426. [CrossRef]

17. Frattini, A.; Orchard, P.J.; Sobacchi, C.; Giliani, S.; Abinun, M.; Mattsson, J.P.; Keeling, D.J.; Andersson, A.K.; Wallbrandt, P.; Zecca, L.; et al. Defects in TCIRG1 subunit of the vacuolar proton pump are responsible for a subset of human autosomal recessive osteopetrosis. *Nat. Genet.* **2000**, *25*, 343–346. [CrossRef]

18. Cotter, K.; Capecci, J.; Sennoune, S.; Huss, M.; Maier, M.; Martinez-Zaguilan, R.; Forgac, M. Activity of plasma membrane V-ATPases is critical for the invasion of MDA-MB231 breast cancer cells. *J. Biol. Chem.* **2015**, *290*, 3680–3692. [CrossRef]

19. Sennoune, S.R.; Bakunts, K.; Martinez, G.M.; Chua-Tuan, J.L.; Kebir, Y.; Attaya, M.N.; Martinez-Zaguilan, R. Vacuolar H+-ATPase in human breast cancer cells with distinct metastatic potential: Distribution and functional activity. *Am. J. Physiol. Cell Physiol.* **2004**, *286*, C1443–C1452. [CrossRef]

20. Nishisho, T.; Hata, K.; Nakanishi, M.; Morita, Y.; Sun-Wada, G.H.; Wada, Y.; Yasui, N.; Yoneda, T. The a3 isoform vacuolar type H$^+$-ATPase promotes distant metastasis in the mouse B16 melanoma cells. *Mol. Cancer Res.* **2011**, *9*, 845–855. [CrossRef]

21. Hinton, A.; Sennoune, S.R.; Bond, S.; Fang, M.; Reuveni, M.; Sahagian, G.G.; Jay, D.; Martinez-Zaguilan, R.; Forgac, M. Function of a subunit isoforms of the V-ATPase in pH homeostasis and in vitro invasion of MDA-MB231 human breast cancer cells. *J. Biol. Chem.* **2009**, *284*, 16400–16408. [CrossRef] [PubMed]

22. Capecci, J.; Forgac, M. The function of vacuolar ATPase (V-ATPase) a subunit isoforms in invasiveness of MCF10a and MCF10CA1a human breast cancer cells. *J. Biol. Chem.* **2013**, *288*, 32731–32741. [CrossRef] [PubMed]

23. McGuire, C.M.; Collins, M.P.; Sun-Wada, G.; Wada, Y.; Forgac, M. Isoform-specific gene disruptions reveal a role for the V-ATPase subunit a4 isoform in the invasiveness of 4T1-12B breast cancer cells. *J. Biol. Chem.* **2019**, *294*, 11248–11258. [CrossRef] [PubMed]

24. Yang, A.; Rajeshkumar, N.V.; Wang, X.; Yabuuchi, S.; Alexander, B.M.; Chu, G.C.; Von Hoff, D.D.; Maitra, A.; Kimmelman, A.C. Autophagy is critical for pancreatic tumor growth and progression in tumors with p53 alterations. *Cancer Discov.* **2014**, *4*, 905–913. [CrossRef] [PubMed]

25. Kamphorst, J.J.; Nofal, M.; Commisso, C.; Hackett, S.R.; Lu, W.; Grabocka, E.; Vander Heiden, M.G.; Miller, G.; Drebin, J.A.; Bar-Sagi, D.; et al. Human pancreatic cancer tumors are nutrient poor and tumor cells actively scavenge extracellular protein. *Cancer Res.* **2015**, *75*, 544–553. [CrossRef] [PubMed]

26. Perera, R.M.; Stoykova, S.; Nicolay, B.N.; Ross, K.N.; Fitamant, J.; Boukhali, M.; Lengrand, J.; Deshpande, V.; Selig, M.K.; Ferrone, C.R.; et al. Transcriptional control of autophagy-lysosome function drives pancreatic cancer metabolism. *Nature* **2015**, *524*, 361–365. [CrossRef]

27. Chung, C.; Mader, C.C.; Schmitz, J.C.; Atladottir, J.; Fitchev, P.; Cornwell, M.L.; Koleske, A.J.; Crawford, S.E.; Gorelick, F. The vacuolar-ATPase modulates matrix metalloproteinase isoforms in human pancreatic cancer. *Lab. Investig.* **2011**, *91*, 732–743. [CrossRef]

28. Ohta, T.; Numata, M.; Yagishita, H.; Futagami, F.; Tsukioka, Y.; Kitagawa, H.; Kayahara, M.; Nagakawa, T.; Miyazaki, I.; Yamamoto, M.; et al. Expression of 16 kDa proteolipid of vacuolar-type H(+)-ATPase in human pancreatic cancer. *Br. J. Cancer* **1996**, *73*, 1511–1517. [CrossRef]

29. Sreekumar, B.K.; Belinsky, G.S.; Einwachter, H.; Rhim, A.D.; Schmid, R.; Chung, C. Polarization of the vacuolar adenosine triphosphatase delineates a transition to high-grade pancreatic intraepithelial neoplasm lesions. *Pancreas* **2014**, *43*, 1256–1263. [CrossRef]

30. Koshiba, T.; Hosotani, R.; Wada, M.; Miyamoto, Y.; Fujimoto, K.; Lee, J.U.; Doi, R.; Arii, S.; Imamura, M. Involvement of matrix metalloproteinase-2 activity in invasion and metastasis of pancreatic carcinoma. *Cancer* **1998**, *82*, 642–650. [CrossRef]

31. Knapinska, A.M.; Estrada, C.A.; Fields, G.B. The Roles of Matrix Metalloproteinases in Pancreatic Cancer. *Prog. Mol. Biol. Transl. Sci.* **2017**, *148*, 339–354. [PubMed]

32. Maatta, M.; Soini, Y.; Liakka, A.; Autio-Harmainen, H. Differential expression of matrix metalloproteinase (MMP)-2, MMP-9, and membrane type 1-MMP in hepatocellular and pancreatic adenocarcinoma: Implications for tumor progression and clinical prognosis. *Clin. Cancer Res.* **2000**, *6*, 2726–2734. [PubMed]

33. Imamura, T.; Ohshio, G.; Mise, M.; Harada, T.; Suwa, H.; Okada, N.; Wang, Z.; Yoshitomi, S.; Tanaka, T.; Sato, H.; et al. Expression of membrane-type matrix metalloproteinase-1 in human pancreatic adenocarcinomas. *J. Cancer Res. Clin. Oncol.* **1998**, *124*, 65–72. [CrossRef] [PubMed]

34. Furukawa, T.; Duguid, W.P.; Rosenberg, L.; Viallet, J.; Galloway, D.A.; Tsao, M.S. Long-term culture and immortalization of epithelial cells from normal adult human pancreatic ducts transfected by the E6E7 gene of human papilloma virus 16. *Am. J. Pathol.* **1996**, *148*, 1763–1770.

35. Ouyang, H.; Mou, L.; Luk, C.; Liu, N.; Karaskova, J.; Squire, J.; Tsao, M.S. Immortal human pancreatic duct epithelial cell lines with near normal genotype and phenotype. *Am. J. Pathol.* **2000**, *157*, 1623–1631. [CrossRef]

36. Andersen, A.P.; Samsoe-Petersen, J.; Oernbo, E.K.; Boedtkjer, E.; Moreira, J.M.A.; Kveiborg, M.; Pedersen, S.F. The net acid extruders NHE1, NBCn1 and MCT4 promote mammary tumor growth through distinct but overlapping mechanisms. *Int. J. Cancer* **2018**, *142*, 2529–2542. [CrossRef]

37. Deer, E.L.; Gonzalez-Hernandez, J.; Coursen, J.D.; Shea, J.E.; Ngatia, J.; Scaife, C.L.; Firpo, M.A.; Mulvihill, S.J. Phenotype and genotype of pancreatic cancer cell lines. *Pancreas* **2010**, *39*, 425–435. [CrossRef]

38. McGuire, C.M.; Forgac, M. Glucose starvation increases V-ATPase assembly and activity in mammalian cells through AMP kinase and phosphatidylinositide 3-kinase/Akt signaling. *J. Biol. Chem.* **2018**, *293*, 9113–9123. [CrossRef]

39. Matsumoto, N.; Sekiya, M.; Tohyama, K.; Ishiyama-Matsuura, E.; Sun-Wada, G.H.; Wada, Y.; Futai, M.; Nakanishi-Matsui, M. Essential Role of the a3 Isoform of V-ATPase in Secretory Lysosome Trafficking via Rab7 Recruitment. *Sci. Rep.* **2018**, *8*, 6701. [CrossRef]

40. Pastor-Soler, N.M.; Hallows, K.R.; Smolak, C.; Gong, F.; Brown, D.; Breton, S. Alkaline pH- and cAMP-induced V-ATPase membrane accumulation is mediated by protein kinase A in epididymal clear cells. *Am. J. Physiol. Cell Physiol.* **2008**, *294*, C488–C494. [CrossRef]

41. Malinda, R.R.Z.; Sharki, P.C.; Ludwig, M.Q.; Pedersen, L.B.; Christensen, S.T.; Pedersen, S.F. TGFβ signaling increases net acid extrusion, proliferation and invasion in Panc-1 pancreatic cancer cells: SMAD4 dependence and link to Merlin/NF2 signaling. *Front. Oncol.* **2019**. in review.

42. Lim, J.H.; Park, J.W.; Kim, M.S.; Park, S.K.; Johnson, R.S.; Chun, Y.S. Bafilomycin induces the p21-mediated growth inhibition of cancer cells under hypoxic conditions by expressing hypoxia-inducible factor-1alpha. *Mol. Pharmacol.* **2006**, *70*, 1856–1865. [CrossRef] [PubMed]

43. Mihaylova, M.M.; Shaw, R.J. The AMPK signalling pathway coordinates cell growth, autophagy and metabolism. *Nat. Cell Biol.* **2011**, *13*, 1016–1023. [CrossRef]

44. Klionsky, D.J.; Abdelmohsen, K.; Abe, A.; Abedin, M.J.; Abeliovich, H.; Acevedo Arozena, A.; Adachi, H.; Adams, C.M.; Adams, P.D.; Adeli, K.; et al. Guidelines for the use and interpretation of assays for monitoring autophagy (3rd edition). *Autophagy* **2016**, *12*, 1–222. [CrossRef] [PubMed]

45. Marino, M.L.; Pellegrini, P.; Di, L.G.; Djavaheri-Mergny, M.; Brnjic, S.; Zhang, X.; Hagg, M.; Linder, S.; Fais, S.; Codogno, P.; et al. Autophagy is a protective mechanism for human melanoma cells under acidic stress. *J. Biol. Chem.* **2012**, *287*, 30664–30676. [CrossRef] [PubMed]

46. Klionsky, D.J.; Abeliovich, H.; Agostinis, P.; Agrawal, D.K.; Aliev, G.; Askew, D.S.; Baba, M.; Baehrecke, E.H.; Bahr, B.A.; Ballabio, A.; et al. Guidelines for the use and interpretation of assays for monitoring autophagy in higher eukaryotes. *Autophagy* **2008**, *4*, 151–175. [CrossRef] [PubMed]

47. Vakifahmetoglu-Norberg, H.; Xia, H.G.; Yuan, J. Pharmacologic agents targeting autophagy. *J. Clin. Investig.* **2015**, *125*, 5–13. [CrossRef]

48. Gao, Y.; Liu, Y.; Hong, L.; Yang, Z.; Cai, X.; Chen, X.; Fu, Y.; Lin, Y.; Wen, W.; Li, S.; et al. Golgi-associated LC3 lipidation requires V-ATPase in noncanonical autophagy. *Cell Death Dis.* **2016**, *7*, e2330. [CrossRef]

49. Hashimoto, D.; Blauer, M.; Hirota, M.; Ikonen, N.H.; Sand, J.; Laukkarinen, J. Autophagy is needed for the growth of pancreatic adenocarcinoma and has a cytoprotective effect against anticancer drugs. *Eur. J. Cancer* **2014**, *50*, 1382–1390. [CrossRef] [PubMed]

50. Mollenhauer, J.; Roether, I.; Kern, H.F. Distribution of extracellular matrix proteins in pancreatic ductal adenocarcinoma and its influence on tumor cell proliferation in vitro. *Pancreas* **1987**, *2*, 14–24. [CrossRef]

51. Toth, M.; Chvyrkova, I.; Bernardo, M.M.; Hernandez-Barrantes, S.; Fridman, R. Pro-MMP-9 activation by the MT1-MMP/MMP-2 axis and MMP-3: Role of TIMP-2 and plasma membranes. *Biochem. Biophys. Res. Commun.* **2003**, *308*, 386–395. [CrossRef]

52. Von Schwarzenberg, K.; Lajtos, T.; Simon, L.; Muller, R.; Vereb, G.; Vollmar, A.M. V-ATPase inhibition overcomes trastuzumab resistance in breast cancer. *Mol. Oncol.* **2014**, *8*, 9–19. [CrossRef] [PubMed]

53. Villanger, O.; Veel, T.; Raeder, M.G. Secretin causes H+/HCO3- Secretion From Pig Pancreatic Ductules by Vacuolar-Type H+-Adenosine Triphosphatase. *Gastroenterology* **1995**, *108*, 850–859. [CrossRef]

54. Hulikova, A.; Harris, A.L.; Vaughan-Jones, R.D.; Swietach, P. Regulation of intracellular pH in cancer cell lines under normoxia and hypoxia 2. *J. Cell Physiol.* **2013**, *228*, 743–752. [CrossRef] [PubMed]

55. Lauritzen, G.; Jensen, M.B.; Boedtkjer, E.; Dybboe, R.; Aalkjaer, C.; Nylandsted, J.; Pedersen, S.F. NBCn1 and NHE1 expression and activity in DeltaNErbB2 receptor-expressing MCF-7 breast cancer cells: Contributions to pHi regulation and chemotherapy resistance. *Exp. Cell Res.* **2010**, *316*, 2538–2553. [CrossRef]

56. Zoncu, R.; Bar-Peled, L.; Efeyan, A.; Wang, S.; Sancak, Y.; Sabatini, D.M. mTORC1 senses lysosomal amino acids through an inside-out mechanism that requires the vacuolar H$^+$-ATPase. *Science* **2011**, *334*, 678–683. [CrossRef] [PubMed]

57. Wu, Y.C.; Wu, W.K.; Li, Y.; Yu, L.; Li, Z.J.; Wong, C.C.; Li, H.T.; Sung, J.J.; Cho, C.H. Inhibition of macroautophagy by bafilomycin A1 lowers proliferation and induces apoptosis in colon cancer cells. *Biochem. Biophys. Res. Commun.* **2009**, *382*, 451–456. [CrossRef]

58. Manabe, T.; Yoshimori, T.; Henomatsu, N.; Tashiro, Y. Inhibitors of vacuolar-type H$^+$-ATPase suppresses proliferation of cultured cells. *J. Cell Physiol.* **1993**, *157*, 445–452. [CrossRef]

59. Von Schwarzenberg, K.; Wiedmann, R.M.; Oak, P.; Schulz, S.; Zischka, H.; Wanner, G.; Efferth, T.; Trauner, D.; Vollmar, A.M. Mode of cell death induction by pharmacological vacuolar H$^+$-ATPase (V-ATPase) inhibition. *J. Biol. Chem.* **2013**, *288*, 1385–1396. [CrossRef]

60. Lim, J.H.; Park, J.W.; Kim, S.J.; Kim, M.S.; Park, S.K.; Johnson, R.S.; Chun, Y.S. ATP6V0C competes with von Hippel-Lindau protein in hypoxia-inducible factor 1alpha (HIF-1alpha) binding and mediates HIF-1alpha expression by bafilomycin A1. *Mol. Pharmacol.* **2007**, *71*, 942–948. [CrossRef]

61. Hubbi, M.E.; Hu, H.; Kshitiz; Ahmed, I.; Levchenko, A.; Semenza, G.L. Chaperone-mediated autophagy targets hypoxia-inducible factor-1alpha (HIF-1alpha) for lysosomal degradation. *J. Biol. Chem.* **2013**, *288*, 10703–10714. [CrossRef] [PubMed]

62. Schneider, L.S.; von, S.K.; Lehr, T.; Ulrich, M.; Kubisch-Dohmen, R.; Liebl, J.; Trauner, D.; Menche, D.; Vollmar, A.M. Vacuolar-ATPase Inhibition Blocks Iron Metabolism to Mediate Therapeutic Effects in Breast Cancer. *Cancer Res.* **2015**, *75*, 2863–2874. [CrossRef] [PubMed]

63. Kissing, S.; Hermsen, C.; Repnik, U.; Nesset, C.K.; von, B.K.; Griffiths, G.; Ichihara, A.; Lee, B.S.; Schwake, M.; De, B.J.; et al. Vacuolar ATPase in phagosome-lysosome fusion. *J. Biol. Chem.* **2015**, *290*, 14166–14180. [CrossRef] [PubMed]

64. Mauvezin, C.; Nagy, P.; Juhasz, G.; Neufeld, T.P. Autophagosome-lysosome fusion is independent of V-ATPase-mediated acidification. *Nat. Commun.* **2015**, *6*, 7007. [CrossRef]

65. Maquoi, E.; Peyrollier, K.; Noel, A.; Foidart, J.M.; Frankenne, F. Regulation of membrane-type 1 matrix metalloproteinase activity by vacuolar H$^+$-ATPases. *Biochem. J.* **2003**, *373*, 19–24. [CrossRef]

66. Pratt, J.; Roy, R.; Annabi, B. Concanavalin-A-induced autophagy biomarkers requires membrane type-1 matrix metalloproteinase intracellular signaling in glioblastoma cells. *Glycobiology* **2012**, *22*, 1245–1255. [CrossRef]

67. Vogelmann, R.; Nguyen-Tat, M.D.; Giehl, K.; Adler, G.; Wedlich, D.; Menke, A. TGFbeta-induced downregulation of E-cadherin-based cell-cell adhesion depends on PI3-kinase and PTEN. *J. Cell Sci.* **2005**, *118*, 4901–4912. [CrossRef]

68. Xu, W.; Wang, Z.; Zhang, W.; Qian, K.; Li, H.; Kong, D.; Li, Y.; Tang, Y. Mutated K-ras activates CDK8 to stimulate the epithelial-to-mesenchymal transition in pancreatic cancer in part via the Wnt/beta-catenin signaling pathway. *Cancer Lett.* **2015**, *356*, 613–627. [CrossRef]

Review

Ferlin Overview: From Membrane to Cancer Biology

Olivier Peulen [1,*], Gilles Rademaker [1], Sandy Anania [1], Andrei Turtoi [2,3,4], Akeila Bellahcène [1] and Vincent Castronovo [1]

[1]   Metastasis Research Laboratory, Giga Cancer, University of Liège, B4000 Liège, Belgium
[2]   Tumor Microenvironment Laboratory, Institut de Recherche en Cancérologie de Montpellier, INSERM U1194, 34000 Montpellier, France
[3]   Institut du Cancer de Montpeiller, 34000 Montpellier, France
[4]   Université de Montpellier, 34000 Montpellier, France
*   Correspondence: olivier.peulen@uliege.be; Tel.: +32-4-366-37-92

Received: 8 August 2019; Accepted: 21 August 2019; Published: 22 August 2019

Abstract: In mammal myocytes, endothelial cells and inner ear cells, ferlins are proteins involved in membrane processes such as fusion, recycling, endo- and exocytosis. They harbour several C2 domains allowing their interaction with phospholipids. The expression of several Ferlin genes was described as altered in several tumoural tissues. Intriguingly, beyond a simple alteration, myoferlin, otoferlin and Fer1L4 expressions were negatively correlated with patient survival in some cancer types. Therefore, it can be assumed that membrane biology is of extreme importance for cell survival and signalling, making Ferlin proteins core machinery indispensable for cancer cell adaptation to hostile environments. The evidences suggest that myoferlin, when overexpressed, enhances cancer cell proliferation, migration and metabolism by affecting various aspects of membrane biology. Targeting myoferlin using pharmacological compounds, gene transfer technology, or interfering RNA is now considered as an emerging therapeutic strategy.

Keywords: ferlin; myoferlin; dysferlin; otoferlin; C2 domain; plasma membrane

1. Introduction

Ferlin is a family of proteins involved in vesicle fusions. To date, more than 760 articles in Pubmed refer to one of its members. Most of these publications are related to muscle biology, while less than 50 are directly related to cancer. However, the emerging idea of targeting plasma membranes [1] and the discovery of a significant correlation between Ferlin gene expression and cancer patient survival, brings attention to cancer. This review focused attention on the roles of these proteins, first in a healthy context, then in cancer.

During the maturation of spermatids to motile spermatozoa in *Caenorhabditis elegans* worm, large vesicles called membranous organelles fuse with the spermatid plasma membrane. This step requires a functional FER-1 protein encoded by the fer-1 gene (*fertilization defective-1*) [2]. When FER-1 was identified and sequenced, no other known proteins had strong resemblance to it. Subsequently, homologs were found by sequence similarity in mammals, forming a family of similar proteins now called ferlins. In humans, a first *C. elegans* fer-1 homolog gene was discovered and the protein encoded by this gene was named dysferlin [3]. Shortly after, a second human FER-1-Like gene was identified. The product of the gene was named otoferlin [4]. The human EST database mining revealed a dysferlin paralog called myoferlin [5,6]. Three new members joined the ferlin gene family: FER1L4, a pseudogene; FER1L5; and FER1L6. The main features of ferlins are summarized in Table 1.

Table 1. Short description of *C. elegans* and human ferlin genes and proteins.

Protein Name (Uniprot Number)	Gene Name	Chromosome Mapping	Main Protein Size
Sperm vesicle fusion protein FER-1 (Q17388)	fer-1		2034 AA (235 KDa)
Dysferlin (O75923)	Fer1-Like 1 Fer1L1	2p13.2	2080 AA (237 KDa)
Otoferlin (Q9HC10)	Fer1-Like Fer1L2	2p23.3	1997 AA (227 KDa)
Myoferlin (Q9NZM1)	Fer1-Like 3 Fer1L3	10q23.33	2061 AA (230 KDa)
FER1L4 (A9Z1Z3)	Fer1-Like 4 Fer1L4	20q11.22	pseudogene
FER1L5 (A0AVI2)	Fer1-Like 5 Fer1L5	2q11.2	2057 AA (238 KDa)
FER1L6 (Q2WGJ9)	Fer1-Like 6 Fer1L6	8q24.13	1857 AA (209 KDa)

The dysferlin mutations were involved in Limb-Girdle muscular dystrophy 2B (LGMD2B), a autosomal recessive degenerative myopathy, and in Miyoshi muscular dystrophy 1 (MMD1), a late-onset muscular dystrophy [3,7]. The otoferlin mutations were described in the non-syndromic prelingual deafness (DFNB9) and in the auditory neuropathy autosomal recessive 1 (AUNB1) [4,8,9]. Nowadays, myoferlin and the 3 last members of the ferlin family are still not linked to human genetic diseases. However, myoferlin was proposed as a modifier protein for muscular dystrophy phenotype [5] and studies of myoferlin-null mice demonstrated impaired myoblast fusion and myofiber formation during muscle development and regeneration [10]. More recently, a truncated variant of myoferlin was associated with Limb-Girdle type muscular dystrophy and cardiomyopathy [11]. Here under, this review discusses that ferlins, mainly myoferlin, are involved in neoplastic diseases and are potential therapeutic targets.

2. Genomic Organization of Ferlin Gene Family

Ferlin genomic organization has not been extensively investigated. Nonetheless, valuable information was obtained from sequencing and subsequent gene annotation (www.ensembl.org). In *C. elegans*, fer-1 gene is approximately 8.6 kb in length and composed of 21 exons [2]. In humans, dysferlin gene (DYSF) is composed of 55 exons [12], and encodes 19 splice variant transcripts. Otoferlin gene (OTOF) contains 47 exons and encodes 7 splice variants. One of them is retaining an intronic sequence from other locus and is not coding for protein. An alternate splicing results in a neuronal-specific domain for otoferlin, regulated by the inclusion of exon 47 [8]. Myoferlin gene (MYOF), is composed of 54 exons and encodes for 9 splice variants. Four of them are not translated to protein and the shortest retains an intronic sequence. Myoferlin promoter includes several consensus-binding sites, such as for Myc, MEF2, CEBP, Sp1, AP1, and NFAT. The latter is able to bind endogenous NFATc1 and NFATc3 [13]. FER1L5 encodes 7 splice variants obtained by the arrangement of 53 exons. Five transcripts are known to encode proteins when the 2 shortest are retaining intronic sequences and do not encode protein. FER1L6 gene is composed of 41 exons and encodes a unique transcript. The main features of ferlin genes are summarized in Table 2.

Table 2. Short description of *C. elegans* and human ferlin genes and transcripts.

Gene Name	Gene Length	Number of Exons	Transcript Size	Number of Variants
Fer-1	8.6 kb	21	6.2 kb	3
Fer1-Like 1 Fer1L1 (DYSF)	233 kb	55	0.5–6.7 kb	19
Fer1-Like 2 Fer1L2 (OTOF)	121 kb	47	0.5–7.2 kb	7
Fer1-Like 3 Fer1L3 (MYOF)	180 kb	54	0.4–6.7 kb	9
Fer1-Like 4 Fer1L4	48 kb	43	0.2–5.9 kb	13
Fer1-Like 5 Fer1L5	64 kb	53	3.5–6.5 kb	7
Fer1-Like 6 Fer1L6	278 kb	41	6 kb	1

3. Ferlin's Structure and Localization

Caenorrhabditis elegans FER-1 is a large protein rich in charged residues. Charged amino acids are distributed throughout the whole protein length such that no particularly acidic or basic domains are observed. The hydrophobicity plot described a 35 amino acid long hydrophobic region at the C-terminal end [2]. To the authors' knowledge, it has never been experimentally demonstrated. Similarity studies suggest that this region might be a transmembrane domain. FER-1 sequence analysis with Pfam protein families database [14] revealed the existence of 4 C2 domains and several other domains.

Ferlins are proteins harboring multi-C2 domains. These structural domains are ~130 amino acid long independently folded modules found in several eukaryotic proteins. They were identified in classical Protein Kinase C (PKC) as the second conserved domain out of four. The typical C2 domain is composed of a beta-sandwich made of 8 beta-strands coordinating calcium ions, participating to their ability to bind phospholipids (for review [15]). However, some C2 domains have lost their capacity to bind calcium but still bind membranes [16]. A large variety of proteins containing C2 domains have been identified, and most of them are involved in membrane biology, such as vesicular transport (synaptotagmin), GTPase regulation (Ras GTPase activating protein) or lipid modification (phospholipase C) (for review [17]).

Human ferlin proteins harbour 5 to 7 C2 domains as described in the Pfam database (Figure 1A). According to this database, in humans, 342 proteins harbour C2 domains. However, the occurrence of multiple tandem C2 domains is uncommon. Only three vertebrate protein families contain more than two C2 domains: The multiple C2 domain and transmembrane region proteins (MCTP) [18], the E-Syt (extended synaptotagmins) [19], and the ferlins. The typical feature of a C2 domain is its ability to interact with two or three calcium ions. The prototype of this domain is the C2A contained in PKC that binds phospholipids in a calcium-dependent manner. Several other distinct C2 domain subtypes, e.g. those found in PI3K and in PTEN, do not have calcium binding abilities and instead specialize in protein-protein interactions [16,17]. In classical Ca^{2+}-binding C2 domains, 5 aspartate residues are involved in the ion binding [20]. Clustal omega alignment of ferlin C2 domains with PKC and synaptotagmin I C2 domains revealed that the 5 Ca^{2+}-binding aspartic acids were conserved or substituted by a glutamic acid in the C2E and C2F domains of all human paralogs (Figure 1B). The aspartic acid to glutamic acid substitution is considered as highly conservative and observed in some non-ferlin Ca^{2+}-binding C2 domains [21]. Some ferlins showed more C2 domains with Ca^{2+}-binding potential, e.g. dysferlin and myoferlin C2C and C2D, otoferlin C2D and fer1L6 C2D [22]. The phylogenic tree created by neighbour-joining of a Clustal omega alignment of C2 domain sequences shows that a C2 domain is more similar to others at a similar position in ortholog proteins than it is to the other C2 domains within the same protein [23]. A Clustal omega alignment reveals an evolutionary distribution of the ferlin proteins into two main subgroups (Figure 1C): The type 1 ferlins containing a DysF domain and the type-2 ferlins without the DysF domain [22]. This domain is present in yeast peroxisomal proteins where its established function is to regulate the peroxisome size and number [24]. In mammals, despite the fact that its solution structure was resolved [25] and that many pathogenic point mutations occur in this region [26,27], the function of this domain remains unknown.

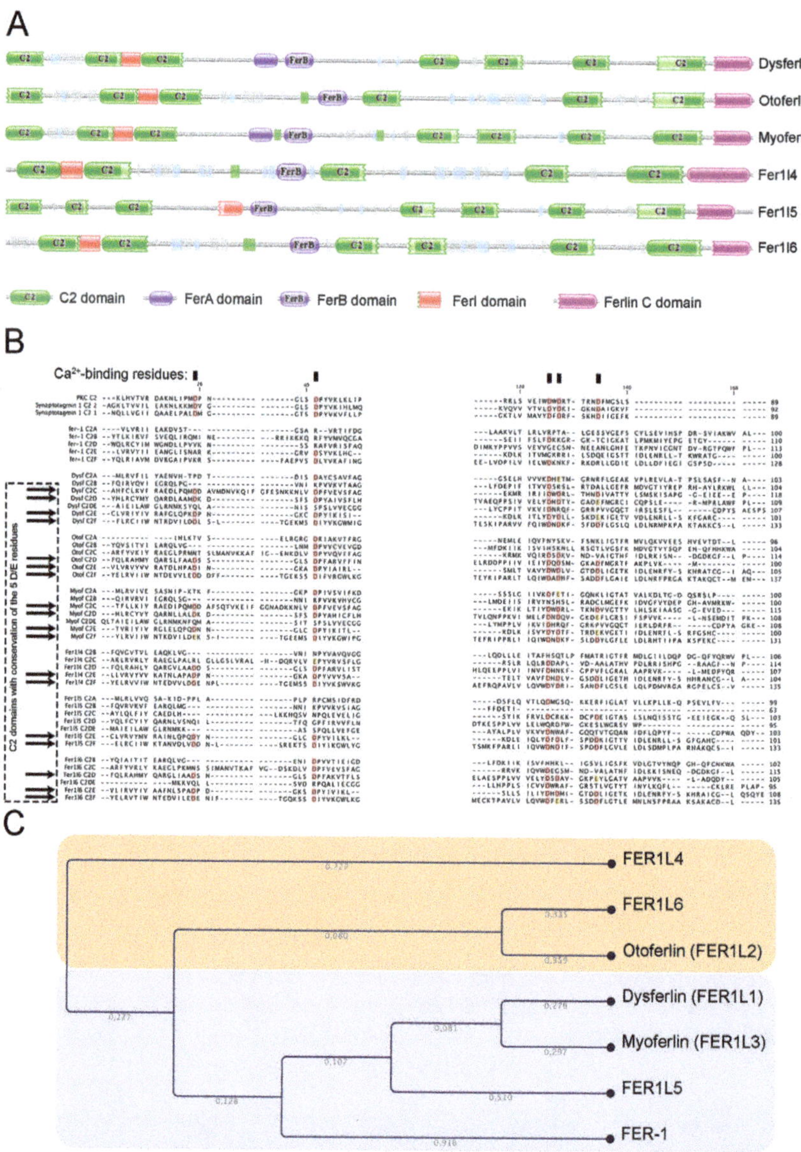

Figure 1. Structure and phylogenic relation of ferlin proteins. (**A**) Schematic structure of FER-1 human homologs as produced by Pfam protein families' database. (**B**) Clustal omega multiple alignment of ferlin C2 domains. Conserved Ca^{2+}-binding site are highlighted in red (aspartic acid—D) or yellow (glutamic acid—E). (**C**) Cladogram of clustal omega alignment indicating type 1 ferlins in blue and type 2 ferlins in yellow. The branch length is indicated in grey.

Immunodetection of a myoferlin-haemagglutinin fusion protein in non-permeabilised COS-7 cells confirmed the presence of the C-terminal domain of the protein in the extracellular compartment [28], supporting the functionality of the putative trans-membrane region. The sublocalisation of ferlins was further studied, indicating robust membrane localisation for dysferlin, myoferlin and Fer1L6 while only low levels of otoferlin were at the plasma membrane and Fer1L5 was intracellular. Dysferlin and myoferlin were localised within the endo-lysosomal pathway accumulating in late endosomes and in recycling compartment. GFP-myoferlin fusion protein revealed that myoferlin was colocalized with lysosomal markers in NIH3T3 cells [29]. Otoferlin has been shown to move from the trans-Golgi network to the plasma membrane and inversely. Fer1L5 was cytosolic while Fer1L6 was detected in a specific sub-compartment of the trans-Golgi network compartment [30].

4. Ferlin's Interactions with Phospholipids

Ferlins are regarded as intrinsic membrane proteins through their putative transmembrane region. However, they can also interact with membranes by other domains. Experimentally, myoferlin C2A was the single C2 domain able to bind to phospholipid vesicles. A significant presence of the negatively charged phosphatidylserine (PS) was required for this interaction. Myoferlin C2A binding to PS-containing vesicles did not occur with calcium concentration similar to the one observed in the basal physiological condition (0.1 μM). Indeed, the half-maximal binding was observed at 1 μM [31], suggesting that the C2A domain is involved in specific processes inside the cell requiring Ca^{2+} release from intracellular stock, like in Ca^{2+}-regulated exocytosis. When cells are stimulated by various means, including depolarization and ligand binding, the cytosolic Ca^{2+} concentration increases to the concentration up to 1 μM or more [32], similar to the one required by myoferlin C2A domain to bind lipids. It appears that dysferlin C2A domain has the same binding properties as myoferlin C2A domain. However, its half-maximal lipid binding was higher (4.5 μM) [31]. A recent publication confirmed that myoferlin and dysferlin C2A domains exhibit different Ca^{2+} affinities. However, they describe myoferlin C2A domain with a lower Ca^{2+} affinity than the dysferlin homolog C2 domain, and a marginal binding of myoferlin C2A domain to phospholipid mixture containing PS [33]. The binding of dysferlin C2A to PS was confirmed and extended to several phosphoinositide monophosphates in a Ca^{2+}-dependent fashion. Therrien et al. observed that all remaining dysferlin C2 domains were able to bind to PS but independently of Ca^{2+} [34]. The laurdan fluorescence emission experiments suggest that dysferlin and myoferlin contribute to increase the lipid order in lipid vesicles. The magnitude of this observation was calcium-enhanced and C2 domains within both N- and C-termini of ferlins influenced lipid packing. The experiments conducted with individual recombinant ferlin's C2A-C domains demonstrated that all of them are able to increase lipid order [35].

The authors described in the first part of this review the conservation of the 5 Ca^{2+}-binding aspartate residues in the C2D-F domains of otoferlin making them putative Ca^{2+}-binding sites. In addition to its C2D-F domains, otoferlin is also able to bind Ca^{2+} via its C2B and C2C domains [36]. Despite the fact that C2A domain from otoferlin does not possess all five aspartate residues, its ability to bind Ca^{2+} is still under debate. Therrien and colleagues showed that otoferlin C2A domain can bind PS in a Ca^{2+}-dependent fashion, suggesting an interaction with this ion [34]. This interaction was confirmed by a direct measure of otoferlin-binding to liposomes in the presence of Ca^{2+} (1 mM). Moreover, C2A-C domains seem to bind lipids also under calcium free conditions [36]. At the opposite, a spectroscopy analysis indicates that otoferlin C2A domain is unable to coordinate Ca^{2+} ion [37].

Floatation assays were unable to confirm the interaction between otoferlin C2A and lipids. This may be due to the presence of a shorter membrane-interacting loop at the top of the domain [37]. As for dysferlin and myoferlin, otoferlin increases lipid order in vesicles. However, its C2A does not participate to the phenomenon [35].

Ferlin proteins contain also a FerA domain recently described as a four-helix bundle fold with its own Ca^{2+}-dependent phospholipid-binding activity [38].

5. Ferlin's Main Functions in Non-Neoplastic Cells and Tissues

5.1. In Mammal Muscle Cells

Dysferlin and myoferlin have a specific temporal pattern of expression in an in vitro model of muscle development. Myoferlin was highly expressed in myoblasts that have elongated prior to fusion to syncytial myotubes. After fusion, myoferlin expression was decreased. The dysferlin expression increased concomitantly with the fusion and maturation of myotubes [31]. A proteomic analysis revealed the interacting partners of dysferlin during muscle differentiation [39]. It appeared that the number of partners decreases during the differentiation process, while the core-set of partners is large (115 proteins). Surprisingly, the dysferlin homolog myoferlin was consistently co-immunoprecipitated with dysferlin. The gene ontology analysis of the core-set proteins indicates that the highest ranked clusters are related to vesicle trafficking. In the C2C12 myoblast model, immunoprecipitation experiments showed that myoferlin interacts with the Eps15 Homology Domain 2 (EHD2) apparently through a NPF (asparagine-proline-phenylalanine) motif in its C2B domain [40]. EHD2 has been implicated in endocytic recycling. It was inferred that the interaction between EHD2 and myoferlin might indirectly regulate disassembly or reorganization of the cytoskeleton that accompanies myoblast fusion.

Dysferlin-null mice develop a slowly progressive muscular dystrophy with a loss of plasma membrane integrity. The presence of a stable and functional dystrophin–glycoprotein complex (DGC), involved in muscle injury-susceptibility when altered, suggests that dysferlin has a role in sarcolemma repair process. This was confirmed in dysferlin-null mice by a markedly delayed membrane resealing, even in the presence of Ca^{2+} [41]. Pharmacological experiments conducted in skeletal muscles demonstrated that dysferlin modulates smooth reticulum Ca^{2+} release and that in its absence injuries cause an increased ryanodine receptor (RyR1)-mediated Ca^{2+} leak from the smooth reticulum into the cytoplasm [42]. In the SJL/J mice model of dysferlinopathy, annexin-1 and -2 co-precipitate with muscle dysferlin and co-localise at sarcolemma in an injury-dependent manner [43]. An immunofluorescence analysis of mitochondrial respiratory chain complexes in the muscles from the patients with dysferlinopathy revealed complex I- and complex IV-deficient myofibers [44]. This report is particularly interesting in light of the dysferlin_v1 alternate transcript discovered in skeletal muscle [45] and harboring a mitochondrial importation signal [39].

Intriguingly, at the site of membrane injury, only the C-terminal extremity of dysferlin was immunodetected. It was reported than dysferlin was cleaved by calpain [46], one of its interacting proteins [39]. The cleavage generate a C-terminal fragment called mini-dysferlin$_{C72}$ bearing two cytoplasmic C2 domains anchored by a transmembrane domain [46]. Myoferlin expression is also up regulated in damaged myofibers and in surrounding mononuclear muscle and inflammatory cells [13]. As it was observed for dysferlin, myoferlin can be cleaved by calpain to produce a mini-myoferlin module composed of the C2E and C2F domains [47].

Membrane repair requires the accumulation and fusion of vesicles with each other and with plasma membrane at the disruption point. A role for dysferlin and myoferlin in these processes is consistent with the presence of several C2 domains and with their homology with FER-1 having a role in vesicle fusion. Moreover, mini-dysferlin and mini-myoferlin bear structural resemblance to synaptotagmin, a well-known actor in synaptic vesicle fusion with the presynaptic membrane [48].

In mouse skeletal muscle, myoferlin was found at the nuclear and plasma membrane [5]. It is highly expressed in myoblasts before their fusion to myotubes [10,31] and found to be highly concentrated at the site of apposed myoblast and myotube membranes, and at site of contact between two myotubes [10]. Myoblast fusion requires a Ca^{2+} concentration increase to 1.4 μM [49], similar to the one reported for myoferlin C2A binding to phospholipids [31]. Myoferlin-null mice myoblasts show impaired fusion in vitro, producing mice with smaller muscles and smaller myofibers in vivo [10]. All together, these observations support a role for myoferlin in the maturation of myotubes and the formation of large myotubes that arise from the fusion of myoblasts to multinucleated myotubes.

Interestingly, myoferlin-null mice are unresponsive to IGF-1 for the myoblast fusion to the pre-existing myofibers. Mechanistic experiments indicate a defect in IGF-1 internalization and a redirection of the IGF1R to the lysosomal degradation pathway instead of recycling. As expected, myoferlin-null myoblasts lacked the IGF1-induced increase in AKT and MAPK activity downstream to IGFR [50].

The defects in myoblast fusion and muscle repair observed in myoferlin-null mice are reminiscent of what was reported in muscle lacking nuclear factor of activated T-cells (NFAT). Demonbreun and colleagues suggested that in injured myofibers, the membrane damages induce an intracellular increase of Ca^{2+} concentration producing a calcineurin-dependent NFAT activation and subsequent translocation to the nucleus. The activated NFAT can therefore bind to its response element on the myoferlin promoter [13].

Using HeLa and HEK293T cell lines overexpressing ADAM-12, it was discovered that myoferlin was one of the ten most abundant interacting partners of ADAM-12 [51]. Though this was discovered in an artificial overexpressing model using cancer cells, it can be considered as pertinent in the context of muscle cell repair. Indeed, ADAM-12 is a marker of skeletal muscle regeneration interacting with the actin-binding protein α-actinin-2 in the context of myoblast fusion [52].

The differentiating myoblast C2C12 expressed Fer1L5 at the protein level with an expression pattern similar to dysferlin throughout myoblast differentiation. Fer1L5 shares with myoferlin a NPF motif in its C2B domain. As in myoferlin, this motif was described as interacting with EHD2, but also with EHD1 [53].

5.2. In mammal Inner Ear Cells

In adult mouse cochlea, otoferlin gene expression is limited to inner hair cells (IHC) [4]. In these cells, the strongest immunostaining of otoferlin was associated with the basolateral region, where the afferent synaptic contacts are located, suggesting that otoferlin is a component of the IHC presynaptic machinery. Ultrastructural observations confirmed the association of otoferlin with the synaptic vesicles. It appears that otoferlin is not necessary for the synapse formation [54], but rather regulates the Ca^{2+}-induced synaptic vesicle exocytosis [36].

At molecular level, otoferlin interacts with plasma membrane t-SNARE (*soluble N-ethylmaleimide-sensitive-factor attachment protein receptor*) proteins (syntaxin 1 and SNAP-25) in a Ca^{2+}-dependent manner [54]. Supporting this discovery, both t-SNARE proteins are known to interact with synaptotagmin I, a C2 domain harbouring protein, in the context of the classical synaptic vesicles docking [55,56]. It was reported that otoferlin relies on C2F domain for its Ca^{2+}-dependent interaction with t-SNARE [57–59]. However, others suggest a Ca^{2+}-dependent interaction through the C2C, C2D, C2E and C2F domains and a Ca^{2+}-independent interaction via the C2A and C2B domains. The SNARE-mediated membrane fusion was reconstituted with proteoliposomes. This assay indicates that in presence of Ca^{2+}, otoferlin accelerates the fusion process [36], suggesting that otoferlin operates as a calcium-sensor for SNARE-mediated membrane fusion.

5.3. In Mammal Endothelial Cells

Bernatchez and colleagues reported that dysferlin and myoferlin are abundant in caveolae-enriched membrane microdomains/lipid rafts (CEM/LR) isolated from human endothelial cells and are highly expressed in mouse blood vessels [28,60]. As observed for dysferlin in muscle cells, myoferlin regulates the endothelial cell membrane resealing after physical damage. In endothelial cells, myoferlin silencing reduced or abolished the ERK-1/2, JNK or PLCγ phosphorylation by VEGF, resulting from a loss of VEGFR-2 stabilization at the membrane. Indeed, myoferlin silencing caused an increase in VEGFR2 polyubiquitination, which leads to its degradation [28]. In contrast to what was observed in myoferlin-silenced endothelial cells, dysferlin gene silencing decrease neither VEGFR2 expression nor its downstream signalling. However, dysferlin-siRNA treated endothelial cells showed a near-complete inhibition of proliferation when they were sub-confluent. The proliferation decrease seems to be due to an impaired attachment rather than to cell death, as supported by adhesion assays and PECAM-1 poly-ubiquitination that leads to its degradation. Co-immunoprecipitation and co-localisation experiments support the formation of a molecular complex between dysferlin and PECAM-1. This PECAM-1 degradation leads, in dysferlin-null mice, to a blunted VEGF-induced angiogenesis [60]. Another angiogenic tyrosine kinase receptor Tie-2 (tyrosine kinase with Ig and epidermal growth factor homology domains-2) is significantly less expressed at the plasma membrane when myoferlin is silenced in endothelial cells [61]. In this case, it appears that proteasomal degradation plays a minor role in the down regulation of the receptor. Strikingly, G-protein coupled receptors (GCPR) were unaffected by the decrease of myoferlin expression, suggesting a selective effect on receptor tyrosine kinases (RTK).

It was also reported that in endothelial cells, myoferlin is required for an efficient clathrin and caveolae/raft-dependent endocytosis, is co-localized with Dynamin-2 protein [62] and that the FASL-induced lysosome fusion to plasma membrane is mediated by dysferlin C2A domain [63].

5.4. Other Mammal's Cells

Dysferlin and myoferlin are expressed in both basal and ciliated airway epithelial cells from healthy human lungs [64]. In the airway epithelial cell line (16HBE), dysferlin and myoferlin were immuno-detected at the plasma membrane, Golgi membrane and in cytoplasm but not in the nuclei. The silencing of myoferlin in these cells induces the loss of zonula occludens (ZO)-1, inducing apoptosis [64].

Myoferlin was also detected in exosomes from human eye trabecular meshwork cells [65] and in phagocytes where it participates to the fusion between lysosomes and the plasma membrane, thus promoting the release of lysosomal contents [29].

The Fer1L5 gene expression was largely restricted to the pancreas, where it was alternatively spliced by removing exon 51 [30].

6. Ferlins in Cancer, Potential Targets to Kill Cancer

It is clear from the data above that ferlins are consistently involved in membrane processes requiring membrane fusion, including endocytosis, exocytosis, membrane repair, recycling and remodelling. Membrane processes are of extreme importance for cell survival and signalling, making them core machinery for cancer cell adaptation to hostile environments.

Considering that ferlins have been only scarcely investigated in cancer, the authors next sought to mine publicly available databases and gain information regarding ferlin's expression or mutation in tumors. Using the FireBrowse gene expression viewer (firebrowse.org), The Cancer Genome Atlas (TCGA) RNAseq data of all ferlin's genes in neoplastic tissues were investigated in order to obtain a differential expression in comparison to their normal counterparts. It appears that all ferlin genes are modulated in several cancer types. Myoferlin and fer1l4 genes are more frequently up regulated than down regulated, while dysferlin, fer1l5, and fer1l6 are more frequently down regulated (Figure 2).

Experimentally, a myoferlin gene was discovered as highly expressed in several tumour tissues including the pancreas [66,67], breast [68], kidneys [68], and head and neck squamous cell carcinoma (HNSCC) [69]. This expression was confirmed at a protein level in tumour tissue and/or cell lines from the pancreas [70–73], breast [74,75], lungs [75], melanoma [75], hepatocellular carcinoma [76], HNSCC [77], clear cell renal carcinoma [78,79], and endometroid carcinoma [80]. Myoferlin was also detected at a protein level in microvesicles/exosomes derived from several cancer cells including the bladder [81], colon [82–85], ovary [86], prostate [87], breast and pancreas, where it plays a role in vesicle fusion with the recipient endothelial cells [88].

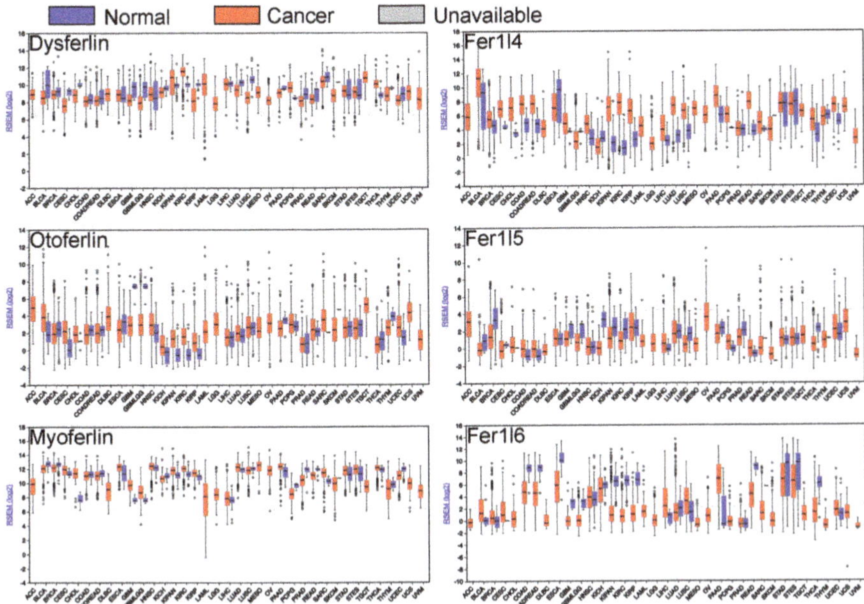

Figure 2. Ferlin gene expression in several cancers (red) and their normal counterparts (blue). Cancer tissues from adrenocortical carcinoma (ACC), bladder urothelial carcinoma (BLCA), breast invasive carcinoma (BRCA), cervical squamous cell carcinoma and endocervical adenocarcinoma (CESC), cholangiocarcinoma (CHOL), colon adenocarcinoma (COAD), colorectal adenocarcinoma (COADREAD), lymphoid neoplasm diffuse large B-cell lymphoma (DLBC), esophageal carcinoma (ESCA), glioblastoma multiforme (GBM), glioma (GBMLGG), head and neck squamous cell carcinoma (HNSC), kidney chromophobe (KICH), pan-kidney cohort (KIPAN), kidney renal clear cell carcinoma (KIRC), kidney renal papillary cell carcinoma (KIRP), acute myeloid leukemia (LAML), brain lower grade glioma (LGG), liver hepatocellular carcinoma (LIHC), lung adenocarcinoma (LUAD), lung squamous cell carcinoma (LUBC), mesothelioma (MESO), ovarian serous cystadenocarcinoma (OV), pancreatic adenocarcinoma (PAAD), pheochromocytoma and paraganglioma (PCPG), prostate adenocarcinoma (PRAD), rectum adenocarcinoma (READ), sarcoma (SARC), skin cutaneous melanoma (SKCM), stomach adenocarcinoma (STAD), stomach and esophageal carcinoma (STES), testicular germ cell tumours (TGCT), thyroid carcinoma (THCA), thymoma (THYM), uterine corpus endometrial carcinoma (UCEC), uterine carcinosarcoma (UCS), uveal melanoma (UVM).

This review then explored the mutations occurring in ferlin genes in tumours using Tumorportal (http://www.tumorportal.org) [89]. Several mutations were reported in ferlin genes in a few cancer types. However, none of them were considered as significant. Survival was also analysed (Table 3) using a pan-cancer method available online (OncoLnc–http://www.oncolnc.org) and combining mRNAs, miRNAs, and lncRNAs expression [90]. Noticeably, otoferlin expression was strongly significantly correlated with survival in renal clear cell carcinoma (KIRC–$p < 10^{-5}$); myoferlin expression was strongly significantly correlated with survival in brain lower grade glioma (LGG–$p < 10^{-4}$) and pancreatic adenocarcinoma (PAAD–$p < 10^{-4}$), and Fer1l4 expression was strongly significantly correlated with survival in bladder urothelial carcinoma (BLCA–$p < 10^{-5}$) and kidney renal clear cell carcinoma (KIRC–$p < 10^{-5}$). The 5 more significant correlations between ferlin's expression and the overall survival were represented as Kaplan-Meier curves with their associated log-rank p-value (Figure 3).

Table 3. Survival analysis by a Cox regression.

Positive Association			Negative Association		
Cohort	Cox Coefficient	p-Value	Cohort	Cox Coefficient	p-Value
DYSF EXPRESSION					
CESC	0.266	$4.20e^{-02}$	SARC	−0.277	$1.00e^{-02}$
STAD	0.171	$4.80e^{-02}$	KIRC	−0.220	$1.00e^{-02}$
OTOF EXPRESSION					
KIRC	**0.377**	**$1.50e^{-06}$**	BLCA	−0.275	$4.50e^{-04}$
KIRP	0.413	$4.90e^{-03}$	SKCM	−0.169	$1.40e^{-02}$
MYOF EXPRESSION					
LGG	**0.441**	**$1.40e^{-05}$**	SKCM	−0.163	$1.90e^{-02}$
PAAD	**0.561**	**$1.70e^{-05}$**			
LAML	0.215	$4.70e^{-02}$			
FER1L4 EXPRESSION					
KIRC	**0.356**	**$5.20e^{-06}$**	**BLCA**	**−0.383**	**$2.90e^{-06}$**
KIRP	0.492	$1.10e^{-03}$	SKCM	−0.225	$1.10e^{-03}$
LGG	0.244	$4.00e^{-03}$			
FER1L5 EXPRESSION					
LUAD	−0.199	$1.30e^{-02}$			
FER1L6 EXPRESSION					
KIRC	−0.160	$4.80e^{-02}$			
READ	−0.401	$4.90e^{-02}$			

Ferlin gene expression from cohorts with cancer was submitted to a survival analysis with a Cox regression. The red rows indicate a negative Cox coefficient, the green rows indicate positive Cox coefficient. The bold p-values were considered as highly significant ($p < 10^{-4}$). Bladder urothelial carcinoma (BLCA), cervical squamous cell carcinoma and endocervical adenocarcinoma (CESC), kidney renal clear cell carcinoma (KIRC), kidney renal papillary cell carcinoma (KIRP), acute myeloid leukemia (LAML), brain lower grade glioma (LGG), lung adenocarcinoma (LUAD), pancreatic adenocarcinoma (PAAD), rectum adenocarcinoma (READ), sarcoma (SARC), skin cutaneous melanoma (SKCM), stomach adenocarcinoma (STAD).

Interestingly, a recent publication points out specific single nucleotide polymorphisms in dysferlin genes as significantly associated with pancreas cancer patient survival [91]. Mining the TCGA database, a high Fer1L4 expression was reported as a predictor of a poor prognosis in glioma [92,93] and as an oncogenic driver in several human cancers [94]. However, several other publications pointed it out as a predictor of good prognosis in osteosarcoma [95], gastric cancer [96], endometrial carcinoma [97].

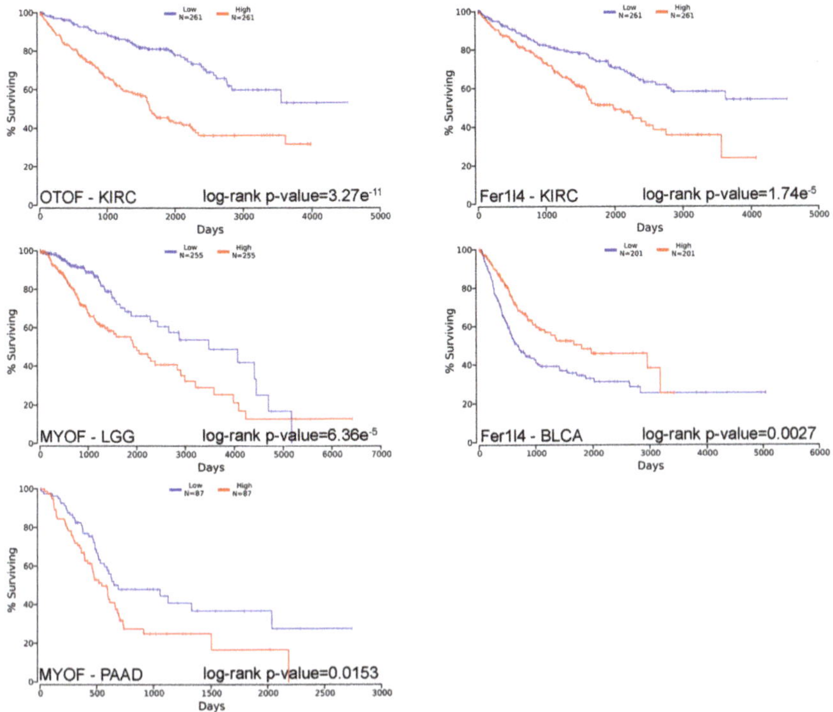

Figure 3. Kaplan-Meier survival curves of patient cohorts with different cancer types. Ferlin gene expression was segregated in low (blue) and high (red) expression according to median in kidney renal clear cell carcinoma (KIRC), brain lower grade glioma (LGG), bladder urothelial carcinoma (BLCA), and pancreatic adenocarcinoma (PAAD).

6.1. Breast Cancer and Melanoma

A mathematical model was proposed to examine the role of myoferlin in cancer cell invasion. This model confirms the experimental observation of decreased invasion of the myoferlin-null breast MDA-MB-231 cell line, and predicts that the pro-invasion effect of myoferlin may be in large partly mediated by MMPs [98]. The model was further validated in vitro suggesting a mesenchymal to epithelial transition (MET) when myoferlin was knockdown [99,100]. Using the same cell model, Blackstone and colleagues showed that myoferlin depletion increased cell adhesion to PET substrate by enhancing focal adhesion kinase (FAK) and its associated protein paxillin (PAX) phosphorylation [101]. Interestingly, myoferlin was reported as regulating the cell migration through a TGF-β1 autocrine loop [102]. Recently, related results were reported in melanoma [103]. Myoferlin expression was first correlated with vasculogenic mimicry (VM) in patients, then its in vitro depletion in A375 cell line impaired VM, migration, and invasion by decreasing MMP-2 production.

Several evidences, obtained from normal endothelial cells, indicate that myoferlin is involved in RTKs recycling (see above). Our group showed that MDA-MB-231 and -468 cells depleted for myoferlin were unable to migrate and to undergo EMT upon EGF stimulation. The authors discovered that myoferlin depletion altered the EGFR fate after ligand binding, most probably by inhibiting the non-clathrin mediated endocytosis [104]. Unexpectedly, myoferlin seemed to be physically associated with lysosomal fraction in MCF-7 cells [105], supporting its involvement in the membrane receptor recycling.

The co-localisation of myoferlin with caveolin-1 [104], the main component of caveolae considered as a metabolic hub [106] prompted our group to investigate the implication of myoferlin in energy metabolism.

In this context, the authors showed in triple-negative breast cancer cells that myoferlin-silencing produces an accumulation of monounsaturated fatty acids (C16:1). Its depletion further decreased oxygen consumption switching the cell metabolism toward glycolysis [107]. This was the first report of the role of myoferlin in mitochondrial function and cell metabolism. A recent report describing the link between dysferlin mutations and mitochondrial respiratory complexes in muscular dysferlinopathy emerged (see above) [44]. It is also intriguing that dysferlin_v1 alternate transcript discovered in skeletal muscle [45] harbours a mitochondrial importation signal [39].

Several breast cancer cell lines and tissues showed a calpain-independent myoferlin cleavage, regardless of cell injuries and subsequent Ca^{2+} influx [108]. The resulting cleaved myoferlin increases ERK phosphorylation in an overexpressing HEK293 system. It would be of interest to further study the link between mini-myoferlin and KRAS mutated cancers as ERK is a mid-pathway signalling protein in this context.

6.2. Pancreas and Colon Cancers

In pancreas adenocarcinoma (PAAD), myoferlin is overexpressed in high grade PAAD in comparison to low grade [73]. The patients with high myoferlin PAAD had a significantly worse prognosis than those with low myoferlin PAAD, with myoferlin appearing as an independent prognosis factor. The experiments undertaken with pancreatic cell lines and siRNA-mediated silencing demonstrated that myoferlin requested to maintain a high proliferation rate. The authors reported that myoferlin is a key element in VEGF exocytosis by PAAD cell lines, correlating with microvessel density in PAAD tissue [109]. Recently, it was demonstrated that myoferlin is critical to maintain mitochondrial structure and oxidative phosphorylation [110]. This discovery was extended to colon cancer where myoferlin seemed also to protect cells from p53-driven apoptosis [111]. The concept claiming that metastatic dissemination relies on oxidative phosphorylation is broadly accepted [112,113]. Based on these reports, the authors discovered that myoferlin was overexpressed in PAAD cells with a high metastatic potential, where it controls mitochondrial respiration [114].

Recently, FER1L4 methylated DNA marker in pancreatic juice has been strongly associated with pancreatic ductal adenocarcinoma suggesting its use as a biomarker for early detection [115].

6.3. Lung Cancer

In mice bearing solid LLC lung tumours, the intratumoral injection of myoferlin siRNA mixed with a lipidic vector reduced the tumour volume by 73%. The observed reduction was neither the consequence of a difference in blood vessel density nor of VEGF secretion. However, a significant reduction of the proportion of the Ki67-positive cells indicated a decrease in cell proliferation [75]. Myoferlin was reported as expressed in human non-small cell lung cancer tissues where it was correlated with VEGFR2, thyroid transcription factor (TTF)-1 and transformation-related protein (p63), especially in the low stage tumours [116].

Recently, it was suggested that long non-coding RNA Fer1L4 negatively controlled proliferation and migration of lung cancer cells, probably through the PI3K/AKT pathway [117]. The same observation was made in osteosarcoma cells [118], esophageal squamous cell carcinoma [119], and hepatocellular carcinoma [120].

6.4. Liver Cancer

In the hepatocellular carcinoma (HCC) cell line, the silencing of the transcriptional coactivator of the serum response factor (SRF), Megakaryoblastic Leukemia 1/2 (MKL1/2), induced a reduction of myoferlin gene expression. It was shown by chromatin immunoprecipitation that MKL1/2 binds effectively to the myoferlin promoter [76]. As in other cancer types, HCC required myoferlin to proliferate and perform invasion or anchorage-independent cell growth. Its depletion enhanced EGFR phosphorylation, in agreement with the concept of myoferlin being a regulator of RTK recycling.

6.5. Head and Neck Cancer

A myoferlin expression pattern was investigated in oropharyngeal squamous carcinoma (OPSCC). It was reported that myoferlin was overexpressed in 50% of the cases and significantly associated with worse survival. Moreover, HPV-negative patients had significantly higher expressions of myoferlin. A subgroup survival analysis indicates the interaction between these two parameters as HPV-negative has the worst prognosis when myoferlin is highly expressed. Nuclear myoferlin expression appeared to be highly predictive of the clinical outcome and associated with IL-6 and nanog overexpression [77]. Upon HNSCC cell line stimulation with IL-6, myoferlin dissociates from EHD2 and binds activated STAT3 to drive it in the nucleus. The observation was extended to breast cancer cell lines [69].

6.6. Gastric Cancer

Recently, a profiling study reported that FER1L4 was a long non-coding RNA (lncRNA) strongly downregulated in gastric cancer tissue [96], in plasma from gastric cancer patients [121] and in human gastric cancer cell lines [122]. In gastric cancer tissues, FER1L4 lncRNA was associated with the tumour diameter, differentiation state, tumour classification, invasion, metastasis, TNM stage and serum CA72-4. Interestingly, the abundance of this lncRNA decreases in plasma shortly after surgery [121]. The same team reported that the FER1L4 lncRNA is a target of miR-106a-5p [122,123]. The cell depletion in FER1L4 lncRNA resulted in an increase in miR-106a-5p and in a decrease of its endogenous target PTEN, suggesting a competing endogenous RNA (ceRNA) [124] role for FER1L4 lncRNA [122]. The control of miR-106a-5p by FER1L4 lncRNA was extended to colon cancer [125] and HCC [126], while it was described over miR-18a-5p in osteosarcoma [127].

6.7. Gynecological Cancers

Lnc Fer1L4 was briefly investigated in ovarian cancer where it was described as downregulated in cancer cells in comparison to normal ovarian epithelial cells [128]. Interestingly, the Fer1L4 expression correlates negatively with the paclitaxel resistance and its re-expression restore the paclitaxel sensitivity through the inhibition of a MAPK signalling pathway.

7. Conclusions

This review clearly shows that all ferlin proteins are membrane-based molecular actors sharing structural similarities. Far beyond their well-described involvement in physiological membrane fusion, several correlations apparently link ferlins, and most particularly myoferlin, to cancer prognosis. However, further investigations are still needed to discover the direct link between myoferlin and cancer biology. Encouragingly, there are many indications that myoferlin depletion interferes with growth factor exocytosis, surface receptor fate determination, exosome composition, and metabolism, indicating the future research axes.

Self-sufficiency in growth factor signalling is a hallmark of cancer cells. Cancer cells overproduce the growth factor to stimulate unregulated proliferation in an autocrine, juxtacrine or paracrine fashions. In this context, myoferlin could be considered as a cancer growth promoter as it helps the exocytosis of the growth factors, at least VEGF. In normal cells, myoferlin was described as involved in receptor tyrosine kinase (EGFR and VEGFR) recycling or expression, allowing as such, the cell response to the growth factors. Knowing that some cancer cells exhibit mutations in tyrosine kinase receptors, which lead to a constitutive receptor activation triggering the downstream pathways, it can be speculated that myoferlin depletion could impede cell proliferation in these cases. This role was indeed described in breast cancer cells [104].

Exosomes are small extracellular vesicles released on exocytosis of multivesicular bodies filled with intraluminal vesicles. They represent an important role in intercellular communication, serving as carrier for the transfer of miRNA and proteins between cells. The exosomes are increasingly described as cancer biomarkers [129] and involved in the preparation of the tumour microenvironment [130].

Interestingly, myoferlin was demonstrated to be present in exosomes isolated from several cancer cell types. However, the biological significance of this localization has still to be investigated.

Metabolism recently integrated the hallmarks of cancer [131], and mitochondria were recognised as key players in cancer metabolism [132]. The indications that myoferlin is necessary for optimal mitochondrial function is a promising avenue in the search for an innovative therapy.

Myoferlin, being overexpressed in several cancer types, offers very promising advantages for cancer diagnosis and targeting. Targeting myoferlin at the expression or functional levels remains, however, the next challenge. Interestingly, recent studies identified new small compounds interacting with the myoferlin C2D domain and demonstrating promising anti-tumoral/metastasis properties in breast and pancreas cancer [133,134].

Gene transfer strategies have undergone profound development in recent years and this is particularly applicable for recessive disorders. The adeno-associated virus (AAV) is a non-pathogenic vector used in a treatment strategy aiming at delivering full-length dysferlin or shorter variants to skeletal muscle in dysferlin-null mice. Several well documented reports demonstrate an improvement in the outcome measures after dysferlin gene therapy [135–138]. Similar AAV vectors were used as a gene delivery system in cancer [139,140], allowing the dream of myoferlin negative-dominant delivery to cancer cells. Moreover, the sleeping beauty transposon system [141] may overcome some of the limitations associated with viral gene transfer vectors and transient non-viral gene delivery approaches that are being used in the majority of ongoing clinical trials.

8. Statistical Methods

The multivariate Cox regressions (Table 3) were performed with the coxph function from the R survival library. For each cancer and data type, OncoLnc attempted to construct a model with gene expression, sex, age, and grade or histology as multivariates [90]. The clinical information was obtained from TCGA and only patients who contained all the necessary clinical information were included in the analysis. The patients were split into low and high expressing according to the median gene expression.

Funding: This research was funded by "Fonds Léon Fredericq" and by the "Patrimoine de l'Université de Liège".

Acknowledgments: The results published here are in whole or part based upon data generated by the TCGA Research Network: http://cancergenome.nih.gov/. AB is a Research Director at the National Fund for Scientific Research (FNRS), Belgium. SA is supported by a FNRS FRIA grant. AT acknowledges LabEx MAbImprove for financial support.

Conflicts of Interest: The authors declare no conflicts of interest. The funders had no role in the design of the study; in the collection, analyses, or interpretation of data; in the writing of the manuscript, or in the decision to publish the results.

References

1. Bernardes, N.; Fialho, A.M. Perturbing the Dynamics and Organization of Cell Membrane Components: A New Paradigm for Cancer-Targeted Therapies. *Int. J. Mol. Sci.* **2018**, *19*, 3871. [CrossRef] [PubMed]
2. Achanzar, W.E.; Ward, S. A nematode gene required for sperm vesicle fusion. *J. Cell Sci.* **1997**, *110*, 1073–1081. [PubMed]
3. Bashir, R.; Britton, S.; Strachan, T.; Keers, S.; Vafiadaki, E.; Lako, M.; Richard, I.; Marchand, S.; Bourg, N.; Argov, Z.; et al. A gene related to caenorhabditis elegans spermatogenesis factor fer-1 is mutated in limb-girdle muscular dystrophy type 2B. *Nat. Genet.* **1998**, *20*, 37–42. [CrossRef] [PubMed]
4. Yasunaga, S.; Grati, M.; Cohen-Salmon, M.; El-Amraoui, A.; Mustapha, M.; Salem, N.; El-Zir, E.; Loiselet, J.; Petit, C. A mutation in OTOF, encoding otoferlin, a FER-1-like protein, causes DFNB9, a nonsyndromic form of deafness. *Nat. Genet.* **1999**, *21*, 363–369. [CrossRef] [PubMed]
5. Davis, D.B.; Delmonte, A.J.; Ly, C.T.; McNally, E.M. Myoferlin, a candidate gene and potential modifier of muscular dystrophy. *Hum. Mol. Genet.* **2000**, *9*, 217–226. [CrossRef] [PubMed]
6. Britton, S.; Freeman, T.; Vafiadaki, E.; Keers, S.; Harrison, R.; Bushby, K.; Bashir, R. The third human FER-1-like protein is highly similar to dysferlin. *Genomics* **2000**, *68*, 313–321. [CrossRef]

7. Liu, J.; Aoki, M.; Illa, I.; Wu, C.; Fardeau, M.; Angelini, C.; Serrano, C.; Urtizberea, J.A.; Hentati, F.; Hamida, M.B.; et al. Dysferlin, a novel skeletal muscle gene, is mutated in Miyoshi myopathy and limb girdle muscular dystrophy. *Nat. Genet.* **1998**, *20*, 31–36. [CrossRef] [PubMed]

8. Choi, B.Y.; Ahmed, Z.M.; Riazuddin, S.; Bhinder, M.A.; Shahzad, M.; Husnain, T.; Griffith, A.J.; Friedman, T.B. Identities and frequencies of mutations of the otoferlin gene (OTOF) causing DFNB9 deafness in Pakistan. *Clin. Genet.* **2009**, *75*, 237–243. [CrossRef]

9. Tekin, M.; Akcayoz, D.; Incesulu, A. A novel missense mutation in a C2 domain of OTOF results in autosomal recessive auditory neuropathy. *Am. J. Med. Genet. A* **2005**, *138*, 6–10. [CrossRef]

10. Doherty, K.R.; Cave, A.; Davis, D.B.; Delmonte, A.J.; Posey, A.; Earley, J.U.; Hadhazy, M.; McNally, E.M. Normal myoblast fusion requires myoferlin. *Development* **2005**, *132*, 5565–5575. [CrossRef]

11. Kiselev, A.; Vaz, R.; Knyazeva, A.; Sergushichev, A.; Dmitrieva, R.; Khudiakov, A.; Jorholt, J.; Smolina, N.; Sukhareva, K.; Fomicheva, Y.; et al. Truncating variant in myof gene is associated with limb-girdle type muscular dystrophy and cardiomyopathy. *Front. Genet.* **2019**, *10*, 608. [CrossRef] [PubMed]

12. Aoki, M.; Liu, J.; Richard, I.; Bashir, R.; Britton, S.; Keers, S.M.; Oeltjen, J.; Brown, H.E.; Marchand, S.; Bourg, N.; et al. Genomic organization of the dysferlin gene and novel mutations in Miyoshi myopathy. *Neurology* **2001**, *57*, 271–278. [CrossRef] [PubMed]

13. Demonbreun, A.R.; Lapidos, K.A.; Heretis, K.; Levin, S.; Dale, R.; Pytel, P.; Svensson, E.C.; McNally, E.M. Myoferlin regulation by NFAT in muscle injury, regeneration and repair. *J. Cell Sci.* **2010**, *123*, 2413–2422. [CrossRef] [PubMed]

14. Finn, R.D.; Coggill, P.; Eberhardt, R.Y.; Eddy, S.R.; Mistry, J.; Mitchell, A.L.; Potter, S.C.; Punta, M.; Qureshi, M.; Sangrador-Vegas, A.; et al. The Pfam protein families database: Towards a more sustainable future. *Nucleic Acids Res.* **2016**, *44*, D279–D285. [CrossRef] [PubMed]

15. Corbalan-Garcia, S.; Gómez-Fernández, J.C. Signaling through C2 domains: More than one lipid target. *Biochim. Biophys. Acta* **2014**, *1838*, 1536–1547. [CrossRef] [PubMed]

16. Zhang, D.; Aravind, L. Identification of novel families and classification of the C2 domain superfamily elucidate the origin and evolution of membrane targeting activities in eukaryotes. *Gene* **2010**, *469*, 18–30. [CrossRef]

17. Nalefski, E.A.; Falke, J.J. The C2 domain calcium-binding motif: Structural and functional diversity. *Protein Sci.* **1996**, *5*, 2375–2390. [CrossRef]

18. Shin, O.-H.; Han, W.; Wang, Y.; Südhof, T.C. Evolutionarily conserved multiple C2 domain proteins with two transmembrane regions (MCTPs) and unusual Ca^{2+} binding properties. *J. Biol. Chem.* **2005**, *280*, 1641–1651. [CrossRef]

19. Min, S.-W.; Chang, W.-P.; Südhof, T.C. E-Syts, a family of membranous Ca^{2+}-sensor proteins with multiple C2 domains. *Proc. Natl. Acad. Sci. USA* **2007**, *104*, 3823–3828. [CrossRef]

20. Rizo, J.; Sudhof, T.C. C2-domains, structure and function of a universal Ca^{2+}-binding domain. *J. Biol. Chem.* **1998**, *273*, 15879–15882. [CrossRef]

21. Von Poser, C.; Ichtchenko, K.; Shao, X.; Rizo, J.; Sudhof, T.C. The evolutionary pressure to inactivate. A subclass of synaptotagmins with an amino acid substitution that abolishes Ca^{2+} binding. *J. Biol. Chem.* **1997**, *272*, 14314–14319. [CrossRef] [PubMed]

22. Lek, A.; Lek, M.; North, K.N.; Cooper, S.T. Phylogenetic analysis of ferlin genes reveals ancient eukaryotic origins. *BMC Evol. Biol.* **2010**, *10*, 231. [CrossRef] [PubMed]

23. Washington, N.L.; Ward, S. FER-1 regulates Ca^{2+}-mediated membrane fusion during C. elegans spermatogenesis. *J. Cell Sci.* **2006**, *119*, 2552–2562. [CrossRef] [PubMed]

24. Yan, M.; Rachubinski, D.A.; Joshi, S.; Rachubinski, R.A.; Subramani, S. Dysferlin domain-containing proteins, Pex30p and Pex31p, localized to two compartments, control the number and size of oleate-induced peroxisomes in Pichia pastoris. *Mol. Biol. Cell.* **2008**, *19*, 885–898. [CrossRef] [PubMed]

25. Patel, P.; Harris, R.; Geddes, S.M.; Strehle, E.-M.; Watson, J.D.; Bashir, R.; Bushby, K.; Driscoll, P.C.; Keep, N.H. Solution Structure of the Inner DysF Domain of Myoferlin and Implications for Limb Girdle Muscular Dystrophy Type 2B. *J. Mol. Biol.* **2008**, *379*, 981–990. [CrossRef] [PubMed]

26. Fuson, K.; Rice, A.; Mahling, R.; Snow, A.; Nayak, K.; Shanbhogue, P.; Meyer, A.G.; Redpath, G.M.I.; Hinderliter, A.; Cooper, S.T.; et al. Alternate splicing of dysferlin C2A confers Ca^{2+}-dependent and Ca^{2+}-independent binding for membrane repair. *Structure* **2014**, *22*, 104–115. [CrossRef] [PubMed]

27. Aartsma-Rus, A.; Van Deutekom, J.C.T.; Fokkema, I.F.; Van Ommen, G.-J.B.; Den Dunnen, J.T. Entries in the Leiden Duchenne muscular dystrophy mutation database: An overview of mutation types and paradoxical cases that confirm the reading-frame rule. *Muscle Nerve* **2006**, *34*, 135–144. [CrossRef] [PubMed]

28. Bernatchez, P.N.; Acevedo, L.; Fernandez-Hernando, C.; Murata, T.; Chalouni, C.; Kim, J.; Erdjument-Bromage, H.; Shah, V.; Gratton, J.-P.; McNally, E.M.; et al. Myoferlin regulates vascular endothelial growth factor receptor-2 stability and function. *J. Biol. Chem.* **2007**, *282*, 30745–30753. [CrossRef]

29. Miyatake, Y.; Yamano, T.; Hanayama, R. Myoferlin-Mediated Lysosomal Exocytosis Regulates Cytotoxicity by Phagocytes. *J. Immunol.* **2018**, *201*, 3051–3057. [CrossRef]

30. Redpath, G.M.I.; Sophocleous, R.A.; Turnbull, L.; Whitchurch, C.B.; Cooper, S.T. Ferlins Show Tissue-Specific Expression and Segregate as Plasma Membrane/Late Endosomal or Trans-Golgi/Recycling Ferlins. *Traffic* **2016**, *17*, 245–266. [CrossRef]

31. Davis, D.B.; Doherty, K.R.; Delmonte, A.J.; McNally, E.M. Calcium-sensitive phospholipid binding properties of normal and mutant ferlin C2 domains. *J. Biol. Chem.* **2002**, *277*, 22883–22888. [CrossRef] [PubMed]

32. Bootman, M.D.; Rietdorf, K.; Hardy, H.; Dautova, Y.; Corps, E.; Pierro, C.; Stapleton, E.; Kang, E.; Proudfoot, D. Calcium Signalling and Regulation of Cell Function. In *eLS*; John Wiley & Sons, Ltd.: Chichester, UK, 2012; pp. 1–14.

33. Harsini, F.M.; Bui, A.A.; Rice, A.M.; Chebrolu, S.; Fuson, K.L.; Turtoi, A.; Bradberry, M.; Chapman, E.R.; Sutton, R.B. Structural Basis for the Distinct Membrane Binding Activity of the Homologous C2A Domains of Myoferlin and Dysferlin. *J. Mol. Biol.* **2019**, *431*, 2112–2126. [CrossRef] [PubMed]

34. Therrien, C.; Fulvio, S.D.; Pickles, S.; Sinnreich, M. Characterization of lipid binding specificities of dysferlin C2 domains reveals novel interactions with phosphoinositides. *Biochemistry* **2009**, *48*, 2377–2384. [CrossRef] [PubMed]

35. Marty, N.J.; Holman, C.L.; Abdullah, N.; Johnson, C.P. The C2 domains of otoferlin, dysferlin, and myoferlin alter the packing of lipid bilayers. *Biochemistry* **2013**, *52*, 5585–5592. [CrossRef] [PubMed]

36. Johnson, C.P.; Chapman, E.R. Otoferlin is a calcium sensor that directly regulates SNARE-mediated membrane fusion. *J. Cell Biol.* **2010**, *191*, 187–197. [CrossRef] [PubMed]

37. Helfmann, S.; Neumann, P.; Tittmann, K.; Moser, T.; Ficner, R.; Reisinger, E. The Crystal Structure of the C2A Domain of Otoferlin Reveals an Unconventional Top Loop Region. *J. Mol. Biol.* **2011**, *406*, 479–490. [CrossRef]

38. Harsini, F.M.; Chebrolu, S.; Fuson, K.L.; White, M.A.; Rice, A.M.; Sutton, R.B. FerA is a Membrane-Associating Four-Helix Bundle Domain in the Ferlin Family of Membrane-Fusion Proteins. *Sci. Rep.* **2018**, *8*, 10949. [CrossRef]

39. De Morrée, A.; Hensbergen, P.J.; van Haagen, H.H.; Dragan, I.; Deelder, A.M.; AC't Hoen, P.; Frants, R.R.; van der Maarel, S.M. Proteomic analysis of the dysferlin protein complex unveils its importance for sarcolemmal maintenance and integrity. *PLoS ONE* **2010**, *5*, e13854. [CrossRef]

40. Doherty, K.R.; Demonbreun, A.R.; Wallace, G.Q.; Cave, A.; Posey, A.D.; Heretis, K.; Pytel, P.; McNally, E.M. The endocytic recycling protein EHD2 interacts with myoferlin to regulate myoblast fusion. *J. Biol. Chem.* **2008**, *283*, 20252–20260. [CrossRef]

41. Bansal, D.; Miyake, K.; Vogel, S.S.; Groh, S.; Chen, C.-C.; Williamson, R.; McNeil, P.L.; Campbell, K.P. Defective membrane repair in dysferlin-deficient muscular dystrophy. *Nature* **2003**, *423*, 168–172. [CrossRef]

42. Lukyanenko, V.; Muriel, J.M.; Bloch, R.J. Coupling of excitation to Ca^{2+} release is modulated by dysferlin. *J. Physiol.* **2017**, *595*, 5191–5207. [CrossRef] [PubMed]

43. Lennon, N.J.; Kho, A.; Bacskai, B.J.; Perlmutter, S.L.; Hyman, B.T.; Brown, R.H. Dysferlin Interacts with Annexins A1 and A2 and Mediates Sarcolemmal Wound-healing. *J. Biol. Chem.* **2003**, *278*, 50466–50473. [CrossRef] [PubMed]

44. Vincent, A.E.; Rosa, H.S.; Alston, C.L.; Grady, J.P.; Rygiel, K.A.; Rocha, M.C.; Barresi, R.; Taylor, R.W.; Turnbull, D.M. Dysferlin mutations and mitochondrial dysfunction. *Neuromuscul. Disord.* **2016**, *26*, 782–788. [CrossRef]

45. Pramono, Z.A.D.; Lai, P.S.; Tan, C.L.; Takeda, S.; Yee, W.C. Identification and characterization of a novel human dysferlin transcript: Dysferlin_v1. *Hum. Genet.* **2006**, *120*, 410–419. [CrossRef] [PubMed]

46. Lek, A.; Evesson, F.J.; Lemckert, F.A.; Redpath, G.M.I.; Lueders, A.-K.; Turnbull, L.; Whitchurch, C.B.; North, K.N.; Cooper, S.T. Calpains, cleaved mini-dysferlinC72, and L-type channels underpin calcium-dependent muscle membrane repair. *J. Neurosci.* **2013**, *33*, 5085–5094. [CrossRef]

47. Redpath, G.M.I.; Woolger, N.; Piper, A.K.; Lemckert, F.A.; Lek, A.; Greer, P.A.; North, K.N.; Cooper, S.T. Calpain cleavage within dysferlin exon 40a releases a synaptotagmin-like module for membrane repair. *Mol. Biol. Cell.* **2014**, *25*, 3037–3048. [CrossRef] [PubMed]

48. O'Connor, V.; Lee, A.G. Synaptic vesicle fusion and synaptotagmin: 2B or not 2B? *Nat. Neurosci.* **2002**, *5*, 823–824. [CrossRef]

49. Przybylski, R.J.; Szigeti, V.; Davidheiser, S.; Kirby, A.C. Calcium regulation of skeletal myogenesis. II. Extracellular and cell surface effects. *Cell Calcium* **1994**, *15*, 132–142. [CrossRef]

50. Demonbreun, A.R.; Posey, A.D.; Heretis, K.; Swaggart, K.A.; Earley, J.U.; Pytel, P.; McNally, E.M. Myoferlin is required for insulin-like growth factor response and muscle growth. *FASEB J.* **2010**, *24*, 1284–1295. [CrossRef]

51. Zhou, Y.; Xiong, L.; Zhang, Y.; Yu, R.; Jiang, X.; Xu, G. Quantitative proteomics identifies myoferlin as a novel regulator of A Disintegrin and Metalloproteinase 12 in HeLa cells. *J. Proteom.* **2016**, *148*, 94–104. [CrossRef]

52. Galliano, M.F.; Huet, C.; Frygelius, J.; Polgren, A.; Wewer, U.M.; Engvall, E. Binding of ADAM12, a marker of skeletal muscle regeneration, to the muscle-specific actin-binding protein, α-actinin-2, is required for myoblast fusion. *J. Biol. Chem.* **2000**, *275*, 13933–13939. [CrossRef] [PubMed]

53. Posey, A.D.; Pytel, P.; Gardikiotes, K.; Demonbreun, A.R.; Rainey, M.; George, M.; Band, H.; McNally, E.M. Endocytic recycling proteins EHD1 and EHD2 interact with fer-1-like-5 (Fer1L5) and mediate myoblast fusion. *J. Biol. Chem.* **2011**, *286*, 7379–7388. [CrossRef] [PubMed]

54. Roux, I.; Safieddine, S.; Nouvian, R.; Grati, M.; Simmler, M.-C.; Bahloul, A.; Perfettini, I.; Le Gall, M.; Rostaing, P.; Hamard, G.; et al. Otoferlin, defective in a human deafness form, is essential for exocytosis at the auditory ribbon synapse. *Cell* **2006**, *127*, 277–289. [CrossRef] [PubMed]

55. Chapman, E.R.; Hanson, P.I.; An, S.; Jahn, R. Ca^{2+} regulates the interaction between synaptotagmin and syntaxin 1. *J. Biol. Chem.* **1995**, *270*, 23667–23671. [CrossRef] [PubMed]

56. Mohrmann, R.; de Wit, H.; Connell, E.; Pinheiro, P.S.; Leese, C.; Bruns, D.; Davletov, B.; Verhage, M.; Sørensen, J.B. Synaptotagmin interaction with SNAP-25 governs vesicle docking, priming, and fusion triggering. *J. Neurosci.* **2013**, *33*, 14417–14430. [CrossRef] [PubMed]

57. Ramakrishnan, N.A.; Drescher, M.J.; Drescher, D.G. Direct interaction of otoferlin with syntaxin 1A, SNAP-25, and the L-type voltage-gated calcium channel Cav1.3. *J. Biol. Chem.* **2009**, *284*, 1364–1372. [CrossRef] [PubMed]

58. Ramakrishnan, N.A.; Drescher, M.J.; Morley, B.J.; Kelley, P.M.; Drescher, D.G. Calcium regulates molecular interactions of otoferlin with soluble NSF attachment protein receptor (SNARE) proteins required for hair cell exocytosis. *J. Biol. Chem.* **2014**, *289*, 8750–8766. [CrossRef]

59. Hams, N.; Padmanarayana, M.; Qiu, W.; Johnson, C.P. Otoferlin is a multivalent calcium-sensitive scaffold linking SNAREs and calcium channels. *Proc. Natl. Acad. Sci. USA* **2017**, *114*, 8023–8028. [CrossRef] [PubMed]

60. Sharma, A.; Yu, C.; Leung, C.; Trane, A.; Lau, M.; Utokaparch, S.; Shaheen, F.; Sheibani, N.; Bernatchez, P. A new role for the muscle repair protein dysferlin in endothelial cell adhesion and angiogenesis. *Arterioscler. Thromb. Vasc. Biol.* **2010**, *30*, 2196–2204. [CrossRef]

61. Yu, C.; Sharma, A.; Trane, A.; Utokaparch, S.; Leung, C.; Bernatchez, P. Myoferlin gene silencing decreases Tie-2 expression in vitro and angiogenesis in vivo. *Vascul. Pharmacol.* **2011**, *55*, 26–33. [CrossRef]

62. Bernatchez, P.N.; Sharma, A.; Kodaman, P.; Sessa, W.C. Myoferlin is critical for endocytosis in endothelial cells. *Am. J. Physiol. Cell. Physiol.* **2009**, *297*, C484–C492. [CrossRef] [PubMed]

63. Han, W.Q.; Xia, M.; Xu, M.; Boini, K.M.; Ritter, J.K.; Li, N.J.; Li, P.L. Lysosome fusion to the cell membrane is mediated by the dysferlin C2A domain in coronary arterial endothelial cells. *J. Cell Sci.* **2012**, *125*, 1225–1234. [CrossRef] [PubMed]

64. Leung, C.; Shaheen, F.; Bernatchez, P.; Hackett, T.-L. Expression of myoferlin in human airway epithelium and its role in cell adhesion and zonula occludens-1 expression. *PLoS ONE* **2012**, *7*, e40478. [CrossRef] [PubMed]

65. Stamer, W.D.; Hoffman, E.A.; Luther, J.M.; Hachey, D.L.; Schey, K.L. Protein profile of exosomes from trabecular meshwork cells. *J. Proteom.* **2011**, *74*, 796–804. [CrossRef] [PubMed]

66. Iacobuzio-Donahue, C.A.; Maitra, A.; Shen-Ong, G.L.; van Heek, T.; Ashfaq, R.; Meyer, R.; Walter, K.; Berg, K.; Hollingsworth, M.A.; Cameron, J.L.; et al. Discovery of Novel Tumor Markers of Pancreatic Cancer using Global Gene Expression Technology. *Am. J. Pathol.* **2002**, *160*, 1239–1249. [CrossRef]

67. Han, H.; Bearss, D.J.; Browne, L.W.; Calaluce, R.; Nagle, R.B.; Von Hoff, D.D. Identification of differentially expressed genes in pancreatic cancer cells using cDNA microarray. *Cancer Res.* **2002**, *62*, 2890–2896. [PubMed]

68. Amatschek, S.; Koenig, U.; Auer, H.; Steinlein, P.; Pacher, M.; Gruenfelder, A.; Dekan, G.; Vogl, S.; Kubista, E.; Heider, K.-H.; et al. Tissue-wide expression profiling using cDNA subtraction and microarrays to identify tumor-specific genes. *Cancer Res.* **2004**, *64*, 844–856. [CrossRef]

69. Yadav, A.; Kumar, B.; Lang, J.C.; Teknos, T.N.; Kumar, P. A muscle-specific protein "myoferlin" modulates IL-6/STAT3 signaling by chaperoning activated STAT3 to nucleus. *Oncogene* **2017**, *36*, 6374–6382. [CrossRef]

70. McKinney, K.Q.; Lee, Y.Y.; Choi, H.S.; Groseclose, G.; Iannitti, D.A.; Martinie, J.B.; Russo, M.W.; Lundgren, D.H.; Han, D.K.; Bonkovsky, H.L.; et al. Discovery of putative pancreatic cancer biomarkers using subcellular proteomics. *J. Proteom.* **2011**, *74*, 79–88. [CrossRef]

71. Turtoi, A.; Musmeci, D.; Wang, Y.; Dumont, B.; Somja, J.; Bevilacqua, G.; De Pauw, E.; Delvenne, P.; Castronovo, V. Identification of novel accessible proteins bearing diagnostic and therapeutic potential in human pancreatic ductal adenocarcinoma. *J. Proteome Res.* **2011**, *10*, 4302–4313. [CrossRef]

72. McKinney, K.Q.; Lee, J.-G.; Sindram, D.; Russo, M.W.; Han, D.K.; Bonkovsky, H.L.; Hwang, S.-I. Identification of differentially expressed proteins from primary versus metastatic pancreatic cancer cells using subcellular proteomics. *Cancer Genom. Proteom.* **2012**, *9*, 257–263.

73. Wang, W.S.; Liu, X.H.; Liu, L.X.; Lou, W.H.; Jin, D.Y.; Yang, P.Y.; Wang, X.L. ITRAQ-based quantitative proteomics reveals myoferlin as a novel prognostic predictor in pancreatic adenocarcinoma. *J. Proteom.* **2013**, *91*, 453–465. [CrossRef] [PubMed]

74. Adam, P.J.; Boyd, R.; Tyson, K.L.; Fletcher, G.C.; Stamps, A.; Hudson, L.; Poyser, H.R.; Redpath, N.; Griffiths, M.; Steers, G.; et al. Comprehensive proteomic analysis of breast cancer cell membranes reveals unique proteins with potential roles in clinical cancer. *J. Biol. Chem.* **2003**, *278*, 6482–6489. [CrossRef] [PubMed]

75. Leung, C.; Yu, C.; Lin, M.I.; Tognon, C.; Bernatchez, P. Expression of myoferlin in human and murine carcinoma tumors: Role in membrane repair, cell proliferation, and tumorigenesis. *Am. J. Pathol.* **2013**, *182*, 1900–1909. [CrossRef] [PubMed]

76. Hermanns, C.; Hampl, V.; Holzer, K.; Aigner, A.; Penkava, J.; Frank, N.; Martin, D.E.; Maier, K.C.; Waldburger, N.; Roessler, S.; et al. The novel MKL target gene myoferlin modulates expansion and senescence of hepatocellular carcinoma. *Oncogene* **2017**, *36*, 3464–3476. [CrossRef] [PubMed]

77. Kumar, B.; Brown, N.V.; Swanson, B.J.; Schmitt, A.C.; Old, M.; Ozer, E.; Agrawal, A.; Schuller, D.E.; Teknos, T.N.; Kumar, P. High expression of myoferlin is associated with poor outcome in oropharyngeal squamous cell carcinoma patients and is inversely associated with HPV-status. *Oncotarget* **2016**, *7*, 18665–18677. [CrossRef] [PubMed]

78. Song, D.H.; Ko, G.H.; Lee, J.H.; Lee, J.S.; Yang, J.W.; Kim, M.H.; An, H.J.; Kang, M.H.; Jeon, K.N.; Kim, D.C. Prognostic role of myoferlin expression in patients with clear cell renal cell carcinoma. *Oncotarget* **2017**, *8*, 89033–89039. [CrossRef]

79. Koh, H.M.; An, H.J.; Ko, G.H.; Lee, J.H.; Lee, J.S.; Kim, D.C.; Seo, D.H.; Song, D.H. Identification of Myoferlin Expression for Prediction of Subsequent Primary Malignancy in Patients With Clear Cell Renal Cell Carcinoma. *In Vivo* **2019**, *33*, 1103–1108. [CrossRef]

80. Kim, M.H.; Song, D.H.; Ko, G.H.; Lee, J.H.; Kim, D.C.; Yang, J.W.; Lee, H.I.; An, H.J.; Lee, J.S. Myoferlin expression and its correlation with FIGO histologic grading in early-stage endometrioid carcinoma. *J. Pathol. Transl. Med.* **2018**, *52*, 93–97. [CrossRef]

81. Welton, J.L.; Khanna, S.; Giles, P.J.; Brennan, P.; Brewis, I.A.; Staffurth, J.; Mason, M.D.; Clayton, A. Proteomics analysis of bladder cancer exosomes. *Mol. Cell. Proteom.* **2010**, *9*, 1324–1338. [CrossRef]

82. Mathivanan, S.; Lim, J.W.E.; Tauro, B.J.; Ji, H.; Moritz, R.L.; Simpson, R.J. Proteomics analysis of A33 immunoaffinity-purified exosomes released from the human colon tumor cell line LIM1215 reveals a tissue-specific protein signature. *Mol. Cell. Proteom.* **2010**, *9*, 197–208. [CrossRef] [PubMed]

83. Beckler, M.D.; Higginbotham, J.N.; Franklin, J.L.; Ham, A.-J.; Halvey, P.J.; Imasuen, I.E.; Whitwell, C.; Li, M.; Liebler, D.C.; Coffey, R.J. Proteomic analysis of exosomes from mutant KRAS colon cancer cells identifies intercellular transfer of mutant KRAS. *Mol. Cell. Proteom.* **2013**, *12*, 343–355. [CrossRef] [PubMed]

84. Ji, H.; Greening, D.W.; Barnes, T.W.; Lim, J.W.; Tauro, B.J.; Rai, A.; Xu, R.; Adda, C.; Mathivanan, S.; Zhao, W.; et al. Proteome profiling of exosomes derived from human primary and metastatic colorectal cancer cells reveal differential expression of key metastatic factors and signal transduction components. *Proteomics* **2013**, *13*, 1672–1686. [CrossRef] [PubMed]

85. Choi, D.-S.; Choi, D.-Y.; Hong, B.S.; Jang, S.C.; Kim, D.-K.; Lee, J.; Kim, Y.-K.; Kim, K.P.; Gho, Y.S. Quantitative proteomics of extracellular vesicles derived from human primary and metastatic colorectal cancer cells. *J. Extracell. Vesicles* **2012**, *1*, 18704. [CrossRef] [PubMed]

86. Liang, B.; Peng, P.; Chen, S.; Li, L.; Zhang, M.; Cao, D.; Yang, J.; Li, H.; Gui, T.; Li, X.; et al. Characterization and proteomic analysis of ovarian cancer-derived exosomes. *J. Proteom.* **2013**, *80*, 171–182. [CrossRef] [PubMed]

87. Sandvig, K.; Llorente, A. Proteomic analysis of microvesicles released by the human prostate cancer cell line PC-3. *Mol. Cell. Proteom.* **2012**, *11*, M111-012914. [CrossRef]

88. Blomme, A.; Fahmy, K.; Peulen, O.J.; Costanza, B.; Fontaine, M.; Struman, I.; Baiwir, D.; De Pauw, E.; Thiry, M.; Bellahcène, A.; et al. Myoferlin is a novel exosomal protein and functional regulator of cancer-derived exosomes. *Oncotarget* **2016**, *7*, 83669–83683. [CrossRef]

89. Lawrence, M.S.; Stojanov, P.; Mermel, C.H.; Robinson, J.T.; Garraway, L.A.; Golub, T.R.; Meyerson, M.; Gabriel, S.B.; Lander, E.S.; Getz, G. Discovery and saturation analysis of cancer genes across 21 tumour types. *Nature* **2014**, *505*, 495–501. [CrossRef]

90. Anaya, J. OncoLnc: Linking TCGA survival data to mRNAs, miRNAs, and lncRNAs. *PeerJ Comp. Sci.* **2016**, *2*, e67. [CrossRef]

91. Tang, H.; Wei, P.; Chang, P.; Li, Y.; Yan, D.; Liu, C.; Hassan, M.; Li, D. Genetic polymorphisms associated with pancreatic cancer survival: A genome-wide association study. *Int. J. Cancer* **2017**, *141*, 678–686. [CrossRef]

92. Ding, F.; Tang, H.; Nie, D.; Xia, L. Long non-coding RNA Fer-1-like family member 4 is overexpressed in human glioblastoma and regulates the tumorigenicity of glioma cells. *Oncol. Lett.* **2017**, *14*, 2379–2384. [CrossRef] [PubMed]

93. Xia, L.; Nie, D.; Wang, G.; Sun, C.; Chen, G. FER1L4/miR-372/E2F1 works as a ceRNA system to regulate the proliferation and cell cycle of glioma cells. *J. Cell. Mol. Med.* **2019**, *23*, 3224–3233. [CrossRef] [PubMed]

94. You, Z.; Ge, A.; Pang, D.; Zhao, Y.; Xu, S. Long noncoding RNA FER1L4 acts as an oncogenic driver in human pan-cancer. *J. Cell. Physiol.* **2019**, *1859*, 46. [CrossRef] [PubMed]

95. Chen, Z.-X.; Chen, C.-P.; Zhang, N.; Wang, T.-X. Low-expression of lncRNA FER1L4 might be a prognostic marker in osteosarcoma. *Eur. Rev. Med. Pharmacol. Sci.* **2018**, *22*, 2310–2314.

96. Song, H.; Sun, W.; Ye, G.; Ding, X.; Liu, Z.; Zhang, S.; Xia, T.; Xiao, B.; Xi, Y.; Guo, J. Long non-coding RNA expression profile in human gastric cancer and its clinical significances. *J. Transl. Med.* **2013**, *11*, 225. [CrossRef] [PubMed]

97. Kong, Y.; Ren, Z. Overexpression of LncRNA FER1L4 in endometrial carcinoma is associated with favorable survival outcome. *Eur. Rev. Med. Pharmacol. Sci.* **2018**, *22*, 8113–8118.

98. Eisenberg, M.C.; Kim, Y.; Li, R.; Ackerman, W.E.; Kniss, D.A.; Friedman, A. Mechanistic modeling of the effects of myoferlin on tumor cell invasion. *Proc. Natl. Acad. Sci. USA* **2011**, *108*, 20078–20083. [CrossRef]

99. Li, R.; Ackerman, W.E.; Mihai, C.; Volakis, L.I.; Ghadiali, S.; Kniss, D.A. Myoferlin depletion in breast cancer cells promotes mesenchymal to epithelial shape change and stalls invasion. *PLoS ONE* **2012**, *7*, e39766. [CrossRef]

100. Volakis, L.I.; Li, R.; Ackerman, W.E.; Mihai, C.; Bechel, M.; Summerfield, T.L.; Ahn, C.S.; Powell, H.M.; Zielinski, R.; Rosol, T.J.; et al. Loss of myoferlin redirects breast cancer cell motility towards collective migration. *PLoS ONE* **2014**, *9*, e86110. [CrossRef]

101. Blackstone, B.N.; Li, R.; Ackerman, W.E.; Ghadiali, S.N.; Powell, H.M.; Kniss, D.A. Myoferlin depletion elevates focal adhesion kinase and paxillin phosphorylation and enhances cell-matrix adhesion in breast cancer cells. *Am. J. Physiol. Cell. Physiol.* **2015**, *308*, C642–C649. [CrossRef]

102. Barnhouse, V.R.; Weist, J.L.; Shukla, V.C.; Ghadiali, S.N.; Kniss, D.A.; Leight, J.L. Myoferlin regulates epithelial cancer cell plasticity and migration through autocrine TGF-β1 signaling. *Oncotarget* **2018**, *9*, 19209–19222. [CrossRef] [PubMed]

103. Zhang, W.; Zhou, P.; Meng, A.; Zhang, R.; Zhou, Y. Down-regulating Myoferlin inhibits the vasculogenic mimicry of melanoma via decreasing MMP-2 and inducing mesenchymal-to-epithelial transition. *J. Cell. Mol. Med.* **2017**, *155*, 739. [CrossRef] [PubMed]

104. Turtoi, A.; Blomme, A.; Bellahcène, A.; Gilles, C.; Hennequière, V.; Peixoto, P.; Bianchi, E.; Noël, A.; De Pauw, E.; Lifrange, E.; et al. Myoferlin is a key regulator of EGFR activity in breast cancer. *Cancer Res.* **2013**, *73*, 5438–5448. [CrossRef] [PubMed]

105. Nylandsted, J.; Becker, A.C.; Bunkenborg, J.; Andersen, J.S.; Dengjel, J.; Jäättelä, M. ErbB2-associated changes in the lysosomal proteome. *Proteomics* **2011**, *11*, 2830–2838. [CrossRef] [PubMed]

106. Örtegren, U.; Aboulaich, N.; Öst, A.; Strålfors, P. A new role for caveolae as metabolic platforms. *Trends Endocrinol. Metab.* **2007**, *18*, 344–349. [CrossRef] [PubMed]

107. Blomme, A.; Costanza, B.; de Tullio, P.; Thiry, M.; Van Simaeys, G.; Boutry, S.; Doumont, G.; Di Valentin, E.; Hirano, T.; Yokobori, T.; et al. Myoferlin regulates cellular lipid metabolism and promotes metastases in triple-negative breast cancer. *Oncogene* **2017**, *36*, 2116–2130. [CrossRef] [PubMed]

108. Piper, A.-K.; Ross, S.E.; Redpath, G.M.; Lemckert, F.A.; Woolger, N.; Bournazos, A.; Greer, P.A.; Sutton, R.B.; Cooper, S.T. Enzymatic cleavage of myoferlin releases a dual C2-domain module linked to ERK signalling. *Cell. Signal.* **2017**, *33*, 30–40. [CrossRef]

109. Fahmy, K.; Gonzalez, A.; Arafa, M.; Peixoto, P.; Bellahcène, A.; Turtoi, A.; Delvenne, P.; Thiry, M.; Castronovo, V.; Peulen, O.J. Myoferlin plays a key role in VEGFA secretion and impacts tumor-associated angiogenesis in human pancreas cancer. *Int. J. Cancer* **2016**, *138*, 652–663. [CrossRef]

110. Rademaker, G.; Hennequière, V.; Brohée, L.; Nokin, M.-J.; Lovinfosse, P.; Durieux, F.; Gofflot, S.; Bellier, J.; Costanza, B.; Herfs, M.; et al. Myoferlin controls mitochondrial structure and activity in pancreatic ductal adenocarcinoma, and affects tumor aggressiveness. *Oncogene* **2018**, *66*, 1–15. [CrossRef]

111. Rademaker, G.; Costanza, B.; Bellier, J.; Herfs, M.; Peiffer, R.; Agirman, F.; Maloujahmoum, N.; Habraken, Y.; Delvenne, P.; Bellahcène, A.; et al. Human colon cancer cells highly express myoferlin to maintain a fit mitochondrial network and escape p53-driven apoptosis. *Oncogenesis* **2019**, *8*, 21. [CrossRef]

112. LeBleu, V.S.; O'Connell, J.T.; Gonzalez Herrera, K.N.; Wikman, H.; Pantel, K.; Haigis, M.C.; de Carvalho, F.M.; Damascena, A.; Domingos Chinen, L.T.; Rocha, R.M.; et al. PGC-1α mediates mitochondrial biogenesis and oxidative phosphorylation in cancer cells to promote metastasis. *Nat. Cell Biol.* **2014**, *16*, 992–1003. [CrossRef] [PubMed]

113. Porporato, P.E.; Payen, V.L.; Pérez-Escuredo, J.; De Saedeleer, C.J.; Danhier, P.; Copetti, T.; Dhup, S.; Tardy, M.; Vazeille, T.; Bouzin, C.; et al. A mitochondrial switch promotes tumor metastasis. *Cell Rep.* **2014**, *8*, 754–766. [CrossRef] [PubMed]

114. Rademaker, G.; Costanza, B.; Anania, S.; Agirman, F.; Maloujahmoum, N.; Di Valentin, E.; Goval, J.J.; Bellahcène, A.; Castronovo, V.; Peulen, O.J. Myoferlin Contributes to the Metastatic Phenotype of Pancreatic Cancer Cells by Enhancing Their Migratory Capacity through the Control of Oxidative Phosphorylation. *Cancers* **2019**, *11*, 853. [CrossRef] [PubMed]

115. Majumder, S.; Raimondo, M.; Taylor, W.R.; Yab, T.C.; Berger, C.K.; Dukek, B.A.; Cao, X.; Foote, P.H.; Wu, C.W.; Devens, M.E.; et al. Methylated DNA in Pancreatic Juice Distinguishes Patients with Pancreatic Cancer from Controls. *Clin. Gastroenterol. Hepatol.* **2019**. [CrossRef] [PubMed]

116. Song, D.H.; Ko, G.H.; Lee, J.H.; Lee, J.S.; Lee, G.-W.; Kim, H.C.; Yang, J.W.; Heo, R.W.; Roh, G.S.; Han, S.-Y.; et al. Myoferlin expression in non-small cell lung cancer: Prognostic role and correlation with VEGFR-2 expression. *Oncol. Lett.* **2016**, *11*, 998–1006. [CrossRef] [PubMed]

117. Gao, X.; Wang, N.; Wu, S.; Cui, H.; An, X.; Yang, Y. Long non-coding RNA FER1L4 inhibits cell proliferation and metastasis through regulation of the PI3K/AKT signaling pathway in lung cancer cells. *Mol. Med. Rep.* **2019**, *20*, 182–190. [CrossRef] [PubMed]

118. Ma, L.; Zhang, L.; Guo, A.; Liu, L.C.; Yu, F.; Diao, N.; Xu, C.; Wang, D. Overexpression of FER1L4 promotes the apoptosis and suppresses epithelial-mesenchymal transition and stemness markers via activating PI3K/AKT signaling pathway in osteosarcoma cells. *Pathol. Res. Pract.* **2019**, *215*, 152412. [CrossRef] [PubMed]

119. Ma, W.; Zhang, C.-Q.; Li, H.-L.; Gu, J.; Miao, G.-Y.; Cai, H.-Y.; Wang, J.-K.; Zhang, L.-J.; Song, Y.-M.; Tian, Y.-H.; et al. LncRNA FER1L4 suppressed cancer cell growth and invasion in esophageal squamous cell carcinoma. *Eur. Rev. Med. Pharmacol. Sci.* **2018**, *22*, 2638–2645. [PubMed]

120. Wang, X.; Dong, K.; Jin, Q.; Ma, Y.; Yin, S.; Wang, S. Upregulation of lncRNA FER1L4 suppresses the proliferation and migration of the hepatocellular carcinoma via regulating PI3K/AKT signal pathway. *J. Cell. Biochem.* **2019**, *120*, 6781–6788. [CrossRef]

121. Liu, Z.; Shao, Y.; Tan, L.; Shi, H.; Chen, S.; Guo, J. Clinical significance of the low expression of FER1L4 in gastric cancer patients. *Tumour Biol.* **2014**, *35*, 9613–9617. [CrossRef] [PubMed]

122. Xia, T.; Chen, S.; Jiang, Z.; Shao, Y.; Jiang, X.; Li, P.; Xiao, B.; Guo, J. Long noncoding RNA FER1L4 suppresses cancer cell growth by acting as a competing endogenous RNA and regulating PTEN expression. *Sci. Rep.* **2015**, *5*, 13445. [CrossRef] [PubMed]

123. Xia, T.; Liao, Q.; Jiang, X.; Shao, Y.; Xiao, B.; Xi, Y.; Guo, J. Long noncoding RNA associated-competing endogenous RNAs in gastric cancer. *Sci. Rep.* **2014**, *4*, 6088. [CrossRef] [PubMed]

124. Salmena, L.; Poliseno, L.; Tay, Y.; Kats, L.; Pandolfi, P.P. A ceRNA hypothesis: The rosetta stone of a hidden RNA language? *Cell* **2011**, *146*, 353–358. [CrossRef] [PubMed]

125. Yue, B.; Sun, B.; Liu, C.; Zhao, S.; Zhang, D.; Yu, F.; Yan, D. Long non-coding RNA Fer-1-like protein 4 suppresses oncogenesis and exhibits prognostic value by associating with miR-106a-5p in colon cancer. *Cancer Sci.* **2015**, *106*, 1323–1332. [CrossRef] [PubMed]

126. Wu, J.; Huang, J.; Wang, W.; Xu, J.; Yin, M.; Cheng, N.; Yin, J. Long non-coding RNA Fer-1-like protein 4 acts as a tumor suppressor via miR-106a-5p and predicts good prognosis in hepatocellular carcinoma. *Cancer Biomark.* **2017**, *20*, 55–65. [CrossRef]

127. Fei, D.; Zhang, X.; Liu, J.; Tan, L.; Xing, J.; Zhao, D.; Zhang, Y. Long Noncoding RNA FER1L4 Suppresses Tumorigenesis by Regulating the Expression of PTEN Targeting miR-18a-5p in Osteosarcoma. *Cell. Physiol. Biochem.* **2018**, *51*, 1364–1375. [CrossRef] [PubMed]

128. Liu, S.; Zou, B.; Tian, T.; Luo, X.; Mao, B.; Zhang, X.; Lei, H. Overexpression of the lncRNA FER1L4 inhibits paclitaxel tolerance of ovarian cancer cells via the regulation of the MAPK signaling pathway. *J. Cell. Biochem.* **2018**, *120*, 7581–7589. [CrossRef]

129. Théry, C. Cancer: Diagnosis by extracellular vesicles. *Nature* **2015**, *523*, 161–162. [CrossRef]

130. Ciardiello, C.; Cavallini, L.; Spinelli, C.; Yang, J.; Reis-Sobreiro, M.; de Candia, P.; Minciacchi, V.R.; Di Vizio, D. Focus on Extracellular Vesicles: New Frontiers of Cell-to-Cell Communication in Cancer. *Int. J. Mol. Sci.* **2016**, *17*, 175. [CrossRef]

131. Hanahan, D.; Weinberg, R.A. Hallmarks of cancer: The next generation. *Cell* **2011**, *144*, 646–674. [CrossRef]

132. Anderson, R.G.; Ghiraldeli, L.P.; Pardee, T.S. Mitochondria in cancer metabolism, an organelle whose time has come? *Biochim. Biophys. Acta Rev. Cancer* **2018**, *1870*, 96–102. [CrossRef] [PubMed]

133. Zhang, T.; Jingjie, L.; He, Y.; Yang, F.; Hao, Y.; Jin, W.; Wu, J.; Sun, Z.; Li, Y.; Chen, Y.; et al. A small molecule targeting myoferlin exerts promising anti-tumor effects on breast cancer. *Nat. Commun.* **2018**, *9*, 3726. [CrossRef] [PubMed]

134. Li, Y.; He, Y.; Shao, T.; Pei, H.; Guo, W.; Mi, D.; Krimm, I.; Zhang, Y.; Wang, P.; Wang, X.; et al. Modification and Biological Evaluation of a Series of 1,5-Diaryl-1,2,4-triazole Compounds as Novel Agents against Pancreatic Cancer Metastasis through Targeting Myoferlin. *J. Med. Chem.* **2019**, *62*, 4949–4966. [CrossRef] [PubMed]

135. Sondergaard, P.C.; Griffin, D.A.; Pozsgai, E.R.; Johnson, R.W.; Grose, W.E.; Heller, K.N.; Shontz, K.M.; Montgomery, C.L.; Liu, J.; Clark, K.R.; et al. AAV.Dysferlin Overlap Vectors Restore Function in Dysferlinopathy Animal Models. *Ann. Clin. Transl. Neurol.* **2015**, *2*, 256–270. [CrossRef] [PubMed]

136. Escobar, H.; Schöwel, V.; Spuler, S.; Marg, A.; Izsvák, Z. Full-length Dysferlin Transfer by the Hyperactive Sleeping Beauty Transposase Restores Dysferlin-deficient Muscle. *Mol. Ther. Nucleic Acids* **2016**, *5*, e277. [CrossRef] [PubMed]

137. Potter, R.A.; Griffin, D.A.; Sondergaard, P.C.; Johnson, R.W.; Pozsgai, E.R.; Heller, K.N.; Peterson, E.L.; Lehtimäki, K.K.; Windish, H.P.; Mittal, P.J.; et al. Systemic Delivery of Dysferlin Overlap Vectors Provides Long-Term Gene Expression and Functional Improvement for Dysferlinopathy. *Hum. Gene Ther.* **2017**. hum.2017.062. [CrossRef]

138. Llanga, T.; Nagy, N.; Conatser, L.; Dial, C.; Sutton, R.B.; Hirsch, M.L. Structure-Based Designed Nano-Dysferlin Significantly Improves Dysferlinopathy in BLA/J Mice. *Mol. Ther.* **2017**, *25*, 2150–2162. [CrossRef]

139. Lee, J.H.; Kim, Y.; Yoon, Y.-E.; Kim, Y.-J.; Oh, S.-G.; Jang, J.-H.; Kim, E. Development of efficient adeno-associated virus (AAV)-mediated gene delivery system with a phytoactive material for targeting human melanoma cells. *New Biotechnol.* **2017**, *37*, 194–199. [CrossRef]

140. Chow, R.D.; Guzman, C.D.; Wang, G.; Schmidt, F.; Youngblood, M.W.; Ye, L.; Errami, Y.; Dong, M.B.; Martinez, M.A.; Zhang, S.; et al. AAV-mediated direct in vivo CRISPR screen identifies functional suppressors in glioblastoma. *Nat. Neurosci.* **2017**, *20*, 1329–1341. [CrossRef]

141. Hodge, R.; Narayanavari, S.A.; Izsvák, Z.; Ivics, Z. Wide Awake and Ready to Move: 20 Years of Non-Viral Therapeutic Genome Engineering with the Sleeping Beauty Transposon System. *Hum. Gene Ther.* **2017**, *28*, 842–855. [CrossRef]

Review

Zinc Finger Transcription Factor MZF1—A Specific Regulator of Cancer Invasion

Ditte Marie Brix [1,2], Knut Kristoffer Bundgaard Clemmensen [1] and Tuula Kallunki [1,3,*]

[1] Cell Death and Metabolism, Center for Autophagy, Recycling and Disease, Danish Cancer Society Research Center, 2100 Copenhagen, Denmark; dittemariebrix@gmail.com (D.M.B.); knucle@cancer.dk (K.K.B.C.)

[2] Danish Medicines Council, Dampfærgevej 27-29, 2100 Copenhagen, Denmark

[3] Department of Drug Design and Pharmacology, Faculty of Health Sciences, University of Copenhagen, 2200 Copenhagen, Denmark

* Correspondence: tk@cancer.dk; Tel.: +45-35-25-7746

Received: 12 December 2019; Accepted: 14 January 2020; Published: 16 January 2020

Abstract: Over 90% of cancer deaths are due to cancer cells metastasizing into other organs. Invasion is a prerequisite for metastasis formation. Thus, inhibition of invasion can be an efficient way to prevent disease progression in these patients. This could be achieved by targeting the molecules regulating invasion. One of these is an oncogenic transcription factor, Myeloid Zinc Finger 1 (MZF1). Dysregulated transcription factors represent a unique, increasing group of drug targets that are responsible for aberrant gene expression in cancer and are important nodes driving cancer malignancy. Recent studies report of a central involvement of MZF1 in the invasion and metastasis of various solid cancers. In this review, we summarize the research on MZF1 in cancer including its function and role in lysosome-mediated invasion and in the expression of genes involved in epithelial to mesenchymal transition. We also discuss possible means to target it on the basis of the current knowledge of its function in cancer.

Keywords: cancer therapy; EMT; lysosome; lysosome-mediated invasion; MZF1; phosphorylation; PAK4; SUMOylation; transcription factor; zinc finger

---

1. Transcription Factors as Drug Targets in Cancer

For a long time, steroid receptors, which are also known as ligand-activated transcription factors, have been the main group of transcription factors targeted in anti-cancer treatments. During recent years, other sequence-specific transcription factors have emerged as promising anti-cancer drug targets and consequently, transcription factors have lost their status of being "undruggable". This is especially the case for zinc finger transcription factors, which is a large group of proteins with their own specific DNA binding sequences. A good example of a cancer drug that targets a transcription factor is thalidomide, an antiemetic drug from the 1950s that has been repurposed as a novel treatment against hematological malignancies and which functions by inactivating zinc finger transcription factors Ikarios (IKZF1) and Aiolos (IKZF3) through their destabilization [1,2]. Here we will summarize the recent literature on the role and function of another cancer-relevant zinc finger transcription factor, Myeloid Zinc Finger 1 (MZF1), and present reasoning for its potential targeting in cancer and discuss the possibilities of how to target it.

2. What Is MZF1?

2.1. MZF1 Is a Sequence-Specific, Oncogenic Transcription Factor Involved in Myeloid Differentiation

MZF1 is a member of the SCAN domain-containing zinc finger transcription factor (SCAN-ZFP) family, a subfamily of zinc finger proteins (ZFPs)[3]. SCAN-ZFPs represent a class of DNA-binding proteins, many of which are known to regulate transcription during different developmental processes. MZF1 was first isolated from the peripheral blood from a patient with chronic myeloid leukemia and was described as a novel zinc finger protein involved in transcriptional regulation of hematopoietic development [4]. A few years later it was shown to regulate the expression of hematopoiesis-specific genes influencing differentiation, proliferation and programmed cell death and its aberrant expression was found to result in the development of hematopoietic cancers [5,6]. During the last decade, MZF1 was shown to be implicated in the development of various types of solid cancers by enhancing cancer cell growth, migration and invasion [7–15]. Knowledge on the detailed mechanisms by which MZF1 activity is regulated and the central target genes it activates has steadily increased and is still emerging due to active research on the topic.

2.2. MZF1 Transcript Variants and Functional Domains

MZF1 is encoded by a single-copy gene located at chromosome 19q13.4, which is the sub-telomeric region of the chromosome 19q, containing a large number zinc finger genes [4]. Full-length MZF1 protein is estimated to be about 82 kD without post translational modifications. *MZF1* gene supposedly encodes three transcript variants, which are predicted to result from alternative use of two transcription initiation sites and from alternative splicing [16,17]. MZF1 gene is composed of six exons [16,17]. Early work on *MZF1* transcripts lead to the identification of two human MZF1 isoforms: one full-length 734 amino acid isoform containing a SCAN domain in the N-terminus; 13 zinc finger DNA-binding domains in the C-terminus; and one N-terminally truncated, 485 amino acid isoform containing the 13 zinc finger DNA-binding domains and a short N-terminal fragment [16,17] (Figure 1).

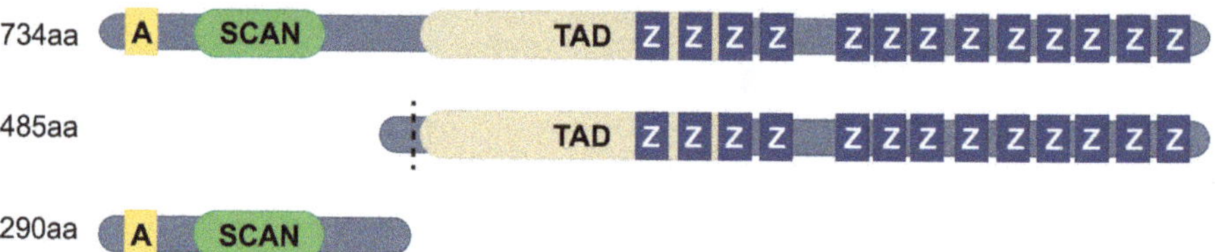

Figure 1. MZF1 protein isoforms. Top: Domain structure of the full-length (734 amino acid) MZF1 isoform containing five distinct domains: acidic domain (A), SCAN domain (SCAN), transactivation domain (TAD), and 13 highly conserved Krüppel-like zinc finger motifs (Z) arranged in two domains. Middle: Domain structure of the putative (485 amino acid) "zinc finger only"-form of MZF1, that in addition to 13 zinc fingers also has the TAD domain. The amino terminus of the new, recently identified 450 kD zinc finger only isoform is marked with a dashed black line. Bottom: Domain structure of the 290 amino acids "SCAN domain only" form of MZF1 that in addition to the SCAN domain also has the acidic domain (A).

Because the smaller 485 amino acid MZF1 isoform (the so-called zinc finger only isoform) was isolated and characterized first, it was for some years believed to be the full-length MZF1. Only when a novel mouse isoform of MZF1, then denoted as Mzf-2, was identified [16,18], was it discovered that also human MZF1 exists in a larger form, containing several additional structural and functional domains [17]. Soon after, these full-length MZF1 isoforms, Mzf-2a (mouse) and MZF1B/C (human), were collectively denoted as MZF1. A third MZF1 isoform of 290 amino acids containing only the SCAN domain in the N-terminus was later identified by the National Institutes of Health Mammalian Gene Collection Program (http://genecollections.nci.nih.gov/MGC/) (Figure 1). This so-called "SCAN domain only" isoform belongs to a group of proteins known as SCAND proteins. SCAND proteins are expected to function as regulators of other SCAN domain-containing proteins [19,20]. The tissue-specific expression and function of this isoform has not been elucidated. It also needs to be noted that the 485 amino acid "zinc finger only" isoform of MZF1 has been recently deleted from the human NBCI sequences and replaced with the predicted sequence of 450 amino acids "zinc finger only" form, where the N-terminus is slightly shorter than in the 485 amino acids form (Figure 1).

Full-length MZF1 consists of five distinctive domains (Figure 1). The domain furthest towards the N-terminus is called an acidic domain (A), which is rich in glutamic and aspartic acid [16,21]. This domain is involved in transcriptional activation and contains regulatory SUMO and phospho-sites [22]. It can also mediate interactions between MZF1 and other transcription factors [22–24]. Downstream of the acidic domain is the SCAN domain of 84 amino acids, a leucine-rich region known to mediate protein–protein interactions [16,17,25,26]. The SCAN domain is found in the SCAN-ZFP family of zinc finger proteins and it mediates homo- and heterodimer formation between SCAN domain containing zinc finger proteins [25–28]. Following the highly conserved SCAN domain there is a so-called transcriptional activation domain (TAD). It is a serine and threonine rich region that is involved in the transcriptional activation of MZF1 [18,21]. This MZF1 domain was identified as a TAD by a classical study by Ogawa and co-workers [21]. In the study, they showed that in murine MZF1 this domain is phosphorylated by mitogen-activated protein kinase ERK and p38 in three of its serines, serine 257, 275 and 295, leading to transcriptional inactivation of Mzf-2a. Consequently, substitution of all of these serines with alanines resulted in strong enhancement of its transcriptional activity in murine myeloid cell line LGM-1[21]. The corresponding sites are conserved in human MZF1, where they are represented as serine 256, 274 and 294. Later on, post-translational modifications in both the acidic domain and the SCAN domain were found to contribute to the transcriptional activity of human MZF1 [22]. In the C-terminal region of MZF1 are the 13 highly conserved Krüppel-like zinc finger motifs arranged in two domains. The first four zinc-finger motifs form one zinc-finger domain and the last nine motifs form another zinc-finger domain, separated from the first one by a glycine-proline-rich region of 24 amino acid residues [4] (Figure 1). The two zinc-finger domains of MZF1 bind to two distinct, yet similar DNA consensus sequences with a common core sequence of four or five guanines, which allows MZF1 to bind more than one DNA sequence at the same time [29], or to bind stronger in genomic sites containing binding sites for both motifs.

3. MZF1 and Cancer

3.1. MZF1 and Hematological Malignancies

MZF1 was originally isolated from chronic myeloid leukemia and was shown to be involved in hematopoietic differentiation due to its ability to control the expression of genes involved in hematopoiesis such as *CD34* and *MYB* [4–6]. Due to these reasons, most of the earliest studies on the function of MZF1 were done in hematopoietic cells. Some of the results concerning the actual role and function of MZF1 in hematopoietic malignancies are contradictory. This is because during the earliest studies of MZF1, the knowledge of MZF1 isoforms was not complete. Thus, many studies were done using overexpression of the so-called zinc-finger-only isoform of MZF1, that was the 485 amino acid isoform (Figure 1), which is practically missing most of the N-terminal regulatory domains. As mentioned, this zinc-finger-only isoform was later deleted from the human NBCI sequences, suggesting that it may be a cloning artifact. It has, however, been replaced with a slightly shorter, 450 amino acid isoform, which is a predicted alternative MZF1 transcript that can theoretically exist. In brief, early experiments involving overexpression of the 485 amino acid zinc-finger-only isoform in myeloid cells showed that it inhibits hematopoietic differentiation of mouse embryonic stem cells and delays retinoic acid-induced granulocytic differentiation and apoptosis by inducing proliferation in human promyeloblasts [6,30]. Contrary to what would be expected from these overexpression studies, silencing of MZF1 with antisense oligodeoxynucleotides (AOS) significantly inhibited granulocyte development in vitro from granulocyte colony-stimulating factor-induced cells originating from normal bone marrow, which was evident from granulocyte colony formation assays [31]. This result coincides with the results obtained from Mzf1 knockout mice, which accumulate highly proliferating myeloid cells in their bone marrow and liver, disturbing the tissue architecture, indicating that Mzf1 may function as a tumor suppressor in the hematopoietic compartment [32]. Since AOS and knockout studies target the full-length MZF1, it is understandable that the results obtained when downregulating or inhibiting MZF1 expression would not necessarily be the opposite of the results obtained when overexpressing the MZF1 zinc-finger-only isoform. The zinc-finger-only-isoform, by lacking the N-terminus that contains most of the regulatory domains of MZF1, would have impaired regulation of its activity and would be unable to dimerize with SCAN domain-containing proteins. However, it could mimic MZF1 in a potential state where it is void of its upstream regulation. This could be achieved for example via cancer-induced aberrant expression of MZF1. Thus, it can be concluded that aberrant expression and regulation of MZF1 can make it oncogenic, which is also supported by the fact that mouse embryonic fibroblasts that overexpress the zinc-finger-only MZF1 isoform form aggressive tumors in athymic mice and lose their contact inhibition and upregulate their growth ex vivo [33].

3.2. MZF1 Acts as an Oncogene in Solid Cancers

Several studies demonstrate that MZF1 can promote tumorigenesis of various solid cancers. These include breast, cervical, colorectal, liver, lung, and prostate cancer [7–9,11–15,34]. Many MZF1 target genes have a central role in cancer, and increased expression and/or activation of MZF1 induces cell growth, migration and invasion [7–15]. Below we will summarize most of the results obtained on MZF1 in some common solid cancers.

3.2.1. MZF1 in Breast Cancer

MZF1 is needed for the invasion of ErbB2-expressing breast cancer cells [7]. In a study by Rafn and co-workers, ErbB2 activation was shown to induce the invasion of breast cancer cell spheroids via activation of a signaling network that involves TGFβ receptor 1 and 2 (*TGFBR1* and 2), ERK2 (*MAPK1*), PAK4, PAK5 and PAK6 (*PAK4*, *PAK5* and *PAK6*), cdc42 binding protein kinase beta (*CDC42BPB*), and protein kinase Cα (PKCα; *PRKCA*). Activation of this signaling network leads to activation of MZF1 and MZF1-mediated induction of expression of lysosomal cysteine cathepsins B and L (*CTSB* and *CTSL1*). This work implied for the first-time involvement of lysosomes in the invasion of ErbB2-expressing cancer cells. It showed that MZF1, upon activation by ErbB2 signaling, can induce the pericellular accumulation of lysosomes at the invadosome-like cellular protrusions in invasive ErbB2 expressing breast cancer cells, thereby initiating and promoting their invasion [7]. Once lysosomes have travelled to the cell periphery, their hydrolytic content can be secreted into the extracellular space (lysosomal exocytosis). This can initiate and induce invasion mainly via cathepsin B, which cleaves and thereby activates matrix metalloproteases (MMP) 2 and 3 and the urokinase plasminogen activator [35–37]. Consistently, ErbB2-positive primary breast tumors exhibit increased mRNA and protein expression of cathepsins B and L. Supporting the in vivo connection of ErbB2 activation and cathepsins B and L, the positive correlation between ErbB2 and cathepsin B and L expression in invasive breast cancer was found to be significant [7].

Interestingly, MZF1 regulates the expression of TGFβ1 gene (*TGFB1*) in response to osteopontin-induced integrin signaling in human mesenchymal stem cells, where increased TGFβ signaling induces them to differentiate and adapt a cancer-associated fibroblast phenotype, a process that leads to increased tumor growth and metastasis [14]. TGFβ1 is considered as one of the main regulators of epithelial mesenchymal transition (EMT) [38] and ErbB2 overexpression is connected to TGFβ overexpression, secretion and activation of TGFβ signaling [39]. TGFβ signaling amplifies oncogenic ErbB2 signaling and promotes invasion and metastasis of ErbB2 positive cancer cells [40–42]. Since ErbB2-induced activation of MZF1 is enhanced by TGFβ receptor signaling and *TGFB1* is a MZF1 target gene, increased TGFβ signaling can further induce ErbB2 signaling via a feedback loop involving MZF1 activation. This may additionally lead to enhanced activation of other MZF1 target genes that are important for amplification of breast cancer signaling networks and promoting breast cancer cell migration and invasion, such as *PRKCA* [11]. Interestingly, a complex formation between Elk-1 and MZF1 has been shown to enhance *PRKCA* expression in a synergistic manner and its expression correlates positively with the expression of Elk1 and MZF1 in various breast cancer cell lines [11]. Moreover, a high level of MZF1 in triple-negative breast cancer cell lines Hs578T and MDA-MB-231 is associated with a mesenchymal phenotype with increased cell migration and invasion, which is mediated via insulin-like growth factor receptor (IGF1) [24]. Consequently, destabilization of MZF1 by the IGF1R-driven p38MAPK-Erα-SLUG-E-cadherin pathway leads to conversion of the invasion-promoting mesenchymal phenotype to the less invasive epithelial phenotype. In osteoblasts, which are involved in osteolytic breast cancer metastasis, MZF1 has been shown to upregulate the expression of another EMT regulator, N-cadherin (*CDH2*) [43]. Moreover, a MZF1 target gene *AXL*, which can be activated upon lapatinib resistance in ErbB2 positive breast cancer cells [44], has been shown to induce EMT in breast cancer cells [45]. This all suggests that MZF1 has a role in the development of aggressive breast cancer.

3.2.2. MZF1 in Cervical and Colorectal Cancers

MZF1 activation has been implicated in the progression of cervical and colorectal cancer, where it increases invasion and metastasis, at least partially, via increased expression and activity of receptor tyrosine kinase AXL [8]. Increased expression of *AXL* has been connected to invasion and metastasis of many types of cancers [45–47]. Supportively, both MZF1 and AXL protein levels are considerably higher in colorectal tumors than in corresponding normal tissue [8]. MZF1 binds directly to the *AXL* promoter, leading to increased *AXL* mRNA and protein expression [8]. However, depletion of *AXL* by RNA interference only partially inhibits MZF1-induced migration and invasion of colorectal cancer cells, suggesting that additional MZF1-regulated genes are involved in this process. MZF1 is also central for the activation of the expression of Phosphoinositide -3-Kinase Regulatory Subunit 3 Gamma (*PIK3R3*), which is a regulatory subunit of PI3 kinase (PI3K) needed for PI3K signaling and is important for cancer cell proliferation [15]. In human papillomavirus infected cervical cancer, MZF1 induces the expression of another transcription factor, *NKX2-1*, which in turn upregulates a cancer stem cell regulator *FOXM1*, resulting in increased tumor growth and invasion [48]. In another study with SiHa human cervical cancer cells, MZF1 was shown to bind the matrix metalloprotease 2 (MMP2) promoter, and a bit surprisingly to suppress its expression, and thus was reported to function as a tumor suppressor in these cells [49].

3.2.3. MZF1 in Liver and Lung Cancer

MZF1 regulates the expression of the PKCα gene, *PRKCA*, in human hepatocellular carcinoma cells, where it binds directly to the MZF1 binding site in the *PRKCA* promoter region [9,12]. Depletion of MZF1 with specific antisense oligonucleotides reduces proliferation, migration and invasion of hepatocellular carcinoma cells [9,12] and suppresses the growth of the corresponding xenografts [10,12]. In lung cancer, MZF1 activates the expression of the c-Myc gene (*MYC*) upon loss of the liver kinase B1 (*LKB1*) [13]. This results in enhanced migration and invasion of lung cancer cells and facilitates their growth in soft agar [13]. In tumors from lung adenocarcinoma patients there is a positive correlation between high MYC and MZF1 and low LKB1 expression. Importantly, lung adenocarcinoma patients with low *LKB1* expression have a shorter overall survival than patients with high *LKB1* expression [13]. In lung cancer, MZF1 can upregulate the expression of *NKX2-1*, which in turn increases the expression of *FOXM1* resulting in facilitated tumor growth and invasion [48]. On the other hand, in lung adenocarcinomas, the loss of *LKB1* is associated with *NKX2-1* expression [50].

3.2.4. MZF1 in Prostate Cancer

The role of MZF1 in prostate cancer is somewhat more complicated. The expression of the cell division control 37 (*CDC37*) gene is increased in prostate cancer cells. Here, MZF1 was shown to bind to the promoter of *CDC37* and upregulate its expression [51]. As expected, depletion of MZF1 in prostate cancer cells decreases *CDC37* expression and reduces their tumorigenesis. Interestingly, SCAND1, a SCAN domain protein that can inhibit MZF1 by dimerizing with it, can upon overexpression accumulate at the MZF1 binding sites at the *CDC37* promoter and downregulate its expression-inhibiting tumorigenesis [51]. On the contrary, MZF1 was shown to have the opposite effect in PC3 and DU145 prostate cancer cells, where expression of MZF1 upregulated ferroportin (*FPN*), the only known mammalian iron exporter [52]. Depletion of MZF1 was found to decrease the expression of *FPN*, as expected, but in turn this was shown to result in enhanced cancer cell growth in addition to increased cytoplasmic iron retention [52]. Consequently, increase in the expression of MZF1 inhibited tumor growth, suggesting that in respect to *FPN* regulation in these prostate cancer cells, MZF1 can exhibit a tumor suppressor type of function.

3.2.5. MZF1 in Other Type of Cancers

In glioma cell lines, MZF1 binds directly to the LIM-only protein 3 (*LMO3*) promoter and induces the expression of *LMO3* [53], which is a transcriptional co-activator that can act as an oncogene in glioma, one of the most aggressive and most common tumors of the central nervous system. *LMO3* is often overexpressed in gliomas and its expression correlates positively with poor prognosis [53]. The 19q chromosomal deletions together with the deletion of 1p are used to define the oligodendroglioma, which is a specific type of glioma with favorable prognosis and good response to chemotherapy [54]. Interestingly, the 19q chromosomal deletions in oligodendroglioma include the MZF1 locus as well as the locus of genes coding for many other zinc-finger proteins. In esophageal squamous cell carcinoma samples of 13 patients, MZF1 was found to be co-activated with three other transcription factors, SPIB, MAFG and NFE2L1 when compared to their paired non-cancerous tissues using microarray analysis, where the expression of 17 other transcription factors was suppressed [55]. In gastric cancer cells, MZF1 upregulates *MMP14* expression by directly binding to its promoter [56]. In the same study it was shown that in the clinical samples, *MZF1* expression correlated positively with *MMP14* expression in gastric cancer. On the contrary to this, another study where human gastric cancer samples were analyzed indicated that MZF1 expression was decreased during gastric cancer progression, which correlated with increased invasiveness of gastric cancer [57].

3.2.6. Many MZF1 Target Genes Have a Role in Cancer

MZF1 exerts its activity via modulating the expression of its target genes. Aberrant MZF1 expression and activation results in transcriptional changes that increase cell growth, migration and invasion (see above and reviewed by Eguchi and co-workers [58]). In summary, MZF1 may promote invasion and migration partially by controlling the expression of kinases that are controlling these processes such as *AXL* and *PRKCA*. It can also increase expression of lysosomal, invasion-inducing and promoting hydrolases *CTSB* and *CTSL1*, which facilitate intra- and extracellular degradation of extracellular matrix components by their direct cleavage or by indirectly cleaving and activating matrix metalloproteases MMP2 and MMP3 and urokinase plasminogen activator, which in turn degrade the extracellular matrix [35,36]. MZF1 is also expected to have a role in EMT by controlling the expression of *TGFB1*, *CDH2* and *FOXM1*, and several other EMT-related genes. In Table 1, we have listed the known MZF1 target genes. However, it needs to be noticed that only the ones that are verified by chromatin immunoprecipitation can be considered definite direct targets of MZF1.

Table 1. MZF1 target genes; their method of identification (ChIP: chromatin immunoprecipitation; EMSA: electrophoretic mobility shift assay), reference, function (the role of MZF1) and their involvement generally in EMT (yes, if involvement has been reported).

Gene	Method	Reference	Function	EMT
AXL	ChIP, Luciferase	[8]	Activator	yes
CD14	EMSA,luciferase	[59]	Activator	yes
CD34	EMSA, Acetyltransferase activity	[6]	Activator	
CDC37	ChIP, Luciferase	[51]	Activator	

Table 1. *Cont.*

Gene	Method	Reference	Function	EMT
CDH2 (N-Cadherin)	EMSA	[43]	Activator	yes
CDH2 (N-Cadherin)	ChIP, Luciferase	[60]	Activator	
CK17	Luciferase, qPCR	[61]	Activator	
CTGF	ChIP	[62]	Activator	yes
CTSB	ChIP, Luciferase	[7]	Activator	
GAPDH	ChIP	[63]	Activator	yes
HK2	ChIP	[64]	Suppressor	
IGFIR	ChIP, Luciferase	[65]	Suppressor	yes
ITGAM (CD11b)	EMSA, luciferase	[59]	Activator	
LMO3	ChIP	[53]	Activator	
MMP2	ChIP, Luciferase	[49]	Suppressor	yes
Mtor	ChIP, EMSA, Luciferase	[66]	Suppressor	yes
MYB (c-myb)	EMSA, Acetyltransferase activity	[6]	Activator	
MYC	ChIP, Luciferase	[13]	Activator	
NFKB1A	ChIP	[67]	Activator	yes
NKX2-1	ChIP, Luciferase	[48]	Activator	yes
NKX2-5	ChIP, Luciferase	[68]	Activator	yes
NOV	ChIP	[61]	Activator	
OOCT4	Luciferase	[69]	Activator	yes
PAX2	Luciferase	[70]	Suppressor	yes
PIK3R3 (p55PIK)	ChIP, Luciferase	[15]	Activator	yes
PRAME	ChIP	[71]	Activator	yes
PRKCA (PKC alpha)	ChIP, Luciferase	[12]	Activator	
SLC40A1 (FPN)	ChIP, Luciferase	[52]	Activator	
SMAD4	ChIP, EMSA, Luciferase	[72]	Activator	
TGFB1	ChIP, Luciferase	[73]	Activator	yes
TNFRSF10B (DR5)	Luciferase	[74]	Activator	
YAP1	ChIP, EMSA, Luciferase	[75]	Activator	yes

4. How Does MZF1 Function?

In order to target the oncogenic functions of MZF1, we need to understand how MZF1 is regulated. The key to MZF1 function in cancer lies in its domain structure and in its post-translational modifications (Figure 2) that are regulating its association with other factors, its activation status and its availability.

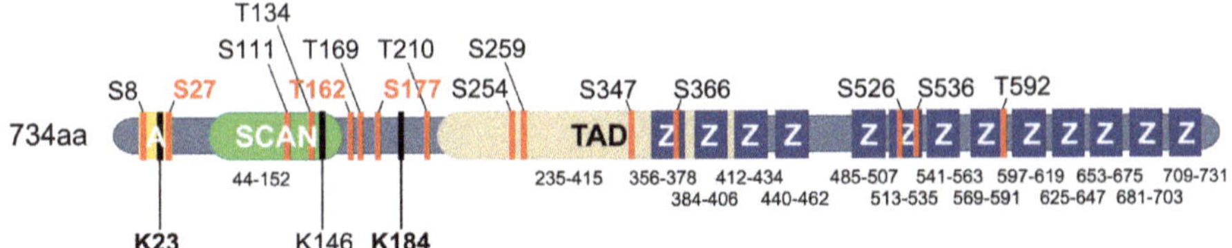

Figure 2. Schematic representation of the protein structure of full-length human MZF1 with reported SUMO-sites (K) and serine (S) and threonine (T) phosphorylation sites. The domain structure of MZF1 is presented as in Figure 1. The location of each indicated SUMO- and phospho-site is shown. The verified SUMO-sites are marked with bold font and the predicted SUMO-site (K146) is marked with regular font. The phospho-sites that are highlighted with red have been identified as ErbB2-responsive sites. Note that the serines 256, 274 and 294 corresponding to the ERK phosphorylation sites in the TAD of murine MZF1 have not yet been reported as phospho-sites in humans.

4.1. Regulation of MZF1 Expression

Relatively little is known about the transcriptional regulation of *MZF1*. Originally, MZF1 was identified as an important transcriptional regulator of myeloid differentiation, and its expression was believed to be myeloid-specific [76]. Later on, it was identified as an oncogenic transcription factor responsible for migratory and invasive phenotypes of various cancer cell lines. Analysis of TCGA website data indicates that there is a significant amplification of the *MZF1* gene in various solid cancers [58]. These include breast, bladder, lung, and uterine cancers. Indeed, increased expression of MZF1 protein levels has been detected in the study of 321 tissue microarray samples containing primary breast cancer and normal breast samples [77]. In these samples, MZF1 levels were shown to significantly increase from normal breast tissue to grade 1–2 tumors, which define invasive ductal carcinoma.

Non-coding RNAs have arisen as important regulators of gene expression. Several microRNAs (miRNAs) can regulate *MZF1* expression. Let-7 miRNAs belong to the group of miRNAs whose aberrant expression is most frequently associated with cancer [78]. Let-7 is upregulated during differentiation, and its expression is systematically downregulated in malignant cancers including breast cancer [79]. Let-7 binds to the 3′-untranslated region of *MZF1*, and ectopic expression of let-7 microRNAs let-7d and let-7e can efficiently downregulate *MZF1* and invasion of constitutively active ErbB2-expressing breast cancer cells [77]. MiR-492 is another microRNA that can bind to the 3′-untranslated region of *MZF1* [52]. MiR-492 regulates *MZF1* expression in prostate cancer cells, and in prostate tumors, miR-492 levels correlate reversibly with the levels of MZF1. Another study with glioma cell lines indicates that overexpression of miR-101 leads to a decrease in *MZF1* expression, without going further into detail in regard to its potential binding sites in *MZF1* [53]. MiRNA-337-3p inhibits gastric cancer progression by downregulating MZF1 activity via a specific mechanism, where miRNA-337-3p binds to the promoter region of *MMP14* adjacent to its MZF1 binding site and represses the MZF1-induced expression of *MMP14* [56]. Consequently, in the same study, miRNA-337-3p was shown to inhibit growth, invasion, metastasis, and angiogenesis of gastric cancer cells in vitro and in vivo via repression of MZF1 activity. Furthermore, miRNA-337-3p expression was found to be an independent prognostic factor for a favorable outcome in gastric cancer.

Interestingly, according to UCSC genome browser (https://genome.ucsc.edu), a validated long non-coding RNA exists that contains the whole MZF1 coding sequence resulting in 15,573 base pair antisense RNA (LOC100131691). Thus far, no studies exist of its actual regulation, expression or function, although it is tempting to speculate that it can have a regulatory role in the expression of MZF1.

4.2. Regulation of the Transcriptional Activity of MZF1

4.2.1. Interaction with Other Transcription Factors

MZF1 has to dimerize to function as a transcription factor. MZF1 utilizes its SCAN domain to form homo- and heterodimers with other SCAN-domain transcription factors [25–28]. The possibility of heterodimerization via the SCAN domain exposes MZF1 to an additional level of regulation, since depending on its dimerization partner, MZF1 may function as a transcriptional activator or repressor. Known SCAN domain-containing MZF1 dimerization partners include SCAN-ZFP family members RAZ1, ZNF24, ZNF174, and ZNF202 [3,26,80], which are all heterodimerizing with MZF1 via a SCAN–SCAN interaction. A recent computational study has tried to shed new light on MZF1 SCAN domain interactions by identification and analysis of cancer-specific mutations in the MZF1 SCAN domain [81]. In this study, 23 cancer-specific mutations were identified in the MZF1 SCAN domain, which could affect MZF1 function by changing its dimerization capacity directly or indirectly via gain or loss of possible post-translational modifications (Nygaard et al., 2016). This work identified cysteine 69 as a potential regulator of MZF1 SCAN–SCAN interactions. Moreover, simultaneous expression and appearance of other SCAN and SCAND domain-containing proteins and possible cancer-inducing mutations in them could also affect MZF1 function for example by directly or indirectly replacing the binding partners of MZF1. However, this type of exiting regulation scheme is still mostly theoretical, especially in the case of MZF1 heterodimers, since detailed biological information on the specific regulation of the transcriptional activity of MZF1 via heterodimeric SCAN–SCAN interactions is still missing.

In addition to SCAN-domain proteins, MZF1 can interact with proteins without the classical SCAN domain, which complicates the scenario of its regulation by binding partners. Moreover, MZF1 interaction with other proteins can even occur via other domains than the SCAN domain. Recent work has indicated that the acidic domain of MZF1 is an additional protein–protein interaction domain. The acidic domain of MZF1 is involved in its association with Elk1 in triple negative breast cancer [12,23,24]. Association of MZF1 and Elk1 via the acidic domain of MZF1 and the heparin-binding domain of Elk1 increases invasion, migration and mesenchymal phenotype of breast cancer cells. This occurs via increasing the expression of *PRKCA* and *IGF1R* by direct binding of MZF1 to their promoter regions. In non-invasive MCF7 breast cancer cells, MZF1 interacts with the CCCTC-binding factor (CTCF) via its acidic domain, which results in downregulation of the transcriptional activity of MZF1 [22]. Activation of ectopic ErbB2 signaling results in SUMO-directed (SUMOylation of lysine 23) phosphorylation of MZF1 serine 27 at its acidic domain, which dissociates MZF1 from its transcriptional repressor CTCF, allowing transcriptional activation of MZF1 [22].

4.2.2. SUMOylation of MZF1

The activity of transcription factors can be modulated by covalent attachment of small ubiquitin-related modifier (SUMO) proteins: SUMO1, SUMO2, SUMO3, and SUMO4 in their SUMO acceptor sites [82]. According to current knowledge of consensus SUMOylation sites, MZF1 has three predicted SUMOylation sites: lysine 23, 184 and 146 (Figure 2) [22,58]. An earlier study that was the first to report MZF1 SUMOylation, suggested that a SUMOylation site would reside in the amino terminus of MZF1 between amino acids 15–27 [80], which is a conserved sequence found in a subset of SCAN-ZFPs [19,80]. This site was identified by showing that overexpressed full-length MZF1 has the ability to accumulate into promyelocytic leukemia nuclear bodies (PML-NBs), a function which requires SUMOylation, and which could be abolished when this area was deleted from MZF1 [80].

SUMOylation of transcription factors and their cofactors may lead to transcriptional activation or inactivation [82,83]. Several studies suggest an important role of PML-NBs in transcriptional regulation [84,85]. Especially, PML-NBs has been presented as a site where SUMO-conjugation occurs and where SUMOylated nuclear proteins reside and accumulate [86]. PML-NBs usually associate to the areas of genomic regions with high transcriptional activity, and many transcription factors can be transiently recruited to PML-NBs. Since SUMOylation usually occurs in the nucleus [83], introduction of mutations in the nuclear SUMO-modified proteins that interfere with their SUMOylation can promote their translocation into the cytoplasm. Thus, MZF1 SUMOylation may be involved in its nuclear retention and its ability to function as a transcription factor.

MZF1 lysine 23 has been identified as a SUMOylation site that directly regulates the transcriptional activity of MZF1 [22]. Its occupation by SUMO groups exposes a nearby serine 27 for phosphorylation by PAK4 in response to ErbB2 activation, which in turn results in increased transcriptional activity of MZF1, indicating its importance for the transcriptional activation of MZF1 as well as defining a new mechanistic type of post-translational regulation, "SUMO-directed phosphorylation". In the same study, lysine 184 was mapped as an additional functional SUMO acceptor site. However, no biological function has yet been identified for it. It is tempting to speculate that MZF1 SUMOylation may be regulating its stability. However, in our studies of MZF1 post-translational modification and their effect on protein stability, we found that in ErbB2-expressing breast cancer cells, MZF1 is a very stable protein and its stability was not affected by mutating its SUMO sites (Brix and Kallunki, unpublished observations). Moreover, in the ErbB2-expressing breast cancer cells, no evidence of SUMOylation of lysine 146 was found [22]. It is possible that lysine 146 is not a functional SUMO acceptor site in MZF1. Supporting this, its probability as a SUMO acceptor site is much lower than that of lysine 23 and 184 [22,58].

4.2.3. Phosphorylation of MZF1

MZF1 is a phosphoprotein that contains several potential as well as functional phosphorylation sites. Even though its massive phosphorylation has been known for some time, thanks to the large amount of phosphorylation analysis done by Mass Spectrometry that has been deposited on the web (http://www.phosphosite.org), thus far no biological function has been shown for the majority of these sites. In a study conducted in MCF7 breast cancer cells with inducible expression of constitutively active ErbB2, ErbB2 activation was shown to increase the transcriptional activity of MZF1 via a signaling network that involves TGFβ receptors 1 and 2 (TGFBR1 and 2), ERK2, PAK4 (PAK5 and PAK6), cdc42 binding protein beta kinase (CDC42BPB), and PKCα (PRKCA) [7]. In a recent study that utilized ErbB2 positive breast cancer cells for phosphorylation analysis by Mass Spectrometry, 13 MZF1 phosphorylation sites were identified [22]. Only three of these were phosphorylated in response to ErbB2 activation. These were serine 27, serine 162 and threonine 177, other sites being constitutively phosphorylated. Of these three sites, only phosphorylation of serine 27 was shown to increase the transcriptional activity of MZF1, the two others having no significant effect on it [22]. Supporting the importance of this phospho-site in vivo, serine 27 phosphorylation was found in an ErbB2-positive breast tumor sample in a proteomics study covering 105 breast tumors that were characterized for TCGA [87]. Furthermore, MZF1 serine 27 phosphorylation was found to correlate positively and significantly with ErbB2 status in a breast tumor tissue microarray containing 225 tissue cores embedded as duplicates and enriched with primary invasive breast cancer samples [22]. The phosphorylation of serine 27 is tightly connected to the SUMOylation of lysine 23 through a mechanism where SUMOylation of lysine 23 is needed as a prerequisite for the phosphorylation of serine 27 by PAK4. In silico modelling of this activation mechanism suggests that SUMOylation of lysine 23 opens up and exposes the serine 27, which otherwise is masked and not approachable for PAK4 to dock to it and phosphorylate it (Figure 3). Phosphorylation of serine 27 will then dissociate MZF1 from the transcriptional repressor CTCF, allowing MZF1 to activate transcription of *CTSB* needed for the invasion of ErbB2-expressing cells. Interestingly, a recent study in human esophageal cancer cell lines demonstrated that phosphorylation of MZF1 serine 27 by constitutively active casein kinase 2 (CK2), which is often upregulated in cancers, mediates their epithelial to mesenchymal transition by inducing the expression of N-cadherin [60].

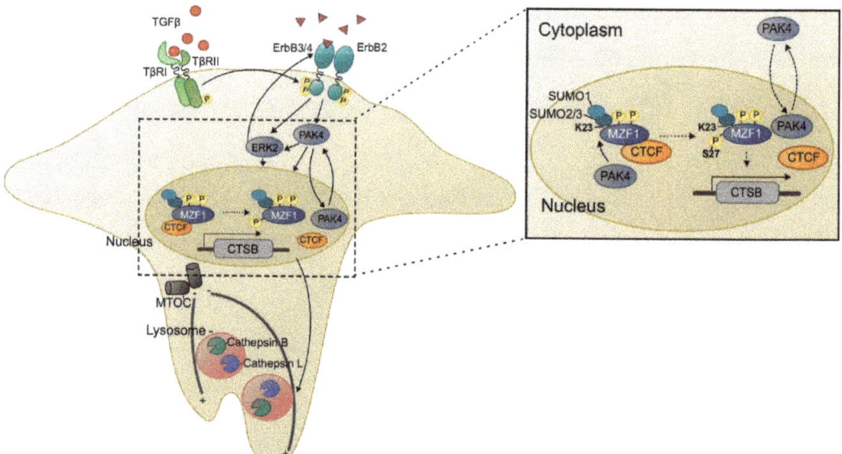

Figure 3. Graphical presentation of MZF1 activation and induction of *CTSB* expression in lysosome-mediated invasion as a response to ErbB2 signaling. ErbB2 activation, further supported by activation of TGFβ signaling, activates ERK2 and PAK4. Active PAK4 will phosphorylate MZF1 serine 27, if its adjacent lysine 23 is SUMOylated, which exposes MZF1 serine 27 to PAK4 phosphorylation. As a response to phosphorylation of serine 27, MZF1 association of its transcriptional repressors, e.g., CTCF, is prevented and MZF1 can now activate *CTSB* expression and lysosome redistribution, which leads to lysosome-mediated invasion.

4.2.4. Mutations Creating New MZF1 Binding Sites in the Genome

Cancer-induced somatic mutations can create new transcription factor binding sites at the regulatory regions of cancer genes [88]. Mutations at the promoter regions of genes that are important for cancer progression can create new transcription factor binding sites that can contribute to the overexpression of that particular gene. Tian and co-workers have identified a mechanism by which MZF1 can affect gene expression via cancer-induced allelic mutations that result in a novel transcription factor co-operation at the promoter of the hepatocyte growth factor gene *HGF* [89]. They have identified single-nucleotide polyformism (SNP) and single nucleotide variants (SNV) in multiple myeloma at the promoter region of *HGF* that result in its increased expression. These mutations resemble the wild-type sequences of the binding motifs of MZF1, nuclear factor kappa-B (NFκB) and nuclear factor erythroid 2-related factor 2 (NFR-2), which together can contribute to increased expression of *HGF*. Whether this is a single type of cancer and a gene where new MZF1 binding sites are gained through a cancer-induced mutation or if multiple cancers and promoters are involved, is not yet known.

5. How Does MZF1 Promote Cancer Invasion and Metastasis?

The mechanistical explanations of how MZF1 promotes cancer progression must rely on the activation of its specific target genes in cancer. Currently known MZF1 target genes have been mapped in individual functional studies, however, no genome-wide studies on MZF1 transcriptional targets have been reported. The majority of the known MZF1 target genes are known cancer genes, whose activation is expected to promote cancer progression (Table 1). Two of the invasive processes activated by MZF1 have been described in more detail. We will briefly present these below.

5.1. Lysosomes, MZF1 and Invasion

Lysosomes have a central role in the induction of invasion by ErbB2 in breast cancer cells [90]. Invasion of the MCF7 breast cancer spheroids expressing the trastuzumab-resistant p95 form of ErbB2 depends on the activation of a signaling network that culminates in the activation of MZF1 [7]. Here MZF1 regulates the function and activity of lysosomes by mediating ErbB2-induced, increased expression of lysosomal cysteine cathepsins B and L (*CTSB* and *CTSL1*), which is necessary for the invasion of these cells. Increased expression of *CTSB* and *CTSL1* leads to increased activity of cathepsins B and L, whose expression correlates positively ($p < 0.0001$) with high ErbB2 status in primary invasive breast cancer [7]. This is connected to the redistribution of lysosomes from a perinuclear to a peripheral position in invadosome-like cellular protrusions adjacent to the cell membrane, which is induced by phosphorylation of MZF1 serine 27 [22]. The appearance of the peripheral population of lysosomes correlates positively with the invasiveness of ErbB2 positive ovarian and breast cancer cells [7,91] and can contribute to extracellular matrix (ECM) degradation both internally and externally [34,36,37]. Peripheral lysosomes degrade the ECM components that have been internalized by the cell. Moreover, they can secrete their hydrolytic content, including cathepsin B, into the extracellular space to initiate and promote invasion. Secreted cathepsin B degrades the ECM components type IV collagen, laminin, and fibronectin and initiates the activation of the extracellular degradome by cleaving the pro-forms of urokinase plasminogen activator and MMP2 and MMP3 [92], which are activators of MMP9 and MMP13 (Figure 4). MZF1 seems to be a central regulator of invasion-associated pericellular lysosome distribution and lysosome-mediated invasion of ErbB2 expressing highly invasive cancer cells [7,22,90] (Figure 3).

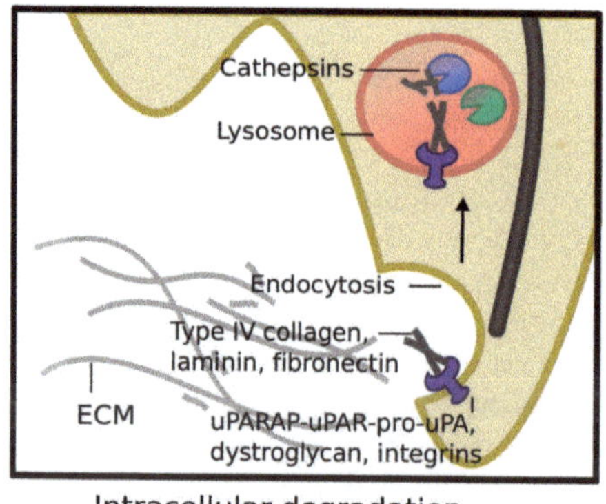

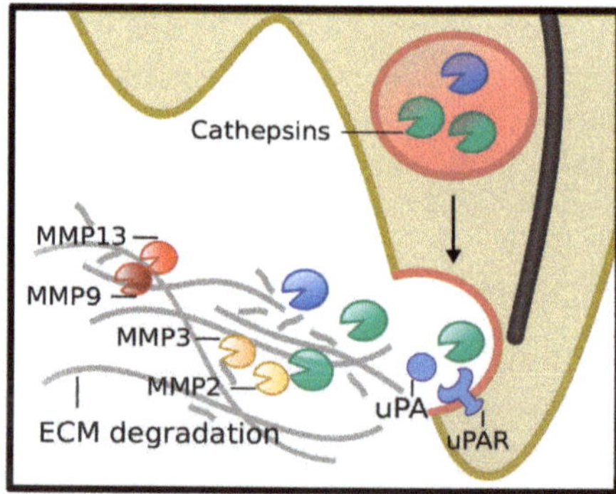

Figure 4. Graphical presentation of cellular mechanisms activated in lysosome-mediated invasion. Peripheral lysosomes contribute to extracellular matrix (ECM) degradation both internally (**left**) and externally (**right**). Peripheral lysosomes degrade the ECM components that have been internalized by the cell e.g., via endocytosis. Peripheral lysosomes can secrete their contents, including cathepsin B, into the extracellular space via lysosomal exocytosis, a process where the lysosome membrane fuses with the plasma membrane, which allows the secretion of the lysosomal contents to the extracellular space. Secreted cathepsin B degrades the ECM components: type IV collagen, laminin and fibronectin and initiates the activation of the extracellular degradome by cleaving the pro-forms of urokinase plasminogen activator and MMP2 and MMP3, which are activators of MMP9 and MMP13.

5.2. MZF1 and EMT

Recently, MZF1 has been connected to EMT, a biological process where epithelial cells lose their polarity and cell–cell adhesion capability and gain invasive and migratory properties by adapting a mesenchymal phenotype. In human esophageal cancer cell lines, phosphorylation of MZF1 serine 27 by CK2 initiates EMT by inducing the transcription of N-cadherin during the EMT-inducing switch from E-cadherin to N-cadherin [60]. Knockdown of MZF1 by specific shRNA reverses the mesenchymal phenotype of these cells into epithelial and downregulates the expression of N-cadherin. In triple-negative breast cancer cells, MZF1 activation can maintain the mesenchymal phenotype by interacting with Elk1 at the promoter region of *IGF1R* [24]. Even though evidence of the connection between MZF1 and EMT is increasing, it is still not clear if the role of MZF1 in EMT is cancer type-specific, or if MZF1 can have a more general role in the initiation and/or maintenance of EMT. Intriguingly, 17 of the 31 (55%) reported MZF1 target genes (Table 1) are somehow involved in EMT in other cancer studies.

6. Conclusions and Future Directions

The majority of studies on MZF1 in cancer report that MZF1 functions as an oncogene in various solid cancers by regulating the expression of genes involved in cancer progression, EMT, extracellular matrix degradation, invasion, and angiogenesis. Inhibition of MZF1 function could be a way to inhibit these processes. Different efficient approaches to inhibit transcription factor activity in cancer exist, including transcription factor destabilization by affecting the post-translational modifications that regulate stability or activity. Regarding MZF1, probably one of the most successful scenarios could be to inhibit its association with its specific co-transcription factors such as Elk1, which is needed for MZF1-induced activation of the expression of *PRKCA* and *IGF1R*, and which contributes to the stability of MZF1 in triple-negative breast cancer [24].

Specific post-translational modifications of MZF1 are induced in invasive cancer, as is the case for breast cancer harboring ErbB2 activation, and these are necessary for the invasive signaling mediated via MZF1 in response to ErbB2 activation in breast cancer cells [22]. Thus, another valid possibility would be to target the enzymes responsible for these post-translational modifications, namely SUMOylation of lysine 23 and/or phosphorylation of serine 27. It is not known what regulates the SUMOylation of lysine 23, which is a prerequisite of the phosphorylation of serine 27 by PAK4 [22]. However, a theory exists according to which the generally high phosphorylation status of MZF1, and especially the phosphorylation of serine 8 can bend the MZF1 molecule to a position where lysine 23 is exposed to SUMOylation [22,81]. Interestingly, increased SUMOylation is generally connected to cancer progression, and in breast cancer it is associated with poor prognosis [93,94]. The expression of SUMOylation-associated enzymes is often increased in cancer, and thus, numerous SUMO-pathway-targeting inhibitors have been developed, many of which can be considered as promising anti-cancer agents [95]. These could also target SUMOylation of MZF1 lysine 23 and thus prevent the activation of MZF1 by hindering the phosphorylation of serine 27.

PAK4, a kinase that can phosphorylate MZF1 serine 27 in response to ErbB2 activation, is considered as a good target for the treatment of a variety of solid cancers including breast cancer, and its inhibition for this purpose has been patented by Hoffman-La Roche and Genentech [96]. Although the resulting PAK4 inhibitor PF-3758309 failed in phase I clinical trials [97], a new PAK4 inhibitor, KPT-9274, has been developed (Karyopharm Therapeutics, USA), which is currently in phase I clinical trials [98]. Identification of MZF1 as an oncogenic target of PAK4, whose activity is important for invasiveness of ErbB2 positive breast cancer cells, suggests that PAK4 inhibitors might be useful for the treatment of cancers whose aggressiveness depends on MZF1.

Another possible way to target MZF1 could be by preventing its binding to the regulatory regions of its cancerous target genes. For this, more understanding of its DNA-binding specificity would be needed. For example, since it has two distinct zinc finger domains with divergent binding sequences, it would be useful to find out if either of them is a preferred binding domain for its target genes that are important in cancer. Interestingly, by using CRISPR-Cas9 gene editing technology, we have experienced that ErbB2-expressing breast cancer cell lines have developed dependency on *MZF1*, so that these cancer cells harboring full knockout of *MZF1* are not viable [22], suggesting that they could have developed oncogene addiction towards MZF1. If this is the case, efficient inhibition of MZF1 could result not only in inhibition of invasion but could also be lethal for them.

An increasing number of studies point to a central role for enhanced MZF1 expression and activation in the invasiveness of different solid cancers, making it an attractive therapeutic target. Several probabilities already exist for how its activity could be controlled, and at the same, interesting possibilities still remain to be studied. One potentially useful future approach would be to carry out an in silico screen to identify compounds that interfere with MZF1 DNA binding, dimerization with specific partners or with post-translational modifications that are important for its activation. Especially, its DNA binding domain as well as SUMOylation of lysine 23 and phosphorylation of serine 27 are well characterized. These modification sites are located in domains for which crystal structures are available and would thus already be suitable for such an approach. To use this approach to identify molecules that can prevent MZF1 heterodimerization, more research would be needed to understand which associations are beneficial for cancer. In general, more research is still needed to increase the understanding of the detailed function of MZF1 in cancer, of the cellular cancer-promoting programs it regulates, the cancers where its inhibition would be most beneficial, and how it should be achieved.

Funding: The research concerning MZF1 function in cancer in TK group has been financially supported by Novo Nordisk Foundation (NNF15OC0017324), the Danish Medical Research Council (0602-02386B), the Danish Cancer Society Scientific Committee (KBVU) (R124-A7854-15-S2 and R56-A3108-12-S2), and the Danish National Research Foundation (DNRF125).

Conflicts of Interest: The authors declare no conflict of interest.

References

1. Gandhi, A.K.; Kang, J.; Havens, C.G.; Conklin, T.; Ning, Y.; Wu, L.; Ito, T.; Ando, H.; Waldman, M.F.; Thakurta, A.; et al. Immunomodulatory agents lenalidomide and pomalidomide co-stimulate T cells by inducing degradation of T cell repressors Ikaros and Aiolos via modulation of the E3 ubiquitin ligase complex CRL4(CRBN.). *Br. J. Haematol.* **2014**, *164*, 811–821. [CrossRef]
2. Licht, J.D.; Shortt, J.; Johnstone, R. From anecdote to targeted therapy: The curious case of thalidomide in multiple myeloma. *Cancer Cell* **2014**, *25*, 9–11. [CrossRef]
3. Peterson, F.C.; Hayes, P.L.; Waltner, J.K.; Heisner, A.K.; Jensen, D.R.; Sander, T.L.; Volkman, B.F. Structure of the SCAN domain from the tumor suppressor protein MZF1. *J. Mol. Biol.* **2006**, *363*, 137–147. [CrossRef]
4. Hromas, R.; Collins, S.J.; Hickstein, D.; Raskind, W.; Deaven, L.L.; O'Hara, P.; Hagen, F.S.; Kaushansky, K. A retinoic acid-responsive human zinc finger gene, MZF-1, preferentially expressed in myeloid cells. *J. Biol. Chem.* **1991**, *266*, 14183–14187.
5. Morris, J.F.; Rauscher, F.J., 3rd; Davis, B.; Klemsz, M.; Xu, D.; Tenen, D.; Hromas, R. The myeloid zinc finger gene, MZF-1, regulates the CD34 promoter in vitro. *Blood* **1995**, *86*, 3640–3647.
6. Perrotti, D.; Melotti, P.; Skorski, T.; Casella, I.; Peschle, C.; Calabretta, B. Overexpression of the zinc finger protein MZF1 inhibits hematopoietic development from embryonic stem cells: Correlation with negative regulation of CD34 and c-myb promoter activity. *Mol. Cell Biol.* **1995**, *15*, 6075–6087. [CrossRef]
7. Rafn, B.; Nielsen, C.F.; Andersen, S.H.; Szyniarowski, P.; Corcelle-Termeau, E.; Valo, E.; Fehrenbacher, N.; Olsen, C.J.; Daugaard, M.; Egebjerg, C.; et al. ErbB2-driven breast cancer cell invasion depends on a complex signaling network activating myeloid zinc finger-1-dependent cathepsin B expression. *Mol. Cell* **2012**, *45*, 764–776. [CrossRef] [PubMed]
8. Mudduluru, G.; Vajkoczy, P.; Allgayer, H. Myeloid zinc finger 1 induces migration, invasion, and in vivo metastasis through Axl gene expression in solid cancer. *Mol. Cancer Res.* **2010**, *8*, 159–169. [CrossRef] [PubMed]
9. Hsieh, Y.H.; Wu, T.T.; Tsai, J.H.; Huang, C.Y.; Hsieh, Y.S.; Liu, J.Y. PKCalpha expression regulated by Elk-1 and MZF-1 in human HCC cells. *Biochem. Biophys. Res. Commun.* **2006**, *339*, 217–225. [CrossRef] [PubMed]
10. Hsieh, Y.H.; Wu, T.T.; Huang, C.Y.; Hsieh, Y.S.; Liu, J.Y. Suppression of tumorigenicity of human hepatocellular carcinoma cells by antisense oligonucleotide MZF-1. *Chin. J. Physiol.* **2007**, *50*, 9–15. [PubMed]
11. Yue, C.H.; Chiu, Y.W.; Tung, J.N.; Tzang, B.S.; Shiu, J.J.; Huang, W.H.; Liu, J.Y.; Hwang, J.M. Expression of protein kinase C alpha and the MZF-1 and Elk-1 transcription factors in human breast cancer cells. *Chin. J. Physiol.* **2012**, *55*, 31–36. [CrossRef]
12. Yue, C.H.; Huang, C.Y.; Tsai, J.H.; Hsu, C.W.; Hsieh, Y.H.; Lin, H.; Liu, J.Y. MZF-1/Elk-1 Complex Binds to Protein Kinase Calpha Promoter and Is Involved in Hepatocellular Carcinoma. *PLoS ONE* **2015**, *10*, e0127420. [CrossRef]
13. Tsai, L.H.; Wu, J.Y.; Cheng, Y.W.; Chen, C.Y.; Sheu, G.T.; Wu, T.C.; Lee, H. The MZF1/c-MYC axis mediates lung adenocarcinoma progression caused by wild-type lkb1 loss. *Oncogene* **2015**, *34*, 1641–1649. [CrossRef] [PubMed]
14. Weber, C.E.; Kothari, A.N.; Wai, P.Y.; Li, N.Y.; Driver, J.; Zapf, M.A.; Franzen, C.A.; Gupta, G.N.; Osipo, C.; Zlobin, A.; et al. Osteopontin mediates an MZF1-TGF-beta1-dependent transformation of mesenchymal stem cells into cancer-associated fibroblasts in breast cancer. *Oncogene* **2015**, *34*, 4821–4833. [CrossRef] [PubMed]
15. Deng, Y.; Wang, J.; Wang, G.; Jin, Y.; Luo, X.; Xia, X.; Gong, J.; Hu, J. p55PIK transcriptionally activated by MZF1 promotes colorectal cancer cell proliferation. *Biomed. Res. Int.* **2013**, *2013*, 868131. [CrossRef] [PubMed]
16. Murai, K.; Murakami, H.; Nagata, S. A novel form of the myeloid-specific zinc finger protein (MZF-2). *Genes Cells* **1997**, *2*, 581–591. [CrossRef] [PubMed]
17. Peterson, M.J.; Morris, J.F. Human myeloid zinc finger gene MZF produces multiple transcripts and encodes a SCAN box protein. *Gene* **2000**, *254*, 105–118. [CrossRef]
18. Murai, K.; Murakami, H.; Nagata, S. Myeloid-specific transcriptional activation by murine myeloid zinc-finger protein 2. *Proc. Natl. Acad. Sci. USA* **1998**, *95*, 3461–3466. [CrossRef]
19. Sander, T.L.; Stringer, K.F.; Maki, J.L.; Szauter, P.; Stone, J.R.; Collins, T. The SCAN domain defines a large family of zinc finger transcription factors. *Gene* **2003**, *310*, 29–38. [CrossRef]

20. Edelstein, L.C.; Collins, T. The SCAN domain family of zinc finger transcription factors. *Gene* **2005**, *359*, 1–17. [CrossRef]

21. Ogawa, H.; Murayama, A.; Nagata, S.; Fukunaga, R. Regulation of myeloid zinc finger protein 2A transactivation activity through phosphorylation by mitogen-activated protein kinases. *J. Biol. Chem.* **2003**, *278*, 2921–2927. [CrossRef] [PubMed]

22. Brix, D.M.; Tvingsholm, S.A.; Hansen, M.B.; Clemmensen, K.B.; Ohman, T.; Siino, V.; Lambrughi, M.; Hansen, K.; Puustinen, P.; Gromova, I.; et al. Release of transcriptional repression via ErbB2-induced, SUMO-directed phosphorylation of myeloid zinc finger-1 serine 27 activates lysosome redistribution and invasion. *Oncogene* **2019**, *38*, 3170–3184. [CrossRef] [PubMed]

23. Lee, C.J.; Hsu, L.S.; Yue, C.H.; Lin, H.; Chiu, Y.W.; Lin, Y.Y.; Huang, C.Y.; Hung, M.C.; Liu, J.Y. MZF-1/Elk-1 interaction domain as therapeutic target for protein kinase Calpha-based triple-negative breast cancer cells. *Oncotarget* **2016**, *7*, 59845–59859. [CrossRef] [PubMed]

24. Yue, C.H.; Liu, J.Y.; Chi, C.S.; Hu, C.W.; Tan, K.T.; Huang, F.M.; Pan, Y.R.; Lin, K.I.; Lee, C.J. Myeloid Zinc Finger 1 (MZF1) Maintains the Mesenchymal Phenotype by Down-regulating IGF1R/p38 MAPK/ERalpha Signaling Pathway in High-level MZF1-expressing TNBC cells. *Anticancer Res.* **2019**, *39*, 4149–4164. [CrossRef]

25. Pengue, G.; Calabro, V.; Bartoli, P.C.; Pagliuca, A.; Lania, L. Repression of transcriptional activity at a distance by the evolutionarily conserved KRAB domain present in a subfamily of zinc finger proteins. *Nucleic Acids Res.* **1994**, *22*, 2908–2914. [CrossRef]

26. Sander, T.L.; Haas, A.L.; Peterson, M.J.; Morris, J.F. Identification of a novel SCAN box-related protein that interacts with MZF1B. The leucine-rich SCAN box mediates hetero- and homoprotein associations. *J. Biol. Chem.* **2000**, *275*, 12857–12867. [CrossRef]

27. Williams, A.J.; Khachigian, L.M.; Shows, T.; Collins, T. Isolation and characterization of a novel zinc-finger protein with transcription repressor activity. *J. Biol. Chem.* **1995**, *270*, 22143–22152. [CrossRef]

28. Williams, A.J.; Blacklow, S.C.; Collins, T. The zinc finger-associated SCAN box is a conserved oligomerization domain. *Mol. Cell Biol.* **1999**, *19*, 8526–8535. [CrossRef]

29. Morris, J.F.; Hromas, R.; Rauscher, F.J., 3rd. Characterization of the DNA-binding properties of the myeloid zinc finger protein MZF1: Two independent DNA-binding domains recognize two DNA consensus sequences with a common G-rich core. *Mol. Cell Biol.* **1994**, *14*, 1786–1795. [CrossRef]

30. Robertson, K.A.; Hill, D.P.; Kelley, M.R.; Tritt, R.; Crum, B.; Van Epps, S.; Srour, E.; Rice, S.; Hromas, R. The myeloid zinc finger gene (MZF-1) delays retinoic acid-induced apoptosis and differentiation in myeloid leukemia cells. *Leukemia* **1998**, *12*, 690–698. [CrossRef]

31. Bavisotto, L.; Kaushansky, K.; Lin, N.; Hromas, R. Antisense oligonucleotides from the stage-specific myeloid zinc finger gene MZF-1 inhibit granulopoiesis in vitro. *J. Exp. Med.* **1991**, *174*, 1097–1101. [CrossRef] [PubMed]

32. Gaboli, M.; Kotsi, P.A.; Gurrieri, C.; Cattoretti, G.; Ronchetti, S.; Cordon-Cardo, C.; Broxmeyer, H.E.; Hromas, R.; Pandolfi, P.P. Mzf1 controls cell proliferation and tumorigenesis. *Genes Dev.* **2001**, *15*, 1625–1630. [CrossRef] [PubMed]

33. Hromas, R.; Morris, J.; Cornetta, K.; Berebitsky, D.; Davidson, A.; Sha, M.; Sledge, G.; Rauscher, F., 3rd. Aberrant expression of the myeloid zinc finger gene, MZF-1, is oncogenic. *Cancer Res.* **1995**, *55*, 3610–3614.

34. Brix, D.M.; Clemmensen, K.K.; Kallunki, T. When Good Turns Bad: Regulation of Invasion and Metastasis by ErbB2 Receptor Tyrosine Kinase. *Cells* **2014**, *3*, 53–78. [CrossRef]

35. Mason, S.D.; Joyce, J.A. Proteolytic networks in cancer. *Trends Cell Biol.* **2011**, *21*, 228–237. [CrossRef]

36. Kallunki, T.; Olsen, O.D.; Jaattela, M. Cancer-associated lysosomal changes: Friends or foes? *Oncogene* **2013**, *32*, 1995–2004. [CrossRef]

37. Hamalisto, S.; Jaattela, M. Lysosomes in cancer-living on the edge (of the cell). *Curr. Opin. Cell Biol.* **2016**, *39*, 69–76. [CrossRef]

38. Imamura, T.; Hikita, A.; Inoue, Y. The roles of TGF-beta signaling in carcinogenesis and breast cancer metastasis. *Breast Cancer* **2012**, *19*, 118–124. [CrossRef]

39. Gupta, P.; Srivastava, S.K. HER2 mediated de novo production of TGFbeta leads to SNAIL driven epithelial-to-mesenchymal transition and metastasis of breast cancer. *Mol. Oncol.* **2014**, *8*, 1532–1547. [CrossRef]

40. Pradeep, C.R.; Zeisel, A.; Kostler, W.J.; Lauriola, M.; Jacob-Hirsch, J.; Haibe-Kains, B.; Amariglio, N.; Ben-Chetrit, N.; Emde, A.; Solomonov, I.; et al. Modeling invasive breast cancer: Growth factors propel progression of HER2-positive premalignant lesions. *Oncogene* **2012**, *31*, 3569–3583. [CrossRef]

41. Seton-Rogers, S.E.; Lu, Y.; Hines, L.M.; Koundinya, M.; LaBaer, J.; Muthuswamy, S.K.; Brugge, J.S. Cooperation of the ErbB2 receptor and transforming growth factor beta in induction of migration and invasion in mammary epithelial cells. *Proc. Natl. Acad. Sci. USA* **2004**, *101*, 1257–1262. [CrossRef] [PubMed]

42. Ueda, Y.; Wang, S.; Dumont, N.; Yi, J.Y.; Koh, Y.; Arteaga, C.L. Overexpression of HER2 (erbB2) in human breast epithelial cells unmasks transforming growth factor beta-induced cell motility. *J. Biol. Chem.* **2004**, *279*, 24505–24513. [CrossRef] [PubMed]

43. Le Mee, S.; Fromigue, O.; Marie, P.J. Sp1/Sp3 and the myeloid zinc finger gene MZF1 regulate the human N-cadherin promoter in osteoblasts. *Exp. Cell Res.* **2005**, *302*, 129–142. [CrossRef] [PubMed]

44. Liu, L.; Greger, J.; Shi, H.; Liu, Y.; Greshock, J.; Annan, R.; Halsey, W.; Sathe, G.M.; Martin, A.M.; Gilmer, T.M. Novel mechanism of lapatinib resistance in HER2-positive breast tumor cells: Activation of AXL. *Cancer Res.* **2009**, *69*, 6871–6878. [CrossRef]

45. Asiedu, M.K.; Beauchamp-Perez, F.D.; Ingle, J.N.; Behrens, M.D.; Radisky, D.C.; Knutson, K.L. AXL induces epithelial-to-mesenchymal transition and regulates the function of breast cancer stem cells. *Oncogene* **2014**, *33*, 1316–1324. [CrossRef]

46. Paccez, J.D.; Vogelsang, M.; Parker, M.I.; Zerbini, L.F. The receptor tyrosine kinase Axl in cancer: Biological functions and therapeutic implications. *Int. J. Cancer* **2014**, *134*, 1024–1033. [CrossRef]

47. Arteaga, C.L.; Engelman, J.A. ERBB receptors: From oncogene discovery to basic science to mechanism-based cancer therapeutics. *Cancer Cell* **2014**, *25*, 282–303. [CrossRef]

48. Chen, P.M.; Cheng, Y.W.; Wang, Y.C.; Wu, T.C.; Chen, C.Y.; Lee, H. Up-regulation of FOXM1 by E6 oncoprotein through the MZF1/NKX2-1 axis is required for human papillomavirus-associated tumorigenesis. *Neoplasia* **2014**, *16*, 961–971. [CrossRef]

49. Tsai, S.J.; Hwang, J.M.; Hsieh, S.C.; Ying, T.H.; Hsieh, Y.H. Overexpression of myeloid zinc finger 1 suppresses matrix metalloproteinase-2 expression and reduces invasiveness of SiHa human cervical cancer cells. *Biochem. Biophys. Res. Commun.* **2012**, *425*, 462–467. [CrossRef]

50. Tsai, L.H.; Chen, P.M.; Cheng, Y.W.; Chen, C.Y.; Sheu, G.T.; Wu, T.C.; Lee, H. LKB1 loss by alteration of the NKX2-1/p53 pathway promotes tumor malignancy and predicts poor survival and relapse in lung adenocarcinomas. *Oncogene* **2014**, *33*, 3851–3860. [CrossRef]

51. Eguchi, T.; Prince, T.L.; Tran, M.T.; Sogawa, C.; Lang, B.J.; Calderwood, S.K. MZF1 and SCAND1 Reciprocally Regulate CDC37 Gene Expression in Prostate Cancer. *Cancers* **2019**, *11*, 792. [CrossRef] [PubMed]

52. Chen, Y.; Zhang, Z.; Yang, K.; Du, J.; Xu, Y.; Liu, S. Myeloid zinc-finger 1 (MZF-1) suppresses prostate tumor growth through enforcing ferroportin-conducted iron egress. *Oncogene* **2015**, *34*, 3839–3847. [CrossRef] [PubMed]

53. Liu, X.; Lei, Q.; Yu, Z.; Xu, G.; Tang, H.; Wang, W.; Wang, Z.; Li, G.; Wu, M. MiR-101 reverses the hypomethylation of the LMO3 promoter in glioma cells. *Oncotarget* **2015**, *6*, 7930–7943. [CrossRef]

54. Sonabend, A.M.; Lesniak, M.S. Oligodendrogliomas: Clinical significance of 1p and 19q chromosomal deletions. *Expert. Rev. Neurother.* **2005**, *5*, 25–32. [CrossRef] [PubMed]

55. Zhao, Y.; Min, L.; Xu, C.; Shao, L.; Guo, S.; Cheng, R.; Xing, J.; Zhu, S.; Zhang, S. Construction of disease-specific transcriptional regulatory networks identifies co-activation of four gene in esophageal squamous cell carcinoma. *Oncol. Rep.* **2017**, *38*, 411–417. [CrossRef] [PubMed]

56. Zheng, L.; Jiao, W.; Mei, H.; Song, H.; Li, D.; Xiang, X.; Chen, Y.; Yang, F.; Li, H.; Huang, K.; et al. miRNA-337-3p inhibits gastric cancer progression through repressing myeloid zinc finger 1-facilitated expression of matrix metalloproteinase 14. *Oncotarget* **2016**, *7*, 40314–40328. [CrossRef] [PubMed]

57. Li, G.Q.; He, Q.; Yang, L.; Wang, S.B.; Yu, D.D.; He, Y.Q.; Hu, J.; Pan, Y.M.; Wu, Y. Clinical Significance of Myeloid Zinc Finger 1 Expression in the Progression of Gastric Tumourigenesis. *Cell Physiol. Biochem.* **2017**, *44*, 1242–1250. [CrossRef]

58. Eguchi, T.; Prince, T.; Wegiel, B.; Calderwood, S.K. Role and Regulation of Myeloid Zinc Finger Protein 1 in Cancer. *J. Cell Biochem.* **2015**, *116*, 2146–2154. [CrossRef]

59. Moeenrezakhanlou, A.; Shephard, L.; Lam, L.; Reiner, N.E. Myeloid cell differentiation in response to calcitriol for expression CD11b and CD14 is regulated by myeloid zinc finger-1 protein downstream of phosphatidylinositol 3-kinase. *J. Leukoc. Biol.* **2008**, *84*, 519–528. [CrossRef] [PubMed]

60. Ko, H.; Kim, S.; Yang, K.; Kim, K. Phosphorylation-dependent stabilization of MZF1 upregulates N-cadherin expression during protein kinase CK2-mediated epithelial-mesenchymal transition. *Oncogenesis* **2018**, *7*, 27. [CrossRef] [PubMed]

61. Wu, L.; Han, L.; Zhou, C.; Wei, W.; Chen, X.; Yi, H.; Wu, X.; Bai, X.; Guo, S.; Yu, Y.; et al. TGF-beta1-induced CK17 enhances cancer stem cell-like properties rather than EMT in promoting cervical cancer metastasis via the ERK1/2-MZF1 signaling pathway. *FEBS J.* **2017**, *284*, 3000–3017. [CrossRef] [PubMed]

62. Piszczatowski, R.T.; Rafferty, B.J.; Rozado, A.; Parziale, J.V.; Lents, N.H. Myeloid Zinc Finger 1 (MZF-1) Regulates Expression of the CCN2/CTGF and CCN3/NOV Genes in the Hematopoietic Compartment. *J. Cell Physiol.* **2015**, *230*, 2634–2639. [CrossRef] [PubMed]

63. Piszczatowski, R.T.; Rafferty, B.J.; Rozado, A.; Tobak, S.; Lents, N.H. The glyceraldehyde 3-phosphate dehydrogenase gene (GAPDH) is regulated by myeloid zinc finger 1 (MZF-1) and is induced by calcitriol. *Biochem. Biophys. Res. Commun.* **2014**, *451*, 137–141. [CrossRef] [PubMed]

64. Gupta, P.; Sheikh, T.; Sen, E. SIRT6 regulated nucleosomal occupancy affects Hexokinase 2 expression. *Exp. Cell Res.* **2017**, *357*, 98–106. [CrossRef]

65. Vishwamitra, D.; Curry, C.V.; Alkan, S.; Song, Y.H.; Gallick, G.E.; Kaseb, A.O.; Shi, P.; Amin, H.M. The transcription factors Ik-1 and MZF1 downregulate IGF-IR expression in NPM-ALK(+) T-cell lymphoma. *Mol. Cancer* **2015**, *14*, 53. [CrossRef]

66. Zhang, S.; Shi, W.; Ramsay, E.S.; Bliskovsky, V.; Eiden, A.M.; Connors, D.; Steinsaltz, M.; DuBois, W.; Mock, B.A. The transcription factor MZF1 differentially regulates murine Mtor promoter variants linked to tumor susceptibility. *J. Biol. Chem.* **2019**, *294*, 16756–16764. [CrossRef]

67. Lin, S.; Wang, X.; Pan, Y.; Tian, R.; Lin, B.; Jiang, G.; Chen, K.; He, Y.; Zhang, L.; Zhai, W.; et al. Transcription Factor Myeloid Zinc-Finger 1 Suppresses Human Gastric Carcinogenesis by Interacting with Metallothionein 2A. *Clin. Cancer Res.* **2019**, *25*, 1050–1062. [CrossRef]

68. Doppler, S.A.; Werner, A.; Barz, M.; Lahm, H.; Deutsch, M.A.; Dressen, M.; Schiemann, M.; Voss, B.; Gregoire, S.; Kuppusamy, R.; et al. Myeloid zinc finger 1 (Mzf1) differentially modulates murine cardiogenesis by interacting with an Nkx2.5 cardiac enhancer. *PLoS ONE* **2014**, *9*, e113775. [CrossRef]

69. Chen, H.; Zuo, Q.; Wang, Y.; Song, J.; Yang, H.; Zhang, Y.; Li, B. Inducing goat pluripotent stem cells with four transcription factor mRNAs that activate endogenous promoters. *BMC Biotechnol.* **2017**, *17*, 11. [CrossRef]

70. Jia, N.; Wang, J.; Li, Q.; Tao, X.; Chang, K.; Hua, K.; Yu, Y.; Wong, K.K.; Feng, W. DNA methylation promotes paired box 2 expression via myeloid zinc finger 1 in endometrial cancer. *Oncotarget* **2016**, *7*, 84785–84797. [CrossRef]

71. Lee, Y.K.; Park, U.H.; Kim, E.J.; Hwang, J.T.; Jeong, J.C.; Um, S.J. Tumor antigen PRAME is up-regulated by MZF1 in cooperation with DNA hypomethylation in melanoma cells. *Cancer Lett.* **2017**, *403*, 144–151. [CrossRef] [PubMed]

72. Lee, J.H.; Kim, S.S.; Lee, H.S.; Hong, S.; Rajasekaran, N.; Wang, L.H.; Choi, J.S.; Shin, Y.K. Upregulation of SMAD4 by MZF1 inhibits migration of human gastric cancer cells. *Int. J. Oncol.* **2017**, *50*, 272–282. [CrossRef] [PubMed]

73. Driver, J.; Weber, C.E.; Callaci, J.J.; Kothari, A.N.; Zapf, M.A.; Roper, P.M.; Borys, D.; Franzen, C.A.; Gupta, G.N.; Wai, P.Y.; et al. Alcohol inhibits osteopontin-dependent transforming growth factor-beta1 expression in human mesenchymal stem cells. *J. Biol. Chem.* **2015**, *290*, 9959–9973. [CrossRef] [PubMed]

74. Horinaka, M.; Yoshida, T.; Tomosugi, M.; Yasuda, S.; Sowa, Y.; Sakai, T. Myeloid zinc finger 1 mediates sulindac sulfide-induced upregulation of death receptor 5 of human colon cancer cells. *Sci. Rep.* **2014**, *4*, 6000. [CrossRef]

75. Verma, N.K.; Gadi, A.; Maurizi, G.; Roy, U.B.; Mansukhani, A.; Basilico, C. Myeloid Zinc Finger 1 and GA Binding Protein Co-Operate with Sox2 in Regulating the Expression of Yes-Associated Protein 1 in Cancer Cells. *Stem Cells* **2017**, *35*, 2340–2350. [CrossRef] [PubMed]

76. Hromas, R.; Davis, B.; Rauscher, F.J., 3rd; Klemsz, M.; Tenen, D.; Hoffman, S.; Xu, D.; Morris, J.F. Hematopoietic transcriptional regulation by the myeloid zinc finger gene, MZF-1. *Curr. Top. Microbiol. Immunol.* **1996**, *211*, 159–164. [CrossRef]

77. Tvingsholm, S.A.; Hansen, M.B.; Clemmensen, K.K.B.; Brix, D.M.; Rafn, B.; Frankel, L.B.; Louhimo, R.; Moreira, J.; Hautaniemi, S.; Gromova, I.; et al. Let-7 microRNA controls invasion-promoting lysosomal changes via the oncogenic transcription factor myeloid zinc finger-1. *Oncogenesis* **2018**, *7*, 14. [CrossRef]

78. Nana-Sinkam, S.P.; Croce, C.M. MicroRNAs as therapeutic targets in cancer. *Transl. Res.* **2011**, *157*, 216–225. [CrossRef]

79. Yu, F.; Yao, H.; Zhu, P.; Zhang, X.; Pan, Q.; Gong, C.; Huang, Y.; Hu, X.; Su, F.; Lieberman, J.; et al. let-7 regulates self renewal and tumorigenicity of breast cancer cells. *Cell* **2007**, *131*, 1109–1123. [CrossRef]

80. Noll, L.; Peterson, F.C.; Hayes, P.L.; Volkman, B.F.; Sander, T. Heterodimer formation of the myeloid zinc finger 1 SCAN domain and association with promyelocytic leukemia nuclear bodies. *Leuk. Res.* **2008**, *32*, 1582–1592. [CrossRef]

81. Nygaard, M.; Terkelsen, T.; Vidas Olsen, A.; Sora, V.; Salamanca Viloria, J.; Rizza, F.; Bergstrand-Poulsen, S.; Di Marco, M.; Vistesen, M.; Tiberti, M.; et al. The Mutational Landscape of the Oncogenic MZF1 SCAN Domain in Cancer. *Front. Mol. Biosci.* **2016**, *3*, 78. [CrossRef] [PubMed]

82. Raman, N.; Nayak, A.; Muller, S. The SUMO system: A master organizer of nuclear protein assemblies. *Chromosoma* **2013**, *122*, 475–485. [CrossRef] [PubMed]

83. Hay, R.T. SUMO: A history of modification. *Mol. Cell* **2005**, *18*, 1–12. [CrossRef]

84. Zhong, S.; Salomoni, P.; Pandolfi, P.P. The transcriptional role of PML and the nuclear body. *Nat. Cell Biol.* **2000**, *2*, E85–E90. [CrossRef] [PubMed]

85. Bernardi, R.; Pandolfi, P.P. Structure, dynamics and functions of promyelocytic leukaemia nuclear bodies. *Nat. Rev. Mol. Cell Biol.* **2007**, *8*, 1006–1016. [CrossRef] [PubMed]

86. Lallemand-Breitenbach, V.; de The, H. PML nuclear bodies: From architecture to function. *Curr. Opin. Cell Biol.* **2018**, *52*, 154–161. [CrossRef] [PubMed]

87. Mertins, P.; Mani, D.R.; Ruggles, K.V.; Gillette, M.A.; Clauser, K.R.; Wang, P.; Wang, X.; Qiao, J.W.; Cao, S.; Petralia, F.; et al. Proteogenomics connects somatic mutations to signalling in breast cancer. *Nature* **2016**, *534*, 55–62. [CrossRef] [PubMed]

88. Melton, C.; Reuter, J.A.; Spacek, D.V.; Snyder, M. Recurrent somatic mutations in regulatory regions of human cancer genomes. *Nat. Genet.* **2015**, *47*, 710–716. [CrossRef]

89. Tian, E.; Borset, M.; Sawyer, J.R.; Brede, G.; Vatsveen, T.K.; Hov, H.; Waage, A.; Barlogie, B.; Shaughnessy, J.D., Jr.; Epstein, J.; et al. Allelic mutations in noncoding genomic sequences construct novel transcription factor binding sites that promote gene overexpression. *Genes Chromosomes Cancer* **2015**, *54*, 692–701. [CrossRef]

90. Rafn, B.; Kallunki, T. A way to invade: A story of ErbB2 and lysosomes. *Cell Cycle* **2012**, *11*, 2415–2416. [CrossRef]

91. Brix, D.M.; Rafn, B.; Bundgaard Clemmensen, K.; Andersen, S.H.; Ambartsumian, N.; Jaattela, M.; Kallunki, T. Screening and identification of small molecule inhibitors of ErbB2-induced invasion. *Mol. Oncol.* **2014**, *8*, 1703–1718. [CrossRef] [PubMed]

92. Olson, O.C.; Joyce, J.A. Cysteine cathepsin proteases: Regulators of cancer progression and therapeutic response. *Nat. Rev. Cancer* **2015**, *15*, 712–729. [CrossRef] [PubMed]

93. Bawa-Khalfe, T.; Yeh, E.T. SUMO Losing Balance: SUMO Proteases Disrupt SUMO Homeostasis to Facilitate Cancer Development and Progression. *Genes Cancer* **2010**, *1*, 748–752. [CrossRef] [PubMed]

94. Kim, K.I.; Baek, S.H. SUMOylation code in cancer development and metastasis. *Mol. Cells* **2006**, *22*, 247–253.

95. Yang, Y.; Xia, Z.; Wang, X.; Zhao, X.; Sheng, Z.; Ye, Y.; He, G.; Zhou, L.; Zhu, H.; Xu, N.; et al. Small-Molecule Inhibitors Targeting Protein SUMOylation as Novel Anticancer Compounds. *Mol. Pharmacol.* **2018**, *94*, 885–894. [CrossRef]

96. P21-Activated Kinase 4 (PAK4) Inhibitors as Potential Cancer Therapy. Available online: https://pubs.acs.org/doi/pdf/10.1021/ml500445c (accessed on 13 January 2020).

97. This Is the First Study Using Escalating Doses of PF-03758309, an Oral Compound, in Patients with Advanced Solid Tumors. Available online: https://clinicaltrials.gov/ct2/show/NCT00932126 (accessed on 13 January 2020).

98. PAK4 and NAMPT in Patients with Solid Malignancies or NHL (PANAMA) (PANAMA). Available online: https://clinicaltrials.gov/ct2/show/NCT02702492 (accessed on 13 January 2020).

Article

Targeting the NPL4 Adaptor of p97/VCP Segregase by Disulfiram as an Emerging Cancer Vulnerability Evokes Replication Stress and DNA Damage while Silencing the ATR Pathway

Dusana Majera [1,†], Zdenek Skrott [1,†], Katarina Chroma [1], Joanna Maria Merchut-Maya [2], Martin Mistrik [1,*] and Jiri Bartek [1,2,3,*]

1   Laboratory of Genome Integrity, Institute of Molecular and Translational Medicine, Faculty of Medicine and Dentistry, Palacky University, 77 147 Olomouc, Czech Republic; dusana.majera@upol.cz (D.M.); zdenek.skrott@upol.cz (Z.S.); katarina.chroma@upol.cz (K.C.)
2   Danish Cancer Society Research Center, 2100 Copenhagen, Denmark; jomema@cancer.dk
3   Division of Genome Biology, Department of Medical Biochemistry and Biophysics, Science for Life Laboratory, Karolinska Institute, 171 77 Stockholm, Sweden
*   Correspondence: martin.mistrik@upol.cz (M.M.); jb@cancer.dk (J.B.)
†   These authors contributed equally to this work.

Received: 21 January 2020; Accepted: 17 February 2020; Published: 18 February 2020

Abstract: Research on repurposing the old alcohol-aversion drug disulfiram (DSF) for cancer treatment has identified inhibition of NPL4, an adaptor of the p97/VCP segregase essential for turnover of proteins involved in multiple pathways, as an unsuspected cancer cell vulnerability. While we reported that NPL4 is targeted by the anticancer metabolite of DSF, the bis-diethyldithiocarbamate-copper complex (CuET), the exact, apparently multifaceted mechanism(s) through which the CuET-induced aggregation of NPL4 kills cancer cells remains to be fully elucidated. Given the pronounced sensitivity to CuET in tumor cell lines lacking the genome integrity caretaker proteins BRCA1 and BRCA2, here we investigated the impact of NPL4 targeting by CuET on DNA replication dynamics and DNA damage response pathways in human cancer cell models. Our results show that CuET treatment interferes with DNA replication, slows down replication fork progression and causes accumulation of single-stranded DNA (ssDNA). Such a replication stress (RS) scenario is associated with DNA damage, preferentially in the S phase, and activates the homologous recombination (HR) DNA repair pathway. At the same time, we find that cellular responses to the CuET-triggered RS are seriously impaired due to concomitant malfunction of the ATRIP-ATR-CHK1 signaling pathway that reflects an unorthodox checkpoint silencing mode through ATR (Ataxia telangiectasia and Rad3 related) kinase sequestration within the CuET-evoked NPL4 protein aggregates.

Keywords: targeted cancer therapy; disulfiram; NPL4; replication stress; DNA damage; BRCA1; BRCA2; ATR pathway

---

1. Introduction

Recent advances in understanding of the altered wiring of cancer cell regulatory pathways, and hence vulnerabilities and dependencies of tumor cells have led to discoveries of new molecular targets potentially exploitable in cancer therapy. As the development of a new drug is time-consuming, very expensive, and prone to frequent failure, drug repurposing as a possible alternative approach to cancer treatment is currently undergoing serious consideration [1]. One of the candidate drugs for repurposing in oncology is disulfiram (tetraethylthiuram disulfide, DSF, commercially known as Antabuse), a cheap and well-tolerated generic drug that has been used for decades to treat alcohol dependency. DSF has shown anticancer activity in preclinical models, and multiple clinical trials to treat various types of human malignancies by DSF are currently underway [2]. We have recently published that DSF is metabolized in vivo into the bis-diethyldithiocarbamate-copper complex (CuET), in a process that requires copper ions, and demonstrated that CuET represents the ultimate anticancer metabolite of DSF in-vivo [3]. Furthermore, our nationwide epidemiological study in Denmark yielded results consistent with the emerging anticancer effects of DSF, documenting a lower risk of death from cancer in those cancer patients who were treated by DSF after their cancer diagnosis [3]. Mechanistically, we reported that CuET causes aggregation and thereby immobilization and dysfunction of NPL4, an essential cofactor of the p97/VCP segregase. This otherwise highly mobile protein complex is involved in the regulation of protein turnover upstream of the proteasome, with important roles in a wide range of cellular processes including fundamental pro-survival stress-tolerance pathways [3].

In a follow-up study devoted to target validation and further mechanistic insights into CuET effects [4], we explored the reported exceptional sensitivity to DSF of human cancer cell lines defective in BRCA1 or BRCA2 tumor suppressors, key components of the genome integrity maintenance machinery [4,5]. We found that CuET spontaneously forms from DSF and available copper ions also in cell culture media, and our experiments confirmed NPL4 as the molecular target while excluding the proposed inhibition of aldehyde dehydrogenase (ALDH) [5] and accumulation of toxic acetaldehydes causing DNA-protein and DNA interstrand cross-links [6], as the potential mechanistic explanation for the reported sensitivity of BRCA-defective tumors [4]. In addition to ALDH, we also excluded the proteasome, another previously suggested candidate target of DSF's anticancer effects, as a valid target. Indeed, we showed that the observed 'proteasome-inhibition-like features' triggered by DSF/CuET turned out to be fully attributable to the disabled NPL4 acting upstream of the proteasome [3]. Collectively, these mechanistic studies identified and validated NPL4 as the genuine, and possibly the only or dominant direct molecular target of DSF/CuET responsible for the widely appreciated tumor-inhibitory effects of DSF [3,4]. Indeed, the available evidence in the field now points to CuET-induced aggregation of NPL4 as the key anticancer mechanism of DSF under both in vitro and in vivo conditions, and a promising cancer vulnerability.

Relevant to the present study and the sensitivity of the BRCA-defective cancers to DSF/CuET, we and others previously discovered enhanced replication stress and endogenous DNA damage as a candidate hallmark of cancer [7–10], thereby pioneering the concept of the ATM-Chk2- and ATR-Chk1-mediated DNA damage response (DDR) checkpoints as important cell-intrinsic barriers against oncogene activation and tumor progression [10–12]. Currently, replication stress is recognized to play a prominent role in driving genomic instability and tumorigenesis, while further drug-mediated enhancement of replication stress or inhibition of replication stress-tolerance pathways such as ATR-Chk1 signaling may provide additional targetable vulnerabilities of cancer [13,14]. The main mechanistic consequence of replication stress is the accumulation of ssDNA and stalling of replication forks [15]. The ssDNA stretches become rapidly coated with replication protein A (RPA), thereby facilitating activation of the ATR-Chk1 signaling module and subsequent phosphorylation of hundreds of cellular proteins as substrates of ATR and Chk1 kinases [15,16]. These phosphorylation cascades also involve BRCA1 and BRCA2 and help to stabilize the stalled forks, thereby preventing fork collapse, while in parallel limiting the cellular entry into mitosis by activation of the S-M checkpoint [17]. Under inhibition or genetic deficiency of ATR, stalled replication forks tend to collapse, leading to a generation of DNA double-strand breaks (DSBs), which, if unrepaired or misrepaired, can cause chromosomal instability, severe pathologies or cell death [14,18].

With the above-mentioned knowledge as the starting point, here we examined potential mechanistic links between cancer-associated replication stress, DNA damage checkpoint signaling and the functional impact of DSF/CuET treatment on DNA replication and genome integrity maintenance, searching for possible explanations of the overall sensitivity of tumor cells, and the observed preferential sensitivity of cancer cells lacking BRCA1 and BRCA2, to treatment with DSF/CuET.

2. Materials and Methods

2.1. Cell Culture

Human non-small cell lung carcinoma H1299 cells expressing a doxycycline (DOX)-inducible BRCA1 and BRCA2 shRNAs, U2OS, MDA-MB-231, MDA-MB-436, U2OS cells expressing NPL4-GFP, U2OS expressing DOX-inducible MUT-NPL4-GFP [3] and U2OS cells expressing ATR-GFP [19] were cultured and maintained in DMEM (Dulbecco's Modified Eagle Medium, Lonza, Basel, Switzerland), supplemented with 10% fetal bovine serum (Thermo Fisher Scientific, Waltham, MA, USA) and 1% penicillin/streptomycin (Sigma-Aldrich, St. Louis, MO, USA). CAPAN-1 cells were grown in DMEM medium, supplemented with 20% fetal bovine serum and 1% penicillin/streptomycin. H1299 expressing a DOX-inducible BRCA1 and BRCA2 shRNA were kindly provided [5]. For efficient BRCA1 and BRCA2 knockdown cells were cultivated in the presence of 2 µg/mL DOX for at least three days.

2.2. Immunoblotting

Equal amounts of cell lysates were separated by SDS-PAGE on hand casted gels and then transferred onto the nitrocellulose membrane. The membrane was blocked in Tris-buffered saline containing 5% milk and 0.1% Tween 20 for 1 h at room temperature and then incubated 1 h at room temperature with primary antibodies, followed by detection with secondary antibodies. Secondary antibodies were visualized by ELC detection reagent (Thermo Fisher Scientific, Waltham, MA, USA).

2.3. Immunofluorescence

Cells were seeded on plastic inserts in 12-well dishes. The next day, cells were treated with compounds at indicated concentrations and subsequently either pre-extracted (0.1% Triton X 100 in Phosphate-Buffered Saline(PBS) for 2 min or fixed with formaldehyde for 15 min at room temperature, washed with PBS and permeabilized with 0.5% Triton X-100 in PBS for 5 min. After PBS washes, the cells on the plastic inserts were immunostained with primary antibody for 1 h at room temperature, followed by PBS washes and staining with fluorescently-conjugated secondary antibody for 60 min at room temperature. Nuclei were visualized by 4′,6-diamidino-2-phenylindole (DAPI, 1 µg/mL) staining at room temperature for 2 min. For NPL4 staining, the cells were pre-extracted (0.1% Triton X 100 in PBS, for 2 min) and fixed with −20 °C methanol for 15 min at room temperature, washed with PBS and permeabilized with 0.5% Triton X-100 in PBS for 5 min. After PBS washes, the cells on the plastic inserts were immunostained with primary antibody for 120 min at room temperature, followed by PBS washes and staining with fluorescently-conjugated secondary antibody for 60 min at room temperature. Dried plastic inserts with cells were mounted using Vectashield mounting medium (Vector Laboratories, Burlingame, CA, USA), and images were acquired using the Zeiss Axioimager Z.1 platform.

2.4. Ethynyldeoxyuridine (EdU) and Bromodeoxyuridine (BrdU) Incorporation and Detection

To detect active DNA replication, cells were incubated with 10 µM EdU (Life Technologies, Carlsbad, CA, USA) for 30 min, fixed, permeabilized and stained using Click-iT reaction (100 mM Tris pH 8.5, 1 mM copper sulfate, 100 µM ascorbic acid, 1 µM azide Alexa fluor 488 (Life Technologies, Carlsbad, CA, USA) for 30 min at room temperature. To detect ssDNA, cells were incubated with 10 µM BrdU (Sigma) for 24 h, then BrdU was washed out, and cells were incubated with the tested compounds as indicated. After pre-extraction and fixation in buffered formol, the incorporated BrdU was detected by an anti-BrdU antibody (BD Biosciences, San Jose, CA, USA) without denaturation.

2.5. Image Quantification

Images were acquired using the Olympus IX81 fluorescence microscope and ScanR Acquisition software. The scans were quantified in automated image and data analysis software ScanR Analysis.

The data was further analyzed in the STATISTICA 13 software tool (Dell Software, Round Rock, TX, USA).

2.6. DNA Combing

H1299 cells were treated with 125 nM CuET for 5 h and subsequently pulsed with 5-Iodo-2'-deoxyuridine (IdU, 20 µM) for 30 min, washed and pulsed with 5-Chloro-2'-deoxyuridine (CIdU, 200 µM) for additional 30 min. DNA replication was stopped by ice-cold PBS. Cells were collected and embedded in 0,5% insert agarose plugs. The plugs were incubated for 32 h in buffer containing proteinase K at 50 °C. Plugs were then washed with TE buffer and melted at 68 °C. The obtained solution was further digested overnight with Agarase I at 42 °C. The next day, the concentration of DNA was measured on nanodrop and combed on silanized cover glasses (Matsunami, Japan) with a speed of 0,3 mm/s. The cover glasses with combed DNA were baked at 60 °C, dehydrated with 70%, 90%, and 100% ethanol series for 3 min each. DNA was denatured at 75 °C in 2xSSC, 50% formamide for 2 min. Next, the cover glasses were dehydrated with a 70%, 90% and 100% ice-cold ethanol series for 5 min each, air dried, blocked using 1% BSA in PBS-Tween for 1 h at 37 °C and subsequently incubated with primary antibodies, mouse anti-BrdU for IdU detection (1:5) and rat anti-BrdU for CIdU detection (1:25) for 1 h at 37 °C. After several washes with PBS-Tween, cover glasses were incubated with secondary antibodies goat anti-mouse A488 (1:100) and goat anti-rat A549 (1:100) for 30 min at 37 °C. After several washes with PBS-Tween, cover glasses were air-dried, mounted, and images of DNA fibers were acquired using the Zeiss Axioimager Z.1 platform.

2.7. Estimation of DNA Replication Origin Density

After the treatment by tested compounds, cells were pulsed with EdU (10 µM) for 20 min, then harvested and resuspended in cold PBS (1 million of cells per 1 mL). 2 µL of cell suspension was applied on glass slides (Superfrost Plus, Thermo Fisher) and allowed to partially evaporate for 5 min, then mixed with a lysis buffer (50 mM EDTA and 0.5% SDS in 200 mM Tris-HCl, pH 7.5) and incubated for 2 min. Slides were tilted to 15° to allow the spreading of fibers. After drying, the samples were fixed in methanol/acetic acid solution for 15 min and thoroughly washed. EdU was detected by click reaction using Alexafluor 488 azide. The signal was further enhanced by anti-Alexa fluor 488 antibody (A-11094, Thermo Fisher) and secondary antibody. DNA was visualized by YOYO-1 (Molecular Probes) staining (1 µM for 20 min). Fiber images were acquired using the Zeiss Axioimager Z.1 platform, and the number of DNA replication origins was calculated on single well-stretched DNA fibers. A conversion factor of 2.59 kb/µm was used in calculations [20].

2.8. Cell Fractionation for Triton X Insoluble Pellets

Cells were treated as indicated, washed in cold PBS and lysed in lysis buffer (50 mM Tris-HCl, pH 7.5, 150 mM NaCl, 2 mM MgCl2, 10% glycerol, 0.5% Triton-X100, protease inhibitor cocktail by Roche) for 2 min, under gentle agitation at 4 °C. Then, cells were scraped to Eppendorf tubes and kept for another 10 min on ice with vortex steps. Next, the lysate was centrifuged at 20,000× *g* for 10 min at 4 °C. Insoluble fraction and supernatant were re-suspended in Laemmli Sample Buffer (1X final concentration; 10% glycerol, 60 mM Tris-HCl, pH 6.8, 2% SDS, 0.01% bromophenol blue, 50 mM dithiothreitol).

2.9. Laser Micro-Irradiation

U2OS cells stably expressing GFP-ATR were seeded into 24-well plates with a glass-bottom (Cellvis) 24 h before laser micro-irradiation in a density of 6 × 105 cells/mL. After seeding the cells into the 24 well plates, the specimen was first placed on an equilibrated bench for 20 min at room temperature (RT) to ensure equal cell distribution and then placed into an incubator. CuET was added to cells 5 h before micro-irradiation in final concentrations of 250 nM and 500 nM. Twenty minutes before laser micro-irradiation, cells were pre-sensitized towards UV-A wavelength by 20 µM 8-Methoxypsoralen

(8-MOP) and placed inside Zeiss Axioimager Z.1 inverted microscope combined with the LSM 780 confocal module. Laser micro-irradiation was performed at 37 °C via X 40 water immersion objective (Zeiss C-Apo 403/1.2WDICIII), using a 355 nm 65 mW laser set on 100% power to induce the DNA damage. The total laser dose that can be further manipulated by the number of irradiation cycles was empirically set to two irradiation cycles. Subsequent immunofluorescence detection and quantitative analysis of the striation pattern in photo-manipulated samples were essentially performed as described previously [21].

2.10. Antibodies and Chemicals

The following antibodies were used for immunoblotting: BRCA1 antibody (Santa Cruz Biotechnology, Dallas, TX, USA, D-9), rabbit polyclonal antibody against BRCA2 (Bethyl, Montgomery, TX, USA, A300-005A) antibody and mouse monoclonal antibody against β-actin (Santa Cruz Biotechnology, C4), lamin B (Santa Cruz Biotechnology, sc-6217), α-Tubulin (Santa Cruz Biotechnology, sc-5286), anti-ubiquitin lys48-specific (Merck Millipore, Burlington, MA, USA, clone Apu2) Chk1 (Santa Cruz, Biotechnology, sc-8404), phospho-Chk1 S317 (Cell Signalling, Danvers, MA, USA, 2344), phospho-Chk1 S345 (Cell Signalling, 2348), RPA (Abcam, ab16855, Cambridge, UK), phospho-RPA S33 (Bethyl, A300-246A), ATR (Santa Cruz Biotechnology, N-19). For immunofluorescence were used the following antibodies: γH2AX (Merck Millipore, 05-636), cyclin A (Santa Cruz Biotechnology, H-3, Santa Cruz Biotechnology, sc-239), RPA (Abcam, ab16855), Rad51 (Abcam, ab63801), NPL4 (Santa Cruz Biotechnology, D-1), p97 (Abcam, ab11433), ATR (Santa Cruz Biotechnology, N-19). For DNA combing assay following antibodies were used: anti-BrdU (BD Biosciences, Franklin Lakes, NJ, USA, BD 347580) and rat anti-BrdU (Abcam ab6323).

Chemicals used in this study were as follows: CuET (bis-diethyldithiocarbamate-copper complex, TCI chemicals), disulfiram (Sigma, St. Louis, MO, USA), bortezomib (Velcade, Janssen-Cilag International N.V.), bathocuproinedisulfonic acid (Sigma, St. Louis, MO, USA), CB-5083 (Selleckchem, Houston, TX, USA), hydroxyurea (Sigma, St. Louis, MO, USA), AZD6738 (AstraZeneca, London, UK).

2.11. Field Inversion Gel Electrophoresis (FIGE)

Treated cells, as indicated in the main text, were trypsinized and melted into 1.0% InCert-Agarose inserts. Subsequently, agarose inserts were digested in a mixture of 10 mM Tris-HCl pH 7.5, 50 mM EDTA, 1% N-laurylsarcosyl, and proteinase K (2 mg/mL) at 50 °C for 24 hr and washed five times in Tris-EDTA (TE buffer, 10 mM Tris-HCl pH 8.0, 100 mM EDTA). The inserts were loaded onto a separation gel 1.0% agarose mixed with GelRed® solution (10,000x). Run conditions for the DNA fragments separation were: 110 V, 7.5 V/cm, 16 h, forward pulse 11 s, reverse pulse 5 s in 1X Tris-acetic acid-EDTA (TAE buffer 40 mM Tris, 20 mM acetic acid, 1 mM EDTA).

2.12. Alkaline Comet Assay

The alkaline comet assay was performed essentially as described in [22]. Briefly, CAPAN-1 and MDA-MB-436 cells were treated with 250 nM CuET or 2 mM hydroxyurea (HU) for 5 h, collected and resuspended in PBS (7500 cells/μL). Cells (75000) were then mixed with 37 °C low melting point agarose (Lonza, Basel, Switzerland), spotted on the normal melting point agarose (Invitrogen, Waltham, MA, USA) pre-coated slides and left to sit for 10 min at 4 °C. Slides were then immersed in the cold alkaline lysis buffer for 2 h at 4 °C. Slides were washed three times with the cold alkaline electrophoresis buffer and electrophoresis was performed (25 min, 4 °C, 0.6 V/cm). Slides were then washed with cold PBS and ddH$_2$O, dehydrated in cold graded ethanol, air-dried and stored at room temperature. For staining, slides were rehydrated with ddH$_2$O, stained with Sybr Gold (1:4000 in TE buffer; Thermo Fisher Scientific, Waltham, MA, USA), washed with PBS and mounted with Mowiol (Sigma-Aldrich, St. Louis, MO, USA). Images were acquired using a fluorescent microscope (Carl Zeiss, Oberkochen, Germany), a 20x air immersion objective (Carl Zeiss, Oberkochen, Germany) and Comet Assay IV software (Perceptive Instruments, Haverhill, UK). Presented results are from the technical duplicate.

Alkaline lysis buffer: 1.2 M NaCl, 100 mM Na_2EDTA, 0.1% sodium lauryl sarcosinate, 0.26 M NaOH (pH > 13, 4 °C, prepared fresh); alkaline electrophoresis buffer: 0.03 M NaOH, 2 mM Na_2EDTA (pH 12.3, 4 °C).

3. Results

3.1. CuET Causes DNA Damage Preferentially Detectable in S/G2-Phase Cells

To initiate our current study, we first wished to assess the impact of CuET on DNA damage in cultured human cancer cells, including isogenic cell pairs with experimentally altered components of the DDR machinery. To this end, we employed the established H1299 lung cancer model allowing for DOX-inducible shRNA-mediated depletion of BRCA1 or BRCA2 [4,5]. Indeed, treatment of these cell lines with CuET resulted in an increased formation of γH2AX foci as well as enhanced overall γH2AX signal intensity, established surrogate markers for chromatin response to DSBs and overall DNA damage signaling by the upstream DDR kinases, respectively (Figure 1A,B; Supplementary Figure S1A,B). Notably, the CuET-evoked increase of γH2AX was more pronounced in the BRCA1- and BRCA2-depleted cells compared with their BRCA-proficient counterpart H1299 cells (unexposed to DOX) (Figure 1A,B; Supplementary Figure S1B). To clarify whether such DNA damage could also be caused by DSF itself, we treated the BRCA2-depleted H1299 cells with DSF in cell culture settings where the cells were first pre-treated by the copper chelator bathocuproinedisulfonic acid (BCDS), a manipulation that we previously reported prevents the otherwise spontaneous and rapid formation of CuET from DSF and copper in cell culture media [4]. As expected, when used alone, DSF caused a similar increase in DNA damage formation as CuET, however when DSF was combined with the copper chelator BCDS pre-treatment step, the γH2AX-inducing effect of DSF was completely abrogated (Figure 1E). These results showed that the DNA damage observed after the treatment with DSF depends on the copper-dependent spontaneous formation of CuET in the culture media, thereby establishing that analogous to the anticancer effects, the active DNA damage-inducing compound is the CuET metabolite, rather than DSF itself.

Next, we pursued our observation that the increase of γH2AX was apparent only in a subset of cells in a given exponentially growing cell population, suggesting that the DNA damage could be cell cycle-dependent. To examine this possibility, we again treated the above mentioned H1299 cells with CuET, yet in the subsequent immunofluorescence analysis, we double stained the cells for γH2AX and cyclin A, an approach commonly used to distinguish cells in G1 phase (cyclin A negative) from those in S/G2 phases (cyclin A positive). Notably, the CuET-induced γH2AX was preferentially seen in cyclin A positive cells, and this cell-cycle effect was even more pronounced in the BRCA1- and BRCA2-depleted cells (Figure 1C,D, Supplementary Figure S1C). The preference of elevated γH2AX intensity in cyclin A positive cells was also confirmed in additional human cancer cell lines (Supplementary Figure S1D,E), thereby excluding a possibility that such genotoxic effects of CuET could be restricted to the H1299 cell model.

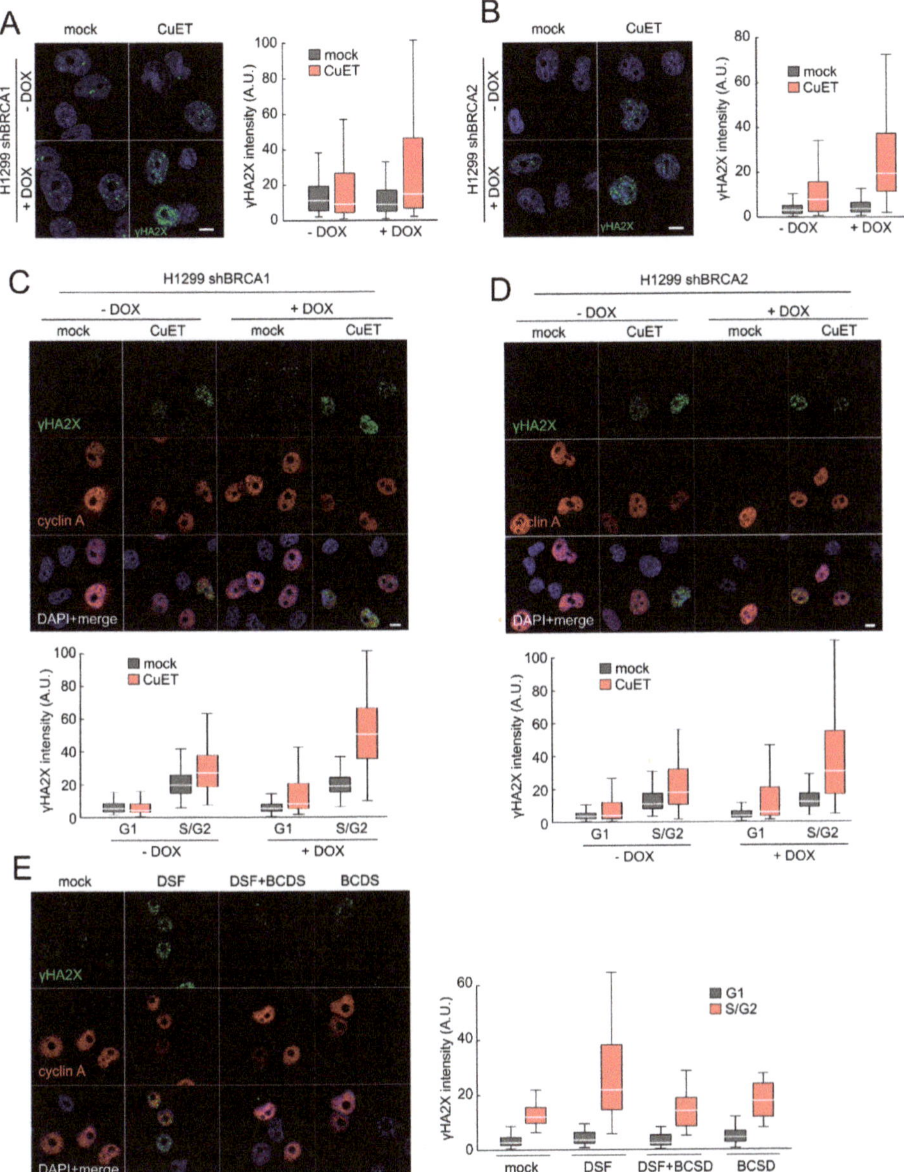

Figure 1. Disulfiram's metabolite bis-diethyldithiocarbamate-copper complex (CuET) causes DNA damage preferentially in S/G2 cells deficient for BRCA1 or BRCA2 proteins. H1299 cells expressing (doxycycline-) DOX-inducible shBRCA1 (**A**) or shBRCA2 (**B**) were cultivated for at least three days in DOX-containing media and then treated with CuET (250 nM) for 5 h, and γH2AX intensity was analyzed by quantitative microscopy. (**C**) H1229 shBRCA1 cells or (**D**) H1299 shBRCA2 cells were treated as in (A) a γH2AX intensity was quantified with respect to cyclin A positivity defining S/G2 phase. (**E**) H1229 shBRCA2 cells pre-incubated with DOX were treated with disulfram (DSF) (500 nM), bathocuproinedisulfonic acid (BCDS) (50 μM), or their combination for 5 h and γH2AX intensity was quantified. Box plot represents 25–75 quartiles, median, and whiskers non-outlier range. Scale bars = 10 μm.

Overall, we conclude from these results that the DNA damage-inducing effects of DSF are attributable to its CuET metabolite, include both elevated γH2AX foci formation and overall γH2AX signal intensity, and occur preferentially in cells traversing S/G2 phases.

3.2. CuET Treatment Decreases DNA Replication Fork Velocity and Increases the Number of Active Replication Origins

Since the CuET-induced DNA damage was more apparent in S/G2 cells, we argued that CuET might preferentially interfere with DNA replication. To examine this possibility, we pre-treated H1299 cells with CuET, followed by a pulse-treatment with the thymine analog EdU that becomes incorporated into newly synthesized DNA, allowing visualization of the rate of ongoing DNA replication using fluorescence readouts. Using this approach, we could indeed confirm severe impairment of DNA replication in CuET treated cells, manifested as a decreased EdU signal in H1299 cells (Figure 2A) and also other cell lines, such as human breast cancer MDA-MB-231 and osteosarcoma U2OS cells (Supplementary Figure S2A,B). DNA replication can be halted by the presence of DNA damage [23] and vice versa; replication interference can be the source of DNA damage [13,14]. To address what is the cause and consequence in this scenario, we performed a kinetic study showing that the decrease of EdU incorporation is an early event, preceding the γH2AX foci formation (Figure 2B). This result indicated that the observed DNA damage most likely results from the CuET-induced impairment of DNA replication. To gain more detailed insights into the observed replication interference phenomenon, we employed DNA combing as an assay enabling us to directly assess the effect of CuET on DNA replication fork velocity. H1299 cells were first pre-treated with a rather low concentration of CuET and then pulsed with IdU and CIdU thymine analogs to detect actively replicating DNA, the length of which can be evaluated by fluorescence microscopy-based measurements [24]. Our analysis of the obtained DNA fibers revealed a robust reduction of DNA replication fork velocity after CuET treatment (Figure 2C). Since such decreased DNA replication fork speed is known to trigger firing of dormant replication origins, we next tested the density of active origins using an established DNA fiber assay [22,25]. We quantified the number of origins per 1 Mb of DNA. Indeed, CuET treatment increased the number of active origins compared to untreated cells, similarly to treatment with the ATR kinase inhibitor AZD6738 (Figure 2D), a known activator of latent replication origin firing used here as a positive control [26].

We interpret these results as documenting a previously unsuspected negative impact of CuET on DNA replication, slowing down the fork velocity and concurrently leading to the firing of more dormant origins.

3.3. CuET-Induced Replication Stress Leads to DNA Damage that Triggers Homologous Recombination Repair Pathway

As replication stress is associated with accumulation of ssDNA stretches detectable by RPA32 protein foci or by staining for DNA-incorporated BrdU under non-denaturating conditions [18,27,28], we next assessed these parameters in human cells treated with CuET. Consistent with the CuET-impaired replication forks (see above), we found enhanced RPA32 foci in several cancer cell lines treated with CuET (Figure 3A,B) and also detected incorporated BrdU under non-denaturing conditions (Figure 3C,D). These data suggest that in CuET-treated cells, DNA helicase becomes uncoupled from DNA polymerases, generating stretches of ssDNA in a manner broadly analogous with effects of the replication stress-inducing drugs such as hydroxyurea or aphidicolin [28]. The RPA-coated ssDNA is known to recruit and activate the ATRIP-ATR-CHK1 signaling pathway [29] to stabilize the stalled replication structures, thereby avoiding fork collapse and formation of DSBs [30]. Importantly, these DNA lesions typically require repair by the homologous recombination (HR) repair pathway that encompasses, among other factors, also BRCA1 and BRCA2, the latter being critical for loading of the Rad51 HR repair protein [31–33]. To test whether Rad51 is involved in the repair process of lesions caused by the CuET treatment, we stained the cells for Rad51 and searched for the typical

DNA-associated Rad51 foci that form within the DSB-flanking chromatin regions under ongoing DNA repair. Indeed, in multiple tested cell lines, the CuET treatment increased the number of Rad51 foci (Figure 3E,F) except for the BRCA2-depleted cells, which are principally incapable of loading Rad51 both after CuET treatment and gamma-irradiation (here used as a positive control) (Figure 3G). The presence of DNA breaks in CuET treated cells was confirmed also by direct physical methods including Field Inversion Gel Electrophoresis (FIGE, detecting largely DSBs) (Figure 3H, Supplementary Figure S3D) and comet assay (Supplementary 3A,B,C, detecting a mixture of single-stranded and double stranded DNA breaks) in BRCA-deficient human cell lines derived from carcinomas of the breast (MDA-MB-436), lung (the H1299 series) and pancreas (CAPAN1), the latter reported by us previously as very sensitive to CuET treatment [3].

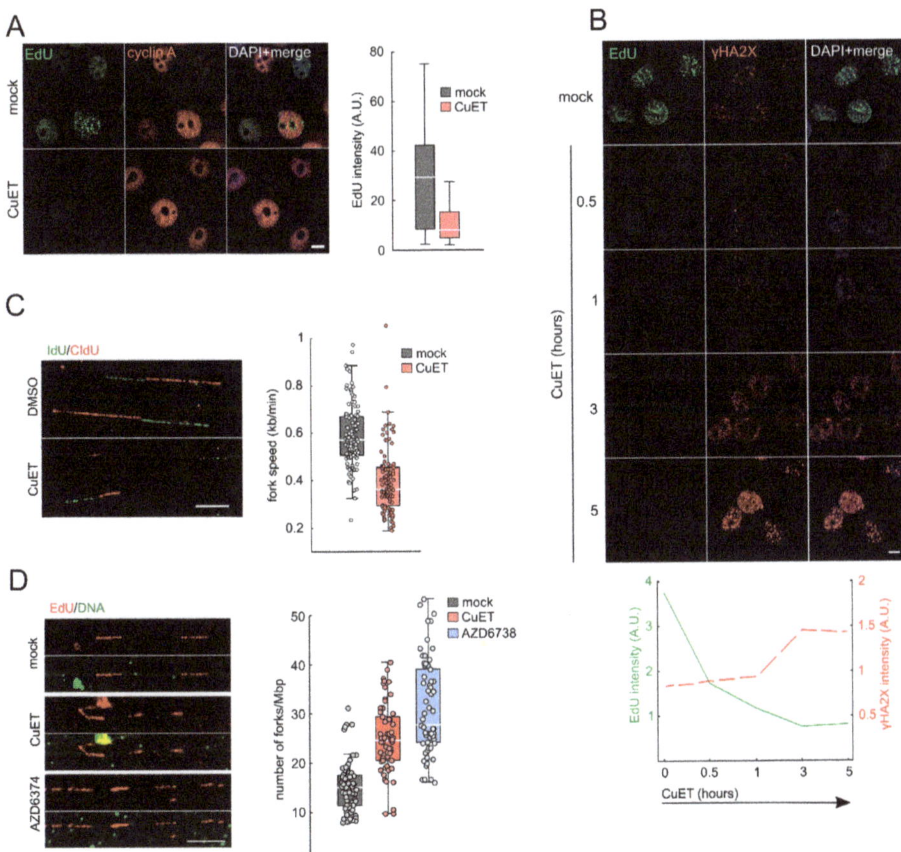

Figure 2. CuET impairs DNA replication. (**A**) H1299 cells were treated with CuET (250 nM) for 3 h, and ethynyldeoxyuridine (EdU) intensity was analyzed in cells positive for cyclin A. (**B**) H1299 cells were treated with CuET (250 nM) for different time points, and EdU and γH2AX intensities were quantified. (**C**) H1299 cells were treated with CuET (125 nM) for 5 h, then pulse-labeled with5-Iodo-2′-deoxyuridine (IdU) and 5-Chloro-2′-deoxyuridine (CIdU) and processed for DNA combing. (**D**) H1299 cells were treated with CuET (250 nM) or AZD6372 (10 μM) for 3 h and then pulsed with EdU and processed for DNA fiber assay. Box plot represents 25–75 quartiles, median, and whiskers non-outlier range. Scale bars = 10 μm.

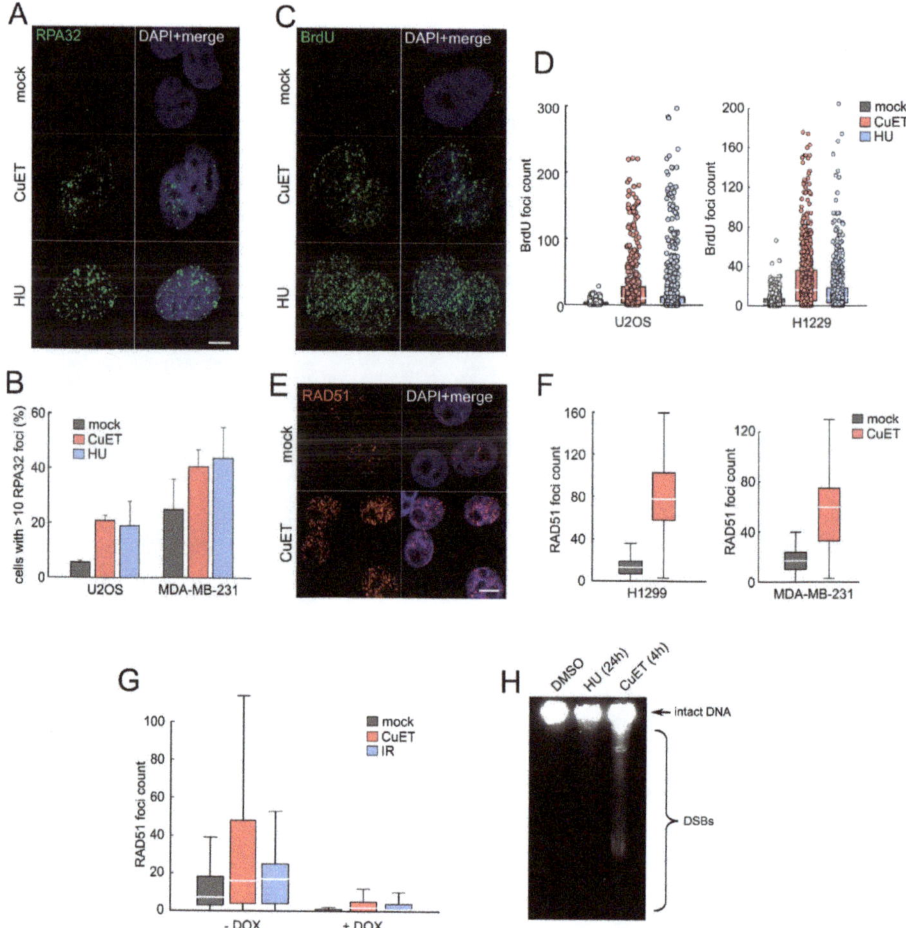

Figure 3. CuET induces replication stress. (**A**) RPA32 foci detection in pre-extracted U2OS cells treated with CuET (250 nM) or hydroxyurea (HU, 2 mM) for 5 h. (**B**) Quantification of cells with more than 10 RPA32 foci treated as in (A) (mean, SD from three independent experiments). (**C**) Formation of single-stranded DNA (ssDNA) visualized by BrdU detected under non-denaturating conditions in U2OS cells treated by CuET (250 nM) and HU (2 mM) for 5 h. (**D**) Quantification of bromodeoxyuridine (BrdU) foci in U2OS and H1299 cells treated as in C. (**E**) Detection of RAD51 foci in pre-extracted H1299 cells treated by CuET (250 nM) for 5 h. (**F**) Quantification of RAD51 foci in cyclin A positive H1299 and MDA-MB-231 cells treated by CuET (250 nM) for 5 h. (**G**) Quantification of Rad51 foci in BRCA2 proficient and deficient H1299 cells after 5-h treatment with 250 nM CuET or 4 Gray (Gy) irradiation. (**H**) FIGE analysis of DSBs in H1299 cells exposed to CuET or HU. Box plot represents 25–75 quartiles, median, and whiskers non-outlier range. Scale bars = 10 μm.

Collectively, these results are consistent with CuET inducing replication stress-associated DNA damage that requires HR repair, including Rad51, a process that is defective in the absence of BRCA1 and BRCA2. Consequently, such DNA damage cannot be properly processed in cells lacking the BRCA factors, which explains the higher amount of DNA damage that contributes to the preferential sensitivity of BRCA-deficient cells to DSF [5] and CuET [4].

3.4. The ATR Signaling Pathway is Compromised in CuET-Treated Cells

In the context of the results obtained so far, we were intrigued by the fact that CuET treatment resulted in DNA breaks relatively quickly within 3–4 h. However, stalled or slowed replication forks should be rather stable for many more hours before turning into DSBs as reported in the U2OS cell line after HU treatment [31] (see also Supplementary Figure S3D). As the prominent role in the stabilization and protection of the stalled forks reflects the function of the RPA-ATRIP-ATR-Chk1 signaling pathway [29,30], we performed immunoblot analysis of extracts from various cell lines treated with CuET, to assess the status of the ATR signaling. In contrast to HU treatment which was used as a positive control, the RPA-ATRIP-ATR-Chk1 signaling pathway was not activated in response to CuET, as manifested by the absence of the ATR-mediated phosphorylations of the effector kinase Chk1: Chk1 S317 and Chk1 S345 (Figure 4A). This result was rather surprising as ssDNA is obviously present in the CuET treated cells (see Figure 3A–D) and also coated by the upstream factor RPA, thereby setting the initial stage for ATR activation and phosphorylation of ATR targets including Chk1. To further investigate whether CuET indeed impairs the RPA-ATRIP-ATR-CHk1 signaling, we treated cells with CuET in the presence of HU. While treatment with HU alone efficiently induced phosphorylation of Chk1 S317 and Ckh1 S345, as expected, the combined treatment with CuET and HU revealed the lack of such Chk1 phosphorylations again, indicating that CuET exerted a dominant effect in suppressing the ATR pathway activity (Figure 4B). These unexpected results were then corroborated by the lack of Serine 33 phosphorylation of yet another ATR substrate, the replication stress marker RPA32, an event seen in the HU-treated control but not in CuET- or combined CuET- and HU-treated cells (Figure 4C).

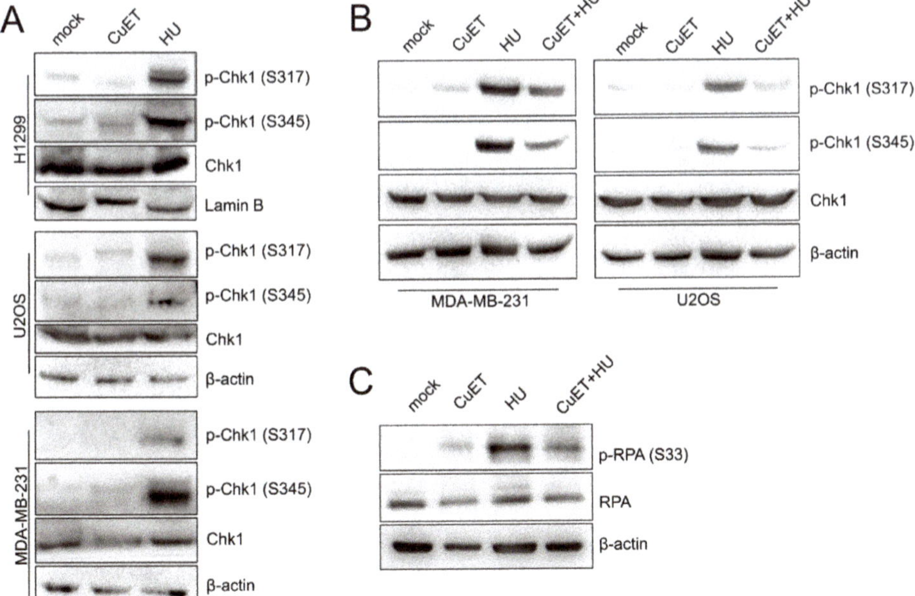

Figure 4. ATR signaling is compromised by CuET. (**A**) Western blot analysis of phosphorylated forms of Chk1 in various cell lines treated by CuET (250 nM) or HU (2 mM) for 5 h. (**B**) WB analysis of Chk1 phosphorylation in U2OS and MDA-MB-231 cells pre-treated by dimethylsulfoxide (DMSO, mock) or CuET (250 nM) for 2 h and then exposed to HU (2 mM) for additional 3 h. (**C**) WB detection of RPA32 phosphorylation in U2OS cells treated as in B.

Together these results suggest that CuET treatment not only causes replication stress by slowing down and/or stalling replication fork progression but at the same time, it also interferes with the activation of the RPA-ATRIP-ATR-Chk1 signaling cascade that is critical for proper cellular responses to replication stress.

3.5. The ATR Signaling Pathway is Compromised in CuET-Treated Cells

The fact that ATR kinase signaling was suppressed after CuET treatment despite ongoing robust replication stress that also included the formation of ssDNA inspired us to focus directly on the ATR protein and its behavior in response to CuET. As a general readout for analysis of ATR abundance, subcellular localization and function we employed the reporter U2OS cells expressing GFP-labeled ATR (U2OS ATR-GFP) that allowed us to directly assess also recruitment of the ATR protein to acutely inflicted DNA lesions induced by laser microirradiation of psoralen pre-sensitized cell nuclei [19,21]. While in control mock-treated cells, the ATR-GFP protein rapidly formed the expected pattern of fluorescent stripes matching the laser tracks, such recruitment of ATR was markedly impaired after CuET exposure (Figure 5A and Supplementary Figure S4). Moreover, we noticed that in CuET-treated cells without any laser exposure, the otherwise pan-nuclear and generally diffuse ATR-GFP fluorescence signal became altered, forming a pattern that was reminiscent of protein aggregates previously reported by us for the NPL4 protein after CuET treatment [3] (Figure 5B). Indeed, further immunofluorescence analysis confirmed co-localization of ATR-GFP with the NPL4/p97 aggregates formed after CuET treatment (Figure 5C) and general immobilization of the ATR protein was then confirmed by two additional complementary approaches: quantitative microscopy on cultured and pre-extracted U2OS ATR-GFP cells (Figure 5D), and immunoblotting identification of protein translocation from the mobile into the immobile (pre-extraction resistant) protein fraction. Notably, unlike the aggregated immobile ATR protein, the downstream component of the ATR cascade, namely the effector kinase Chk1 was not immobilized after CuET treatment (Figure 5E). To distinguish whether or not ATR immobilization was caused by CuET independently of CuET's key reported target, the NPL4 protein [3], we employed our U2OS cell model conditionally expressing a mutated form of NPL4-GFP, a protein which tends to aggregate spontaneously when expressed in cells due to the point mutation in the putative zinc-finger domain involved in the interaction with CuET [3]. We have already shown that such spontaneous aggregation of the NPL4-MUT protein mimics multiple aspects of CuET treatment including association and immobilization of various cellular stress-response proteins including HSP70, p97, SUMO, polyUb, and TDP43 with the NPL4 aggregates [3]. Indeed, using this model, we found the association and immobilization of ATR-GFP within the spontaneously formed NPL4-MUT aggregates (Figure 5F,H).

In summary, these experiments identified NPL4 aggregation, induced by either CuET in the case of wild-type NPL4, or mutation-caused conformational change of the NPL4-MUT protein in the absence of any added CuET, as the primary event and a pre-requisite for the subsequent sequestration of ATR in such NPL4 aggregates, with the ensuing signaling defect of the ATR-Chk1 signaling pathway.

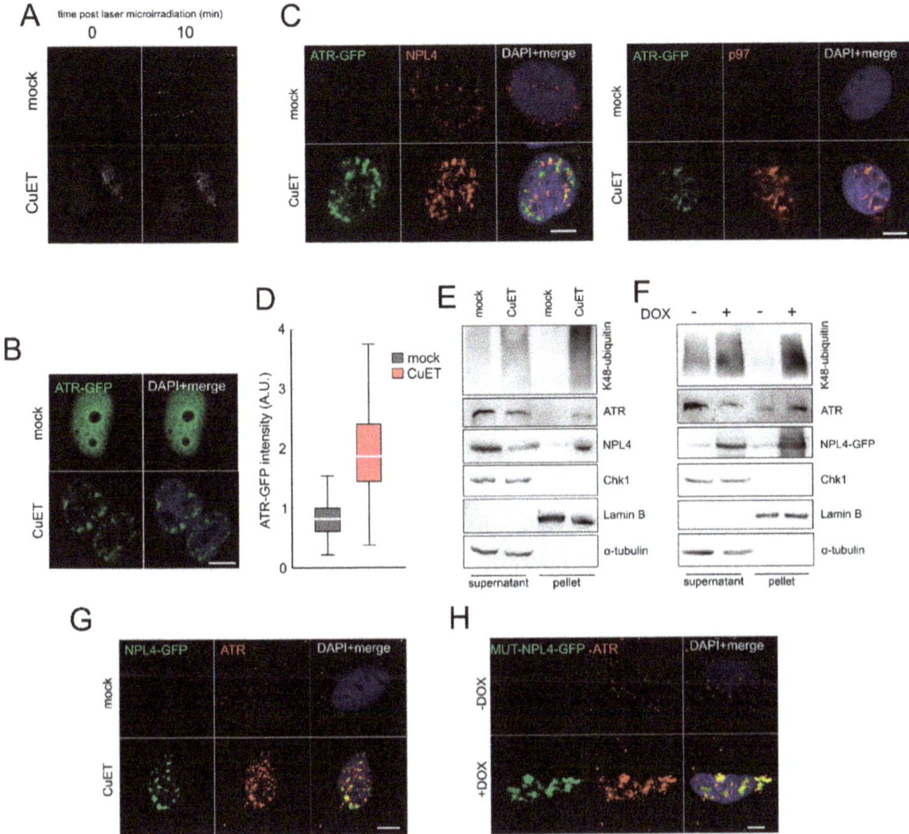

Figure 5. CuET induces immobilization of ATR and its localization to NPL4 aggregates. (**A**) ATR recruitment to sites of damage caused by laser-microirradiation is impaired after CuET treatment (250 nM for 5 h). (**B**) ATR-GFP forms typical nuclear clusters after CuET treatment (250 nM for 5 h). (**C**) Microscopic analysis of co-localization of ATR-GFP with NPL4 and p97 after CuET treatment (250 nM, 3 h) in pre-extracted U2OS cells. (**D**) Quantitative microscopic analysis of pre-extraction resistant ATR-GFP protein in U2OS cells in control and CuET treated cells (250 nM, 5 h). (**E**) WB analysis of immobilized ATR, K48 ubiquitinated proteins, and NPL4 in extracts of CuET-treated (250 nM, 3 h) U2OS cells. (**F**) WB analysis of immobilized ATR, K48 ubiquitinated proteins, and NPL4 in MUT-NPL4-GFP expressing U2OS (Doxycycline induction for 18 h). (**G**) Microscopic analysis of co-localization of NPL4-GFP with ATR after CuET treatment (250 nM, 3 h) in pre-extracted U2OS cells. (**H**) Microscopic analysis of co-localization of MUT-NPL4-GFP with ATR after 18 h doxycycline induction in pre-extracted U2OS cells. Scale bars = 10 μm.

4. Discussion

The major advance provided by the results from our present study is the identification of a new mode of cancer cell cytotoxicity evoked by diethyldithiocarbamate-copper complex, CuET [3,4], the anticancer metabolite of the alcohol aversion drug DSF that is currently tested in clinical trials for repurposing in oncology. Indeed, after years of convoluted efforts to understand the tumor-inhibitory effects of DSF, the field has been aided by our discovery of CuET as the ultimate cancer-killing compound that rapidly forms as DSF becomes metabolized under both in vivo [3], and cell culture [4,34] conditions. At the mechanistic level, we found that CuET impairs the cellular protein degradation machinery upstream of the proteasome, by inducing aggregation and immobilization of NPL4, an essential cofactor of the p97/VCP segregase complex [3]. This mechanism helps explain the observed preferential toxicity in cancer cells experiencing high levels of proteotoxic stress, such as multiple myeloma [3].

Inspired by the recent intriguing observation that human cancer cells lacking the BRCA1/2 DNA damage response genes are particularly sensitive to DSF [4,5], here we focused on potential genotoxic/replication stress as another aberrant cancer-associated trait [7–10] that could be triggered and/or enhanced by CuET. Indeed, we have found that CuET induces DNA damage preferentially in S-phase cells consistent with robust impairment of DNA replication, induction of replication stress, and impairment of ATR signaling. The same effects can be recapitulated with replacing CuET by DSF, as the culture media contain traces of copper that enable the spontaneous formation of CuET [34]. We validated the latter notion by combined treatment of cells with DSF and the copper chelator BCDS (Figure 1E), which efficiently precludes the spontaneous formation of CuET [4] and thereby the cellular phenotypes otherwise shared by CuET and DSF.

The fundamental question that emerges from our present study, and which we address only partially here, is the nature of the precise molecular mechanism behind the CuET-induced replication stress. As CuET impairs the p97/NPL4 pathway that is directly implicated in several processes linked to DNA repair and replication [35], it remains to be seen whether the replication interference could be explained by impacting such processes, including DNA replication, translesion synthesis, DNA-protein crosslinks repair, or termination of replication [36], possibly in a combination. Moreover, p97, together with diverse cofactors, is also directly involved in DSB repair, contributing to the recruitment of the 53BP1 repair factor [37] and also other DDR proteins [38–40]. On the other hand, also indirect effects of NPL4 aggregation, for example, the triggered heat-shock response, could plausibly contribute to the phenotypes observed here. In our previous work, we observed that apart from NPL4/p97, the CuET-induced aggregates contain several proteotoxic stress-related proteins, including HSP70, SUMO2/3, polyubiquitin chains, and TDP-43 [3]. Here, we have surprisingly found that also ATR kinase, a key factor required for proper cellular response to replication stress, is trapped and sequestered in the NPL4 aggregates, thus explaining the dysfunction of ATR signaling in CuET-treated cells. Conceptually, given that ATR dysfunction is known to trigger replication stress, a feature we see also after CuET treatment, one could argue that ATR aggregation could represent the primary and/or major cause of the CuET-induced replication stress. On the other hand, our time-course analysis suggests that DNA replication becomes impaired very quickly upon CuET addition, as judged from the EdU staining (Figure 2B), in fact preceding any detectable ATR aggregation. Therefore, we currently believe that the two processes, replication fork stalling, and ATR aggregation are possibly initiated independent of each other and act rather in a complementary manner to cause the observed robust replication stress phenotype. A related emerging question for future work is what brings ATR to the vicinity of the forming NPL4 aggregates in the first place? This issue is speculative at present, and it remains to be seen whether some structural features of ATR, possibly shared by additional proteins, such as unstructured regions or high dependency on chaperones, could be involved. Alternatively, the recruitment to aggregates might share the mechanism of the reported ATR recruitment into areas of high topological stress within the nuclear envelope [41]. ATR might be sequestrated by the aggregates also through direct interaction with NPL4 or due to the global proteotoxic stress-related changes in the cell. The latter scenario would partially resemble the so-called β-sheets-containing protein aggregates that sequester and mislocalize several proteins involved in RNA metabolism and nucleocytoplasmic transport [42]. Alternatively, liquid–liquid phase separation might also be involved in this process. A recent study [43] revealed that acute hyperosmotic stress induces phase separation of the proteasomes and formation of discrete puncta in the nucleus. Interestingly, these structures also contained K48-ubiquitinated proteins or p97 segregase, the proteins also found in NPL4 clusters, raising the question of whether phase separation plays a role in the case of NPL4 aggregation or attraction of other proteins. These questions need to be addressed by dedicated future studies, to help us better understand the effects of

NPL4 aggregates on cellular physiology, providing clues about why so many seemingly unrelated phenotypes have so far been described after DSF treatment.

Last but not least, our present results are also highly relevant from the clinical point of view, not least because protein aggregation represents an unorthodox and so far largely unexplored mechanism of action for anticancer drugs. This rather unique mechanism may also contribute to the observed synergistic effects of DSF/copper with either ionizing radiation [44] or the DNA damage-inducing drug temozolomide [45] a combination currently tested in several clinical trials focusing on glioblastoma patients [46–48], as well as a combination of DSF with cisplatin [49]. We hope that the data we report here will inspire further research in this rapidly evolving area of biomedicine, and yield additional effective therapies based on combining DSF/copper (CuET) with other currently used DNA damage-related therapeutic modalities.

Overall, based on our present results we suggest that CuET (DSF/copper) evokes and/or exacerbates replication stress in tumor cells while concomitantly precluding the ATR-mediated pro-survival response to such stress, thereby collectively creating a toxic scenario (understandably more severe in BRCA1/2-defective cells) reminiscent of 'killing two birds with one stone.'

Supplementary Materials: The following are available online at http://www.mdpi.com/2073-4409/9/2/469/s1, Figure S1: CuET is causing DNA damage preferentially in S/G2 cells. Figure S2.: CuET impairs DNA replication in MDA-MB-231 and U2OS cells. Figure S3: Detection of DNA breaks after CuET treatment. Figure S4: Microscopy-based quantitative analysis of fluorescence signal in cells.

Author Contributions: D.M., Z.S., M.M., and J.B. conceived the study, D.M. and Z.S. designed and performed most experiments, K.C. contributed the laser microirradiation data. J.M.M.-M. performed alkaline comet assay. D.M., Z.S., M.M., and J.B. interpreted the results and wrote the manuscript, which was approved by all authors. All authors have read and agreed to the published version of the manuscript.

Funding: The study was supported by grants from Grant Agency of Czech Rep. GACR 17-25976S, The work was supported by the MEYS CR (Large RI Project LM2018129 Czech-BioImaging) and ERDF (project No. CZ.02.1.01/0.0/0.0/16_013/0001775), Internal grant of the Palacky University IGA_LF_2019_026), Ministry of School, Education, Youth and Sports of the Czech Republic (ENOCH No. CZ.02.1.01/0.0/0.0/16_019/0000868), the Novo Nordisk Foundation (no. 16854), the Danish National Research Foundation (project CARD: no. DNRF125), the Danish Cancer Society (R204-A12617), the Swedish Research Council (VR-MH 2014-46602-117891-30), and the Swedish Cancer Society (no. 170176).

Acknowledgments: We thank M. Tarsounas (Oxford, UK) for the human H1299 cell lines with the regulatable expression of shBRCA1 and shBRCA2 and Mgr. Tatana Stosova for help with FIGE (Palacky University, Olomouc, Czech Republic).

Conflicts of Interest: The authors declare no conflict of interest.

References

1. Collins, F.S. Mining for therapeutic gold. *Nat. Rev. Drug Discov.* **2011**, *10*, 397. [CrossRef] [PubMed]

2. McMahon, A.; Chen, W.; Li, F. Old wine in new bottles: Advanced drug delivery systems for disulfiram-based cancer therapy. *J. Control. Release* **2020**, *319*, 352–359. [CrossRef] [PubMed]

3. Skrott, Z.; Mistrik, M.; Andersen, K.K.; Friis, S.; Majera, D.; Gursky, J.; Ozdian, T.; Bartkova, J.; Turi, Z.; Moudry, P.; et al. Alcohol-abuse drug disulfiram targets cancer via p97 segregase adaptor NPL4. *Nature* **2017**, *552*, 194–199. [CrossRef] [PubMed]

4. Skrott, Z.; Majera, D.; Gursky, J.; Buchtova, T.; Hajduch, M.; Mistrik, M.; Bartek, J. Disulfiram's anti-cancer activity reflects targeting NPL4, not inhibition of aldehyde dehydrogenase. *Oncogene* **2019**, *38*, 6711–6722. [CrossRef]

5. Tacconi, E.M.; Lai, X.; Folio, C.; Porru, M.; Zonderland, G.; Badie, S.; Michl, J.; Sechi, I.; Rogier, M.; Matía García, V.; et al. BRCA1 and BRCA2 tumor suppressors protect against endogenous acetaldehyde toxicity. *EMBO Mol. Med.* **2017**, *9*, 1398–1414. [CrossRef]

6. Lorenti Garcia, C.; Mechilli, M.; Proietti De Santis, L.; Schinoppi, A.; Katarzyna, K.; Palitti, F. Relationship between DNA lesions, DNA repair and chromosomal damage induced by acetaldehyde. *Mutat. Res. Mol. Mech. Mutagen.* **2009**, *662*, 3–9. [CrossRef]

7. Bartkova, J.; Hořejší, Z.; Koed, K.; Krämer, A.; Tort, F.; Zieger, K.; Guldberg, P.; Sehested, M.; Nesland, J.M.; Lukas, C.; et al. DNA damage response as a candidate anti-cancer barrier in early human tumorigenesis. *Nature* **2005**, *434*, 864–870. [CrossRef]

8. Bartkova, J.; Rezaei, N.; Liontos, M.; Karakaidos, P.; Kletsas, D.; Issaeva, N.; Vassiliou, L.-V.F.; Kolettas, E.; Niforou, K.; Zoumpourlis, V.C.; et al. Oncogene-induced senescence is part of the tumorigenesis barrier imposed by DNA damage checkpoints. *Nature* **2006**, *444*, 633–637. [CrossRef]

9. Gorgoulis, V.G.; Vassiliou, L.-V.F.; Karakaidos, P.; Zacharatos, P.; Kotsinas, A.; Liloglou, T.; Venere, M.; DiTullio, R.A.; Kastrinakis, N.G.; Levy, B.; et al. Activation of the DNA damage checkpoint and genomic instability in human precancerous lesions. *Nature* **2005**, *434*, 907–913. [CrossRef]

10. Halazonetis, T.D.; Gorgoulis, V.G.; Bartek, J. An Oncogene-Induced DNA Damage Model for Cancer Development. *Science* **2008**, *319*, 1352–1355. [CrossRef]

11. Bartek, J.; Bartkova, J.; Lukas, J. DNA damage signalling guards against activated oncogenes and tumour progression. *Oncogene* **2007**, *26*, 7773–7779. [CrossRef] [PubMed]

12. Jackson, S.P.; Bartek, J. The DNA-damage response in human biology and disease. *Nature* **2009**, *461*, 1071–1078. [CrossRef] [PubMed]

13. Gaillard, H.; García-Muse, T.; Aguilera, A. Replication stress and cancer. *Nat. Rev. Cancer* **2015**, *15*, 276–289. [CrossRef]

14. Bartek, J.; Mistrik, M.; Bartkova, J. Thresholds of replication stress signaling in cancer development and treatment. *Nat. Struct. Mol. Biol.* **2012**, *19*, 5–7. [CrossRef]

15. Zeman, M.K.; Cimprich, K.A. Causes and consequences of replication stress. *Nat. Cell Biol.* **2014**, *16*, 2–9. [CrossRef] [PubMed]

16. Berti, M.; Vindigni, A. Replication stress: Getting back on track. *Nat. Struct. Mol. Biol.* **2016**, *23*, 103–109. [CrossRef]

17. Eykelenboom, J.K.; Harte, E.C.; Canavan, L.; Pastor-Peidro, A.; Calvo-Asensio, I.; Llorens-Agost, M.; Lowndes, N.F. ATR Activates the S-M Checkpoint during Unperturbed Growth to Ensure Sufficient Replication Prior to Mitotic Onset. *Cell Rep.* **2013**, *5*, 1095–1107. [CrossRef]

18. Toledo, L.I.; Altmeyer, M.; Rask, M.-B.; Lukas, C.; Larsen, D.H.; Povlsen, L.K.; Bekker-Jensen, S.; Mailand, N.; Bartek, J.; Lukas, J. ATR Prohibits Replication Catastrophe by Preventing Global Exhaustion of RPA. *Cell* **2014**, *156*, 374. [CrossRef]

19. Bekker-Jensen, S.; Lukas, C.; Kitagawa, R.; Melander, F.; Kastan, M.B.; Bartek, J.; Lukas, J. Spatial organization of the mammalian genome surveillance machinery in response to DNA strand breaks. *J. Cell Biol.* **2006**, *173*, 195–206. [CrossRef]

20. Jackson, D.A.; Pombo, A. Replicon Clusters Are Stable Units of Chromosome Structure: Evidence That Nuclear Organization Contributes to the Efficient Activation and Propagation of S Phase in Human Cells. *J. Cell Biol.* **1998**, *140*, 1285–1295. [CrossRef]

21. Mistrik, M.; Vesela, E.; Furst, T.; Hanzlikova, H.; Frydrych, I.; Gursky, J.; Majera, D.; Bartek, J. Cells and Stripes: A novel quantitative photo-manipulation technique. *Sci. Rep.* **2016**, *6*, 19567. [CrossRef] [PubMed]

22. Maya-Mendoza, A.; Moudry, P.; Merchut-Maya, J.M.; Lee, M.; Strauss, R.; Bartek, J. High speed of fork progression induces DNA replication stress and genomic instability. *Nature* **2018**, *559*, 279–284. [CrossRef] [PubMed]

23. Budzowska, M.; Kanaar, R. Mechanisms of Dealing with DNA Damage-Induced Replication Problems. *Cell Biochem. Biophys.* **2009**, *53*, 17–31. [CrossRef] [PubMed]

24. Bianco, J.N.; Poli, J.; Saksouk, J.; Bacal, J.; Silva, M.J.; Yoshida, K.; Lin, Y.-L.; Tourrière, H.; Lengronne, A.; Pasero, P. Analysis of DNA replication profiles in budding yeast and mammalian cells using DNA combing. *Methods* **2012**, *57*, 149–157. [CrossRef]

25. Quinet, A.; Carvajal-Maldonado, D.; Lemacon, D.; Vindigni, A. DNA Fiber Analysis: Mind the Gap! *Methods Enzymology* **2017**, *591*, 55–82.

26. Couch, F.B.; Bansbach, C.E.; Driscoll, R.; Luzwick, J.W.; Glick, G.G.; Betous, R.; Carroll, C.M.; Jung, S.Y.; Qin, J.; Cimprich, K.A.; et al. ATR phosphorylates SMARCAL1 to prevent replication fork collapse. *Genes Dev.* **2013**, *27*, 1610–1623. [CrossRef]

27. Sogo, J.M. Fork Reversal and ssDNA Accumulation at Stalled Replication Forks Owing to Checkpoint Defects. *Science* **2002**, *297*, 599–602. [CrossRef]

28. Byun, T.S.; Pacek, M.; Yee, M.C.; Walter, J.C.; Cimprich, K.A. Functional uncoupling of MCM helicase and DNA polymerase activities activates the ATR-dependent checkpoint. *Genes Dev.* **2005**, *19*, 1040–1052. [CrossRef]

29. Lee, Z.; Elledge, S.J. Sensing DNA Damage Through ATRIP Recognition of RPA-ssDNA Complexes. *Science* **2003**, *300*, 1542–1548. [CrossRef]

30. Liao, H.; Ji, F.; Helleday, T.; Ying, S. Mechanisms for stalled replication fork stabilization: New targets for synthetic lethality strategies in cancer treatments. *EMBO Rep.* **2018**, *19*. [CrossRef]

31. Petermann, E.; Orta, M.L.; Issaeva, N.; Schultz, N.; Helleday, T. Hydroxyurea-Stalled Replication Forks Become Progressively Inactivated and Require Two Different RAD51-Mediated Pathways for Restart and Repair. *Mol. Cell* **2010**, *37*, 492–502. [CrossRef] [PubMed]

32. Whelan, D.R.; Lee, W.T.C.; Yin, Y.; Ofri, D.M.; Bermudez-Hernandez, K.; Keegan, S.; Fenyo, D.; Rothenberg, E. Spatiotemporal dynamics of homologous recombination repair at single collapsed replication forks. *Nat. Commun.* **2018**, *9*, 3882. [CrossRef] [PubMed]

33. Davies, A.A.; Masson, J.Y.; McIlwraith, M.J.; Stasiak, A.Z.; Stasiak, A.; Venkitaraman, A.R.; West, S.C. Role of BRCA2 in control of the RAD51 recombination and DNA repair protein. *Mol. Cell* **2001**, *7*, 273–282. [CrossRef]

34. Majera, D.; Skrott, Z.; Bouchal, J.; Bartkova, J.; Simkova, D.; Gachechiladze, M.; Steigerova, J.; Kurfurstova, D.; Gursky, J.; Korinkova, G.; et al. Targeting genotoxic and proteotoxic stress-response pathways in human prostate cancer by clinically available PARP inhibitors, vorinostat and disulfiram. *Prostate* **2019**, *79*, 352–362. [CrossRef]

35. Ramadan, K. p97/VCP- and Lys48-linked polyubiquitination form a new signaling pathway in DNA damage response. *Cell Cycle* **2012**, *11*, 1062–1069. [CrossRef] [PubMed]

36. Ramadan, K.; Halder, S.; Wiseman, K.; Vaz, B. Strategic role of the ubiquitin-dependent segregase p97 (VCP or Cdc48) in DNA replication. *Chromosoma* **2017**, *126*, 17–32. [CrossRef]

37. Meerang, M.; Ritz, D.; Paliwal, S.; Garajova, Z.; Bosshard, M.; Mailand, N.; Janscak, P.; Hübscher, U.; Meyer, H.; Ramadan, K. The ubiquitin-selective segregase VCP/p97 orchestrates the response to DNA double-strand breaks. *Nat. Cell Biol.* **2011**, *13*, 1376. [CrossRef]

38. Bergink, S.; Ammon, T.; Kern, M.; Schermelleh, L.; Leonhardt, H.; Jentsch, S. Role of Cdc48/p97 as a SUMO-targeted segregase curbing Rad51–Rad52 interaction. *Nat. Cell Biol.* **2013**, *15*, 526–532. [CrossRef]

39. Singh, A.N.; Oehler, J.; Torrecilla, I.; Kilgas, S.; Li, S.; Vaz, B.; Guérillon, C.; Fielden, J.; Hernandez-Carralero, E.; Cabrera, E.; et al. The p97-Ataxin 3 complex regulates homeostasis of the DNA damage response E3 ubiquitin ligase RNF8. *EMBO J.* **2019**, *38*, e102361. [CrossRef]

40. Davis, E.J.; Lachaud, C.; Appleton, P.; Macartney, T.J.; Näthke, I.; Rouse, J. DVC1 (C1orf124) recruits the p97 protein segregase to sites of DNA damage. *Nat. Struct. Mol. Biol.* **2012**, *19*, 1093–1100. [CrossRef]

41. Kumar, A.; Mazzanti, M.; Mistrik, M.; Kosar, M.; Beznoussenko, G.V.; Mironov, A.A.; Garrè, M.; Parazzoli, D.; Shivashankar, G.V.; Scita, G.; et al. ATR Mediates a Checkpoint at the Nuclear Envelope in Response to Mechanical Stress. *Cell* **2014**, *158*, 633–646. [CrossRef] [PubMed]

42. Woerner, A.C.; Frottin, F.; Hornburg, D.; Feng, L.R.; Meissner, F.; Patra, M.; Tatzelt, J.; Mann, M.; Winklhofer, K.F.; Hartl, F.U.; et al. Cytoplasmic protein aggregates interfere with nucleocytoplasmic transport of protein and RNA. *Science* **2016**, *351*, 173–176. [CrossRef] [PubMed]

43. Yasuda, S.; Tsuchiya, H.; Kaiho, A.; Guo, Q.; Ikeuchi, K.; Endo, A.; Arai, N.; Ohtake, F.; Murata, S.; Inada, T.; et al. Stress- and ubiquitylation-dependent phase separation of the proteasome. *Nature* **2020**, *578*, 296–300. [CrossRef] [PubMed]

44. Wang, Y.; Li, W.; Patel, S.S.; Cong, J.; Zhang, N.; Sabbatino, F.; Liu, X.; Qi, Y.; Huang, P.; Lee, H.; et al. Blocking the formation of radiation induced breast cancer stem cells. *Oncotarget* **2014**, *5*, 3743–3755. [CrossRef]

45. Lun, X.; Wells, J.C.; Grinshtein, N.; King, J.C.; Hao, X.; Dang, N.-H.; Wang, X.; Aman, A.; Uehling, D.; Datti, A.; et al. Disulfiram when Combined with Copper Enhances the Therapeutic Effects of Temozolomide for the Treatment of Glioblastoma. *Clin. Cancer Res.* **2016**, *22*, 3860–3875. [CrossRef]

46. Huang, J.; Campian, J.L.; Gujar, A.D.; Tsien, C.; Ansstas, G.; Tran, D.D.; DeWees, T.A.; Lockhart, A.C.; Kim, A.H. Final results of a phase I dose-escalation, dose-expansion study of adding disulfiram with or without copper to adjuvant temozolomide for newly diagnosed glioblastoma. *J. Neurooncol.* **2018**, *138*, 105–111. [CrossRef]

47. Huang, J.; Campian, J.L.; Gujar, A.D.; Tran, D.D.; Lockhart, A.C.; DeWees, T.A.; Tsien, C.I.; Kim, A.H. A phase I study to repurpose disulfiram in combination with temozolomide to treat newly diagnosed glioblastoma after chemoradiotherapy. *J. Neurooncol.* **2016**, *128*, 259–266. [CrossRef]
48. Jakola, A.S.; Werlenius, K.; Mudaisi, M.; Hylin, S.; Kinhult, S.; Bartek, J.J.; Salvesen, O.; Carlsen, S.M.; Strandeus, M.; Lindskog, M.; et al. Disulfiram repurposing combined with nutritional copper supplement as add-on to chemotherapy in recurrent glioblastoma (DIRECT): Study protocol for a randomized controlled trial. *F1000Research* **2018**, *7*, 1797. [CrossRef]
49. Nechushtan, H.; Hamamreh, Y.; Nidal, S.; Gotfried, M.; Baron, A.; Shalev, Y.I.; Nisman, B.; Peretz, T.; Peylan-Ramu, N. A Phase IIb Trial Assessing the Addition of Disulfiram to Chemotherapy for the Treatment of Metastatic Non-Small Cell Lung Cancer. *Oncologist* **2015**, *20*, 366–367. [CrossRef]

Article

Disruption of the NF-κB/IL-8 Signaling Axis by Sulconazole Inhibits Human Breast Cancer Stem Cell Formation

Hack Sun Choi [1,2,†], Ji-Hyang Kim [3,†], Su-Lim Kim [1,3] and Dong-Sun Lee [1,2,3,*]

[1] School of Biomaterials Sciences and Technology, College of Applied Life Science, Jeju National University, Jeju 63243, Korea
[2] Subtropical/tropical Organism Gene Bank, Jeju National University, Jeju 63243, Korea
[3] Interdisciplinary Graduate Program in Advanced Convergence Technology & Science, Jeju National University, Jeju 63243, Korea
* Correspondence: dongsunlee@jejunu.ac.kr
† These authors contributed equally to this work.

Received: 5 August 2019; Accepted: 29 August 2019; Published: 30 August 2019

Abstract: Breast cancer stem cells (BCSCs) are tumor-initiating cells that possess the capacity for self-renewal. Cancer stem cells (CSCs) are responsible for poor outcomes caused by therapeutic resistance. In our study, we found that sulconazole—an antifungal medicine in the imidazole class—inhibited cell proliferation, tumor growth, and CSC formation. This compound also reduced the frequency of cells expressing CSC markers (CD44^{high}/CD24^{low}) as well as the expression of another CSC marker, aldehyde dehydrogenase (ALDH), and other self-renewal-related genes. Sulconazole inhibited mammosphere formation, reduced the protein level of nuclear NF-κB, and reduced extracellular IL-8 levels in mammospheres. Knocking down NF-κB expression using a p65-specific siRNA reduced CSC formation and secreted IL-8 levels in mammospheres. Sulconazole reduced nuclear NF-κB protein levels and secreted IL-8 levels in mammospheres. These new findings show that sulconazole blocks the NF-κB/IL-8 signaling pathway and CSC formation. NF-κB/IL-8 signaling is important for CSC formation and may be an important therapeutic target for BCSC treatment.

Keywords: sulconazole; NF-κB; IL-8; mammosphere; breast cancer stem cells

---

1. Introduction

Breast cancer is a cancer that develops from common breast tissue and a major fatal health problem among females [1]. Patients treated with different therapeutics suffer from cancer relapse and metastasis because of cancer stem cells (CSCs), a subpopulation of tumor cells. CSCs are heterogeneous bulk tumor cells that differentiate into cancer cells. CSCs are resistant to chemotherapies and contribute to tumor heterogeneity [2]. CSCs were first identified in leukemia and found to show properties similar to those of stem cells by Bonnet and Dick [3]. Markers of breast cancer stem cells (BCSCs) include CD44, CD133, and ALDH1. CD44 expression is upregulated in the microenvironment that promotes cancer progression and metastasis [4]. Additionally, CD44 isoforms are reliable markers of CSCs. The CD44 isoform CD44v-xCT regulates redox in cancer stem cells [5]. The signaling pathways regulating CSC stemness and differentiation are the Wnt, Hedgehog, Hippo, and Notch signaling pathways. Molecular targeting of these pathways to inhibit BCSCs may be a useful tool for cancer treatment [6]. Sox2, Nanog, Oct4, and c-Myc are crucial for CSC formation and potential targets for cancer therapy. One report showed that NF-κB was involved in CSCs from primary acute myeloid leukemia (AML) samples [7]. Additionally, BCSCs overexpress NF-κB signaling pathway components and induce NF-κB activity. BCSCs have high protein expression of NF-κB [8]. Inhibiting NF-κB signaling with BMS-345541 in lung cancer reduces the stemness and self-renewal capacity of lung CSCs [9]. The cytokines IL-6 and IL-8 regulate links between CSCs and the microenvironment. The Stat3 and NF-κB pathways regulate the gene expression of IL-6 and IL-8 in breast cancer. Microenvironmental IL-6 and IL-8 regulate BCSC populations [10]. In lung cancer patients, high extracellular IL-6 levels are associated with a poor prognosis [11,12]. IL-6 regulates BCSC formation through the IL-6/IL-6 receptor interaction [13]. The protein expression level of IL-8 is higher in breast cancer cells than in normal breast tissue cells, and IL-8 promotes cancer progression. IL-8 promotes BCSC activity through the CXCR1/IL8 interaction. IL-8/CXCR1 signaling is an important pathway for targeting BCSCs [14]. Azole compounds used as antifungal drugs inhibit the ergosterol biosynthesis pathway through suppression of the enzyme lanosterol 14-α-demethylase, a cytochrome P450 (CYP) enzyme. Azole antifungal drugs consist of an imidazole (clotrimazole and ketoconazole) and a triazole (fluconazole and itraconazole). Recently, antifungal imidazole drugs have well-established pharmacokinetic profiles and known toxicity, which can make these generic drugs strong candidates for repositioning as antitumor therapies [15]. Sulconazole is an antifungal medicine in the imidazole class and has broad-spectrum activity against dermatophytes [16]. We demonstrated that sulconazole had antiproliferative properties in breast cancer and inhibited BCSC formation through a reduction in IL-8 expression induced by disrupting the NF-κB pathway.

2. Materials and Methods

2.1. Cell Lines and Media

MCF-7 and MDA-MB-231 cells were grown in Dulbecco's modified Eagle's medium (DMEM; Gibco, Thermo Fisher Scientific, Waltham, CA, USA) supplemented with 10% (v/v) fetal bovine serum (FBS; Thermo Fisher Scientific, Waltham, CA, USA), 1% penicillin and streptomycin in a humidified 5% CO_2 incubator at 37 °C. Breast cancer cells were cultured at a concentration of 3.5×10^4 or 0.5×10^4 cells/well in an Ultralow Adherent plate containing MammoCult™ medium (STEMCELL Technologies, Vancouver, BC, Canada) supplemented with heparin and hydrocortisone in a humidified 5% CO_2 incubator at 37 °C. A 6-well plate was scanned, and mammosphere counting was performed using the NICE program [17]. A mammosphere formation assay was determined by evaluating mammosphere formation efficiency (MFE) (%) as previously described [18].

2.2. Antibodies, siRNAs, and Plasmids

Anti-pStat3 (Y705) (rabbit monoclonal) antibodies were obtained from Cell Signaling Technology. Anti-p65 (mouse polyclonal), anti-pp65, anti-Stat3 (rabbit monoclonal), anti-β-actin (mouse polyclonal), and anti-Lamin b antibodies were purchased from Santa Cruz Biotechnology. Anti-CD44 FITC-conjugated and anti-CD24 PE-conjugated antibodies were obtained from BD Pharmingen. A human p65-specific siRNA and scrambled siRNA were purchased from Bioneer (Daejeon, Korea).

2.3. Cell Proliferation

We used a previously reported method [19]. Breast cancer cells were incubated in a 96-well plate with sulconazole for 24 h. We followed the manufacturer's protocol for a CellTiter 96® Aqueous One Solution cell kit (Promega), and the optical density at 490 nm (OD_{490}) was determined using a plate reader (SpectraMax, Molecular Devices, San Jose, CA, USA).

2.4. Colony Formation and Migration Assays

MDA-MB-231 cells were cultured at 2×10^3 cells/well with different concentrations of sulconazole in DMEM/10% FBS. The cancer cells were incubated, and colonies were counted. The cancer cells were incubated in a 6-well plate, and a scratch was made using a microtip. After washing with

DMEM, the breast cancer cells were cultured with sulconazole. We followed a previously described method [20].

2.5. Flow Cytometry Analysis of the Expression of CD24 and CD44 and an ALDEFLUOR Assay

We used a previously described method [20]. In total, 1×10^6 cells were incubated with FITC-conjugated anti-CD44 and PE-conjugated anti-CD24 antibodies (BD, San Jose, CA, USA) and incubated on ice for 20 min. The breast cancer cells were washed two times with 1X PBS and assayed by using a flow cytometer (BD curi C6, San Jose, CA, USA). An ALDH1 assay was performed using an ALDEFUOR kit (STEMCELL Technologies, Vancouver, BC, Canada). We followed a previously described method [20]. Breast cancer cells were incubated in ALDH assay buffer at 37 °C for 20 min. ALDH-positive cells were determined by using a personal flow cytometer (BD Accuri C6).

2.6. RNA Isolation and Real-Time RT-qPCR

Total RNA was purified, and RT-qPCR was performed using a one-step RT-qPCR kit (Takara, Tokyo, Japan). We followed a previously described method [19]. The specific primers used can be found in Supplementary Table S1. The β-actin gene was used as an internal control for RT-qPCR.

2.7. Immunoblot Analysis

Proteins isolated from breast cancer cells and mammospheres were separated by 10% SDS-PAGE and transferred to a polyvinylidene fluoride (PVDF) membrane (EMD Millipore, Burlington, MA, USA). The blots were blocked in 5% skim milk in 1X PBS-Tween 20 at room temperature for 60 min and then incubated overnight at 4 °C with primary antibodies. The antibodies were anti-JAK2, anti-Stat3, anti-p65, anti-pp65, anti-lamin B, anti-phospho-Stat3 (Cell Signaling, Danvers, MA, USA), and anti-β-actin (Santa Cruz Biotechnology, Dallas, TA, USA) antibodies. After washing, the blots were detected with IRDye 680 RD and 800 CW secondary antibodies, and images were detected by using ODYSSEY CLx (LI-COR, Lincoln, NE, USA).

2.8. Electrophoretic Mobility Shift Assays (EMSAs)

Nuclear extracts were prepared as described previously [21]. An EMSA for NF-κB binding was performed using an IRDye 800-labeled NF-κB consequence oligonucleotide (LI-COR) for 30 min at room temperature. Samples were run on a nondenaturing 6% PAGE gel, and EMSA data were captured by ODYSSEY CLx (LI-COR). Supershifts were analyzed by incubating nuclear extracts for 30 min before the addition of the IRDye 800-labeled NF-κB consequence oligonucleotide.

2.9. In Vivo Mouse Experiments

Twelve female BALB/C nude mice were injected with MDA-MB-231 cells and treated with/without sulconazole (10 mg/kg). Tumor volume was measured after 1.5 months using a formula (Figure 2). Mouse experiments were performed as described previously [22]. Animal care and animal experiments were conducted in accordance with protocols approved by the Jeju National University Animal Care and Use Committee. Female BALB/C nude mice (5 weeks old) were obtained from OrientBio (Seoul, Korea) and kept in mouse facilities for 7 days. Twelve female BALB/C nude mice injected with MDA-MB-231 cells were monitored. Nude mice ($n = 6$) received sulconazole using mammary fat pad injection with an optimized dosage of 10 mg/kg. The dose of drug used was 10 mg/kg (200 µg/100 µL) once a week. The measurement was made every 3 to 4 days starting from day 10. The solvent used is DMSO. Tumor volumes were measured using the formula: $V = (width^2 \times length)/2$.

2.10. Statistical Analysis

All data from three independent experiments are shown as the mean ± standard deviation (SD). Data were analyzed using one-way ANOVA. A p-value less than 0.05 was considered statistically significant.

3. Results

3.1. Sulconazole Inhibits the Proliferation of Breast Cancer Cells

We examined the antiproliferative effect of sulconazole on human breast cancer cells. Sulconazole inhibited proliferation (Figure 1A,B). Apoptosis in breast cancer cells was induced by sulconazole at a concentration of 20 μM (Figure 1C). Sulconazole induced caspase3/7 activity in breast cancer cells (Figure 1D). The breast cancer cells showed formation of apoptotic bodies in response to sulconazole treatment (Figure 1E). Sulconazole inhibited the migration of cancer cells and reduced the number of colonies (Figure 1F,G). Our data showed that sulconazole effectively inhibited proliferation, migration, and colony formation.

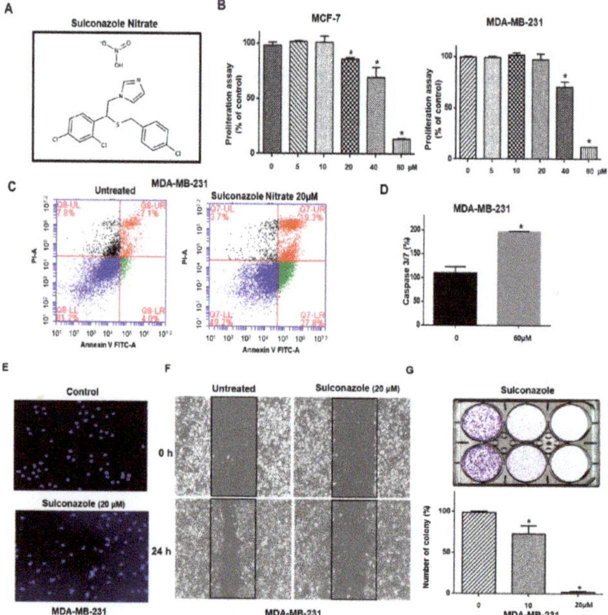

Figure 1. Sulconazole inhibits cell proliferation in breast cancer. (**A**) The molecular structure of sulconazole is shown. (**B**) Breast cancer cells were incubated in a 96-well plate with the indicated concentration of sulconazole. Cell proliferation was measured by an MTS assay. (**C**) Sulconazole induced apoptosis in cancer cells at the indicated concentration. Apoptotic cells were determined using Annexin V/PI staining. (**D**) The caspase3/7 activity of cancer cells was determined using a Caspase-Glo 3/7 assay kit (Promega). The data are presented as the mean ± SD; $n = 3$ independent experiments; * $p < 0.05$ vs. the control (0.3% DMSO). (**E**) Apoptotic cells were analyzed by fluorescence nuclear staining using Hoechst 33,258 dye (magnification, 40×). (**F**) The effect of sulconazole on the migration of cancer cells was evaluated using a scratch assay. The scratch assay was performed with cancer cells treated with sulconazole. (**G**) The effect of sulconazole on colony formation is shown. 1000 cancer cells were incubated in 6-well plates with sulconazole (0.1% DMSO) and 0.1% DMSO. Representative images were recorded. The data are presented as the mean ± SD; $n = 3$ independent experiments; * $p < 0.05$ vs. the control.

3.2. Sulconazole Inhibits Tumor Growth

As sulconazole has cytotoxic activity in breast cancer, we tested whether sulconazole inhibits tumor growth in an in vivo mouse model. The tumor volume in sulconazole-injected mice was smaller than that in control mice (Figure 2A). The tumor weights in the sulconazole-injected mice were lower than those in the control mice (Figure 2B). The sulconazole-treated mice showed body weights similar to those of the control mice (data not shown). Our data showed that sulconazole effectively decreased tumor growth in the xenograft mouse model.

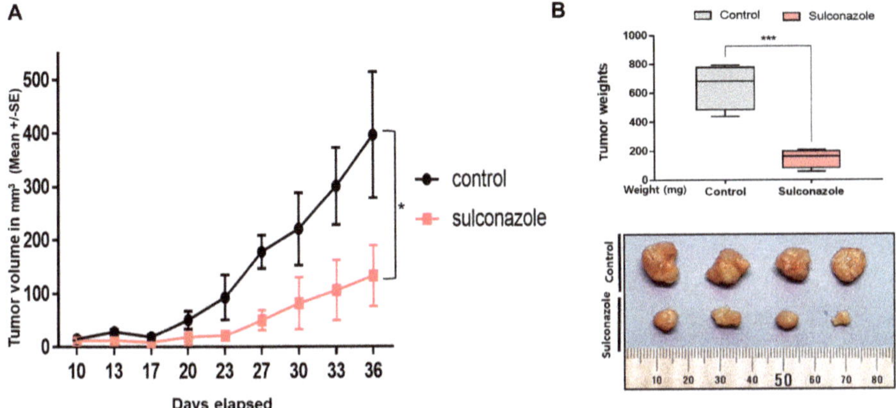

Figure 2. Effect of sulconazole on in vivo tumor growth. (**A**) NOD-SCID nude mice were inoculated with MDA-MB-231 cells and treated with sulconazole or vehicle. The dose of drug used was 10 mg/kg once a week. Tumor volume was measured at the indicated time points using a caliper and calculated as (width^2 × length)/2 and are reported (Mean ± SE). (**B**) The effect of sulconazole on tumor weights was evaluated. Tumor weights were assayed after sacrifice. Photographs were taken of isolated tumors from control or sulconazole-treated mice. * $p < 0.05$ and *** $p < 0.05$ vs. the control.

3.3. Effect of Sulconazole on the Properties of BCSCs

To examine whether sulconazole inhibits mammosphere formation, we treated mammospheres derived from breast cancer cells (MCF-7 and MDA-MB-231) with different concentrations of sulconazole. Sulconazole inhibited mammosphere formation. The number of mammospheres declined by 90%, and mammosphere size also decreased (Figure 3A,B). CD44$^+$/CD24$^-$ cancer cells were assessed under sulconazole treatment. Sulconazole reduced the percentage of CD44$^+$/CD24$^-$ cells from 14.23% to 3.53% (Figure 4A). Additionally, we performed an ALDEFLUOR assay to examine the effect of sulconazole on ALDH-positive cells. Sulconazole reduced the ALDH-positive cell percentage from 3.2% to 1.5% (Figure 4B). Our data show that sulconazole inhibits BCSCs.

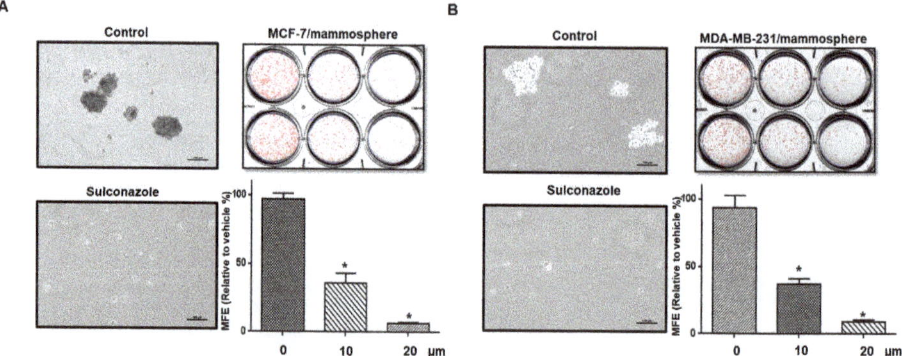

Figure 3. Effect of sulconazole on the mammosphere-forming ability of breast cancer cells. (**A**, **B**) Effect of sulconazole on the mammosphere formation of breast cancer cells. To establish mammospheres, MCF-7 and MDA-MB-231 cells were seeded at a density of 4×10^4 and 1×10^4 cells/well, respectively, in ultralow attachment 6-well plates containing 2 mL of complete MammoCult^{TM} medium (StemCell Technologies) which was supplemented with 4 µg/mL heparin, 0.48 µg/mL hydrocortisone, 100 U/mL penicillin, and 100 µg/mL streptomycin. Mammospheres were cultured with sulconazole (10 or 20 µM) solubilized in 0.05% DMSO or 0. 1% DMSO. The breast cancer cells were incubated with sulconazole in CSC culture medium for 7 days. A mammosphere formation assay evaluated mammosphere formation efficiency (MFE, % of control), which corresponds to the number of mammospheres per well/the number of total cells plated per well ×100 as previously described (scale bar = 100 µm) [22]. The data are presented as the mean ± SD; $n = 3$ independent experiments;* $p < 0.05$ vs. the control (0. 1% DMSO).

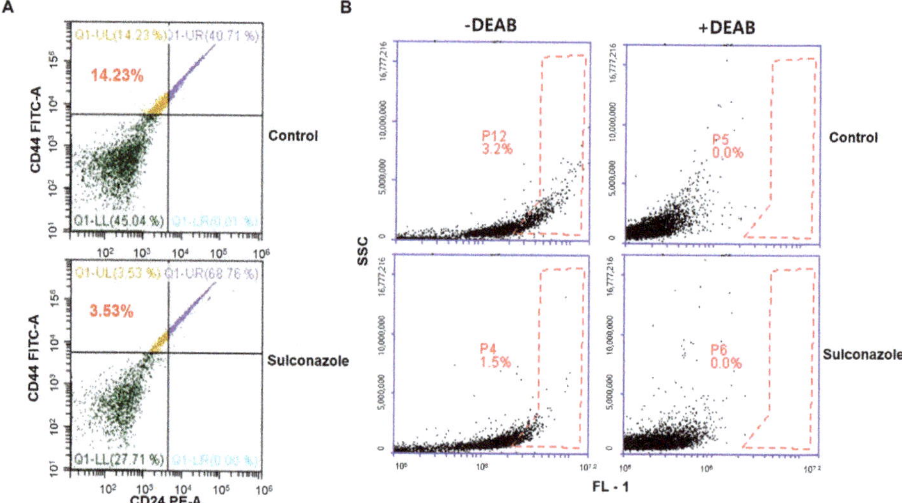

Figure 4. Effect of sulconazole on CD44$^+$/CD24$^-$-and ALDH-positive cell populations. (**A**) The CD44$^+$/CD24$^-$ cell population treated with sulconazole (20 µM) was assayed by flow cytometry. For FACS analysis, 10,000 cells were assayed. Gating was based on binding of the control antibody (Red cross). (**B**) ALDH-positive cells were detected by using an ALDEFLUOR kit. A representative flow cytometry dot plot is shown. The right panel indicates ALDH-positive cells treated with the ALDH inhibitor DEAB (7.5 µM), and the left panel shows ALDH-positive cells without DEAB treatment. The ALDH-positive population was gated in a box (red dot line box).

3.4. Sulconazole Inhibits Mammosphere Formation Through the Inhibition of p65 Nuclear Translocation

To understand the molecular mechanism of sulconazole in mammosphere formation, the nuclear translocation of p65 was evaluated in mammospheres. Our data showed that nuclear phosphor-p65 and p65 levels were reduced significantly in a dose-dependent manner under sulconazole treatment (Figure 5A). Because caffeic acid phenethyl ester (CAPE) inhibits the nuclear translocation of p65 and the activation of the NF-κB signaling pathway [23], we evaluated mammosphere formation after treatment with CAPE. CAPE inhibited mammosphere formation. As a result of the use of sulconazole and CAPE, we showed that the inhibition of p65 nuclear translocation blocked mammosphere formation (Figure 5C). To examine p65 function in BCSCs, we tested the effect of NF-kB using a p65-specific siRNA. siRNA-p65 inhibited mammosphere formation in breast cancer (Figure 5B). In conclusion, we show that NF-κB regulates CSC formation and a CSC survival factor (Figure 5).

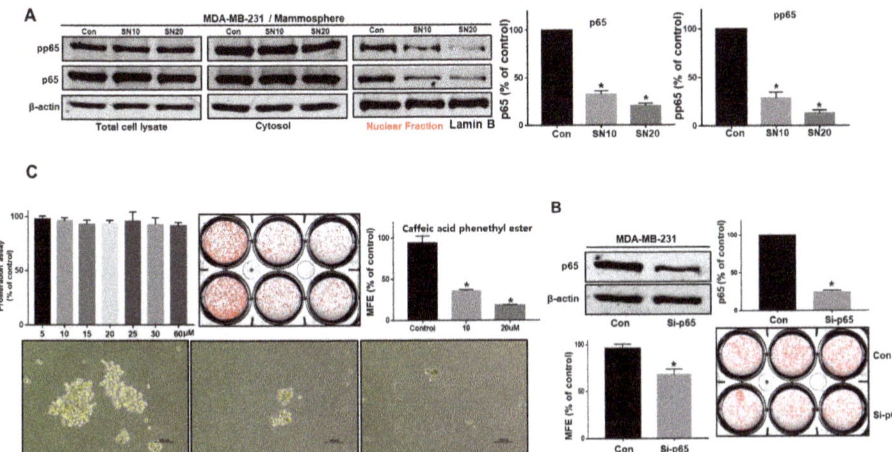

Figure 5. Sulconazole inhibits mammosphere formation through disruption of NF-κB activity. (**A**) Cancer cells were treated with sulconazole for 24 h. Nuclear and cytosolic proteins were run on a 10% SDS-PAGE gel, followed by immunoblotting with anti-p65 and anti-pp65 antibodies. (**B**) The effect of knocking down p65 expression using a siRNA specific for p65 on mammosphere formation was evaluated. The p65-knockdown effect was confirmed by immunoblotting using an anti-p65 antibody. (**C**) The effect of caffeic acid phenethyl ester, an NF-κB-specific inhibitor, on mammosphere formation was evaluated. A mammosphere formation assay evaluated mammosphere formation efficiency (MFE) (scale bar = 100 μm). The data are presented as the mean ± SD; n = 3 independent experiments;* p < 0.05 vs. the control.

3.5. Sulconazole Inhibits the NF-κB Signaling Pathway and Production of Extracellular IL-8 in Mammospheres

To analyze the biological function of sulconazole, we examined NF-κB signaling and the extracellular IL-8 level in mammospheres treated with sulconazole. Compared with a vehicle, sulconazole reduced nuclear p65 protein levels (Figure 6A). We checked NF-κB binding with sulconazole-treated nuclear proteins using an IRDye 800-NF-κB probe that binds an NF-κB oligonucleotide with high affinity. Sulconazole reduced the ability of p65 to bind to the IRDye 800-NF-κB probe (Figure 6B, lane 3). NF-κB/IRDye 800-NF-κB probe specificity was confirmed using a 10-fold increased concentration of self-competitor oligonucleotides (Figure 6B, lane 4). Sulconazole decreased the DNA-binding capacities of NF-κB. Extracellular IL-6 and IL-8 have essential functions in CSC formation [13]. NF-κB regulated the transcription of the IL-6 and IL-8 genes, binding to the promoter regions of the IL-6 and IL-8 genes. To assess the transcriptional levels of IL-6 and IL-8 under sulconazole treatment, we performed real-time RT-qPCR analysis of mammospheres using IL-6- and IL-8-specific primers. The transcript data showed that sulconazole reduced the transcript level of IL-8 but not that of IL-6 (Figure 6C). After using a siRNA targeting p65, the transcript data showed that sulconazole reduced the transcript level of IL-8 (Figure 6D). To test the level of extracellular IL-8, we performed cytokine profiling of the culture medium from mammospheres. After sulconazole treatment, the cytokine profiling data showed that sulconazole reduced the level of extracellular IL-8 but not that of IL-6 (Figure 6E).

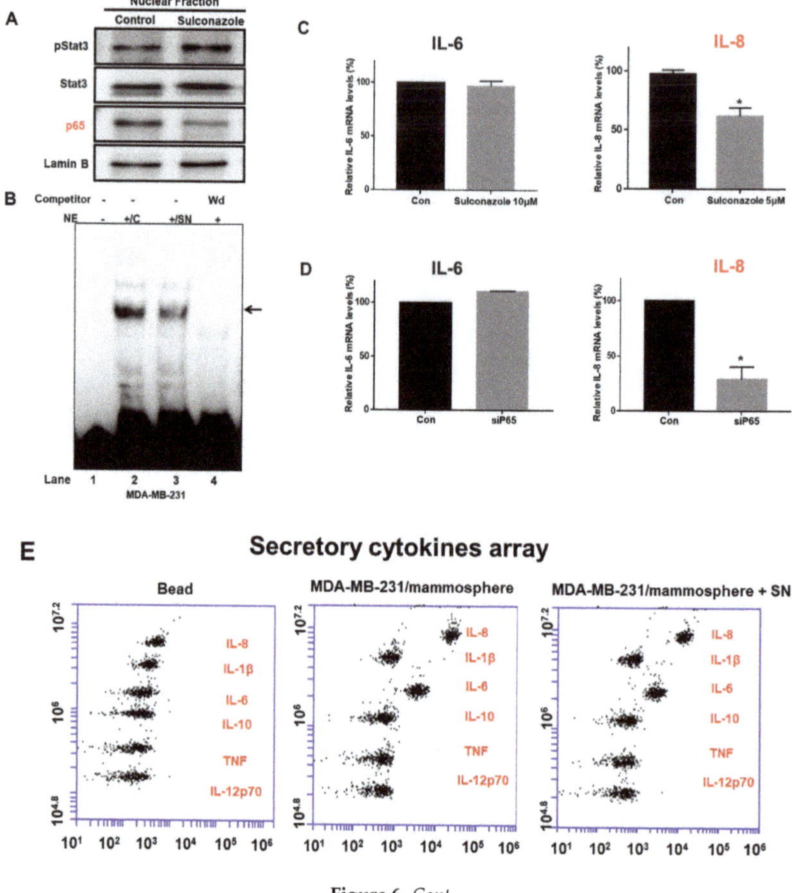

Figure 6. *Cont.*

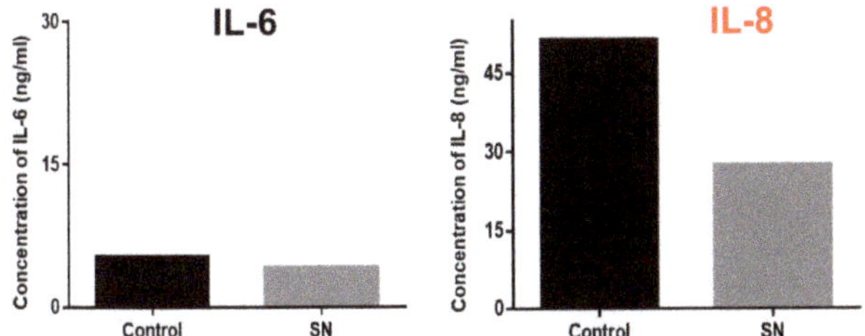

Figure 6. Effect of sulconazole on the NF-κB and IL-8 signaling pathways. (**A**) Cancer cells were treated with sulconazole for 24 h. Nuclear proteins were resolved on a 10% SDS-PAGE gel, followed by western blotting with anti-pStat3, anti-Stat3, anti-p65, and anti-Lamin B antibodies. (**B**) An electrophoretic mobility shift assay (EMSA) was used to assess nuclear lysates from mammospheres treated with sulconazole. The nuclear proteins were incubated with an IRDye 800-NF-κB probe and separated by 6% PAGE. Lane 1: probe only; lane 2: nuclear proteins with probe; lane 3: sulconazole-treated nuclear proteins with probe; lane 4: 10× self-competition. The arrow indicates the DNA/NF-κB interaction in the nuclear lysates. (**C,D**) Transcriptional levels of the IL-6 and IL-8 genes were determined in sulconazole-treated mammospheres and p65-knockout samples treated with a siRNA specific for p65. IL-6- and IL-8-specific primers were used for real-time RT-PCR. β-actin acted as an internal control. The data are presented as the mean ± SD; $n = 3$ independent experiments; * $p < 0.05$ vs. the control. (**E**) The cytokine profiles of conditioned media from mammosphere cultures were determined with cytokine-specific antibodies and cytokine beads. Sulconazole reduced extracellular IL-8 levels in the mammosphere cultures.

3.6. Sulconazole Inhibits Stem Cell Marker Gene Expression and Mammosphere Growth

To determine whether sulconazole regulates stem cell marker genes, we tested the transcription of stem cell marker genes. Sulconazole inhibited the expression of genes such as Nanog, c-Myc, and CD44 in BCSCs (Figure 7A). To verify that sulconazole reduces mammosphere growth, we added sulconazole to a mammosphere culture and counted the mammosphere cells. Sulconazole induced cell death in the mammospheres. These data showed that sulconazole led to a dramatic reduction in mammosphere growth (Figure 7B). These data showed that NF-κB signaling was essential for regulating mammosphere growth and that sulconazole inhibited mammosphere formation through deregulation of the NF-κB/IL-8 signaling pathway.

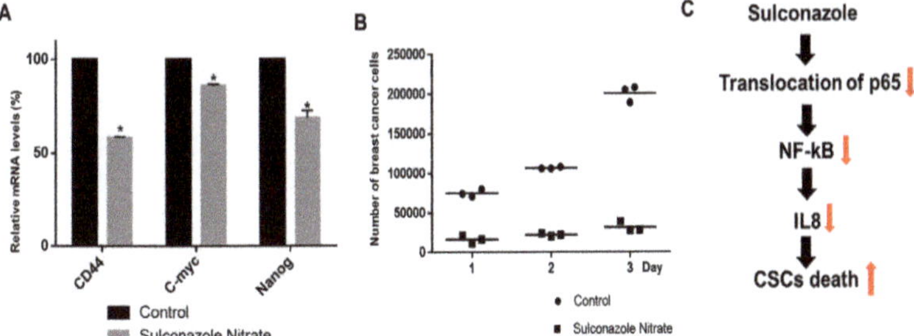

Figure 7. Effects of CSC loads on breast cancer. (**A**) The transcriptional levels of Nanog, C-myc, and CD44 were assayed in sulconazole- and 0.1% DMSO-treated mammospheres using specific primers. β-actin acted as an internal control. The data are presented as the mean ± SD; $n = 3$ independent experiments;* $p < 0.05$ vs. the control. (**B**) Sulconazole prevented mammosphere growth. Sulconazole-treated mammospheres were dissociated into single cells and plated in 6-well plates with equal numbers of cells. The cells were counted in triplicate 1, 2, and 3 days after plating and the mean value was plotted. The data shown represent the mean ± SD of three independent experiments. (**C**) The proposed model for CSC death induced by sulconazole is shown.

4. Discussion

Breast cancer is a female cancer that develops in the breast tissue. Breast cancer treatment using chemotherapy and radiotherapy eradicates the primary tumor, resulting in an increased survival rate in breast cancer patients [24]. Cancer metastasis and relapse have been attributed to CSC existence after chemotherapy [25]. BCSCs remain incompletely understood and are potential targets for breast cancer therapies [26]. The minimum biomarkers for BCSCs are the cell-surface markers CD44+/CD24- and CD44 upregulation is linked to tumor formation [27].

Our data show that sulconazole has potential as an antitumor and anti-CSC agent for breast cancer therapy. Sulconazole inhibits breast cancer hallmarks (Figure 1) and BCSC hallmarks (Figures 3 and 4). It is well known that the maintenance of BCSC properties is regulated by Stat3 [19,28,29]. We checked the Stat3 signaling pathway in the context of sulconazole treatment, but sulconazole did not regulate the Stat3 signaling pathway (Figure 6). The involvement of the NF-κB signaling pathway has been observed in primary AML samples, and elevated or constitutive NF-κB signaling activation is known to be present in many solid tumor types [30]. A high level of nuclear p65 is an essential feature of CSC formation [31]. As the NF-κB signaling pathway is important for BCSC survival, we examined the localization of the p65 subunit. Our results showed that sulconazole inhibited the translocation of p65. The inhibition of p65 translocation induced the inhibition of BCSC formation. CAPE is a strong specific inhibitor of NF-κB and prevents the nuclear translocation of p65 [32]. CAPE inhibited the translocation of p65 and pp65 and induced the inhibition of mammosphere formation. A siRNA specific for p65 inhibited mammosphere formation. Pyrrolidinedithiocarbamate (PDTC), another NF-κB pathway inhibitor, is known to inhibits CSC formation [33]. The imidazole-class drug BMS-345541 IKK inhibitor as an NF-kB inhibitor reduced their stem cell concentrations and self-renewal capacity in lung cancer cells. The nuclear levels of p65 and NF-κB signaling are important for BCSC survival.

Tumor progression and CSC survival can be regulated by the cytokines IL-6 and IL-8 in an autocrine or paracrine manner [10,34,35]. The JAK2/STAT3/IL-6 pathway is hyperactivated in several types of cancer and important to the growth of BCSCs [28]. This hyperactivation is related to a poor prognosis [36]. Extracellular IL-8 is overexpressed in triple-negative breast cancer (TNBC) and is an important therapeutic target in TNBC [37]. IL-8 signaling is an important key for targeting BCSCs [14]. We know that NF-κB can regulate the transcriptional regulation of the IL-6 and IL-8 genes in BCSCs. We assessed IL-6 and IL-8 gene transcripts in BCSCs treated with a p65 translocation inhibitor, sulconazole, and a siRNA specific for p65 that induced p65 downregulation. Both conditions showed that the RNA level of IL-8 was lower in treated samples than in control samples, but the RNA level of IL-6 was not changed between the treated samples and the control samples (Figure 6). Sulconazole reduced CSC formation through downregulation of NF-κB/IL-8 in breast cancer.

Sulconazole is an antifungal and antibacterial medicine in the imidazole class and is available as a cream to treat skin infection. Sulconazole inhibits the growth of common pathogenic dermatophytes by blocking sterol 14α-demethylase (CYP51) [38]. In this study, we first showed that sulconazole has cytotoxicity against breast cancer cells and reduces BCSC characteristics. These results support sulconazole as an important therapeutic agent to inhibit breast cancer and BCSCs.

5. Conclusions

In this study, we found that sulconazole—an antifungal medicine in the imidazole class—inhibited cell proliferation, tumor growth, and CSC formation. This compound also reduced the frequency of cells with the CSC marker phenotype of CD44$^+$/CD24$^-$ as well as the expression of ALDH and self-renewal-related genes. Sulconazole inhibited mammosphere formation and reduced the protein level of nuclear NF-κB. NF-κB knockdown using a p65-specific siRNA reduced CSC formation and secreted IL-8 levels in mammospheres. Sulconazole reduced nuclear NF-κB protein levels and extracellular IL-8 levels. These new findings showed that sulconazole blocked the NF-κB/IL-8 signaling pathway and CSC formation. NF-κB/IL-8 signaling is important for CSC formation and may be an important therapeutic target for BCSC treatment.

Supplementary Materials: The following are available online at http://www.mdpi.com/2073-4409/8/9/1007/s1, Supplementary Table S1. Specific primer sequences for real-time RT-PCR.

Author Contributions: H.S.C. and J.-H.K. designed and performed all the experiments; H.S.C. and D.-S.L. wrote the manuscript; S.-L.K. helped to design and perform the experiments; D.-S.L. supervised the study.

Funding: This research was supported by the Basic Science Research Program through the National Research Foundation of Korea (NRF) funded by the Ministry of Education (NRF-2016R1A6A1A03012862 and NRF-2019R1H1A2039751).

Conflicts of Interest: The authors declare that they have no conflicts of interest.

References

1. Torre, L.A.; Bray, F.; Siegel, R.L.; Ferlay, J.; Lortet-Tieulent, J.; Jemal, A. Global cancer statistics, 2012. *CA Cancer J. Clin.* **2015**, *65*, 87–108. [CrossRef] [PubMed]

2. Kusoglu, A.; Biray Avci, C. Cancer stem cells: A brief review of the current status. *Gene* **2019**, *681*, 80–85. [CrossRef] [PubMed]

3. Bonnet, D.; Dick, J.E. Human acute myeloid leukemia is organized as a hierarchy that originates from a primitive hematopoietic cell. *Nat. Med.* **1997**, *3*, 730–737. [CrossRef] [PubMed]

4. Gotte, M.; Yip, G.W. Heparanase, hyaluronan, and CD44 in cancers: A breast carcinoma perspective. *Cancer Res.* **2006**, *66*, 10233–10237. [CrossRef] [PubMed]

5. Nagano, O.; Okazaki, S.; Saya, H. Redox regulation in stem-like cancer cells by CD44 variant isoforms. *Oncogene* **2013**, *32*, 5191–5198. [CrossRef] [PubMed]

6. Koury, J.; Zhong, L.; Hao, J. Targeting Signaling Pathways in Cancer Stem Cells for Cancer Treatment. *Stem Cells Int.* **2017**, *2017*, 2925869. [CrossRef] [PubMed]

7. Guzman, M.L.; Neering, S.J.; Upchurch, D.; Grimes, B.; Howard, D.S.; Rizzieri, D.A.; Luger, S.M.; Jordan, C.T. Nuclear factor-kappaB is constitutively activated in primitive human acute myelogenous leukemia cells. *Blood* **2001**, *98*, 2301–2307. [CrossRef] [PubMed]

8. Vazquez-Santillan, K.; Melendez-Zajgla, J.; Jimenez-Hernandez, L.E.; Gaytan-Cervantes, J.; Munoz-Galindo, L.; Pina-Sanchez, P.; Martinez-Ruiz, G.; Torres, J.; Garcia-Lopez, P.; Gonzalez-Torres, C.; et al. NF-kappaBeta-inducing kinase regulates stem cell phenotype in breast cancer. *Sci. Rep.* **2016**, *6*, 37340. [CrossRef]

9. Zakaria, N.; Mohd Yusoff, N.; Zakaria, Z.; Widera, D.; Yahaya, B.H. Inhibition of NF-kappaB Signaling Reduces the Stemness Characteristics of Lung Cancer Stem Cells. *Front. Oncol.* **2018**, *8*, 166. [CrossRef]

10. Korkaya, H.; Liu, S.; Wicha, M.S. Regulation of cancer stem cells by cytokine networks: Attacking cancer's inflammatory roots. *Clin. Cancer Res.* **2011**, *17*, 6125–6129. [CrossRef]

11. Silva, E.M.; Mariano, V.S.; Pastrez, P.R.A.; Pinto, M.C.; Castro, A.G.; Syrjanen, K.J.; Longatto-Filho, A. High systemic IL-6 is associated with worse prognosis in patients with non-small cell lung cancer. *PLoS ONE* **2017**, *12*, e0181125. [CrossRef] [PubMed]

12. Lippitz, B.E.; Harris, R.A. Cytokine patterns in cancer patients: A review of the correlation between interleukin 6 and prognosis. *Oncoimmunology* **2016**, *5*, e1093722. [CrossRef] [PubMed]

13. Sansone, P.; Storci, G.; Tavolari, S.; Guarnieri, T.; Giovannini, C.; Taffurelli, M.; Ceccarelli, C.; Santini, D.; Paterini, P.; Marcu, K.B.; et al. IL-6 triggers malignant features in mammospheres from human ductal breast carcinoma and normal mammary gland. *J. Clin. Investig.* **2007**, *117*, 3988–4002. [CrossRef] [PubMed]

14. Singh, J.K.; Simoes, B.M.; Howell, S.J.; Farnie, G.; Clarke, R.B. Recent advances reveal IL-8 signaling as a potential key to targeting breast cancer stem cells. *Breast Cancer Res.* **2013**, *15*, 210. [CrossRef] [PubMed]

15. Bae, S.H.; Park, J.H.; Choi, H.G.; Kim, H.; Kim, S.H. Imidazole Antifungal Drugs Inhibit the Cell Proliferation and Invasion of Human Breast Cancer Cells. *Biomol. Ther.* **2018**, *26*, 494–502. [CrossRef] [PubMed]

16. Benfield, P.; Clissold, S.P. Sulconazole. A review of its antimicrobial activity and therapeutic use in superficial dermatomycoses. *Drugs* **1988**, *35*, 143–153. [CrossRef] [PubMed]

17. Clarke, M.L.; Burton, R.L.; Hill, A.N.; Litorja, M.; Nahm, M.H.; Hwang, J. Low-cost, high-throughput, automated counting of bacterial colonies. *Cytom. Part A* **2010**, *77*, 790–797. [CrossRef] [PubMed]

18. Choi, H.S.; Kim, D.A.; Chung, H.; Park, I.H.; Kim, B.H.; Oh, E.S.; Kang, D.H. Screening of breast cancer stem cell inhibitors using a protein kinase inhibitor library. *Cancer Cell Int.* **2017**, *17*, 25. [CrossRef] [PubMed]

19. Choi, H.S.; Kim, J.H.; Kim, S.L.; Deng, H.Y.; Lee, D.; Kim, C.S.; Yun, B.S.; Lee, D.S. Catechol derived from aronia juice through lactic acid bacteria fermentation inhibits breast cancer stem cell formation via modulation Stat3/IL-6 signaling pathway. *Mol. Carcinog.* **2018**, *57*, 1467–1479. [CrossRef] [PubMed]

20. Choi, H.S.; Kim, S.L.; Kim, J.H.; Deng, H.Y.; Yun, B.S.; Lee, D.S. Triterpene Acid (3-*O*-*p*-Coumaroyltormentic Acid) Isolated From Aronia Extracts Inhibits Breast Cancer Stem Cell Formation through Downregulation of c-Myc Protein. *Int. J. Mol. Sci.* **2018**, *19*, 2528. [CrossRef] [PubMed]

21. Choi, H.S.; Hwang, C.K.; Kim, C.S.; Song, K.Y.; Law, P.Y.; Wei, L.N.; Loh, H.H. Transcriptional regulation of mouse mu opioid receptor gene: Sp3 isoforms (M1, M2) function as repressors in neuronal cells to regulate the mu opioid receptor gene. *Mol. Pharmacol.* **2005**, *67*, 1674–1683. [CrossRef] [PubMed]

22. Kim, S.L.; Choi, H.S.; Kim, J.H.; Jeong, D.K.; Kim, K.S.; Lee, D.S. Dihydrotanshinone-Induced NOX5 Activation Inhibits Breast Cancer Stem Cell through the ROS/Stat3 Signaling Pathway. *Oxid. Med. Cell. Longev.* **2019**, *2019*, 9296439. [CrossRef] [PubMed]

23. Liang, Y.; Feng, G.; Wu, L.; Zhong, S.; Gao, X.; Tong, Y.; Cui, W.; Qin, Y.; Xu, W.; Xiao, X.; et al. Caffeic acid phenethyl ester suppressed growth and metastasis of nasopharyngeal carcinoma cells by inactivating the NF-kappaB pathway. *Drug Des. Dev. Ther.* **2019**, *13*, 1335–1345. [CrossRef] [PubMed]

24. Siegel, R.L.; Miller, K.D.; Jemal, A. Cancer statistics, 2016. *CA Cancer J. Clin.* **2016**, *66*, 7–30. [CrossRef] [PubMed]

25. Yu, Y.; Ramena, G.; Elble, R.C. The role of cancer stem cells in relapse of solid tumors. *Front. Biosci.* **2012**, *4*, 1528–1541. [CrossRef]

26. Saeg, F.; Anbalagan, M. Breast cancer stem cells and the challenges of eradication: A review of novel therapies. *Stem Cell Investig* **2018**, *5*, 39. [CrossRef] [PubMed]

27. Fillmore, C.; Kuperwasser, C. Human breast cancer stem cell markers CD44 and CD24: Enriching for cells with functional properties in mice or in man? *Breast Cancer Res.* **2007**, *9*, 303. [CrossRef] [PubMed]

28. Marotta, L.L.; Almendro, V.; Marusyk, A.; Shipitsin, M.; Schemme, J.; Walker, S.R.; Bloushtain-Qimron, N.; Kim, J.J.; Choudhury, S.A.; Maruyama, R.; et al. The JAK2/STAT3 signaling pathway is required for growth of CD44(+)CD24(−) stem cell-like breast cancer cells in human tumors. *J. Clin. Investig.* **2011**, *121*, 2723–2735. [CrossRef]

29. An, H.; Kim, J.Y.; Oh, E.; Lee, N.; Cho, Y.; Seo, J.H. Salinomycin Promotes Anoikis and Decreases the CD44+/CD24− Stem-Like Population via Inhibition of STAT3 Activation in MDA-MB-231 Cells. *PLoS ONE* **2015**, *10*, e0141919. [CrossRef]

30. Rinkenbaugh, A.L.; Baldwin, A.S. The NF-kappaB Pathway and Cancer Stem Cells. *Cells* **2016**, *5*, 16. [CrossRef]

31. Garner, J.M.; Fan, M.; Yang, C.H.; Du, Z.; Sims, M.; Davidoff, A.M.; Pfeffer, L.M. Constitutive activation of signal transducer and activator of transcription 3 (STAT3) and nuclear factor kappaB signaling in glioblastoma cancer stem cells regulates the Notch pathway. *J. Biol. Chem.* **2013**, *288*, 26167–26176. [CrossRef]

32. Natarajan, K.; Singh, S.; Burke, T.R., Jr.; Grunberger, D.; Aggarwal, B.B. Caffeic acid phenethyl ester is a potent and specific inhibitor of activation of nuclear transcription factor NF-kappa B. *Proc. Natl. Acad. Sci. USA* **1996**, *93*, 9090–9095. [CrossRef]

33. Zhou, J.; Zhang, H.; Gu, P.; Bai, J.; Margolick, J.B.; Zhang, Y. NF-kappaB pathway inhibitors preferentially inhibit breast cancer stem-like cells. *Breast Cancer Res. Treat.* **2008**, *111*, 419–427. [CrossRef]

34. Tan, C.; Hu, W.; He, Y.; Zhang, Y.; Zhang, G.; Xu, Y.; Tang, J. Cytokine-mediated therapeutic resistance in breast cancer. *Cytokine* **2018**, *108*, 151–159. [CrossRef]

35. Zhao, M.; Liu, Y.; Liu, R.; Qi, J.; Hou, Y.; Chang, J.; Ren, L. Upregulation of IL-11, an IL-6 Family Cytokine, Promotes Tumor Progression and Correlates with Poor Prognosis in Non-Small Cell Lung Cancer. *Cell. Physiol. Biochem.* **2018**, *45*, 2213–2224. [CrossRef]

36. Johnson, D.E.; O'Keefe, R.A.; Grandis, J.R. Targeting the IL-6/JAK/STAT3 signalling axis in cancer. *Nat. Rev. Clin. Oncol.* **2018**, *15*, 234–248. [CrossRef]

37. Dominguez, C.; McCampbell, K.K.; David, J.M.; Palena, C. Neutralization of IL-8 decreases tumor PMN-MDSCs and reduces mesenchymalization of claudin-low triple-negative breast cancer. *JCI Insight* **2017**, *2*, e94296. [CrossRef]
38. Thomson, S.; Rice, C.A.; Zhang, T.; Edrada-Ebel, R.; Henriquez, F.L.; Roberts, C.W. Characterisation of sterol biosynthesis and validation of 14alpha-demethylase as a drug target in Acanthamoeba. *Sci. Rep.* **2017**, *7*, 8247. [CrossRef]

Article

Dexamethasone Inhibits Spheroid Formation of Thyroid Cancer Cells Exposed to Simulated Microgravity

Daniela Melnik [1], Jayashree Sahana [2], Thomas J. Corydon [2,3], Sascha Kopp [1,4], Mohamed Zakaria Nassef [1], Markus Wehland [1,4], Manfred Infanger [1,4], Daniela Grimm [2,4,5] and Marcus Krüger [1,4,*]

1   Clinic for Plastic, Aesthetic and Hand Surgery, Otto von Guericke University, Leipziger Str. 44, 39120 Magdeburg, Germany; daniela.melnik@med.ovgu.de (D.M.); sascha.kopp@med.ovgu.de (S.K.); mohamed.nassef@med.ovgu.de (M.Z.N.); markus.wehland@med.ovgu.de (M.W.); manfred.infanger@med.ovgu.de (M.I.)
2   Department of Biomedicine, Aarhus University, Hoegh-Guldbergsgade 10, 8000 Aarhus C, Denmark; jaysaha@biomed.au.dk (J.S.); corydon@biomed.au.dk (T.J.C.); dgg@biomed.au.dk (D.G.)
3   Department of Ophthalmology, Aarhus University Hospital, Palle Juul-Jensens Boulevard 99, 8200 Aarhus N, Denmark
4   Research Group "Magdeburger Arbeitsgemeinschaft für Forschung unter Raumfahrt- und Schwerelosigkeitsbedingungen" (MARS), Otto von Guericke University, Universitätsplatz 2, 39106 Magdeburg, Germany
5   Department of Microgravity and Translational Regenerative Medicine, Otto von Guericke University, Pfälzer Platz, 39106 Magdeburg, Germany
*   Correspondence: marcus.krueger@med.ovgu.de; Tel.: +49-391-6721-267

Received: 9 December 2019; Accepted: 4 February 2020; Published: 5 February 2020

Abstract: Detachment and the formation of spheroids under microgravity conditions can be observed with various types of intrinsically adherent human cells. In particular, for cancer cells this process mimics metastasis and may provide insights into cancer biology and progression that can be used to identify new drug/target combinations for future therapies. By using the synthetic glucocorticoid dexamethasone (DEX), we were able to suppress spheroid formation in a culture of follicular thyroid cancer (FTC)-133 cells that were exposed to altered gravity conditions on a random positioning machine. DEX inhibited the growth of three-dimensional cell aggregates in a dose-dependent manner. In the first approach, we analyzed the expression of several factors that are known to be involved in key processes of cancer progression such as autocrine signaling, proliferation, epithelial–mesenchymal transition, and anoikis. Wnt/β-catenin signaling and expression patterns of important genes in cancer cell growth and survival, which were further suggested to play a role in three-dimensional aggregation, such as *NFKB2*, *VEGFA*, *CTGF*, *CAV1*, *BCL2(L1)*, or *SNAI1*, were clearly affected by DEX. Our data suggest the presence of a more complex regulation network of tumor spheroid formation involving additional signal pathways or individual key players that are also influenced by DEX.

Keywords: glucocorticoids; 3D growth; nuclear factor kappa-light-chain-enhancer of activated B-cells (NF-κB); epithelial–mesenchymal transition; anoikis; proliferation

---

1. Introduction

Glucocorticoids (GCs) are a class of steroid hormones involved in many physiological processes such as metabolism, proliferation, differentiation, and survival of cells [1]. GCs induce their pharmacodynamic effects through binding to glucocorticoid receptors (GRs) [2], which interact downstream with signaling molecules in the cytoplasm or are able to translocate into the nucleus, where they repress the activity of other transcription factors (such as nuclear factor kappa-light-chain-enhancer of activated B-cells, NF-κB, or activator protein 1, AP-1) or initiate transcription of genes associated with anti-inflammatory and immunosuppressive effects (via binding to specific glucocorticoid response elements, GREs) (Figure 1A) [3,4]. Due to these properties, GCs are utilized in the treatment of a variety of immunological disorder treatments to reduce pain and electrolyte imbalance, but also to enhance the anti-tumor effect of chemotherapeutics and prevent adverse effects caused by cytotoxic agents [5–7]. The synthetic GC dexamethasone (DEX; Figure 1B) is commonly administered as a supportive care co-medication to reduce cancer-related fatigue in patients with advanced disease [8]. DEX was further reported to inhibit proliferation of different cancer cells in vitro and in vivo [9–13]. Effective inhibition of tumor growth was suggested to be associated with downregulation of JAK3/STAT3, hypoxia inducible factor 1α, vascular endothelial growth factor (VEGF), and interleukin-6 [12,14,15]. Nevertheless, the exact mechanism by which DEX suppresses cancer cell growth is still unclear.

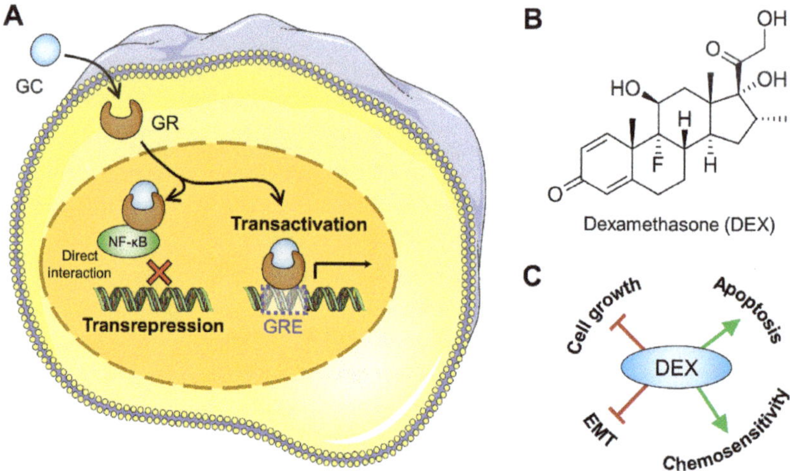

Figure 1. (**A**) Sketch showing the genomic actions of glucocorticoids (GCs) such as dexamethasone (DEX). When bound to DEX, the glucocorticoid receptor (GR) complex translocates into the nucleus and modifies the synthesis of several metabolic proteins. This is done either through directly binding to glucocorticoid response elements (GREs) on the DNA or through influencing the activity of transcription factors (i.e., nuclear factor kappa-light-chain-enhancer of activated B-cells, NF-κB); (**B**) Chemical structure of DEX; (**C**) Described effects of DEX on cancer cells. Parts of the figure are drawn using pictures from Servier Medical Art (https://smart.servier.com), licensed under a Creative Commons Attribution 3.0 Unported License (https://creativecommons.org/licenses/by/3.0).

Over the last 50 years, the incidence of thyroid cancer has increased worldwide and the incidence rate is still on the rise. The result of improved diagnostic procedures, an elevated prevalence of individual risk factors (e.g., obesity), and increased exposure to environmental risk factors (e.g., iodine levels), thyroid cancer is the most common form of endocrine malignancy today [16] and is expected to become the fourth leading type of cancer across the globe [17]. Especially poorly differentiated thyroid tumors are aggressive and tend to metastasize. The prognosis for differentiated thyroid cancer is related to the capability of tumor cells to accumulate radioiodine. Due to de-differentiation, some tumor cells may lose their iodine uptake capability, leaving only extremely limited treatment options, despite intensive searches for new drugs and targets. Therefore, novel approaches to control thyroid cancer progression are required.

Metastasis is the most limiting factor in cancer therapy and responsible for 90% of cancer-related deaths [18]. During the development of metastatic competence, carcinoma cells change their adhesive properties, secrete proteinases, and become motile, which allows them to detach from their primary tumor [19]. Therefore, tumor cells respond to mechanical signals, sensed by integrins or other adhesion receptors [20,21], and chemical signals, sensed by chemokines or growth factor receptors [22] causing changes in their transcriptional profile. The process which enables tumor cells to achieve migration and invasion is called epithelial–mesenchymal transition (EMT) and represents a driving force in tumorigenesis [23]. In the course of EMT, essential proteins for epithelial cell–cell adhesion, such as E-cadherin, are downregulated, thus weakening epithelial tissue integrity and polarization of epithelial cell layers [24]. Under normal circumstances, detached epithelial cells undergo apoptosis, a phenomenon termed anoikis. Cancer cells acquire resistance to anoikis to survive after they have left the primary tumor. In this way, they are able to travel via the circulatory and lymphatic systems disseminating throughout the body. EMT and anoikis resistance are critical steps of the metastatic cascade and potential targets to impact a natural molecular prerequisite for the aggressive metastatic spread of cancer [25,26].

Microgravity (μg) has become a powerful tool in cancer research by enabling metastasis-like cell detachment and formation of three-dimensional (3D) multicellular spheroids (MCS) [27–30]. Experiments in μg contribute to drug discovery by providing an environment which is helpful to detect changes in gene expression and protein synthesis and secretion that occur during the progression from 2D to 3D growth and which might represent new targets for drug development against thyroid cancer. A couple of these proteins were found in follicular thyroid cancer cells by analyzing multiple pilot studies, performed in μg, with the help of semantic methods [31,32]. Some of these potential drugs, including DEX, were recently reviewed [33]. In this study, we investigate the effects of DEX supplementation on the growth of follicular thyroid cancer (FTC) cells exposed to simulated μg produced by a random positioning machine (RPM).

2. Materials and Methods

2.1. Cell Culture

The human follicular thyroid carcinoma cell line FTC-133 was cultured in RPMI-1640 medium (Life Technologies, Carlsbad, CA, USA), supplemented with 10% fetal calf serum (FCS; Sigma-Aldrich, St. Louis, MO, USA), and 1% penicillin/streptomycin (Life Technologies) at 37 °C and 5% CO_2 until use for the experiment. For RPM experiments FTC-133 cells were seeded at a density of 1×10^6 cells per flask either in T25 cell culture flasks (Sarstedt, Nümbrecht, Germany) for mRNA and protein extraction or in slide flasks (Sarstedt) for immunofluorescence staining. Cells were given at least 24 h to attach to the bottom of the flasks.

2.2. Dexamethasone Treatment

Water-soluble DEX (dexamethasone–cyclodextrin complex) was purchased from Sigma-Aldrich. Then, 24 h after seeding, cells were synchronized in RPMI-1640 medium with 0.25% FCS and 1% penicillin/streptomycin for 4 h. Afterwards, the cells were cultured according to Section 2.1, supplemented with DEX concentrations of 10 nM, 100 nM, or 1000 nM [34].

2.3. Random Positioning Machine

The used desktop-RPM (Dutch Space, Leiden, Netherlands) was located in an incubator with 37 °C/5% CO_2 and operated in real random mode, with a constant angular velocity of 60°/s. Before the run, the flasks were filled up completely and air bubble-free with medium to avoid shear stress. The slide and culture flasks were installed on the prewarmed RPM. After 4 h (short-term experiments) or 3 days (long-term experiments), the cells were photographed and fixed with 4% paraformaldehyde (PFA; Carl Roth, Karlsruhe, Germany) for immunostaining. For RNA and protein extraction adherent cells were harvested by adding ice-cold phosphate-buffered saline (PBS; Life Technologies) and using cell scrapers. The suspensions were centrifuged at 3000× *g* for 10 min at 4 °C followed by discarding the PBS and storage of cell pellets at −150 °C. MCS were collected by centrifuging supernatant at

3000× g for 10 min at 4 °C and subsequent storage at −150 °C. Corresponding static controls were prepared in parallel under the same conditions and stored next to the device in an incubator.

2.4. Phase Contrast Microscopy

Cells were observed and photographed using an Axiovert 25 Microscope (Carl Zeiss Microscopy, Jena, Germany) equipped with a Canon EOS 550D camera (Canon, Tokio, Japan).

2.5. Immunofluorescence Microscopy

Immunofluorescence staining was performed to visualize possible translocal alteration of NF-κB proteins and β-catenin by dexamethasone in cells. The PFA-fixed cells were permeabilized with 0.1% Triton™ X-100 for 15 min and blocked with 3% bovine serum albumin (BSA) for 45 min at ambient temperature. Afterwards, the cells were labeled with primary NF-κB p65 rabbit polyclonal antibody #PA1-186 (Invitrogen, Carlsbad, CA, USA) at 1 μg/mL or β-catenin mouse monoclonal antibody #MA1-300 (Invitrogen) at a dilution of 1:200 in 0.1% BSA and incubated overnight at 4 °C in a moist chamber. The next day, cells were washed three times with PBS before incubation with the secondary Alexa Fluor 488 (AF488)-conjugated anti-rabbit (Cell Signaling Technology, Danvers, MA, USA) or anti-mouse antibody (Invitrogen) at a dilution of 1:1000 for 1 h at ambient temperature. Cells were washed again three times with PBS and mounted with Fluoroshield™ with DAPI (4′,6-diamidino-2-phenylindole) (Sigma-Aldrich). The slides were subsequently investigated with a Zeiss LSM 710 confocal laser scanning microscope (Carl Zeiss) [35].

2.6. mRNA Isolation and Quantitative Real-Time PCR

RNA isolation and quantitative real-time PCR were performed according to routine protocols [36–38]. Briefly, RNA was isolated by using the RNeasy Mini Kit (Qiagen, Venlo, Netherlands) according to the manufacturer's protocol and quantified with a spectrophotometer. Afterwards, cDNA was produced with the High Capacity cDNA Reverse Transcription Kit (Applied Biosystems, Foster City, CA, USA) following manufacturer's instructions. To determine the expression level of the target genes shown in Table S1, quantitative real-time PCR was performed applying the Fast SYBR™ Green Master Mix (Applied Biosystems) and the 7500 Fast Real-Time PCR System (Applied Biosystems). Primers were designed to have a $T_m \approx 60$ °C and to span exon/exon boundaries using Primer-BLAST [39] (Supplementary Materials). Primer were synthesized by TIB Molbio (Berlin, Germany). All samples were measured in triplicates and analyzed by the comparative threshold cycle ($\Delta\Delta C_T$) method with 18S rRNA as reference.

2.7. Western Blot Analysis

Western blot analysis was performed with routine protocols as described previously [36]. The control and DEX-treated samples were collected after 4 h and 3 days, solubilized in lysis buffer and compared to the control samples without DEX. Each condition included three batches with a total number of 24 samples (4 h) and 33 samples (3 days), respectively. Following lysis and centrifugation, aliquots of 40 μg total protein were subjected to SDS-PAGE and Western blotting. The samples were loaded onto Criterion XT 4–12% precast gels (Bio-Rad, Hercules, CA, USA), run for 1 h at 150 V and transferred to a polyvinylidene difluoride membrane using TurboBlot (Bio-Rad) (100 V, 30 min). Cyclophilin B was used as a loading control. Membranes were blocked with TBS-T containing 0.3% I-Block (Applied Biosystems) for 2 h at ambient temperature. For detection of the proteins shown in Table S2, the membranes were incubated overnight at 4 °C in TBS-T containing 0.3% I-Block solutions of the antibodies. The next day, membranes were washed three times with TBS-T for 5 min and incubated for 2 h at ambient temperature with a horseradish peroxidase (HRP)-linked antibody (Cell Signaling Technology) diluted 1:1000 in TBS-T with 0.3% I-Block. The respective protein bands were visualized using Clarity ECL Western Blot Substrate (Bio-Rad). Images were captured with Image Quant LAS 4000 mini (GE Healthcare, Chicago, IL, USA) and analyzed using ImageJ software (imagej.net) for densitometric quantification.

2.8. Terminal Deoxynucleotidyl Transferase dUTP Nick-End Labeling (TUNEL) Assay

The Click-iT™ Plus TUNEL assay (Invitrogen) was used for apoptosis detection. FTC-133 cells were cultured in slide flasks (Sarstedt) under static culture conditions or exposed to the RPM, supplemented without or with 1000 nM DEX. After 4 h or 3 days cells were fixed with 4% PFA and prepared for the evaluation of apoptosis. The staining procedure was performed according to the manufacturer's recommendation. A positive control sample was treated with DNase I (Epicentre, Madison, WI, USA) to induce DNA fragmentation. The stained cells were examined using a Zeiss LSM 800 confocal laser scanning microscope (Carl Zeiss) equipped with an external light source and an objective with a calibrated 630× magnification.

2.9. Ki-67 Proliferation Assay

Cells were cultured and prepared as described in Section 2.5. Cells were labeled with an AF488 recombinant anti-Ki-67 antibody #ab197234 (Abcam, Cambridge, UK) at a dilution of 1:100 in 0.1% BSA and incubated overnight at 4 °C. The next day, cells were washed three times with PBS and mounted with Fluoroshield™ with DAPI (Sigma-Aldrich). The cell proliferation was evaluated by a Zeiss LSM 800 confocal laser scanning microscope (Carl Zeiss) and an objective with a calibrated 230× magnification. Five microscopic images for each condition were analyzed using ImageJ (imagej.net). The percentage of Ki-67 positive cells was counted for each condition and normalized to the control.

2.10. Spheroid Formation Assay

Approximately 1×10^6 cells per flasks were seeded into T25 cell culture flasks (Sarstedt). After 24 h the culture flasks were filled up completely (air bubble-free) with media and were installed on the prewarmed RPM (37 °C, 5% CO_2). To investigate the ability of MCS formation, two RPM running time points were considered: media was completely removed from culture flasks after 24 h and after 48 h exposure to the RPM. Flasks were re-filled with fresh media for a further 24 h run on the RPM. After each run, cells were examined and photographed.

2.11. Statistics

Statistical evaluation was performed using GraphPad Prism 7.01 (GraphPad Software, San Diego, CA, USA). The nonparametric Mann–Whitney U test was used to compare DEX-free with DEX-treated samples as well as static and µg conditions. All data are presented as mean ± standard deviation (SD) with a significance level of $p < 0.05$.

3. Results

Based on the knowledge that NF-κB seems to play a crucial role in spheroid formation of MCF-7 breast cancer cells [34] and that NF-κB subunit p65 (RelA) accumulates in thyroid cancer cells on the RPM [40], we decided to target RelA in µg-grown thyroid cancer cells using DEX. Therefore, we cultured the human follicular thyroid cancer cell line FTC-133 on an RPM in the presence of three different DEX concentrations (10, 100, 1000 nM). After three days on the RPM, the cells showed a DEX dose-dependent inhibition of spheroid formation in µg (Figure 2).

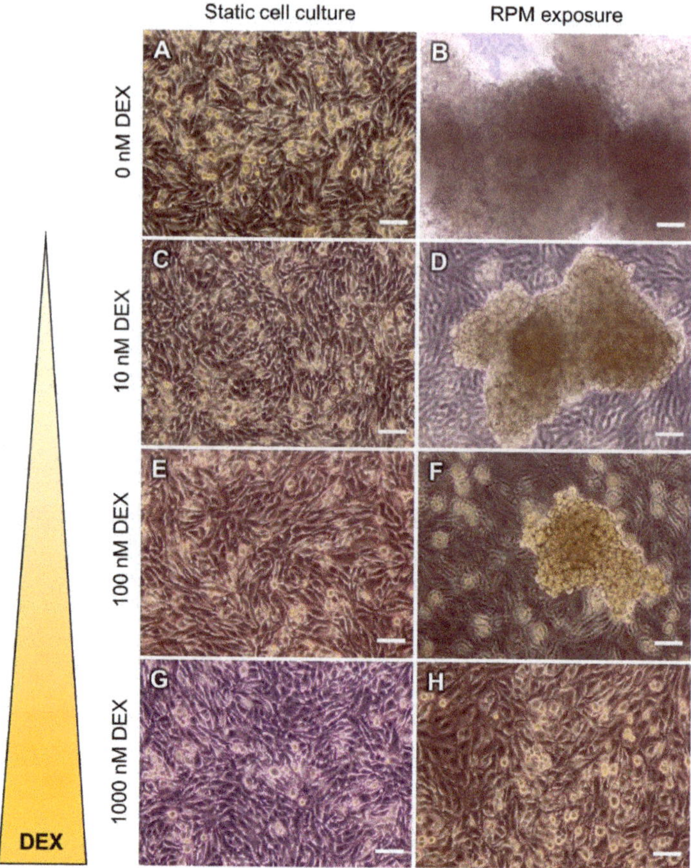

Figure 2. Impact of DEX on the spheroid formation ability of follicular thyroid cancer (FTC)-133 cells exposed to a random positioning machine (RPM). (**A,B**) After three days cells showed a dose-dependent inhibition of spheroid formation when treated with (**C,D**) 10 nM DEX, (**E,F**) 100 nM DEX, or (**G,H**) 1000 nM DEX on the RPM (right column). Scale bars: 100 µm.

3.1. NF-κB Pathway

NF-κB transcription factors play a fundamental role in the tumorigenesis of many cancer types, including thyroid cancer [41,42] and may be a target in the treatment of advanced thyroid cancer [43]. DEX is known to have inhibitory effects on the NF-κB pathway [44].

We investigated the transcription of the NF-κB family members subunit p50 and its precursor p105 (encoded by *NFKB1*) as well as subunit p52 and its precursor p100 (encoded by *NFKB2*). The mRNA levels of both genes were reduced in MCS cells grown for three days on the RPM and *NFKB2* was upregulated in adherently growing (AD) cells, harvested from the RPM (Figure 3A,B). In addition, we found a dose-dependent inhibitory effect of DEX on the mRNA synthesis of *NFKB2* (Figure 3B) and a less pronounced effect on the mRNA synthesis of *NFKB1* (Figure 3A). In contrast to DEX-treated MCF-7 cells [34], RelA was not translocated into the nucleus of FTC-133 cells in a significant amount after DEX supplementation (Figure 3E,F). Furthermore, RelA expression was not significantly altered by DEX on the transcriptional level (Figure 3C). RelA protein was increased after three days on the RPM and seemed to be augmented by DEX treatment in µg-exposed cells (Figure 3D).

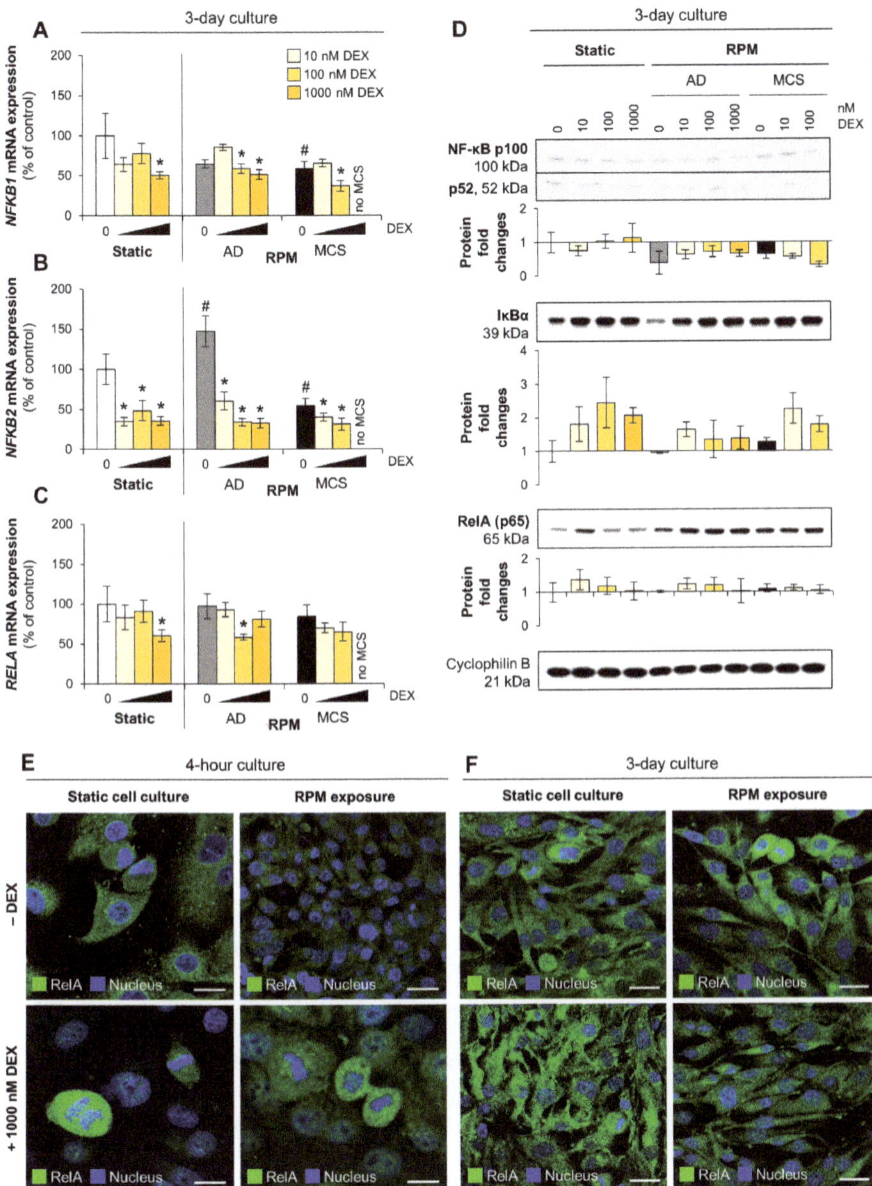

Figure 3. Effect of DEX on NF-κB family members in FTC-133 cells. (**A**) *NFKB1* mRNA expression; (**B**) *NFKB2* mRNA expression; (**C**) *RELA* mRNA expression. Depicted are means of relative mRNA levels ± standard deviations (*n* = 5). *: *p* < 0.05 vs. DEX-free samples. #: *p* < 0.05 vs. static cultures; (**D**) Western blots indicate protein levels of regulated genes after three days. Representatives of each of the three replicates are shown. Diagrams describe relative fold changes to control. AD: adherently growing cells; MCS: multicellular spheroids. (**E,F**) Immunofluorescence shows only minor translocation of RelA (green) into the nucleus (blue; 4′,6-diamidino-2-phenylindole (DAPI)-stained) in FTC-133 cells. Scale bars: 20 μm.

NF-κB dimers can be sequestered in the cytoplasm by the inhibitor of κB (IκB) proteins. Therefore, we analyzed the expression of IκBα (encoded by *NFKBIA*), IκBβ (encoded by *NFKBIB*), and IκBε (encoded by *NFKBIE*). The effect of DEX supplementation on *NFKBIA* expression was limited to RPM-exposed cells (Figure 4A), but it tended to upregulate the IκBα protein level in three-day cultures, independent of gravity (Figure 3D). In addition, DEX lowered *NFKBIB* and *NFKBIE* mRNA in cells cultured under normal conditions. The *NFKBIB* mRNA synthesis seemed to be increased (Figure 4B) whereas *NFKBIE* mRNA synthesis was reduced in adherently growing cells on the RPM (Figure 4C). Overall, the mRNA synthesis of all IκB proteins was reduced by long-term exposure to the RPM.

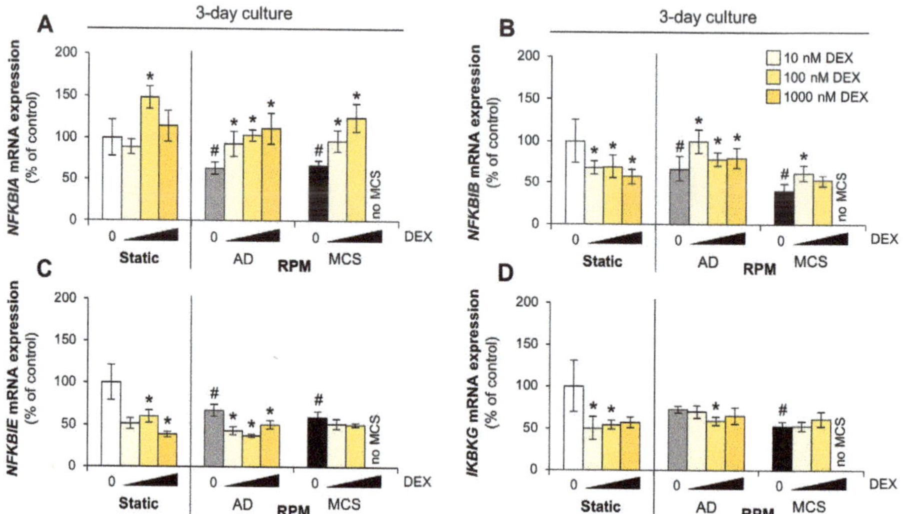

Figure 4. Effect of DEX on NF-κB regulators in FTC-133 cells. (**A**) *NFKBIA* mRNA expression; (**B**) *NFKBIB* mRNA expression; (**C**) *NFKBIE* mRNA expression; (**D**) *IKBKG* mRNA expression. Depicted are means of relative mRNA levels ± standard deviations ($n = 5$). *: $p < 0.05$ vs. DEX-free samples. #: $p < 0.05$ vs. static cultures. AD: adherently growing cells; MCS: multicellular spheroids.

Activation of NF-κB is initiated by the signal-induced degradation of IκB proteins, mainly via activation of IκB kinase (IKK). IKK is composed of the catalytic IKKα/IKKβ heterodimer and the master regulator NEMO (NF-κB essential modulator), also referred as IKKγ (encoded by *IKBKG*). Three-day MCS showed a reduction in *IKBKG* mRNA synthesis. DEX reduced *IKBKG* mRNA synthesis only in static cultured cells after three days. Under all other conditions, transcription was unaffected by DEX supplementation (Figure 4D).

Since NF-κB is obviously not the main target of DEX in suppressing µg-based spheroid formation of FTC-133 cells, we proceeded to illuminate further cancer-related processes which have been reported in connection with DEX (Figure 1C) and that are also involved in the formation of tumor spheroids.

3.2. Growth Factors and Proliferation

Different growth factors are expressed and secreted by cancer cells and contribute to proliferation, survival, and migration. Previous experiments designed to elucidate the growth behavior of cancer cells in µg reveal that especially connective tissue growth factor (CTGF), epidermal growth factor (EGF), transforming growth factor beta (TGF-β), and VEGF were regulated in FTC-133 cells after gravity was omitted [37]. CTGF is a member of the CCN family of secreted, matrix-associated proteins that plays a key role in tumor development, progression, and angiogenesis [45]. CTGF is suggested to regulate cancer cell migration, invasion, angiogenesis, and anoikis [46]. In our experiments, CTGF was upregulated in adherently growing FTC-133 cells after DEX supplementation (Figure 5A). RPM-exposure also enhanced the *CTGF* mRNA level resulting in an additive effect of µg and DEX supplementation. However, the transcription was lower in MCS cells compared to cells in static cultures after three days (Figure 5A).

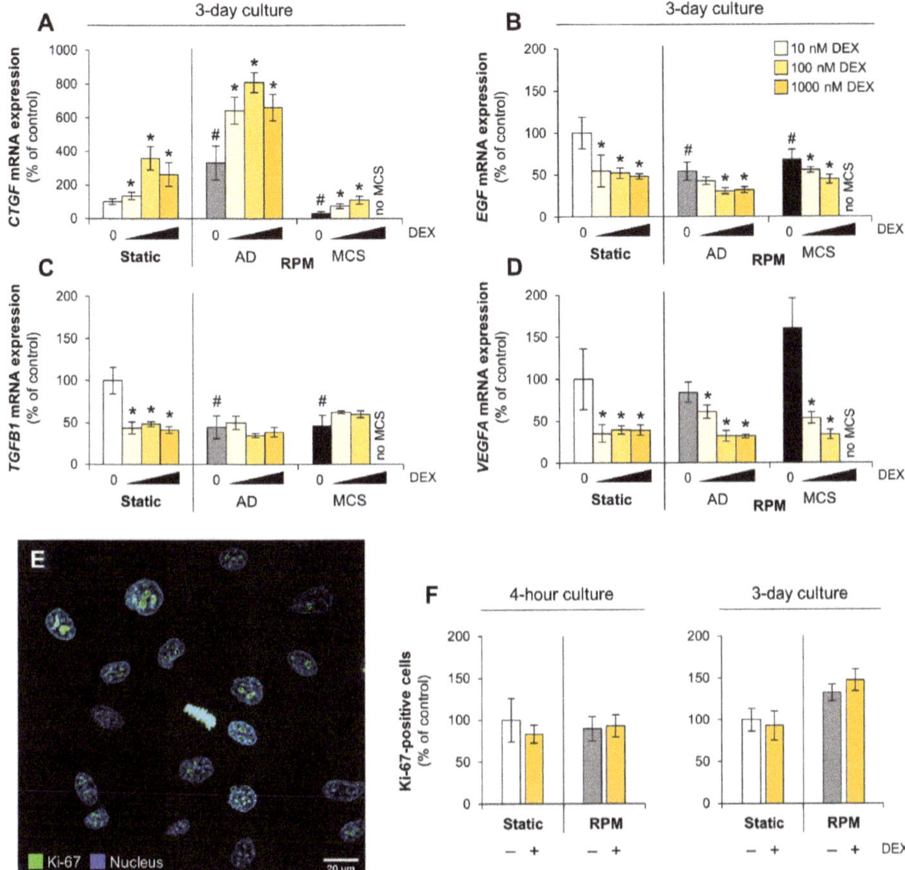

Figure 5. Effect of DEX on autocrine growth factors and proliferation markers in FTC-133 cells. (**A**) *CTGF* mRNA expression; (**B**) *EGF* mRNA expression; (**C**) *TGFB1* mRNA expression; (**D**) *VEGFA* mRNA expression. Depicted are means of relative mRNA levels ± standard deviations ($n = 5$); (**E**) Immunofluorescence. Nuclear expression of Ki-67 indicates proliferating cells; (**F**) Proliferation analysis using Ki-67. *: $p < 0.05$ vs. DEX-free samples. #: $p < 0.05$ vs. static cultures. AD: adherently growing cells; MCS: multicellular spheroids.

TGF-β and EGF represent two physiological regulators of thyroid cell differentiation and proliferation. Whereas EGF is a strong mitogen for follicular thyroid cells [47], TGF-β has a complicated role in cancer. Initially, TGF-β is a tumor suppressor that inhibits the growth of thyrocytes and induces apoptosis [48]. However, at later stages of tumor progression, TGF-β acts as a potent EMT inducer and then it plays a fundamental role in tumor progression and metastasis formation [49–51]. *EGF* mRNA was downregulated both in presence of DEX and in µg (Figure 5B). *TGFB1* mRNA levels were also lower in µg-grown cells, but DEX decreased *TGFB1* mRNA synthesis only in cells from static cultures (Figure 5C). In addition, DEX suppressed VEGF under normal culture conditions and in both cell populations on the

RPM (Figure 5D). In accordance with previous studies that investigated other follicular thyroid cancer cells on the RPM [52], *VEGFA* expression was somewhat increased in MCS cells after three days.

DEX was previously reported to have anti-proliferative effects on human medullary thyroid cancer cells [53]. To prove this effect with follicular thyroid cancer cells and in the context of µg, we searched for cellular markers for proliferation such as the Ki-67 protein (encoded by *MKI67*) [54]. Ki-67 can be detected during all active phases of the cell cycle (G1, S, G2, and M), but not in resting cells (G0). Thus, the nuclear expression of Ki-67 can be evaluated to study tumor proliferation using immunofluorescence microscopy (Figure 5E). In our experiments, neither µg nor DEX had a significant influence on the proliferation of FTC-133 cells (Figure 5F).

3.3. Epithelial and Mesenchymal Characteristics, Wnt/β-catenin Signaling

To find signs for EMT, that is also influenced by µg in cancer cells [55], different epithelial (E-cadherin) and mesenchymal markers (N-cadherin, vimentin, fibronectin, Snail1) were analyzed. In a four-hour culture the E-cadherin mRNA (encoded by *CDH1*) was reduced in cells incubated with DEX (Figure 6A), without significant changes in E-cadherin protein levels (Figure 6H). After three days on the RPM, we found a difference in the *CDH1* gene expression between the two phenotypes: adherently growing cells showed a lower, whereas MCS showed a higher, *CDH1* expression compared to control cells. The elevated *CDH1* expression in spheroids was significantly reduced by DEX supplementation (Figure 6A).

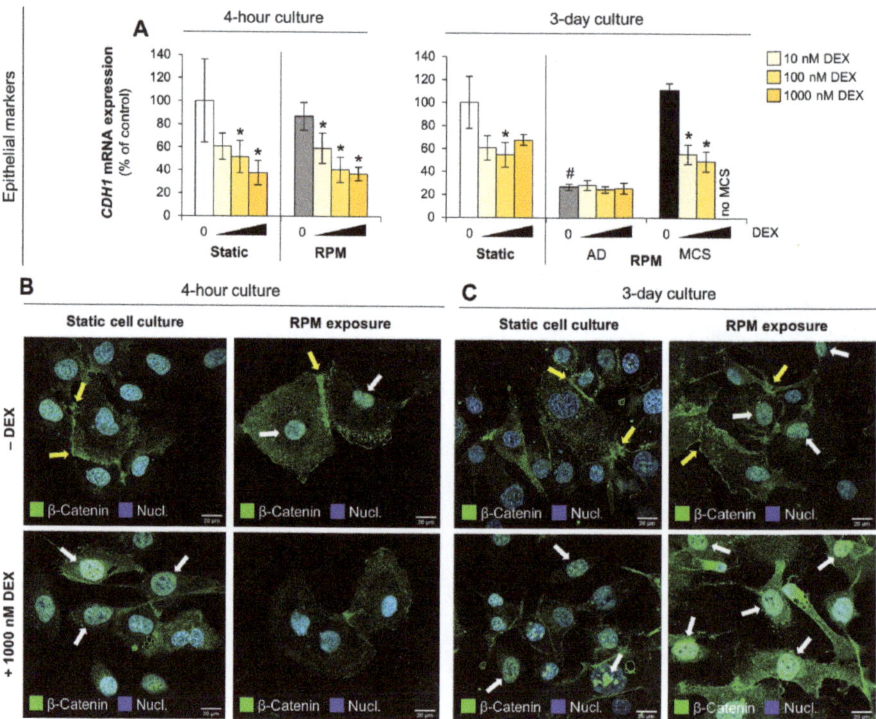

Figure 6. *Cont.*

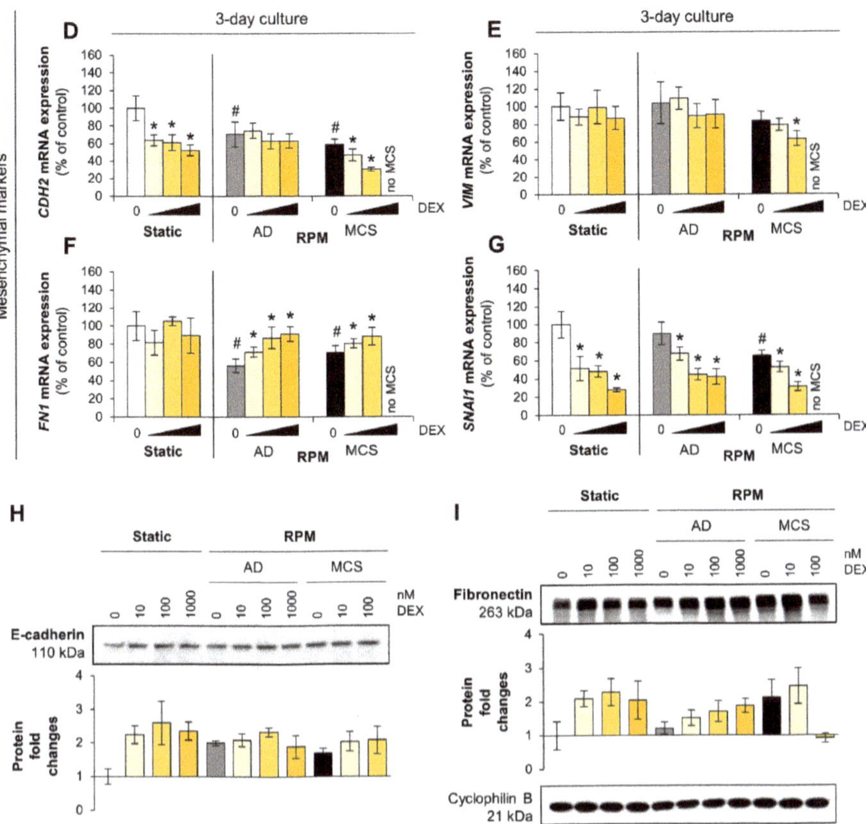

Figure 6. Effect of DEX on the mRNA synthesis of epithelial markers, mesenchymal markers, and other epithelial–mesenchymal transition (EMT) players in FTC-133 cells. (**A**) *CDH1* mRNA expression; (**B,C**) Immunofluorescence. White arrows show translocation of β-catenin (green) into the nucleus (blue; DAPI-stained) both in μg and in the presence of DEX. Yellow arrows indicate an increased occurrence of β-catenin on the plasma membrane in the absence of DEX. Scale bars: 20 μm; (**D**) *CDH2* mRNA expression; (**E**) *VIM* mRNA expression; (**F**) *FN1* mRNA expression; (**G**) *SNAI1* mRNA expression. Depicted are means of relative mRNA levels ± standard deviations ($n = 5$). *: $p < 0.05$ vs. DEX-free samples. #: $p < 0.05$ vs. static cultures; (**H,I**) Western blots indicate protein levels of regulated genes after 3 days. Representatives of each of the three replicates is shown. Diagrams describe relative fold changes to control. AD: adherently growing cells; MCS: multicellular spheroids.

The amount of E-cadherin protein was slightly higher in cells exposed to the RPM than those from static cell cultures and was not influenced by DEX in μg. Under normal culture conditions DEX seemed to increase E-cadherin levels (Figure 6H). Downstream of the cadherin complex, β-catenin mRNA (encoded by *CTNNB1*) was not influenced significantly by DEX (Figure S3). However, β-catenin was translocated from the plasma membrane into the nucleus in the presence of DEX (Figure 6B,C) suggesting an involvement of the Wnt/β-catenin pathway. This is supported by the fact that the transcription of the E-cadherin repressor Snail1 (encoded by *SNAI1*) was also downregulated after DEX treatment (Figure 6G).

N-cadherin (encoded by *CDH2*) and vimentin (encoded by *VIM*) were identified to promote thyroid tumorigenesis [56,57]. *CDH2* mRNA was upregulated in cells after short-term exposure (Figure S3C) and downregulated after long-term exposure to the RPM (Figure 6D). Similar to *CDH1*, *CDH2* expression in spheroids was reduced by DEX supplementation (Figure 6D). Furthermore, DEX elicited the same reducing effects in control cells of a three-day culture. However, adherently growing cells on the RPM were not influenced by DEX. *VIM* expression was not altered by RPM-exposure, but slightly reduced by high DEX concentrations in MCS cells after three days (Figure 6E). RPM-exposure reduced *FN1* mRNA levels in three-day cultures. This effect could be reversed in the presence of DEX (Figure 6F). Protein levels of fibronectin were also slightly increased after DEX supplementation (Figure 6I).

3.4. Anoikis Factors

There is a further possibility that RPM-based spheroid formation of FTC-133 cells in the presence of DEX is abolished through anoikis of detached cells. Cells undergo apoptosis before aggregates are formed. Unfortunately, live/dead staining of detached cells inside the RPM is technically not possible. Adherent cells showed no signs of apoptosis after DEX treatment and after μg-exposure as visualized by a TUNEL staining (Figure 7A). Caspase-3 cleavage tests were negative, both for adherent and spheroid cells in the presence of DEX (Figure 7B). Additionally, we investigated several factors involved in anoikis on the transcriptional level. The cysteine protease caspase-8 (encoded by *CASP8*) is implicated in apoptosis and involved in the induction of NF-κB nuclear translocation [58]. DEX had only a minor effect on *CASP8* gene expression in our experiments (Figure 7C).

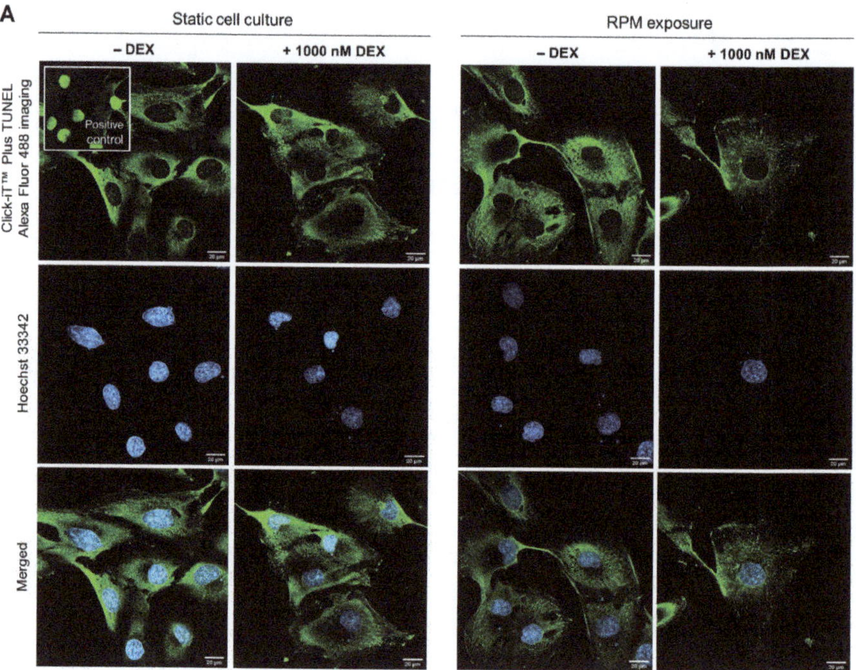

Figure 7. *Cont.*

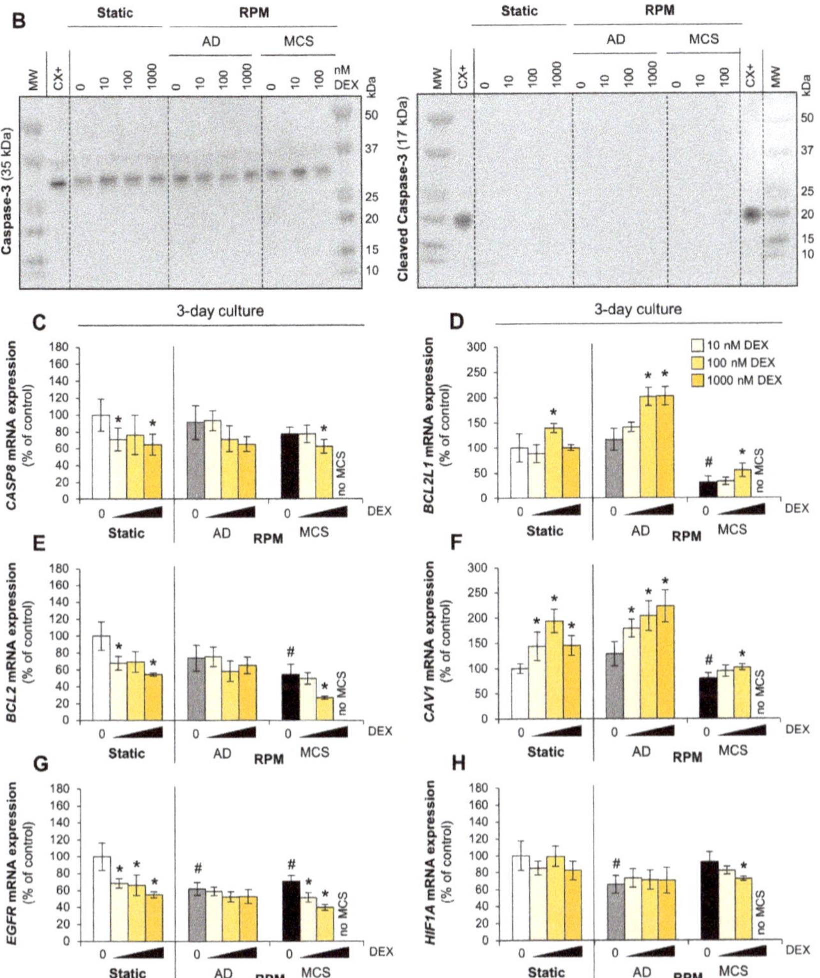

Figure 7. Effect of DEX on apoptosis and anoikis-related proteins in FTC-133 cells. (**A**) No apoptotic cells (green nuclei) were detected by transferase dUTP nick-end labeling (TUNEL) staining after three days. The staining indicates free fluorophores in the cytoplasm in all images except for the positive control. Scale bars: 20 µm; (**B**) Caspase-3 cleavage as an indicator of apoptosis; (**C**) *CASP8* mRNA expression; (**D**) *BCL2L1* mRNA expression; (**E**) *BCL2* mRNA expression; (**F**) *CAV1* mRNA expression; (**G**) *EGFR* mRNA expression; (**H**) *HIF1A* mRNA expression. Depicted are means of relative mRNA levels ± standard deviations ($n = 5$). *: $p < 0.05$ vs. DEX-free samples. #: $p < 0.05$ vs. static cultures. AD: adherently growing cells; MCS: multicellular spheroids.

The anti-apoptotic protein B-cell lymphoma-extra large (Bcl-xL; encoded by *BCL2L1*) has been implicated in the survival of cancer cells by inhibiting the function of the tumor suppressor p53 [59,60]. *BCL2L1* mRNA synthesis was upregulated in FTC-133 cells exposed to the RPM after four hours and reduced by DEX supplementation (Figure S4C). After three days, the *BCL2L1* mRNA synthesis was downregulated in MCS cells, but remained unchanged in adherently growing cells on the RPM. DEX increased *BCL2L1* mRNA after long-term exposure to the RPM (Figure 7D). In contrast, B-cell lymphoma 2 (Bcl-2; encoded by *BCL2*) was further downregulated by DEX in MCS cells (Figure 7E).

A further factor, caveolin-1 (encoded by *CAV1*), was shown to inhibit anchorage-independent growth, anoikis, and invasiveness in human breast cancer cells [61]. Indeed, µg affected the *CAV1* gene expression during spheroid formation: in MCS cells, the *CAV1* mRNA level was reduced. DEX treatment led to an upregulation of caveolin-1 mRNA (Figure 7F).

The loss of coupling between normal integrin and EGF receptor (EGFR) signaling may be further cause for anoikis resistance in tumor cells [62]. We analyzed *EGFR* mRNA synthesis and found a downregulation of EGFR in RPM-grown cells. In addition, DEX decreased *EGFR* transcription in control cells and MCS after three days (Figure 7G).

Hypoxia inducible factor-1 alpha (HIF-1α; encoded by *HIF1A*) is abundantly expressed in most human carcinomas and their metastases. HIF-1α can be induced via EGFR activation and is known to control central metastasis-associated pathways such as angiogenesis, invasion, and resistance to anoikis [63]. Transcription of *HIF1A* was only downregulated in adherently growing cells in three-day RPM cultures and remained unaffected in MCS or by DEX supplementation (Figure 7H).

3.5. Dexamethasone vs. Microgravity—Elucidation of Spheroid Formation Capability

Comparing DEX-induced gene expression data of control cells and transcriptional adaption of FTC-133 cells to µg revealed similar regulation patterns (Figure 8A). We performed an additional two-step RPM culture experiment to check if spheroid formation capability was lost after long-term exposure to µg. Cells that were pre-incubated on the RPM for 48 h showed only marginally reduced spheroid formation during the following 24 h (Figure 8B,C).

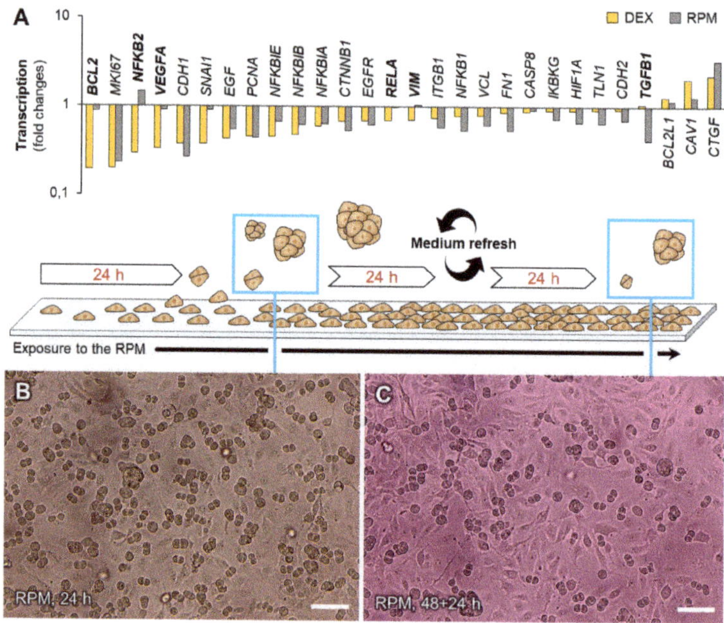

Figure 8. Spheroid formation capability of FTC-133 cells cultured on the RPM. (**A**) Comparison of transcription regulation patterns 4 h after DEX supplementation (yellow bars) and after a three-day RPM-exposure (grey bars). Bold gene symbols indicate fold changes >2.5 or regulation in opposite directions. (**B**) Cells 24 h after the RPM-experiment started; (**C**) Cells 24 h after an initial two-day RPM exposure. Medium was refreshed and spheroids were discarded after the first two days. Although many genes were similarly regulated after DEX supplementation and after a three-day-exposure to the RPM, in contrast to DEX treatment, cells did not lose the ability to form spheroids in µg. Scale bars: 100 µm.

4. Discussion

We investigated the effects of DEX supplementation on the growth of follicular thyroid cancer cells exposed to simulated µg. During a three-day culture on an RPM, cells grew into the form of a large MCS, as it was reported and studied earlier [28,52,64,65]. Previous research revealed that the addition of DEX to spinner flask cultures led to smaller, irregularly shaped spheroids of rat hepatocytes. Higher DEX concentrations inhibited MCS aggregation and promoted MCS disassembly in culture dishes [66]. Kopp et al. [34] described an inhibitory effect of DEX on the MCS formation rate of MCF-7 breast cancer cells cultured on the RPM. However, the authors did not perform any further analyses to elucidate the underlying effects of DEX on MCF-7 cells. After the current study we can confirm similar effects on FTC-133 cells. We found a dose-dependent inhibition of RPM-based spheroid formation by DEX, that was independent from RelA nuclear translocation which was described for DEX-treated breast cancer cells [34]. This finding agrees with the theory of Bauerle et al. [43] that global regulation of thyroid cancer cell growth is not achieved by NF-κB signaling alone and indicates that NF-κB (pathway) may not be the main target of DEX inhibiting the 3D growth of FTC-133 cells in µg. Therefore, we used transcriptional and translational methods to find answers for the changed growth behavior. Interestingly, after DEX supplementation a couple of genes were regulated in the same direction as after a three-day exposure to the RPM. Since the ability of spheroid formation was not suppressed in the RPM-cultures, especially those genes that are of interest, which had a differential expression pattern (Table 1).

Table 1. Significant differences ($p < 0.05$) in mRNA synthesis of adherently growing FTC-133 cells in presence of DEX in static cell culture compared with cells grown without DEX on the RPM.

Process/Pathway	4-Hour Culture	Both Time Points	3-Day Culture
NF-κB pathway	*NFKB1↑, NFKBIA↓, NFKBIB↓*	*NFKB2↓, NFKBIE↓*	*NFKB1↓, RELA↓, NFKBIA↑, IKBKG↓*
Autocrine signaling	*EGF↓, TGFB1↓*	*VEGFA↓*	
EMT	*CDH1↓, CTNNB1↓, VIM↓,*	*CDH2↓, SNAI1↓*	*CDH1↑*
Anoikis	*CASP8↓, BCL2↓, BCL2L1↓, EFGR↓, HIF1A↓*		*HIF1A↑*
Proliferation	*MKI67↓, PCNA↓*		*MKI67↑*

↑: significant upregulation in DEX-treated cells; ↓: significant downregulation in DEX-treated cells.

4.1. Cell Detachment in Microgravity and Epithelial–Mesenchymal Transition

The EMT describes a fundamental process of cancer progression when carcinoma cells lose their epithelial characteristics and acquire a migratory behavior, indicated by mesenchymal markers. This alteration enables them to escape from their epithelial cell community and invade into surrounding tissues, even at distant locations, and contributes to the acquisition and maintenance of stem cell-like properties [49,67]. In previous studies, DEX proved to suppress cell invasion in bladder cancer [68], inhibited hypoxia-induced EMT in colon cancer cells [69], and reduced TGF-β-induced EMT in non-malignant cells [70].

The interaction of DEX-bound GRs with NF-κB affects the expression of several target genes, one of which is TGF-β [71]. TGF-β induces the upregulation mesenchymal markers such as vimentin and downregulation of the epithelial marker E-cadherin, which are considered critical prerequisites for metastasis in numerous human cancers [72]. Therefore, it is not surprising that the expression of E-cadherin and β-catenin in thyroid cancer is associated with better prognosis [73]. Among the set of analyzed genes, *TGFB1* was regulated differently in µg and in the presence of DEX. Hinz [74] suggested that the TGF-β complex functions as an extracellular mechanosensory that can be activated by contractile forces that are transmitted by integrins. Indeed, µg was identified as a possible cause changing TGF-β expression levels [75]. In four-hour cultures stacked on the RPM, the *TGFB1* mRNA was slightly elevated whereas in three-day cultures the mRNA level was attenuated, maybe due to missing forces in µg. DEX supplementation led to a slow downregulation of *TGFB1* expression in static cell cultures or in follicular thyroid cancer cells grown for four hours on the RPM (Figure S2C). This observation could be cell-type specific, as DEX increased *TGFB1* expression in prostate cancer and pancreatic ductal carcinoma cells [76,77]. TGF-β signaling is identified as one of the most altered pathways in ovarian tumor spheroids [78] and cell aggregation proved to be induced by TGF-β in ovarian cancer cells [79]. Therefore, *TGFB1* could be a possible target gene for DEX that may inhibit spheroid formation of FTC-133 cells in an early culture stage at least in part by downregulation of *TGFB1*.

The translocation of β-catenin into the nucleus as well as upregulation of *FN1* mRNAs suggest an activation of the Wnt/β-catenin pathway after DEX supplementation. On the other hand, expression of Snail1 is reduced in the presence of DEX. However, in ovarian adenocarcinoma cells Snail1 is downregulated when TGF-β and Wnt signaling pathways are co-activated [80]. Snail1 acts as an EMT inducer and a potent repressor of E-cadherin [81,82]. The finding that E-cadherin expression correlates with spheroid formation capability suggests that intercellular adhesion plays a key role in 3D growth [83,84]. Sahana et al. [85] found that the blocking of E-cadherin activity with antibodies promoted µg-driven spheroid formation of MCF-7 cells. Budding of ovarian cancer spheroids from monolayers correlated with the expression of vimentin and lack of cortical E-cadherin [86]. In our experiments the *CDH1* gene was downregulated in adherent cells but remained nearly unchanged in MCS cells of a three-day RPM culture. E-cadherin protein increased after DEX supplementation. This finding is consistent with the observations of Sahana et al. [85] and suggests a quantity-dependent influence of E-cadherin on cancer cell aggregation, that is not directly related to its mRNA synthesis. For renal cell carcinoma, N-cadherin was shown not to be an essential molecule for spheroid formation indicating a somewhat different role from cell–cell adhesion. However, anti-N-cadherin antibodies inhibit spheroid formation in a renal cell carcinoma cell line that expressed N-cadherin alone [87]. Tsai et al. [88] suggested that N-cadherin might play a role in the formation and maintenance of spheroid core structures. Due to a higher affinity of N-cadherin to form homodimers [89], cells with higher N-cadherin expression aggregate first. Indeed, in the spheroid-inducing environment of the RPM, the *CDH2* gene was upregulated after four hours (Figure S3B). DEX reduced *CDH2* expression in these cells as well as in spheroids after three days. This explains the reduced spheroid formation in the presence of DEX, but on the other hand, it suggests a destabilization of formed spheroids.

Fibronectin was downregulated in µg, but upregulated by DEX, in three-day cultures. Thus, it was the only mesenchymal factor showing a significantly altered regulation after DEX supplementation. Abu-Absi et al. [66] previously reported an increase in fibronectin and collagen III mRNA when rat hepatocytes were cultured in spinner flasks in the presence of DEX and suggested that a modification of the extracellular matrix (ECM) contributes to the changes in morphology. It is further known that DEX treatment significantly increases the strength of cell–ECM adhesion in glioblastoma cells and thus decreases their motility [90]. Robinson et al. [91] confirmed in different experiments that fibronectin matrix assembly plays a key role in cell aggregation and spheroid formation. So, it is very likely that DEX alters the ECM composition of FTC-133 cells, including fibronectin, in a way that they are no longer susceptible to 3D aggregation in µg.

Our data indicate that FTC-133 cells were not shifted to a typical mesenchymal phenotype during spheroid formation on the RPM and the phenotype was not strongly influenced by DEX. However, the results confirm the involvement of the Wnt/β-catenin axis and TGF-β-induced signaling in µg-triggered spheroid formation ability. These pathways were affected by DEX and can regulate some individual adhesion and matrix proteins (e.g., E-cadherin and fibronectin) which are important for detachment and 3D aggregation.

4.2. Survival of Detached Cells

At least for adherently growing FTC-133 cells, a cell-based assay showed no apoptotic cells after DEX treatment. For osteoblasts it is known that DEX can cause anoikis, probably due to the decreased integrin β1 expression [92]. In our experiments, indeed we found a downregulation of *ITGB1* transcription in DEX-treated cells as well as lower integrin β1 levels in MCS cells that grew in the presence of DEX (Figure S5).

The induction of anoikis occurs through an interplay of the two apoptotic pathways involving activation of caspase-8 and inhibition of Bcl-2 [26]. Caspase-8 expression was only marginally affected by μg and DEX treatment. We observed a counter-regulation for enhanced *CASP8* mRNA synthesis by DEX in cells after four hours on the RPM (Figure S4A). DEX significantly reduced expression of the anti-apoptotic *BCL2* gene in FTC-133 cells but stimulated the expression of the anti-apoptotic Bcl-xL. An upregulation of Bcl-xL expression by DEX was reported earlier for follicular thyroid cancer cells where it promotes survival [93]. That Bcl-2 plays an important role in the efficacy of DEX was confirmed in a study with myeloma cells where Bcl-2 overexpression was associated with resistance to DEX [94]. Overexpression of Bcl-2 correlates with the progression and metastasis of prostate cancer [95] and was shown to inhibit anoikis at least in intestinal epithelial cells [96]. Looking at apoptosis signaling, there are a some, but not all, indications that anoikis can be induced in FTC-133 cells after DEX treatment.

Apart from the apoptotic pathways, there are other proteins playing important roles in the complex network of survival signaling. It has been reported that inhibition of E-cadherin binding prevented cell–cell aggregation and could induce anoikis in epithelial cells [97,98]. In addition, the overexpression of β-catenin, a downstream regulator of cadherin signaling, resulted in anoikis resistance [99]. Indeed, on the transcriptional level we saw a DEX-mediated downregulation of *CDH1* together with an upregulation of *CTNNB1* in MCS cells.

The integral membrane protein caveolin-1 was identified as an important factor in spheroid formation of thyroid cancer cells in an μg environment [65,100] which further inhibits anchorage-independent growth and anoikis, obviously two independent processes, in MCF-7 breast cancer cells [61]. The *CAV1* gene was downregulated when FTC-133 cells formed MCS on an RPM [100]. We were able to confirm this effect in our experiments. The upregulation of *CAV1* expression in the presence of DEX may suppress caveolin's yet undefined effects on 3D aggregation of thyroid carcinoma cells. Interestingly, caveolin-1 is overexpressed in other metastatic carcinoma cells where it promotes growth [101,102] and resistance to anoikis [61]. Caveolin-1 controls the stability of focal adhesions and contributes to mechanosensing and adaptation in response to mechanical stimuli including cell detachment [103]. Moreno-Vicente et al. [104] demonstrated that caveolin-1 regulates yes-associated protein activity which in turn modulates pathophysiological processes such as ECM remodeling. The authors suggested that this regulation could determine the onset and progression of tumor development. A possible explanation for suppressed spheroid formation would be that a more rigid ECM inhibits the growth into 3D aggregates.

In summary, we found regulatory indications but no clear evidence of anoikis after DEX treatment, as both caspase-3 cleavage and TUNEL staining were negative. Therefore, we suggest that apoptosis does not play a (major) role in inhibition of FTC-133 spheroid formation by DEX.

4.3. Autocrine Signaling

Growth factors have been considered to be involved in spheroid formation of thyroid cancer cells for long time. During the Shenzhou-8 space experiment, extraordinarily large 3D aggregates were formed by FTC-133 cells which showed an altered expression of *EGF* and *CTGF* genes under real μg [105]. A decreased expression of *CTGF* in MCS compared to an increased expression in adherent cells was observed after cultivation of FTC-133 on different μg-devices suggesting an important role for CTGF in spheroid formation [65]. DEX supplementation resulted in an upregulation of *CTGF* mRNA synthesis. In vitro, CTGF was identified to stimulate ECM synthesis, proliferation, or integrin expression and has been implicated in different cancer-related processes, comprising migration, invasion, angiogenesis, and anoikis [46]. Elevated *CTGF* expression levels in primary papillary thyroid carcinoma samples were correlated with metastasis [106]. EGF acts as a strong mitogen for follicular thyroid cells [47] and has been shown to increase the spheroid size in various tumor cell lines [107–109]. Both, DEX and μg reduced *EGF* expression. However, CTGF upregulation and EGF downregulation by DEX cannot explain the inhibition of spheroid formation, especially since adherent cells on the RPM are able to form MCS but show the same gene regulations.

VEGF promotes tumor angiogenesis by stimulating proliferation and survival of endothelial cells and can directly modulate cancer cell behavior [110]. Studies have shown that VEGF expression is upregulated in most human tumors and correlates with the risk for the development of metastasis in papillary thyroid cancer [111,112]. VEGF expression can be upregulated in response to hypoxia and was found in the microenvironment of tumor spheroids formed by HT-29 human colon cancer cells [113]. Furthermore, spheroids formed by FTC-133 cells on an RPM showed an increase in *VEGFA* gene expression [52]. Inhibition of VEGF signaling significantly reduced cell viability of thyroid cancer cells and increased apoptosis in the NPA'87 tumor-derived cell line [114]. DEX was reported to reduce VEGF secretion in some head and neck cancer cells via STAT3 [115]. We can confirm a decreasing effect of DEX on *VEGFA* expression in FTC-133 cells, independently of their exposure to μg or their growth behavior on the RPM. A decrease in VEGF-A by treatment using siRNA or anti-VEGF-A, reduced spheroid formation, proliferation, migration, and invasion of epidermal cancer stem cells [116]. This finding highlights the role of VEGF (signaling) in the formation of solid tumors and could also provide an explanation for the effect of DEX on FTC-133 cells. In T47D breast cancer cells, DEX was shown to affect the PI3K/AKT/mTOR pathway [117] that could also be a possible target in thyroid cancer cells responsible for the suppression of spheroid formation [118].

5. Conclusions

In our study, DEX suppressed spheroid formation of FTC-133 cells cultured on an RPM in a dose-dependent manner. Interestingly, DEX did not influence NF-κB in a way that would explain the inhibition of μg-triggered spheroid formation indicating that NF-κB (pathway) may not be the main target of DEX in FTC-133 cells. However, transcriptional regulation of important individual factors in cancer cell biology, which were previously suggested to play a role in spheroid formation, was clearly affected by DEX. Thereby, our data indicate the presence of a more complex regulation network of spheroid formation also involving other signal pathways, such as Wnt/β-catenin and TGF-β, that regulate adhesion and matrix proteins which are important for cell detachment and 3D growth. According to our results, it will be necessary to carry out a broad transcriptome analysis in order to identify the exact influence of DEX on the growth behavior of follicular thyroid cancer cells. Furthermore, it needs to be clarified whether DEX not only inhibits formation of spheroids but also promotes their disassembly in μg.

Supplementary Materials: The following materials are available online at http://www.mdpi.com/2073-4409/9/2/367/s1, Figure S1. Effect of 4 h DEX treatment on NF-κB family members and regulators in FTC-133 cells; Figure S2. Effect of 4 h DEX treatment on growth factors and proliferation markers in FTC-133 cells. Figure S3: Effect of 4 h DEX treatment on epithelial markers, mesenchymal markers and other EMT players in FTC-133 cells; Figure S4: Effect of 4 h DEX treatment on anoikis-related factors in FTC-133 cells; Figure S5: Effect of DEX on cell–cell contacts of FTC-133 cells.; Table S1: Primer sequences used in the quantitative real-time PCR; Table S2: Antibodies used for Western blot analyses.

Author Contributions: Conceptualization, D.M., M.K., and D.G.; methodology, D.M., J.S., and T.J.C.; software, D.M.; validation, D.M., M.K., and S.K.; formal analysis, D.M.; investigation, D.M., M.K., T.J.C., and M.Z.N.; resources, M.I. and T.J.C.; data curation, D.M., M.K., S.K., and D.G.; writing—original draft preparation, M.K. and D.M.; writing—review and editing, M.K., D.G., S.K., T.J.C., and M.W.; visualization, D.M. and M.K.; supervision, D.G. and M.K.; project administration, D.G.; funding acquisition, D.G. and M.I. All authors have read and agreed to the published version of the manuscript.

Funding: This research was funded by Deutsches Zentrum für Luft- und Raumfahrt (DLR), grant numbers 50WB1524 and 50WB1924.

Acknowledgments: The authors would like to thank the Institute of Anatomy (Otto von Guericke University), especially Stefan Kahlert and Andrea Kröber, for the technical support. We also acknowledge financial support by the Open Access Publication Fonds of the Otto von Guericke University Magdeburg.

Conflicts of Interest: The authors declare no conflicts of interest.

References

1. Kadmiel, M.; Cidlowski, J.A. Glucocorticoid receptor signaling in health and disease. *Trends Pharmacol. Sci.* **2013**, *34*, 518–530. [CrossRef] [PubMed]
2. Mangelsdorf, D.J.; Thummel, C.; Beato, M.; Herrlich, P.; Schütz, G.; Umesono, K.; Blumberg, B.; Kastner, P.; Mark, M.; Chambon, P.; et al. The nuclear receptor superfamily: The second decade. *Cell* **1995**, *83*, 835–839. [CrossRef]
3. Coutinho, A.E.; Chapman, K.E. The anti-inflammatory and immunosuppressive effects of glucocorticoids, recent developments and mechanistic insights. *Mol. Cell. Endocrinol.* **2011**, *335*, 2–13. [CrossRef]
4. Vandevyver, S.; Dejager, L.; Tuckermann, J.; Libert, C. New insights into the anti-inflammatory mechanisms of glucocorticoids: An emerging role for glucocorticoid-receptor-mediated transactivation. *Endocrinology* **2013**, *154*, 993–1007. [CrossRef]
5. Vandewalle, J.; Luypaert, A.; De Bosscher, K.; Libert, C. Therapeutic mechanisms of glucocorticoids. *Trends Endocrinol. Metab.* **2018**, *29*, 42–54. [CrossRef]
6. Herr, I.; Pfitzenmaier, J. Glucocorticoid use in prostate cancer and other solid tumours: Implications for effectiveness of cytotoxic treatment and metastases. *Lancet Oncol.* **2006**, *7*, 425–430. [CrossRef]
7. Wang, H.; Wang, Y.; Rayburn, E.R.; Hill, D.L.; Rinehart, J.J.; Zhang, R. Dexamethasone as a chemosensitizer for breast cancer chemotherapy: Potentiation of the antitumor activity of adriamycin, modulation of cytokine expression, and pharmacokinetics. *Int. J. Oncol.* **2007**, *30*, 947–953. [CrossRef]
8. Yennurajalingam, S.; Frisbee-Hume, S.; Palmer, J.L.; Delgado-Guay, M.O.; Bull, J.; Phan, A.T.; Tannir, N.M.; Litton, J.K.; Reddy, A.; Hui, D.; et al. Reduction of cancer-related fatigue with dexamethasone: A double-blind, randomized, placebo-controlled trial in patients with advanced cancer. *J. Clin. Oncol.* **2013**, *31*, 3076–3082. [CrossRef]
9. Wang, L.J.; Li, J.; Hao, F.R.; Yuan, Y.; Li, J.Y.; Lu, W.; Zhou, T.Y. Dexamethasone suppresses the growth of human non-small cell lung cancer via inducing estrogen sulfotransferase and inactivating estrogen. *Acta Pharmacol. Sin.* **2016**, *37*, 845–856. [CrossRef]
10. Lin, K.-T.; Sun, S.-P.; Wu, J.-I.; Wang, L.-H. Low-dose glucocorticoids suppresses ovarian tumor growth and metastasis in an immunocompetent syngeneic mouse model. *PLoS ONE* **2017**, *12*, e0178937. [CrossRef]
11. Gong, H.; Jarzynka, M.J.; Cole, T.J.; Lee, J.H.; Wada, T.; Zhang, B.; Gao, J.; Song, W.C.; DeFranco, D.B.; Cheng, S.Y.; et al. Glucocorticoids antagonize estrogens by glucocorticoid receptor-mediated activation of estrogen sulfotransferase. *Cancer Res.* **2008**, *68*, 7386–7393. [CrossRef] [PubMed]
12. Geng, Y.; Wang, J.; Jing, H.; Wang, H.W.; Bao, Y.X. Inhibitory effect of dexamethasone on lewis mice lung cancer cells. *Genet. Mol. Res.* **2014**, *13*, 6827–6836. [CrossRef] [PubMed]
13. Moon, E.Y.; Ryu, Y.K.; Lee, G.H. Dexamethasone inhibits in vivo tumor growth by the alteration of bone marrow cd11b(+) myeloid cells. *Int. Immunopharmacol.* **2014**, *21*, 494–500. [CrossRef] [PubMed]
14. Sau, S.; Banerjee, R. Cationic lipid-conjugated dexamethasone as a selective antitumor agent. *Eur. J. Med. Chem.* **2014**, *83*, 433–447. [CrossRef] [PubMed]
15. Komiya, A.; Shimbo, M.; Suzuki, H.; Imamoto, T.; Kato, T.; Fukasawa, S.; Kamiya, N.; Naya, Y.; Mori, I.; Ichikawa, T. Oral low-dose dexamethasone for androgen-independent prostate cancer patients. *Oncol. Lett.* **2010**, *1*, 73–79. [CrossRef] [PubMed]
16. Bray, F.; Ferlay, J.; Soerjomataram, I.; Siegel, R.L.; Torre, L.A.; Jemal, A. Global cancer statistics 2018: Globocan estimates of incidence and mortality worldwide for 36 cancers in 185 countries. *CA* **2018**, *68*, 394–424. [CrossRef]
17. Kim, J.; Gosnell, J.E.; Roman, S.A. Geographic influences in the global rise of thyroid cancer. *Nat. Rev. Endocrinol.* **2019**, *16*, 17–29. [CrossRef]
18. Pachmayr, E.; Treese, C.; Stein, U. Underlying mechanisms for distant metastasis—Molecular biology. *Visc. Med.* **2017**, *33*, 11–20. [CrossRef]
19. Yilmaz, M.; Christofori, G. Mechanisms of motility in metastasizing cells. *Mol. Cancer Res.* **2010**, *8*, 629–642. [CrossRef]
20. Paszek, M.J.; Zahir, N.; Johnson, K.R.; Lakins, J.N.; Rozenberg, G.I.; Gefen, A.; Reinhart-King, C.A.; Margulies, S.S.; Dembo, M.; Boettiger, D.; et al. Tensional homeostasis and the malignant phenotype. *Cancer Cell* **2005**, *8*, 241–254. [CrossRef]

21. Moore, S.W.; Roca-Cusachs, P.; Sheetz, M.P. Stretchy proteins on stretchy substrates: The important elements of integrin-mediated rigidity sensing. *Dev. Cell* **2010**, *19*, 194–206. [CrossRef] [PubMed]

22. Roussos, E.T.; Condeelis, J.S.; Patsialou, A. Chemotaxis in cancer. *Nat. Rev. Cancer* **2011**, *11*, 573–587. [CrossRef] [PubMed]

23. Brabletz, T.; Kalluri, R.; Nieto, M.A.; Weinberg, R.A. Emt in cancer. *Nat. Rev. Cancer* **2018**, *18*, 128–134. [CrossRef] [PubMed]

24. Yu, W.; Yang, L.; Li, T.; Zhang, Y. Cadherin signaling in cancer: Its functions and role as a therapeutic target. *Front. Oncol.* **2019**, *9*, 989. [CrossRef]

25. Taddei, M.L.; Giannoni, E.; Fiaschi, T.; Chiarugi, P. Anoikis: An emerging hallmark in health and diseases. *J. Pathol.* **2012**, *226*, 380–393. [CrossRef]

26. Paoli, P.; Giannoni, E.; Chiarugi, P. Anoikis molecular pathways and its role in cancer progression. *Biochim. Biophys. Acta* **2013**, *1833*, 3481–3498. [CrossRef]

27. Chang, T.T.; Hughes-Fulford, M. Molecular mechanisms underlying the enhanced functions of three-dimensional hepatocyte aggregates. *Biomaterials* **2014**, *35*, 2162–2171. [CrossRef]

28. Kopp, S.; Warnke, E.; Wehland, M.; Aleshcheva, G.; Magnusson, N.E.; Hemmersbach, R.; Corydon, T.J.; Bauer, J.; Infanger, M.; Grimm, D. Mechanisms of three-dimensional growth of thyroid cells during long-term simulated microgravity. *Sci. Rep.* **2015**, *5*, 16691. [CrossRef]

29. Kunz-Schughart, L.A. Multicellular tumor spheroids: Intermediates between monolayer culture and in vivo tumor. *Cell Biol. Int.* **1999**, *23*, 157–161. [CrossRef]

30. Martin, A.; Zhou, A.; Gordon, R.E.; Henderson, S.C.; Schwartz, A.E.; Schwartz, A.E.; Friedman, E.W.; Davies, T.F. Thyroid organoid formation in simulated microgravity: Influence of keratinocyte growth factor. *Thyroid* **2000**, *10*, 481–487. [CrossRef] [PubMed]

31. Bauer, J.; Grimm, D.; Gombocz, E. Semantic analysis of thyroid cancer cell proteins obtained from rare research opportunities. *J. Biomed. Inf.* **2017**, *76*, 138–153. [CrossRef] [PubMed]

32. Bauer, J.; Wehland, M.; Infanger, M.; Grimm, D.; Gombocz, E. Semantic analysis of posttranslational modification of proteins accumulated in thyroid cancer cells exposed to simulated microgravity. *Int. J. Mol. Sci.* **2018**, *19*, 2257. [CrossRef] [PubMed]

33. Krüger, M.; Melnik, D.; Kopp, S.; Buken, C.; Sahana, J.; Bauer, J.; Wehland, M.; Hemmersbach, R.; Corydon, T.J.; Infanger, M.; et al. Fighting thyroid cancer with microgravity research. *Int. J. Mol. Sci.* **2019**, *20*, 2553. [CrossRef] [PubMed]

34. Kopp, S.; Sahana, J.; Islam, T.; Petersen, A.G.; Bauer, J.; Corydon, T.J.; Schulz, H.; Saar, K.; Huebner, N.; Slumstrup, L.; et al. The role of nfkappab in spheroid formation of human breast cancer cells cultured on the random positioning machine. *Sci. Rep.* **2018**, *8*, 921. [CrossRef]

35. Corydon, T.J.; Mann, V.; Slumstrup, L.; Kopp, S.; Sahana, J.; Askou, A.L.; Magnusson, N.E.; Echegoyen, D.; Bek, T.; Sundaresan, A.; et al. Reduced expression of cytoskeletal and extracellular matrix genes in human adult retinal pigment epithelium cells exposed to simulated microgravity. *Cell. Physiol. Biochem.* **2016**, *40*, 1–17. [CrossRef]

36. Grosse, J.; Wehland, M.; Pietsch, J.; Ma, X.; Ulbrich, C.; Schulz, H.; Saar, K.; Hübner, N.; Hauslage, J.; Hemmersbach, R.; et al. Short-term weightlessness produced by parabolic flight maneuvers altered gene expression patterns in human endothelial cells. *Faseb J.* **2012**, *26*, 639–655. [CrossRef]

37. Ma, X.; Pietsch, J.; Wehland, M.; Schulz, H.; Saar, K.; Hübner, N.; Bauer, J.; Braun, M.; Schwarzwälder, A.; Segerer, J.; et al. Differential gene expression profile and altered cytokine secretion of thyroid cancer cells in space. *Faseb J.* **2014**, *28*, 813–835. [CrossRef]

38. Ma, X.; Wehland, M.; Schulz, H.; Saar, K.; Hubner, N.; Infanger, M.; Bauer, J.; Grimm, D. Genomic approach to identify factors that drive the formation of three-dimensional structures by ea.Hy926 endothelial cells. *PLoS ONE* **2013**, *8*, e64402. [CrossRef]

39. Ye, J.; Coulouris, G.; Zaretskaya, I.; Cutcutache, I.; Rozen, S.; Madden, T.L. Primer-blast: A tool to design target-specific primers for polymerase chain reaction. *BMC Bioinform.* **2012**, *13*, 134. [CrossRef]

40. Grosse, J.; Wehland, M.; Pietsch, J.; Schulz, H.; Saar, K.; Hübner, N.; Eilles, C.; Bauer, J.; Abou-El-Ardat, K.; Baatout, S.; et al. Gravity-sensitive signaling drives 3-dimensional formation of multicellular thyroid cancer spheroids. *Faseb J.* **2012**, *26*, 5124–5140. [CrossRef]

41. Pacifico, F.; Leonardi, A. Role of nf-kappab in thyroid cancer. *Mol. Cell. Endocrinol.* **2010**, *321*, 29–35. [CrossRef] [PubMed]

42. Giuliani, C.; Bucci, I.; Napolitano, G. The role of the transcription factor nuclear factor-kappa b in thyroid autoimmunity and cancer. *Front. Endocrinol.* **2018**, *9*, 471. [CrossRef] [PubMed]

43. Bauerle, K.T.; Schweppe, R.E.; Haugen, B.R. Inhibition of nuclear factor-kappa b differentially affects thyroid cancer cell growth, apoptosis, and invasion. *Mol. Cancer* **2010**, *9*, 117. [CrossRef] [PubMed]

44. Yamamoto, Y.; Gaynor, R.B. Therapeutic potential of inhibition of the nf-kappab pathway in the treatment of inflammation and cancer. *J. Clin. Investig.* **2001**, *107*, 135–142. [CrossRef]

45. Holbourn, K.P.; Acharya, K.R.; Perbal, B. The ccn family of proteins: Structure-function relationships. *Trends Biochem. Sci.* **2008**, *33*, 461–473. [CrossRef]

46. Chu, C.Y.; Chang, C.C.; Prakash, E.; Kuo, M.L. Connective tissue growth factor (ctgf) and cancer progression. *J. Biomed. Sci.* **2008**, *15*, 675–685. [CrossRef]

47. Asmis, L.M.; Gerber, H.; Kaempf, J.; Studer, H. Epidermal growth factor stimulates cell proliferation and inhibits iodide uptake of frtl-5 cells in vitro. *J. Endocrinol.* **1995**, *145*, 513–520. [CrossRef]

48. Colletta, G.; Cirafici, A.M.; Di Carlo, A. Dual effect of transforming growth factor beta on rat thyroid cells: Inhibition of thyrotropin-induced proliferation and reduction of thyroid-specific differentiation markers. *Cancer Res.* **1989**, *49*, 3457–3462.

49. Xu, J.; Lamouille, S.; Derynck, R. Tgf-β-induced epithelial to mesenchymal transition. *Cell Res.* **2009**, *19*, 156–172. [CrossRef]

50. Gugnoni, M.; Sancisi, V.; Manzotti, G.; Gandolfi, G.; Ciarrocchi, A. Autophagy and epithelial-mesenchymal transition: An intricate interplay in cancer. *Cell Death Dis.* **2016**, *7*, e2520. [CrossRef]

51. Bhatti, M.Z.; Pan, L.; Wang, T.; Shi, P.; Li, L. Reggamma potentiates tgf-beta/smad signal dependent epithelial-mesenchymal transition in thyroid cancer cells. *Cell Signal.* **2019**, *64*, 109412. [CrossRef]

52. Riwaldt, S.; Bauer, J.; Wehland, M.; Slumstrup, L.; Kopp, S.; Warnke, E.; Dittrich, A.; Magnusson, N.E.; Pietsch, J.; Corydon, T.J.; et al. Pathways regulating spheroid formation of human follicular thyroid cancer cells under simulated microgravity conditions: A genetic approach. *Int. J. Mol. Sci.* **2016**, *17*, 528. [CrossRef]

53. Chung, Y.J.; Lee, J.I.; Chong, S.; Seok, J.W.; Park, S.J.; Jang, H.W.; Kim, S.W.; Chung, J.H. Anti-proliferative effect and action mechanism of dexamethasone in human medullary thyroid cancer cell line. *Endocr. Res.* **2011**, *36*, 149–157. [CrossRef]

54. Scholzen, T.; Gerdes, J. The ki-67 protein: From the known and the unknown. *J. Cell. Physiol.* **2000**, *182*, 311–322. [CrossRef]

55. Chen, Z.Y.; Guo, S.; Li, B.B.; Jiang, N.; Li, A.; Yan, H.F.; Yang, H.M.; Zhou, J.L.; Li, C.L.; Cui, Y. Effect of weightlessness on the 3d structure formation and physiologic function of human cancer cells. *Biomed. Res. Int.* **2019**, *2019*, 4894083. [CrossRef]

56. Da, C.; Wu, K.; Yue, C.; Bai, P.; Wang, R.; Wang, G.; Zhao, M.; Lv, Y.; Hou, P. N-cadherin promotes thyroid tumorigenesis through modulating major signaling pathways. *Oncotarget* **2017**, *8*, 8131–8142. [CrossRef]

57. Vasko, V.; Espinosa, A.V.; Scouten, W.; He, H.; Auer, H.; Liyanarachchi, S.; Larin, A.; Savchenko, V.; Francis, G.L.; de la Chapelle, A.; et al. Gene expression and functional evidence of epithelial-to-mesenchymal transition in papillary thyroid carcinoma invasion. *Proc. Natl. Acad. Sci. USA* **2007**, *104*, 2803–2808. [CrossRef]

58. Su, H.; Bidere, N.; Zheng, L.; Cubre, A.; Sakai, K.; Dale, J.; Salmena, L.; Hakem, R.; Straus, S.; Lenardo, M. Requirement for caspase-8 in nf-kappab activation by antigen receptor. *Science* **2005**, *307*, 1465–1468. [CrossRef]

59. Schott, A.F.; Apel, I.J.; Nuñez, G.; Clarke, M.F. Bcl-xl protects cancer cells from p53-mediated apoptosis. *Oncogene* **1995**, *11*, 1389–1394.

60. Li, M.; Wang, D.; He, J.; Chen, L.; Li, H. Bcl-xl: A multifunctional anti-apoptotic protein. *Pharmacol. Res.* **2020**, *151*, 104547. [CrossRef]

61. Fiucci, G.; Ravid, D.; Reich, R.; Liscovitch, M. Caveolin-1 inhibits anchorage-independent growth, anoikis and invasiveness in mcf-7 human breast cancer cells. *Oncogene* **2002**, *21*, 2365–2375. [CrossRef] [PubMed]

62. Reginato, M.J.; Mills, K.R.; Paulus, J.K.; Lynch, D.K.; Sgroi, D.C.; Debnath, J.; Muthuswamy, S.K.; Brugge, J.S. Integrins and egfr coordinately regulate the pro-apoptotic protein bim to prevent anoikis. *Nat. Cell Biol.* **2003**, *5*, 733–740. [CrossRef] [PubMed]

63. Rohwer, N.; Welzel, M.; Daskalow, K.; Pfander, D.; Wiedenmann, B.; Detjen, K.; Cramer, T. Hypoxia-inducible factor 1alpha mediates anoikis resistance via suppression of alpha5 integrin. *Cancer Res.* **2008**, *68*, 10113–10120. [CrossRef] [PubMed]

64. Riwaldt, S.; Pietsch, J.; Sickmann, A.; Bauer, J.; Braun, M.; Segerer, J.; Schwarzwälder, A.; Aleshcheva, G.; Corydon, T.J.; Infanger, M.; et al. Identification of proteins involved in inhibition of spheroid formation under microgravity. *Proteomics* **2015**, *15*, 2945–2952. [CrossRef]

65. Warnke, E.; Pietsch, J.; Wehland, M.; Bauer, J.; Infanger, M.; Görög, M.; Hemmersbach, R.; Braun, M.; Ma, X.; Sahana, J.; et al. Spheroid formation of human thyroid cancer cells under simulated microgravity: A possible role of ctgf and cav1. *Cell Commun. Signal.* **2014**, *12*, 32. [CrossRef]

66. Abu-Absi, S.F.; Hu, W.-S.; Hansen, L.K. Dexamethasone effects on rat hepatocyte spheroid formation and function. *Tissue Eng.* **2005**, *11*, 415–426. [CrossRef]

67. Thiery, J.P. Epithelial–mesenchymal transitions in tumour progression. *Nat. Rev. Cancer* **2002**, *2*, 442–454. [CrossRef]

68. Zheng, Y.; Izumi, K.; Li, Y.; Ishiguro, H.; Miyamoto, H. Contrary regulation of bladder cancer cell proliferation and invasion by dexamethasone-mediated glucocorticoid receptor signals. *Mol. Cancer Ther.* **2012**, *11*, 2621–2632. [CrossRef]

69. Kim, J.H.; Hwang, Y.-J.; Han, S.H.; Lee, Y.E.; Kim, S.; Kim, Y.J.; Cho, J.H.; Kwon, K.A.; Kim, J.H.; Kim, S.-H. Dexamethasone inhibits hypoxia-induced epithelial-mesenchymal transition in colon cancer. *World J. Gastroenterol.* **2015**, *21*, 9887–9899. [CrossRef]

70. Jang, Y.H.; Shin, H.S.; Sun Choi, H.; Ryu, E.S.; Jin Kim, M.; Ki Min, S.; Lee, J.H.; Kook Lee, H.; Kim, K.H.; Kang, D.H. Effects of dexamethasone on the tgf-beta1-induced epithelial-to-mesenchymal transition in human peritoneal mesothelial cells. *Lab. Investig.* **2013**, *93*, 194–206. [CrossRef]

71. Parrelli, J.M.; Meisler, N.; Cutroneo, K.R. Identification of a glucocorticoid response element in the human transforming growth factor beta 1 gene promoter. *Int. J. Biochem. Cell Biol.* **1998**, *30*, 623–627. [CrossRef]

72. Huber, M.A.; Kraut, N.; Beug, H. Molecular requirements for epithelial-mesenchymal transition during tumor progression. *Curr. Opin. Cell Biol.* **2005**, *17*, 548–558. [CrossRef]

73. Ivanova, K.; Ananiev, J.; Aleksandrova, E.; Ignatova, M.M.; Gulubova, M. Expression of e-cadherin/beta-catenin in epithelial carcinomas of the thyroid gland. *Open Access Maced. J. Med. Sci.* **2017**, *5*, 155–159. [CrossRef]

74. Hinz, B. The extracellular matrix and transforming growth factor-β1: Tale of a strained relationship. *Matrix Biol.* **2015**, *47*, 54–65. [CrossRef]

75. Beheshti, A.; Ray, S.; Fogle, H.; Berrios, D.; Costes, S.V. A microrna signature and tgf-β1 response were identified as the key master regulators for spaceflight response. *PLoS ONE* **2018**, *13*, e0199621. [CrossRef]

76. Li, Z.; Chen, Y.; Cao, D.; Wang, Y.; Chen, G.; Zhang, S.; Lu, J. Glucocorticoid up-regulates transforming growth factor-β (tgf-β) type ii receptor and enhances tgf-β signaling in human prostate cancer pc-3 cells. *Endocrinology* **2006**, *147*, 5259–5267. [CrossRef]

77. Liu, L.; Aleksandrowicz, E.; Schonsiegel, F.; Groner, D.; Bauer, N.; Nwaeburu, C.C.; Zhao, Z.; Gladkich, J.; Hoppe-Tichy, T.; Yefenof, E.; et al. Dexamethasone mediates pancreatic cancer progression by glucocorticoid receptor, tgfbeta and jnk/ap-1. *Cell Death Dis.* **2017**, *8*, e3064. [CrossRef]

78. Ameri, W.A.; Ahmed, I.; Al-Dasim, F.M.; Mohamoud, Y.A.; AlAzwani, I.K.; Malek, J.A.; Karedath, T. Tgf-β mediated cell adhesion dynamics and epithelial to mesenchymal transition in 3d and 2d ovarian cancer models. *Biorxiv* **2018**, 465617. [CrossRef]

79. Sodek, K.L.; Ringuette, M.J.; Brown, T.J. Compact spheroid formation by ovarian cancer cells is associated with contractile behavior and an invasive phenotype. *Int. J. Cancer* **2009**, *124*, 2060–2070. [CrossRef]

80. Mitra, T.; Roy, S.S. Co-activation of tgfβ and wnt signalling pathways abrogates emt in ovarian cancer cells. *Cell. Physiol. Biochem.* **2017**, *41*, 1336–1345. [CrossRef]

81. Peinado, H.; Olmeda, D.; Cano, A. Snail, zeb and bhlh factors in tumour progression: An alliance against the epithelial phenotype? *Nat. Rev. Cancer* **2007**, *7*, 415–428. [CrossRef] [PubMed]

82. Yook, J.I.; Li, X.Y.; Ota, I.; Fearon, E.R.; Weiss, S.J. Wnt-dependent regulation of the e-cadherin repressor snail. *J. Biol. Chem.* **2005**, *280*, 11740–11748. [CrossRef]

83. Lin, R.Z.; Chou, L.F.; Chien, C.C.; Chang, H.Y. Dynamic analysis of hepatoma spheroid formation: Roles of e-cadherin and beta1-integrin. *Cell Tissue Res.* **2006**, *324*, 411–422. [CrossRef] [PubMed]

84. Smyrek, I.; Mathew, B.; Fischer, S.C.; Lissek, S.M.; Becker, S.; Stelzer, E.H.K. E-cadherin, actin, microtubules and fak dominate different spheroid formation phases and important elements of tissue integrity. *Biol. Open* **2019**, *8*, bio037051. [CrossRef]

85. Sahana, J.; Nassef, M.Z.; Wehland, M.; Kopp, S.; Krüger, M.; Corydon, T.J.; Infanger, M.; Bauer, J.; Grimm, D. Decreased e-cadherin in mcf7 human breast cancer cells forming multicellular spheroids exposed to simulated microgravity. *Proteomics* **2018**, *18*, e1800015. [CrossRef]

86. Pease, J.C.; Brewer, M.; Tirnauer, J.S. Spontaneous spheroid budding from monolayers: A potential contribution to ovarian cancer dissemination. *Biol. Open* **2012**, *1*, 622–628. [CrossRef]

87. Shimazui, T.; Schalken, J.A.; Kawai, K.; Kawamoto, R.; van Bockhoven, A.; Oosterwijk, E.; Akaza, H. Role of complex cadherins in cell-cell adhesion evaluated by spheroid formation in renal cell carcinoma cell lines. *Oncol. Rep.* **2004**, *11*, 357–360. [CrossRef]

88. Tsai, C.-W.; Wang, J.-H.; Young, T.-H. Core/shell multicellular spheroids on chitosan as in vitro 3d coculture tumor models. *Artif. Cells Nanomed. Biotechnol.* **2018**, *46*, S651–S660. [CrossRef]

89. Katsamba, P.; Carroll, K.; Ahlsen, G.; Bahna, F.; Vendome, J.; Posy, S.; Rajebhosale, M.; Price, S.; Jessell, T.M.; Ben-Shaul, A.; et al. Linking molecular affinity and cellular specificity in cadherin-mediated adhesion. *Proc. Natl. Acad. Sci. USA* **2009**, *106*, 11594–11599. [CrossRef]

90. Shannon, S.; Vaca, C.; Jia, D.; Entersz, I.; Schaer, A.; Carcione, J.; Weaver, M.; Avidar, Y.; Pettit, R.; Nair, M.; et al. Dexamethasone-mediated activation of fibronectin matrix assembly reduces dispersal of primary human glioblastoma cells. *PLoS ONE* **2015**, *10*, e0135951. [CrossRef]

91. Robinson, E.E.; Foty, R.A.; Corbett, S.A. Fibronectin matrix assembly regulates $\alpha5\beta1$-mediated cell cohesion. *Mol. Biol. Cell* **2004**, *15*, 973–981. [CrossRef]

92. Naves, M.A.; Pereira, R.M.; Comodo, A.N.; de Alvarenga, E.L.; Caparbo, V.F.; Teixeira, V.P. Effect of dexamethasone on human osteoblasts in culture: Involvement of beta1 integrin and integrin-linked kinase. *Cell Biol. Int.* **2011**, *35*, 1147–1151. [CrossRef]

93. Petrella, A.; Ercolino, S.F.; Festa, M.; Gentilella, A.; Tosco, A.; Conzen, S.D.; Parente, L. Dexamethasone inhibits trail-induced apoptosis of thyroid cancer cells via bcl-xl induction. *Eur. J. Cancer* **2006**, *42*, 3287–3293. [CrossRef]

94. Gazitt, Y.; Fey, V.; Thomas, C.; Alvarez, R. Bcl-2 overexpression is associated with resistance to dexamethasone, but not melphalan, in multiple myeloma cells. *Int. J. Oncol.* **1998**, *13*, 397–405. [CrossRef]

95. Lin, Y.; Fukuchi, J.; Hiipakka, R.A.; Kokontis, J.M.; Xiang, J. Up-regulation of bcl-2 is required for the progression of prostate cancer cells from an androgen-dependent to an androgen-independent growth stage. *Cell Res.* **2007**, *17*, 531–536. [CrossRef]

96. Toruner, M.; Fernandez-Zapico, M.; Sha, J.J.; Pham, L.; Urrutia, R.; Egan, L.J. Antianoikis effect of nuclear factor-kappab through up-regulated expression of osteoprotegerin, bcl-2, and iap-1. *J. Biol. Chem.* **2006**, *281*, 8686–8696. [CrossRef]

97. Bergin, E.; Levine, J.S.; Koh, J.S.; Lieberthal, W. Mouse proximal tubular cell-cell adhesion inhibits apoptosis by a cadherin-dependent mechanism. *Am. J. Physiol. Renal Physiol.* **2000**, *278*, F758–F768. [CrossRef]

98. Kantak, S.S.; Kramer, R.H. E-cadherin regulates anchorage-independent growth and survival in oral squamous cell carcinoma cells. *J. Biol. Chem.* **1998**, *273*, 16953–16961. [CrossRef]

99. Orford, K.; Orford, C.C.; Byers, S.W. Exogenous expression of beta-catenin regulates contact inhibition, anchorage-independent growth, anoikis, and radiation-induced cell cycle arrest. *J. Cell. Biol.* **1999**, *146*, 855–868. [CrossRef]

100. Riwaldt, S.; Bauer, J.; Pietsch, J.; Braun, M.; Segerer, J.; Schwarzwälder, A.; Corydon, T.J.; Infanger, M.; Grimm, D. The importance of caveolin-1 as key-regulator of three-dimensional growth in thyroid cancer cells cultured under real and simulated microgravity conditions. *Int. J. Mol. Sci.* **2015**, *16*, 28296–28310. [CrossRef]

101. Liu, W.R.; Jin, L.; Tian, M.X.; Jiang, X.F.; Yang, L.X.; Ding, Z.B.; Shen, Y.H.; Peng, Y.F.; Gao, D.M.; Zhou, J.; et al. Caveolin-1 promotes tumor growth and metastasis via autophagy inhibition in hepatocellular carcinoma. *Clin. Res. Hepatol. Gastroenterol.* **2016**, *40*, 169–178. [CrossRef] [PubMed]

102. Chatterjee, M.; Ben-Josef, E.; Thomas, D.G.; Morgan, M.A.; Zalupski, M.M.; Khan, G.; Andrew Robinson, C.; Griffith, K.A.; Chen, C.-S.; Ludwig, T.; et al. Caveolin-1 is associated with tumor progression and confers a multi-modality resistance phenotype in pancreatic cancer. *Sci. Rep.* **2015**, *5*, 10867. [CrossRef] [PubMed]

103. Sinha, B.; Köster, D.; Ruez, R.; Gonnord, P.; Bastiani, M.; Abankwa, D.; Stan, R.V.; Butler-Browne, G.; Vedie, B.; Johannes, L.; et al. Cells respond to mechanical stress by rapid disassembly of caveolae. *Cell* **2011**, *144*, 402–413. [CrossRef] [PubMed]

104. Moreno-Vicente, R.; Pavon, D.M.; Martin-Padura, I.; Catala-Montoro, M.; Diez-Sanchez, A.; Quilez-Alvarez, A.; Lopez, J.A.; Sanchez-Alvarez, M.; Vazquez, J.; Strippoli, R.; et al. Caveolin-1 modulates mechanotransduction

responses to substrate stiffness through actin-dependent control of yap. *Cell Rep.* **2018**, *25*, 1622–1635.e1626. [CrossRef] [PubMed]

105. Pietsch, J.; Ma, X.; Wehland, M.; Aleshcheva, G.; Schwarzwälder, A.; Segerer, J.; Birlem, M.; Horn, A.; Bauer, J.; Infanger, M.; et al. Spheroid formation of human thyroid cancer cells in an automated culturing system during the shenzhou-8 space mission. *Biomaterials* **2013**, *34*, 7694–7705. [CrossRef]

106. Cui, L.; Zhang, Q.; Mao, Z.; Chen, J.; Wang, X.; Qu, J.; Zhang, J.; Jin, D. Ctgf is overexpressed in papillary thyroid carcinoma and promotes the growth of papillary thyroid cancer cells. *Tumour Biol.* **2011**, *32*, 721–728. [CrossRef]

107. Dufau, I.; Frongia, C.; Sicard, F.; Dedieu, L.; Cordelier, P.; Ausseil, F.; Ducommun, B.; Valette, A. Multicellular tumor spheroid model to evaluate spatio-temporal dynamics effect of chemotherapeutics: Application to the gemcitabine/chk1 inhibitor combination in pancreatic cancer. *BMC Cancer* **2012**, *12*, 15. [CrossRef]

108. Engebraaten, O.; Bjerkvig, R.; Pedersen, P.H.; Laerum, O.D. Effects of egf, bfgf, ngf and pdgf(bb) on cell proliferative, migratory and invasive capacities of human brain-tumour biopsies in vitro. *Int. J. Cancer* **1993**, *53*, 209–214. [CrossRef]

109. Mueller-Klieser, W. Three-dimensional cell cultures: From molecular mechanisms to clinical applications. *Am. J. Physiol.* **1997**, *273*, C1109–C1123. [CrossRef]

110. Lichtenberger, B.M.; Tan, P.K.; Niederleithner, H.; Ferrara, N.; Petzelbauer, P.; Sibilia, M. Autocrine vegf signaling synergizes with egfr in tumor cells to promote epithelial cancer development. *Cell* **2010**, *140*, 268–279. [CrossRef]

111. Klein, M.; Vignaud, J.M.; Hennequin, V.; Toussaint, B.; Bresler, L.; Plenat, F.; Leclere, J.; Duprez, A.; Weryha, G. Increased expression of the vascular endothelial growth factor is a pejorative prognosis marker in papillary thyroid carcinoma. *J. Clin. Endocrinol. Metab.* **2001**, *86*, 656–658. [CrossRef] [PubMed]

112. Karaca, Z.; Tanriverdi, F.; Unluhizarci, K.; Ozturk, F.; Gokahmetoglu, S.; Elbuken, G.; Cakir, I.; Bayram, F.; Kelestimur, F. Vegfr1 expression is related to lymph node metastasis and serum vegf may be a marker of progression in the follow-up of patients with differentiated thyroid carcinoma. *Eur. J. Endocrinol.* **2011**, *164*, 277–284. [CrossRef] [PubMed]

113. Waleh, N.S.; Brody, M.D.; Knapp, M.A.; Mendonca, H.L.; Lord, E.M.; Koch, C.J.; Laderoute, K.R.; Sutherland, R.M. Mapping of the vascular endothelial growth factor-producing hypoxic cells in multicellular tumor spheroids using a hypoxia-specific marker. *Cancer Res.* **1995**, *55*, 6222–6226. [PubMed]

114. Vieira, J.M.; Santos, S.C.; Espadinha, C.; Correia, I.; Vag, T.; Casalou, C.; Cavaco, B.M.; Catarino, A.L.; Dias, S.; Leite, V. Expression of vascular endothelial growth factor (vegf) and its receptors in thyroid carcinomas of follicular origin: A potential autocrine loop. *Eur. J. Endocrinol.* **2005**, *153*, 701–709. [CrossRef] [PubMed]

115. Shim, S.H.; Hah, J.H.; Hwang, S.Y.; Heo, D.S.; Sung, M.W. Dexamethasone treatment inhibits vegf production via suppression of stat3 in a head and neck cancer cell line. *Oncol. Rep.* **2010**, *23*, 1139–1143. [CrossRef]

116. Grun, D.; Adhikary, G.; Eckert, R.L. Vegf-a acts via neuropilin-1 to enhance epidermal cancer stem cell survival and formation of aggressive and highly vascularized tumors. *Oncogene* **2016**, *35*, 4379–4387. [CrossRef]

117. Meng, X.G.; Yue, S.W. Dexamethasone disrupts cytoskeleton organization and migration of t47d human breast cancer cells by modulating the akt/mtor/rhoa pathway. *Asian Pac. J. Cancer Prev.* **2014**, *15*, 10245–10250. [CrossRef]

118. Srivastava, A.; Kumar, A.; Giangiobbe, S.; Bonora, E.; Hemminki, K.; Forsti, A.; Bandapalli, O.R. Whole genome sequencing of familial non-medullary thyroid cancer identifies germline alterations in mapk/erk and pi3k/akt signaling pathways. *Biomolecules* **2019**, *9*, 605. [CrossRef]

Article

The Oncogene AF1Q is Associated with WNT and STAT Signaling and Offers a Novel Independent Prognostic Marker in Patients with Resectable Esophageal Cancer

Elisabeth S. Gruber [1], Georg Oberhuber [2,3], Peter Birner [2], Michaela Schlederer [2], Michael Kenn [4], Wolfgang Schreiner [4], Gerd Jomrich [1], Sebastian F. Schoppmann [1], Michael Gnant [1], William Tse [5,6,*] and Lukas Kenner [2,7,8,9,*]

[1] Division of General Surgery, Department of Surgery, Comprehensive Cancer Center, Medical University of Vienna, 1090 Vienna, Austria; elisabeth.s.gruber@meduniwien.ac.at (E.S.G.); gerd.jomrich@meduniwien.ac.at (G.J.); sebastian.schoppmann@meduniwien.ac.at (S.F.S.); michael.gnant@meduniwien.ac.at (M.G.)

[2] Institute of Pathology, Department of Experimental and Translational Pathology, Medical University of Vienna, 1090 Vienna, Austria; georg.oberhuber@patho.at (G.O.); peter.birner@meduniwien.ac.at (P.B.); michaela.schlederer@meduniwien.ac.at (M.S.)

[3] PIZ—Patho im Zentrum GmbH, 3100 St. Poelten, Lower Austria, Austria

[4] Section of Biosimulation and Bioinformatics, Center for Medical Statistics, Informatics and Intelligent Systems (CeMSIIS), Medical University of Vienna, 1090 Vienna, Austria; michael.kenn@meduniwien.ac.at (M.K.); wolfgang.schreiner@meduniwien.ac.at (W.S.)

[5] James Graham Brown Cancer Center, University of Louisville School of Medicine, Louisville, KY 40202, USA

[6] Division of Blood and Bone Marrow Transplantation, Department of Medicine, University of Louisville School of Medicine, Louisville, KY 40202, USA

[7] Christian Doppler Laboratory for Applied Metabolomics (CDL-AM), Medical University of Vienna, 1090 Vienna, Austria

[8] Institute of Laboratory Animal Pathology, University of Veterinary Medicine Vienna, 1210 Vienna, Austria

[9] CBmed Core Lab 2, Medical University of Vienna, 1090 Vienna, Austria

* Correspondence: william.tse@louisville.edu (W.T.); lukas.kenner@meduniwien.ac.at (L.K.); Tel.: +1-502-562-4363 (W.T.); +43-1-40400-51760 (L.K.)

Received: 4 October 2019; Accepted: 29 October 2019; Published: 30 October 2019

Abstract: AF1q impairs survival in hematologic and solid malignancies. AF1q expression is associated with tumor progression, migration and chemoresistance and acts as a transcriptional co-activator in WNT and STAT signaling. This study evaluates the role of AF1q in patients with resectable esophageal cancer (EC). A total of 278 patients operated on for EC were retrospectively included and the expression of AF1q, CD44 and pYSTAT3 was analyzed following immunostaining. Quantified data were processed to correlational and survival analysis. In EC tissue samples, an elevated expression of AF1q was associated with the expression of CD44 ($p = 0.004$) and pYSTAT3 ($p = 0.0002$). High AF1q expression in primary tumors showed high AF1q expression in the corresponding lymph nodes ($p = 0.016$). AF1q expression was higher after neoadjuvant therapy ($p = 0.0002$). Patients with AF1q-positive EC relapsed and died earlier compared to patients with AF1q-negative EC (disease-free survival (DFS), $p = 0.0005$; disease-specific survival (DSS), $p = 0.003$); in the multivariable Cox regression model, AF1q proved to be an independent prognostic marker (DFS, $p = 0.01$; DSS, $p = 0.03$). AF1q is associated with WNT and STAT signaling; it impairs and independently predicts DFS and DSS in patients with resectable EC. Testing AF1q could facilitate prognosis estimation and provide a possibility of identifying the patients responsive to the therapeutic blockade of its oncogenic downstream targets.

Keywords: AF1Q; MLLT11; WNT; STAT; esophageal cancer; prognosis

---

1. Introduction

AF1Q was originally identified as an *MLL* fusion partner in acute myeloid leukemia patients with a t(1; 11)(q21; q23) translocation (*MLLT11*); here, multiple chromosomal translocations of the AF1q locus have been described [1]. Enhanced AF1q expression is associated with poor clinical outcome in hematologic and several solid malignancies, such as breast, thyroid as well as testicular cancer and neuroblastoma [2–9]. Further on, AF1q plays a role in cell differentiation and maintenance during neuronal development [10], and is involved in stem cell differentiation. Here, it promotes T cell development, and at the same time impairs B cell differentiation; in addition, CD34-enriched stem cells show high AF1q levels, which are diminished during cell differentiation [11]. We delineated the precise oncogenic function of AF1q in breast cancer models, where AF1q functions as transcriptional co-factor and interacts with the T-cell factor/lymphoid enhancer factor (TCF/LEF) transcriptional complex in the wingless type MMTV integration site family (WNT) [3] and signal transducer and activator of transcription 3 (STAT3) [12] signaling pathway. These two core cancer pathways play a pivotal role in tumors of the gastrointestinal tract and are involved in tumor initiation, progression, metastases, and chemoresistance [13,14]. In colorectal cancer, WNT signaling is a major driver of oncogenicity, and AF1q has been reported to drive proliferation, migration, and invasion in vitro and in vivo [5,15]. Recently, the suppression of breast cancer dissemination has been subscribed to GATA3-driven miR29b induction, which reportedly inhibits AF1q [16]. Controversly, some studies reported γ-irradiation or doxorubicin-induced pro-apoptotic effects through Blc-2-antagonist of cell death (BAD) in liver and ovarian cancer cells that were mediated by AF1q [7,8]. However, the exact mechanisms that AF1q involves to promote cancer are largely unidentified.

Esophageal cancer (EC) is one of the 10 most common types of cancer and is responsible for more than half a million deaths worldwide [17]. Surgery is the treatment of choice in locally limited resectable disease, and surgical techniques have been extended recently to meet the resection challenges. For locally advanced stage cancers, preoperative treatment is recommended [18]. At the time of diagnosis, half of the patients are metastasized and might benefit from targeted therapies; still, randomized data in EC are scarce [18]. Since EC's genetic landscape is very distinct and driver mutations differ largely among subgroups, the definition of therapeutic targets on the basis of a genetic analysis is challenging. However, several targets for blocking the proposed AF1q downstream pathways WNT and STAT have demonstrated satisfying tumor response in cancer patients [12,13,19–22].

The aim of this study was to elucidate the role of AF1q in resectable EC, which to date remains unclear. By correlating tumoral AF1q expression with the expression of downstream targets CD44 and STAT3 as well as clinciopathological parameters, we wanted to evaluate the oncogenic potential of AF1q and its prognostic value for postoperative survival. Our data provide evidence that AF1q is associated with both WNT and STAT signaling and serves as a prognostic factor in EC patients.

2. Materials and Methods

2.1. Study Population

Patients diagnosed with cancer of the esophagus and the esophagogastric junction who had undergone surgery between 1992 and 2002 at the Division of General Surgery, Medical University of Vienna were included into the study under ethical approval from the local review board ('Ethikkommission') of the Medical University of Vienna, #1197/2019). In case of locally advanced stage cancer, patients received neoadjuvant treatment according to the latest clinical practice guidelines at the time of diagnosis followed by surgery. Histopathological staging was conducted according to the AJCC/UICC staging system [23]. Surgical specimens were processed to tissue microarrays (TMAs), in which each tumor was represented by triplicate core biopsies. The location of the tumors was assessed according to the rules published in the fourth edition of the World Health Organization (WHO) classification of gastrointestinal tumors [24]. In brief, all squamous carcinomas from the tubular esophagus and from the area of the esophagogastric junction were considered to be carcinomas of the esophagus. The proximal extent of the gastric folds was used as the landmark for the esophagogastric junction. Due to similar initiating factors and prognosis of adenocarcinomas of the distal esophagus and adenocarcinomas of the esophagogastric junction (AEG) as well as challenges in clear clinical and histopathological distinction, these two groups were merged before analysis [18]. Histological response to neoadjuvant chemotherapy was graded according to Mandard [25].

2.2. Immunohistochemistry

The expression of AF1q as well as CD44 (as bona fide WNT target gene) as a possible AF1q downstream target was evaluated in resected human EC specimens; pySTAT3 expression was available from previous studies [26]. Immunostainings were performed using a standard protocol. Paraffin sections were de-waxed, and for the antigen retrieval, a citrate buffer pH 6 (CD44) or a Tris/Ethylenediaminetetraacetic acid (EDTA) buffer pH 9 (AF1q) was used. After endogenous peroxidase blocking, avidin and biotin blocking steps were performed. The antibodies were incubated overnight at 4 °C in PBS+1% bovine serum albumin (BSA). The following antibodies were used: CD44 (Santa Cruz, sc-9960) in a 1:200 dilution and AF1q (Abcam, ab109016) in a 1:200 dilution. Slides were washed with phosphate buffered saline (PBS) the following day and incubated with polyvalent-secondary antibody (IDetect Super Stain System HRP, ID laboratories) and horseradish peroxidase (HRP; IDetect Super Stain System HRP, ID laboratories). Signals were visualized with 3-amino-9-ethylcarbazole (ID laboratories). After counterstaining with hemalaun, the slides were mounted. The specimen were analyzed by a board-certified pathologist. A specimen was considered as positive when at least 50% of tumor cells showed moderate or strong cytoplasmic marker expression. Antibody specificity has been confirmed in previous studies [3,12,26–29].

2.3. Statistical Analysis

In terms of experimental characteristics, data were described by mean (± standard deviation). Statistical differences were analyzed by a Student's t-test. A two-sided p-value less than 0.05 was considered statistically significant.

In terms of patient and tumor characteristics, numerical data were described by median (range) and categorical variables were described by frequencies. In order to compare AF1q expression with patient and tumor characteristics, the chi-square test was applied as appropriate. In order to describe the correlation between AF1q and ordinal or continuous variables, the Spearman rank correlation coefficient (r_s) was calculated. The inverse Kaplan–Meier method was used to estimate the median follow-up time [30]. Survival estimates were calculated by the Kaplan–Meier method with the log-rank test for group comparisons. DSS was defined as time from esophageal surgery until death from EC. DFS was defined as time from esophageal surgery until EC recurrence. In order to identify independent predictive factors for survival, established prognostic factors such as neoadjuvant therapy and histological tumor subtype, as well as TNM staging, tumor grading, and resection margin [18,31], were entered into a multivariable Cox regression model in addition to AF1q. MATLAB Version 9.6/R2019a (MathWorks) was used for all statistical calculations. For all analyses, a two-sided p-value less than 0.05 was considered statistically significant.

3. Results

3.1. AF1q Expression Analysis in Human Esophageal Cancer Samples

In total, 278 patients with cancer of the esophagus and the esophagogastric junction as well as corresponding fully annotated tumor samples were retrospectively defined for analysis. The cohort consisted of 118 patients with esophageal squamous cell carcinoma (ESCC, 42.2%), 67 patients with esophageal adenocarcinoma (EAC, 24.1%) and 93 patients with adenocarcinoma of the esophagogastric junction (AEG, 33.5%); out of those, 138 tumor samples (49.6%) showed significant AF1q expression (ESCC, n = 54, 45.8%; EAC, n = 42, 62.7%; AEG, n = 42, 45.2%). An example of positive tumoral AF1q expression in EC is shown in Figure 1. Patient and tumor characteristics compared between AF1q-positive and AF1q-negative tumors are compiled in Table 1. In short, patients who received neoadjuvant therapy showed higher tumoral AF1q expression compared to patients who were resected upfront (p = 0.0002); histological response to neoadjuvant therapy did not correlate with AF1q expression (r_s = 0.22, p = 0.09). Patients with AF1q-positive tumors suffered from a higher rate of positive resection margins (R1) compared to patients with AF1q-negative tumors (p = 0.004).

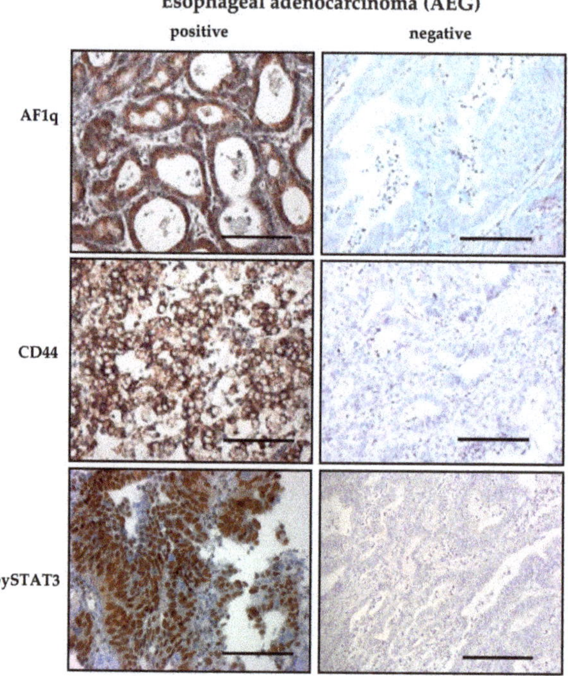

Figure 1. Examples of immunohistochemical (IHC) stained esophageal adenocarcinoma. Examples of AF1q expression in high-risk esophageal adenocarcinoma: Right example showing enhanced AF1q expression in high-risk adenocarcinoma vs. left example showing no AF1q expression in low-risk esophageal adenocarcinoma, size bar 100 μm.

Table 1. Patient and tumor characteristics compared between patients with AF1q-positive and AF1q-negative EC.

Factor	Patients with EC, n = 278 (100.0)	Patients with AF1q-positive EC, n = 138 (49.6)	Patients with AF1q-negative EC, n = 140 (50.4)	*p*-value
Female sex	65 (23.4)	28 (43.1)	37 (56.9)	0.23 *
Male sex	213 (76.6)	110 (51.6)	103 (48.4)	
Age	63.3 (34–90)	63.8 (38–90)	63.9 (34–90)	rs = 0.03 ** (0.65)
Adiposity	16 (5.8)	10 (62.5)	6 (37.5)	0.11 *
Reflux	5 (1.8)	2 (40.0)	3 (60)	0.22 *
Neoadjuvant therapy	68 (24.5)	47 (69.1)	21 (30.9)	0.0002 *
Factor	Patients with EC, n = 278 (100.0)	Patients with AF1q-positive EC, n = 138 (49.6)	Patients with AF1q-negative EC, n = 140 (50.4)	*p*-value
Histological tumor subtype				
ESCC	118 (42.4)	54 (45.8)	64 (54.2)	rs = 0.02 ** (0.88)
AC	160 (57.6)	84 (52.5)	76 (47.5)	
EAC	67 (24.1)	42 (62.7)	25 (37.3)	n.a.
AEG	93 (33.5)	42 (45.2)	51 (54.8)	
AJCC/UICC tumor staging ESCC n = 118 (42.4)				
IB	3 (2.5)	0 (0.0)	3 (100.0)	
IIA	32 (27.1)	15 (46.9)	17 (53.1)	
IIIA	13 (11.0)	6 (46.2)	7 (53.8)	rs = 0.06 ** (0.55)
IIIB	45 (38.1)	20 (44.4)	25 (55.6)	
IVA	19 (15.3)	10 (55.6)	8 (44.4)	
IVB	7 (5.9)	3 (42.9)	4 (57.1)	
AJCC/UICC tumor staging EAC n = 67 (24.1)/AEG n = 93 (33.5)				
IC	8 (5.0)	4 (50.0)	4 (50.0)	
IIA	3 (1.9)	0 (0.0)	3 (100.0)	
IIB	21 (13.1.)	10 (47.6)	11 (32.4)	rs = 0.12 ** (0.14)
IIIA	8 (5.0)	2 (25.0)	6 (75.0)	
IIIB	59 (36.9)	33 (55.9)	26 (44.1)	
IVA	61 (38.1)	35 (57.4)	26 (42.6)	
Regional lymph nodes pN				
N1	119 (44.1)	65 (54.6)	54 (45.4)	0.18 *
Tumor grading G				
G1	11 (4.0)	4 (36.4)	7 (63.6)	
G2	132 (47.5)	66 (50.0)	66 (50.0)	rs = 0.02 ** (0.69)
G3	135 (48.6)	68 (50.4)	67 (49.6)	
Resection margin R				
R0	228 (82.0)	104 (45.6)	124 (54.5)	0.004 *
R1	50 (18.0)	34 (68.0)	16 (32.0)	

Table 1. *Cont.*

Factor	Patients with EC, n = 278 (100.0)	Patients with AF1q-positive EC, n = 138 (49.6)	Patients with AF1q-negative EC, n = 140 (50.4)	*p*-value
Histological response to neoadjuvant therapy				
none	7 (10.3)	7 (10.3)	0 (0.0)	
poor	30 (44.1)	21 (70.0)	9 (30.0)	rs = 0.22 ** (0.09)
moderate	17 (25.0)	11 (64.7)	6 (35.3)	
good	4 (5.9)	2 (50.0)	2 (50.0)	

Note. Continuous variables are shown as median and range, categorical variables are expressed as absolute and relative numbers, n (%); adiposity is defined as BMI >30 kg/m2; EC—esophageal cancer; ESCC—esophageal squamous cell carcinoma; AC—adenocarcinomas; EAC—esophageal adenocarcinoma; AEG—adenocarcinomas of the esophagogastric junction; AJCC/UICC—American Joint Committee on Cancer/Union for International Cancer Control (https://cancerstaging.org/Pages/default.aspx; https://www.uicc.org); n.a.—not applicable; * chi-square test, ** Spearman correlation coefficient.

In order to evaluate an association of AF1q with WNT and STAT signaling, we correlated immunohistochemical (IHC) expression levels of CD44 (as bona fide WNT target gene) and tyrosine phosphorylated STAT3 (pYSTAT3) as possible AF1q downstream targets; we found a strong association of AF1q expression with both CD44 (n = 268, $p = 0.004$) as well as pYSTAT3 (n = 227, $p = 0.0002$) levels in EC (Table 2). Examples of positive tumoral CD44 as well as pYSTAT3 expression in EC are shown in Figure 1.

Table 2. Correlation of AF1q expression with proposed downstream target expression (CD44, pYSTAT3) in tumors of EC patients.

Factor	Patients with AF1q-positive EC, n = 133 (49.6)	Patients with AF1q-negative EC, n = 135 (50.4)	*p*-value
CD44, n = 268 (100)			
positive, n = 94 (35.1)	58 (21.6)	36 (13.5)	0.004
negative, n = 174 (74.9)	75 (28.0)	99 (36.9)	
Factor	**Patients with AF1q-positive EC, n = 117 (51.5)**	**Patients with AF1q-negative EC, n = 110 (48.5)**	***p*-value**
pYSTAT3, n = 227 (100)			
positive, n = 101 (44.5)	66 (29.1)	35 (15.4)	0.0002
negative, n = 126 (55.5)	51 (22.4)	75 (33.1)	

Note. EC—esophageal cancer; variables are expressed as absolute and relative numbers, n (%).

In order to explore AF1q-mediated dissemination features, we compared AF1q expression in lymph node metastases (n = 32) to AF1q expression in corresponding primary tumors and found a significant correlation ($p = 0.016$; Table 3).

Table 3. Correlation of enhanced AF1q expression in primary tumors and lymph node metastases of EC patients.

Factor	AF1q-positive local EC, n = 18 (56.3)	AF1q-negative local EC, n = 14 (43.7)	*p*-value
Lymph node metastases, n = 32 (100)			
AF1q positive, n = 19 (59.4)	14 (43.8)	5 (15.6)	0.016
AF1q negative, n = 13 (40.6)	4 (12.5)	9 (28.1)	

Note. EC—esophageal cancer; variables are expressed as absolute and relative numbers, n (%).

3.2. Survival Analysis

During a median postoperative follow-up period of 71 months (range 52–90 months), 156 out of 278 (56.1%) patients relapsed, and 185 out of 278 (66.5%) patients had died of EC. The median DFS time was 17 months (range 13–20 months), and the median DSS time was 21 months (range 15–25 months). In patients with a high tumoral AF1q expression, the median DFS was 13 months (range 10–16 months) and the median DSS was 22 months (range 18–27 months) compared to 23 months (range 16–31 months) median DFS and 26 months (range 18–24 months) median DSS in patients with a low tumoral AF1q expression. Enhanced AF1q expression in the tumor proper resulted in significantly decreased DFS (Kaplan Meier/log rank; $p = 0.0005$; Figure 2) and DSS (Kaplan Meier/log rank, $p = 0.003$; Figure 3).

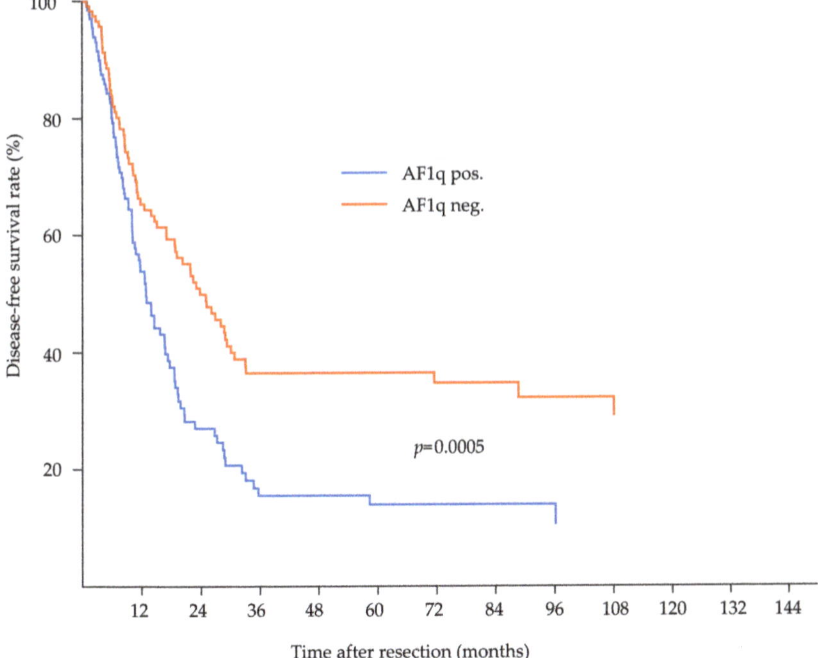

Figure 2. Kaplan–Meier analysis for disease-free survival in esophageal cancer (EC) patients. EC patients with high tumoral AF1q levels relapse earlier compared to patients with low or no tumoral AF1q expression (disease-free survival (DFS), log rank: $p = 0.0005$).

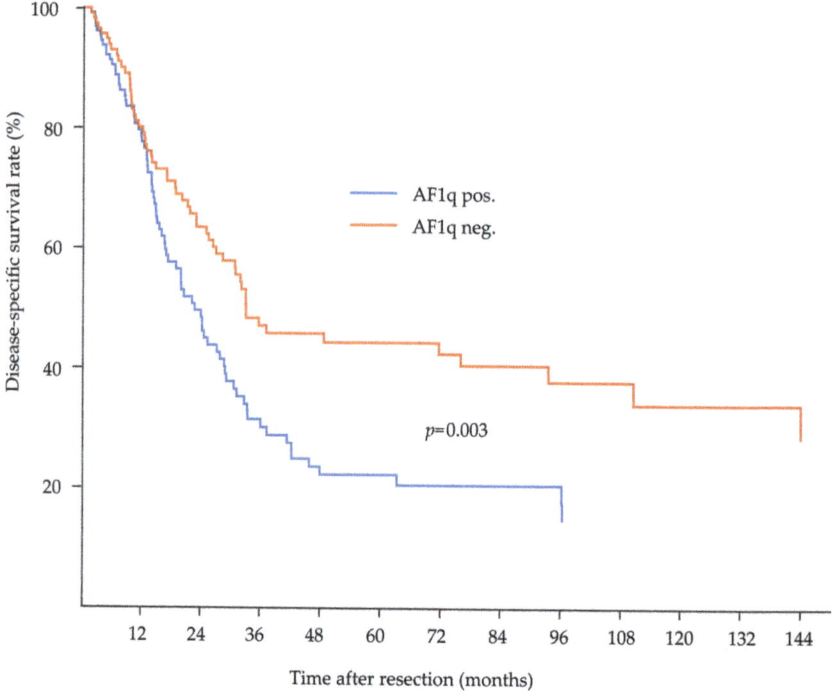

Figure 3. Kaplan–Meier analysis for disease-specific survival in esophageal cancer (EC) patients. EC patients with high tumoral AF1q die earlier compared to patients with low or no tumoral AF1q expression (disease-specific survival (DSS), log rank; $p = 0.003$).

Cox regression analysis showed a 1.5 times higher risk for both disease recurrence and disease-specific death in patients with a high tumoral AF1q expression. Further prognostic factors were neoadjuvant therapy in regard to DFS ($p = 0.0002$), local tumor stage (DFS, $p = 0.001$ and DSS, $p = 0.0004$, respectively), regional lymph node metastases (DFS, $p < 0.0001$ and DSS, $p < 0.0001$, respectively), and resection margin (DFS, $p = 0.02$ and DSS, $p = 0.01$, respectively). In a multivariable Cox regression model, AF1q proved to be an independent factor for survival (DFS $p = 0.01$; DSS $p = 0.03$) next to the local tumor stage in regard to DSS ($p = 0.007$, respectively), regional lymph node metastases (DFS, $p < 0.0001$ and DSS, $p < 0.0001$, respectively), and histological tumor subtype (DFS $p = 0.0004$ and DSS, $p = 0.003$) as well as neoadjuvant therapy in regard to DFS ($p = 0.005$). The data of univariate and multivariable Cox regression analysis are compiled in Table 4.

Table 4. Univariable and multivariable Cox regression analysis in EC patients.

Factor	Univariate *p*-Value	Multivariable *p*-Value	HR	95% CI Low	95% CI High
Disease-free survival					
AF1q	0.0005	0.01	1.5	1.1	2.2
Neoadjuvant therapy	0.0002	0.005	1.7	1.2	2.6
Histological tumor subtype *	0.16	0.0004	1.9	1.3	2.7
Local tumor stage pT	0.001	0.07		n.a.	
Regional lymph nodes pN	<0.0001	<0.0001	2.3	1.6	3.2
Tumor grading G	0.13	0.30		n.a.	
Resection margin R	0.02	0.50		n.a.	
Disease-specific survival					
AF1q	0.003	0.03	1.5	1.0	2.1
Neoadjuvant therapy	0.10	0.83		n.a.	
Histological tumor subtype *	0.30	0.003	1.8	1.2	2.7
Local tumor stage pT	0.0004	0.007	1.9	1.2	3.0
Regional lymph nodes pN	<0.0001	<0.0001	2.2	1.5	3.2
Tumor grading G	0.22	0.48		n.a.	
Resection margin R	0.01	0.31		n.a.	

Note. EC—esophageal cancer; HR—hazard ratio; CI—confidence interval; HR and CR refer to multivariable model; * adenocarcinoma and squamous cell carcinoma.

4. Discussion

We show here for the first time the oncogenic potential as well as the utility of AF1q as a prognostic marker in esophageal cancer (EC). We demonstrate a significant association of AF1q expression with WNT as well as STAT3 signaling pathways, and show that tumoral AF1q expression impairs survival and serves as an independent prognostic factor in patients with resectable EC.

Well known for exerting multiple oncogenic functions, WNT and STAT3 have been shown to be of prognostic value in EC [13,14,26,32]. Both signaling pathways promote intestinal tumor growth and regeneration; interestingly, this can be reversed through gp130-JAK-STAT3 blockade [33]. These data imply that there might be a link between these two core cancer pathways. In fact, what WNT and STAT3 seem to share is AF1q as an activator of their target gene transcription. We previously reported that AF1q physically binds TCF/LEF and STAT3 and boosts their target gene transcription [3,12]. In breast cancer patients, the cooperation of AF1q with TCF7 led to the transcription of CD44 [3], which is a WNT target gene that is essentially involved in tumor progression and epithelial-to-mesenchymal transition [34,35]. In invasive breast cancer cells, AF1q further induced pYSTAT3 levels through the Src kinase-driven PDGFB/PDGFRB cascade [12]. Similarily to the findings in breast cancer, we here demonstrate an association of AF1q with both WNT and STAT signaling in the sense that the patients with AF1q-positive EC show enhanced tumoral levels of both proposed downstream targets CD44 and pYSTAT3.

AF1q expression is reportedly associated with metastatic spread in colorectal, lung, and breast cancer [5,12,36,37]. Since HER-2 is commonly expressed in EC, lung, and breast cancer [38–41], HER-2 might be another signaling pathway AF1q interacts with. However, further studies are needed to prove this concept.

We here demonstrate a correlation of AF1q expression in the local tumor with AF1q expression in the lymph node metastases of EC patients. In addition, AF1q-positive ECs show higher rates of positive resection margins compared to AF1q-negative ECs. We believe that the AF1q-induced co-activation of

WNT and STAT signaling plays a key role in EC initiation, proliferation, and dissemination. Recently, Src kinase and JAK phosphorylation have been shown to facilitate STAT3 signaling [42], which further results in excessive proliferation and the malignant transformation of intestinal epithelial cells [43]. In invasive breast cancer cells, enhanced AF1q expression induced pYSTAT3 levels through the Src kinase-driven platelet-derived growth factor beta (PDGFB)/platelet-derived growth factor receptor beta (PDGFRB) cascade, which was reversible upon Src kinase blockade with protein phosphate 1 (PP1) or PDGFRB blockade using imatinib [12]. Consequently, patients with AF1q-positive EC are at exceptional risk to develop rapid disease progression that might respond to CD44 and STAT3 inhibition [13,19–22].

Conversely, AF1q has been reported to mediate the pro-apoptotic effects of cytotoxic agents such as doxorubicin, retinoic acid, or γ-irradiation through activation of the Blc-2-antagonist of cell death (BAD) [7–9]. In advanced stage EC, neoadjuvant (radio-)chemotherapy is applied to facilitate tumor shrinkage and consecutive R0 resection rates, and prevent recurrence [18,44]. Consistent with other findings, we showed an increased AF1q expression in patients treated with neoadjuvant therapy compared to treatment-naïve patients [7,8,12]. These findings support the fact that AF1q induction can be at least partly triggered by external factors such as cytotoxic agents and/or γ-irradiation [18]. Paradoxically, these data also indiciate that patients with AF1q-positive EC might suffer a disadvantage from currently applied oncologic treatment protocols due to fact that an enhanced induction of AF1q expression not only promotes EC dissemination, but might rather enforce EC's metastic potential.

For a prognosis estimation of esophageal cancer, all staging efforts are recommended to be based on neoadjuvant therapy, histological tumor subtype, and TNM staging, as well as tumor grading and resection margin as important prognostic factors [18,31]. In this cohort of patients, AF1q proved to have a highly significant prognostic impact on both DFS and DSS in the univariate analysis. In fact, neoadjuvant therapy, histological tumor subtype, and regional lymph node status remained as independent prognostic factors for DFS and histological tumor subtype, local tumor stage, as well as regional lymph node status remained as independent prognostic factors for DSS in the model next to AF1q, respectively. Although none of these established prognostic factors correlated with tumoral AF1q expression, they demonstrate predominant importance for survival estimation in this selected cohort of patients with resectable esophageal cancer.

5. Conclusions

Our data underline the potency of AF1q to enforce and link the two major oncogenic pathways WNT and STAT3 involved in tumor initiation, progression, and dissemination. We demonstrate here for the first time a positive correlation between the expression of AF1q and the WNT and the STAT3 target genes CD44 and pYSTAT3 in EC, suggesting that AF1q may act as a cofactor for boosting the transcription of CD44 and pYSTAT3 in EC. Consequently, we demonstrate that patients with AF1q-positive EC relapse and die earlier. The association of AF1q with WNT and STAT3 signaling implicates various (combinatorial) targeting options such as CD44 and STAT3, JAK, and Src, as well as tyrosine kinases. Particularly, blockade of the AF1q/TCF7/CD44 regulatory axis as well as PDGF-B might hold therapeutic promise for patients with EC. Importantly, AF1q serves as an independent prognostic marker, and might add value to the estimation of resectability and disease progression. This study should prompt future clinical studies to validate these findings.

Author Contributions: Conception and design: E.S.G., G.O., M.S., P.B., L.K.; development of methodology: E.S.G., P.B., M.K., W.S.; acquisition of data: P.B., S.F.S., G.J.; analysis and interpretation of data: E.S.G., G.O., M.S., P.B., L.K.; Writing, review and/or revision of the manuscript: E.S.G., G.O., S.F.S., M.G., W.T., L.K.; administrative, technical or material support: none; study supervision: W.T., L.K.

Funding: This study was granted by the "Dr. Anton Oelzelt-Newin Stiftung", Austrian Academy of Sciences. LK is supported by the Austrian Science Fund (FWF, grant no. P26011 and P 29251), the MCEU-ITN ALKATRAS. Network (grant no. 675712) as well as the BM Funds (grant no. 15142) and the Margaretha Hehberger Stiftung (grant no. 15142).

Acknowledgments: We thank Gertrude Krainz for her support in spelling and phraseology.

Conflicts of Interest: M.G. has received institutional research support from AstraZeneca, Roche, Novartis, and Pfizer, and has received lecture fees, honoraria for participation on advisory boards, and travel support from Amgen, AstraZeneca, Celgene, EliLilly, Invectys, Pfizer, Nanostring, Novartis, Roche, and Medison. He has served as a consultant for AstraZeneca and EliLilly, and an immediate family member is employed by Sandoz. None of the other authors have financial and personal relationships with individuals or organizations that could inappropriately influence their work.

References

1. Tse, W.; Zhu, W.; Chen, H.S.; Cohen, A. A novel gene, AF1q, fused to MLL in t(1;11) (q21;q23), is specifically expressed in leukemic and immature hematopoietic cells. *Blood* **1995**, *85*, 650–656. [CrossRef] [PubMed]

2. Strunk, C.J.; Platzbecker, U.; Thiede, C.; Schaich, M.; Illmer, T.; Kang, Z.; Leahy, P.; Li, C.; Xie, X.; Laughlin, M.J.; et al. Elevated AF1q expression is a poor prognostic marker for adult acute myeloid leukemia patients with normal cytogenetics. *Am. J. Hematol.* **2009**, *84*, 308–309. [CrossRef] [PubMed]

3. Park, J.; Schlederer, M.; Schreiber, M.; Ice, R.; Merkel, O.; Bilban, M.; Hofbauer, S.; Kim, S.; Addison, J.; Zou, J.; et al. AF1q is a novel TCF7 co-factor which activates CD44 and promotes breast cancer metastasis. *Oncotarget* **2015**, *6*, 20697–20710. [CrossRef] [PubMed]

4. Tse, W.; Meshinchi, S.; Alonzo, T.A.; Stirewalt, D.L.; Gerbing, R.B.; Woods, W.G.; Appelbaum, F.R.; Radich, J.P. Elevated expression of the AF1q gene, an MLL fusion partner, is an independent adverse prognostic factor in pediatric acute myeloid leukemia. *Blood* **2004**, *104*, 3058–3063. [CrossRef] [PubMed]

5. Hu, J.; Li, G.; Liu, L.; Wang, Y.; Li, X.; Gong, J. AF1q Mediates Tumor Progression in Colorectal Cancer by Regulating AKT Signaling. *Int. J. Mol. Sci.* **2017**, *18*, 987. [CrossRef]

6. Liu, T.; Bohlken, A.; Kuljaca, S.; Lee, M.; Nguyen, T.; Smith, S.; Cheung, B.; Norris, M.D.; Haber, M.; Holloway, A.J.; et al. The retinoid anticancer signal: Mechanisms of target gene regulation. *Br. J. Cancer* **2005**, *93*, 310–318. [CrossRef]

7. Co, N.; Tsang, W.; Wong, T.; Cheung, H.; Tsang, T.; Kong, S.; Kwok, T. Oncogene AF1q enhances doxorubicin-induced apoptosis through BAD-mediated mitochondrial apoptotic pathway. *Mol. Cancer Ther.* **2008**, *7*, 3160–3168. [CrossRef]

8. Co, N.; Tsang, W.; Tsang, T.; Yeung, H.; Yau, P.; Kong, S.K.; Kwok, T.T. AF1q enhancement of γ irradiation-induced apoptosis by up-regulation of BAD expression via NF-κB in human squamous carcinoma A431 cells. *Oncol. Rep.* **2010**, *24*, 547–554. [CrossRef]

9. Tiberio, P.; Cavadini, E.; Callari, M.; Daidone, M.G.; Appierto, V. AF1q: A novel mediator of basal and 4-HPR-induced apoptosis in ovarian cancer cells. *PLoS ONE* **2012**, *7*, e39968. [CrossRef]

10. Lin, H.; Shaffer, K.; Sun, Z.; Jay, G.; He, W.; Ma, W. AF1q, a differentially expressed gene during neuronal differentiation, transforms HEK cells into neuron-like cells. *Mol. Brain Res.* **2004**, *131*, 126–130. [CrossRef]

11. Tse, W.; Deeg, H.; Stirewalt, D.; Appelbaum, F.; Radich, J.; Gooley, T. Increased AF1q gene expression in high-risk myelodysplastic syndrome. *Br. J. Haematol.* **2005**, *128*, 218–220. [CrossRef] [PubMed]

12. Park, J.; Kim, S.; Joh, J.; Remick, S.C.; Miller, D.M.; Yan, J.; Kanaan, Z.; Chao, J.-H.; Krem, M.M.; Basu, S.K.; et al. MLLT11/AF1q boosts oncogenic STAT3 activity through Src-PDGFR tyrosine kinase signaling. *Oncotarget* **2016**, *7*, 43960–43973. [CrossRef] [PubMed]

13. Anastas, J.N.; Moon, R.T. WNT signalling pathways as therapeutic targets in cancer. *Nat. Rev. Cancer* **2013**, *13*, 11–26. [CrossRef] [PubMed]

14. Yu, H.; Lee, H.; Herrmann, A.; Buettner, R.; Jove, R. Revisiting STAT3 signalling in cancer: New and unexpected biological functions. *Nat. Rev. Cancer* **2014**, *14*, 736–746. [CrossRef] [PubMed]

15. Vacante, M.; Borzì, A.M.; Basile, F.; Biondi, A. Biomarkers in colorectal cancer: Current clinical utility and future perspectives. *World J. Clin. Cases* **2018**, *6*, 869–881. [CrossRef] [PubMed]

16. Xiong, Y.; Li, Z.; Ji, M.; Tan, A.; Bemis, J.; Tse, J.; Huang, G.; Park, J.; Ji, C.; Chen, J.; et al. MIR29B regulates expression of MLLT11 (AF1Q), an MLL fusion partner, and low MIR29B expression associates with adverse cytogenetics and poor overall survival in AML. *Br. J. Haematol.* **2011**, *153*, 753–757. [CrossRef]

17. Bray, F.; Ferlay, J.; Soerjomataram, I.; Siegel, R.L.; Torre, L.A.; Jemal, A. Global cancer statistics 2018: GLOBOCAN estimates of incidence and mortality worldwide for 36 cancers in 185 countries. *CA Cancer J. Clin.* **2018**, *68*, 394–424. [CrossRef]

18. Lordick, F.; Mariette, C.; Haustermans, K.; Obermannova, R.; Arnold, D.; Committee, E.G. Oesophageal cancer: ESMO Clinical Practice Guidelines for diagnosis, treatment and follow-up. *Ann. Oncol.* **2016**, *27*, v50–v57. [CrossRef]

19. Buchert, M.; Burns, C.J.; Ernst, M. Targeting JAK kinase in solid tumors: Emerging opportunities and challenges. *Oncogene* **2016**, *35*, 939–951. [CrossRef]

20. Hubbard, J.M.; Grothey, A. Napabucasin: An Update on the First-in-Class Cancer Stemness Inhibitor. *Drugs* **2017**, *77*, 1091–1103. [CrossRef]

21. Dembowsky, K. A Safety and Pharmacokinetic Phase I/Ib Study of AMC303 in Patients With Solid Tumours. 2016. Available online: https://clinicaltrials.gov/ct2/show/NCT03009214 (accessed on 1 August 2016).

22. Menke-van der Houven, C.W.; van Oordt, C.G.R.; van Herpen, C.; Coveler, A.L.; Mahalingam, D.; Verheul, H.M.; van der Graaf, W.T.; Christen, R.; Rüttinger, D.; Weigand, S.; et al. First-in-human phase I clinical trial of RG7356, an anti-CD44 humanized antibody, in patients with advanced, CD44-expressing solid tumors. *Oncotarget* **2016**, *7*, 80046–80058.

23. Rice, T.W.; Patil, D.T.; Blackstone, E.H. 8th edition AJCC/UICC staging of cancers of the esophagus and esophagogastric junction: Application to clinical practice. *Ann. Cardiothorac. Surg.* **2017**, *6*, 119–130. [CrossRef] [PubMed]

24. Odze, R.; Flejou, J.; Boffetta, P.; Hofler, H.; Montgomery, E.; Spechler, S. Adenocarcinoma of the oesophgogastric junction. In *WHO Classification of Tumours of the Digestive System, World Health Organization Classification of Tumours*; Bosman, F., Carneiro, F., Hruban, R., Theise, N., Eds.; IARC Press: Lyon, Paris, 2010; pp. 39–44.

25. Mandard, A.M.; Dalibard, F.; Mandard, J.C.; Marnay, J.; Henry-Amar, M.; Petiot, J.F.; Roussel, A.; Jacob, J.H.; Segol, P.; Samama, G.; et al. Pathologic assessment of tumor regression after preoperative chemoradiotherapy of esophageal carcinoma. Clinicopathologic correlations. *Cancer* **1994**, *73*, 2680–2686. [CrossRef]

26. Schoppmann, S.F.; Jesch, B.; Friedrich, J.; Jomrich, G.; Maroske, F.; Birner, P. Phosphorylation of signal transducer and activator of transcription 3 (STAT3) correlates with Her-2 status, carbonic anhydrase 9 expression and prognosis in esophageal cancer. *Clin. Exp. Metastasis* **2012**, *29*, 615–624. [CrossRef] [PubMed]

27. Addison, J.B.; Koontz, C.; Fugett, J.H.; Creighton, C.J.; Chen, D.; Farrugia, M.K.; Padon, R.R.; Voronkova, M.A.; McLaughlin, S.L.; Livengood, R.H.; et al. KAP1 Promotes Proliferation and Metastatic Progression of Breast Cancer Cells. *Cancer Res.* **2015**, *75*, 344–355. [CrossRef] [PubMed]

28. Chu, T.-H.; Chan, H.-H.; Hu, T.-H.; Wang, E.M.; Ma, Y.-L.; Huang, S.-C.; Wu, J.-C.; Chang, Y.-C.; Weng, W.-T.; Wen, Z.-H.; et al. Celecoxib enhances the therapeutic efficacy of epirubicin for Novikoff hepatoma in rats. *Cancer Med.* **2018**, *7*, 2567–2580. [CrossRef]

29. Jin, H.; Sun, W.; Zhang, Y.; Yan, H.; Liufu, H.; Wang, S.; Chen, C.; Gu, J.; Hua, X.; Zhou, L.; et al. MicroRNA-411 Downregulation Enhances Tumor Growth by Upregulating MLLT11 Expression in Human Bladder Cancer. *Mol. Ther. Nucleic Acids* **2018**, *11*, 312–322. [CrossRef]

30. Schemper, M.; Smith, T.L. A Note on Quantifying Follow-up in Studies of Failure Time. *Control. Clin. Trials* **1996**, *17*, 343–346. [CrossRef]

31. Rice, T.; Kelsen, D.; Blackstone, E.; Ishwaran, H.; Patil, D.; Bass, A.; Erasmus, J.; Gerdes, H.; Hofstetter, W. Esophagus and Esophagogastric Junction. In *AJCC Cancer Staging Manual*, 8th ed.; Springer International Publishing: New York, NY, USA, 2017. [CrossRef]

32. Yu, H.; Kortylewski, M.; Pardoll, D. Crosstalk between cancer and immune cells: Role of STAT3 in the tumour microenvironment. *Nat. Rev. Immunol.* **2007**, *7*, 41–51. [CrossRef]

33. Phesse, T.J.; Buchert, M.; Stuart, E.; Flanagan, D.J.; Faux, M.; Afshar-Sterle, S.; Walker, F.; Zhang, H.H.; Nowell, C.J.; Jorissen, R. Partial inhibition of gp130-Jak-Stat3 signaling prevents Wnt–b-catenin–mediated intestinal tumor growth and regeneration. *Cancer Biol.* **2014**, *7*, 1–11. [CrossRef]

34. Naor, D.; Wallach-Dayan, S.B.; Zahalka, M.A.; Sionov, R.V. Involvement of CD44, a molecule with a thousand faces, in cancer dissemination. *Semin. Cancer Biol.* **2008**, *18*, 260–267. [CrossRef] [PubMed]

35. Zoller, M. CD44: Can a cancer-initiating cell profit from an abundantly expressed molecule? *Nat. Rev. Cancer* **2011**, *11*, 254–267. [CrossRef] [PubMed]

36. Chang, X.Z.; Li, D.Q.; Hou, Y.F.; Wu, J.; Lu, J.S.; Di, G.H.; Jin, W.; Ou, Z.L.; Shen, Z.Z.; Shao, Z.M. Identification of the functional role of AF1Q in the progression of breast cancer. *Breast Cancer Res. Treat.* **2008**, *111*, 65–78. [CrossRef] [PubMed]

37. Yang, S.; Dong, Q.; Yao, M.; Shi, M.; Ye, J.; Zhao, L.; Su, J.; Gu, W.; Xie, W.; Wang, K.; et al. Establishment of an experimental human lung adenocarcinoma cell line SPC-A-1BM with high bone metastases potency by (99m)Tc-MDP bone scintigraphy. *Nucl. Med. Biol.* **2009**, *36*, 313–321. [CrossRef] [PubMed]
38. Bartley, A.N.; Washington, M.K.; Ventura, C.B.; Ismaila, N.; Colasacco, C.; Benson, A.B., III; Carrato, A.; Gulley, M.L.; Jain, D.; Kakar, S.; et al. HER2 Testing and Clinical Decision Making in Gastroesophageal Adenocarcinoma: Guideline From the College of American Pathologists, American Society for Clinical Pathology, and American Society of Clinical Oncology. *Am. J. Clin. Pathol.* **2016**, *146*, 647–669. [CrossRef]
39. Schoppmann, S.F.; Jesch, B.; Friedrich, J.; Wrba, F.; Schultheis, A.; Pluschnig, U.; Maresch, J.; Zacherl, J.; Hejna, M.; Birner, P. Expression of Her-2 in carcinomas of the esophagus. *Am. J. Surg. Pathol.* **2010**, *34*, 1868–1873. [CrossRef]
40. Cardoso, F.; Kyriakides, S.; Ohno, S.; Penault-Llorca, F.; Poortmans, P.; Rubio, I.T.; Zackrisson, S.; Senkus, E. Early Breast Cancer: ESMO Clinical Practice Guidelines. *Ann. Oncol.* **2019**, *30*, 1194–1220. [CrossRef]
41. Kerr, K.; Bubendorf, L.; Edelman, M.; Marchetti, A.; Mok, T.; Novello, S.; O'Byrne, K.; Stahel, R.; Peters, S.; Felip, E. ESMO Consensus Guidelines: Pathology and molecular biomarkers for non-small-cell lung cancer. *Ann. Oncol.* **2014**, *25*, 1681–1690. [CrossRef]
42. Qing, Y.; Stark, G.R. Alternative activation of STAT1 and STAT3 in response to interferon-gamma. *J. Biol. Chem.* **2004**, *279*, 41679–41685. [CrossRef]
43. Peterson, L.W.; Artis, D. Intestinal epithelial cells: Regulators of barrier function and immune homeostasis. *Nat. Rev. Immunol.* **2014**, *14*, 141–153. [CrossRef]
44. Smyth, E.C.; Verheij, M.; Allum, W.; Cunningham, D.; Cervantes, A.; Arnold, D.; Committee, E.G. Gastric cancer: ESMO Clinical Practice Guidelines for diagnosis, treatment and follow-up. *Ann. Oncol.* **2016**, *27*, v38–v49. [CrossRef] [PubMed]

Comment

Comment on "Dexamethasone Inhibits Spheroid Formation of Thyroid Cancer Cells Exposed to Simulated Microgravity"

Joseph J. Bevelacqua [1], James Welsh [2] and S. M. J. Mortazavi [3,*]

[1] Bevelacqua Resources, Richland, WA 99352, USA; bevelresou@aol.com
[2] Department of Radiation Oncology, Edward Hines Jr VA Hospital, Hines, IL 60141, USA; James.Welsh@va.gov
[3] Diagnostic Imaging Department, Fox Chase Cancer Center, Philadelphia, PA 19111, USA
* Correspondence: mortazavismj@gmail.com; Tel.: +1-(215)-214-1769

Received: 20 February 2020; Accepted: 13 July 2020; Published: 21 July 2020

Keywords: dexamethasone; thyroid cancer; microgravity; space environment

---

This letter addresses our concerns about a paper by Melnik et al. entitled "Dexamethasone Inhibits Spheroid Formation of Thyroid Cancer Cells Exposed to Simulated Microgravity" that was published in *Cells* [1]. Melnik et al. have used synthetic glucocorticoid dexamethasone (DEX) to suppress the formation of spheroids in a culture of follicular thyroid cancer (FTC)-133 cells after exposure to simulated microgravity. Despite strengths, this paper suffers from a few shortcomings. The first shortcoming of this paper comes from its simulation protocol. The authors state that they used a random positioning machine (RPM) for simulating microgravity: "The used desktop-RPM (Dutch Space, Leiden, The Netherlands) was located in an incubator with 37 °C/5% CO_2 and operated in real random mode, with a constant angular velocity of 60°/s". RPMs that represent the more sophisticated development of the single-axis clinostats are laboratory instruments that provide a continuous random change in orientation relative to the gravity vector of an accommodated (biological) experiment. Given this consideration, RPMs usually consist of two independently rotating frames (the second frame is positioned inside the first frame) to give a very complex net of orientation change to a biological sample. RPMs cannot really simulate microgravity, and as well as this the cells experience strong shearing forces and microscopic friction that can significantly affect the findings of this study. It is worth noting that the convection and shear stresses that occur inside a cell culture flask during RPM experiments are addressed in previous studies: "The RPM rotation introduces fluid motion in the culture flask, leading to shear forces and enhanced convection" [2]. Along the flask walls, shear stresses can reach up to a few 100 mPa depending on the rotational velocity [3]. Therefore, the fluid dynamic in the culture flasks and its potential effects on cells inside culture flasks turning on an operating RPM should be fully studied.

Moreover, in an actual space environment, the interactions of the key stressors such as radiation and microgravity play a basic role in biological effects, but a single stressor like microgravity does not. The issue of the possible interactions of the space stressors, such as radiation and microgravity, dates back to 1999 [4]. More advanced approaches occurred in the period 2003–2004 [5,6], and this topic still receives great attention [7,8].

In addition, readers of the paper authored by Melnik et al. should be aware of the following points:

1. It is not possible to simulate microgravity on earth. The 9.8 m/s^2 gravitational field directed towards the center of the planet cannot be eliminated and contributed to the experiment. Given this consideration, microgravity simulation studies conducted using RPMs suffer from two major limitations:

I.   Physical Limitations:

Gravity is an interaction that is based on the distance between masses. For Earth, the gravitational interaction and its gradient can be approximated by a direction with respect to the center of the Earth. Since that distance is large, there is not much of a surface gradient. When that distance is shortened, the gradient becomes more severe because the acceleration varies with the distance to the various parts of a body. Acceleration is a vector quantity. This effect increases as the body size increases. It is negligible for a point source, but not so for an extended body.

Given this consideration, while on earth, a preferred direction is defined by the planet's gravitational field of 9.8 m/s^2, whereas in space there is no preferred direction. Thus, simply we cannot do an experiment on Earth to negate this field for an extended body.

A random positioning machine (RPM) can, in principle, establish an oppositely directed acceleration for an instant at a particular body location (or a particular point of a biological sample, cell culture, etc.), but the rest of the body/sample is subjected to the Earth's gravity, or a portion of it. Therefore, parts of the body/sample still experience an acceleration. For an extended body, the RPM does not uniformly negate gravity, and various body parts receive various accelerations. This is not representative of a microgravity environment. Moreover, the duration of the RPM usage is much shorter than a space mission, particularly an extended International Space Station (ISS) or Mars mission.

II.   Biological Limitations:

The RPM device subjects the body to stresses that are notencountered in microgravity. These stresses have an effect on biological processes that differ from a microgravity environment. In particular, cells will experience strong shearing forces and microscopic friction.

In summary, the classic error is assuming that if a vector sum of acceleration is zero, it can be equated to zero gravity. This may be true for a point particle, but not an extended system. The physics does not yield zero, and the biology certainly differs. In other words, the Earth's gravitational force is balanced by a centripetal force, mv^2/r, where r is the distance to the center of the device. This distance variation creates the gradient. It does not lead to a significant gradient for the Earth, but it does for the smaller RPM machines. This gradient is important because the microgravity is so small. Therefore, any gradient is significant compared to microgravity.

2. The preferred direction produced by the field is not equivalent to a microgravity non-directional field that is encountered in space.

3. Individual cells do not necessarily provide specific data regarding cancer in an organism or specific organ. The immune system, DNA repair mechanisms, and whole-body response are not equivalent to individual cellular response data. This is part of the linear non-threshold fallacy that uses single cell data to derive the biological response in an organism. Cell data are important, but also must be properly characterized by including the mitigating measures associated with the collective human biological repair mechanisms.

Readers of [1] should be aware of the limitations of performing Earth-based experiments to simulate microgravity. The experimental simulations and the Earth's inherent gravitational field and associated forces are not equivalent to the space microgravity environment. In addition, the space radiation environment is not equivalent to the Earth-based radiation environment. Evaluating the effects of microgravity in space must also address the space radiation environment to provide a comprehensive evaluation of the presumed biological effects under investigation.

Author Contributions: Conceptualization, J.J.B. and S.M.J.M.; writing—review and editing, all authors; All authors have read and agreed to the published version of the manuscript.

Funding: This research received no external funding.

Conflicts of Interest: The authors declare no conflict of interest.

References

1. Melnik, D.; Sahana, J.; Corydon, T.J.; Kopp, S.; Nassef, M.Z.; Wehland, M.; Infanger, M.; Grimm, D.; Krüger, M. Dexamethasone Inhibits Spheroid Formation of Thyroid Cancer Cells Exposed to Simulated Microgravity. *Cells* **2020**, *9*, 367. [CrossRef]
2. Wuest, S.L.; Richard, S.; Kopp, S.; Grimm, D.; Egli, M. Simulated microgravity: Critical review on the use of random positioning machines for mammalian cell culture. *BioMed Res. Int.* **2015**, *2015*. [CrossRef]
3. Wuest, S.L.; Stern, P.; Casartelli, E.; Egli, M. Fluid dynamics appearing during simulated microgravity using random positioning machines. *PLoS ONE* **2017**, *12*, e0170826. [CrossRef]
4. Kiefer, J.; Pross, H. Space radiation effects and microgravity. *Mutat. Res. Fundam. Mol. Mech. Mutagenesis* **1999**, *430*, 299–305. [CrossRef]
5. Ohnishi, K.; Ohnishi, T. The biological effects of space radiation during long stays in space. *Biol. Sci. Space* **2004**, *18*, 201–205. [CrossRef]
6. Mortazavi, S.J.; Cameron, J.; Niroomand-Rad, A. Adaptive response studies may help choose astronauts for long-term space travel. *Adv. Space Res.* **2003**, *31*, 1543–1551. [CrossRef]
7. Moreno-Villanueva, M.; Wong, M.; Lu, T.; Zhang, Y.; Wu, H. Interplay of space radiation and microgravity in DNA damage and DNA damage response. *Npj Microgravity* **2017**, *3*, 1–8. [CrossRef] [PubMed]
8. Yatagai, F.; Honma, M.; Dohmae, N.; Ishioka, N. Biological effects of space environmental factors: A possible interaction between space radiation and microgravity. *Life Sci. Space Res.* **2019**, *20*, 113–123. [CrossRef] [PubMed]

Reply

Science between Bioreactors and Space Research—Response to Comments by Joseph J. Bevelacqua et al. on "Dexamethasone Inhibits Spheroid Formation of Thyroid Cancer Cells Exposed to Simulated Microgravity"

Marcus Krüger [1,2,*], **Sascha Kopp** [1,2], **Markus Wehland** [1,2], **Thomas J. Corydon** [3,4] and **Daniela Grimm** [1,2,3]

1   Department of Microgravity and Translational Regenerative Medicine, Clinic for Plastic, Aesthetic and Hand Surgery, Otto von Guericke University, Universitätsplatz 2, 39106 Magdeburg, Germany; sascha.kopp@med.ovgu.de (S.K.); markus.wehland@med.ovgu.de (M.W.); dgg@biomed.au.dk (D.G.)

2   Research Group "Magdeburger Arbeitsgemeinschaft für Forschung unter Raumfahrt- und Schwerelosigkeitsbedingungen" (MARS), Otto von Guericke University, Universitätsplatz 2, 39106 Magdeburg, Germany

3   Department of Biomedicine, Aarhus University, Høegh-Guldbergsgade 10, 8000 Aarhus C, Denmark; corydon@biomed.au.dk

4   Department of Ophthalmology, Aarhus University Hospital, Palle Juul-Jensens Boulevard 167, 8200 Aarhus N, Denmark

*   Correspondence: marcus.krueger@med.ovgu.de; Tel.: +49-391-6757471

Received: 14 July 2020; Accepted: 21 July 2020; Published: 23 July 2020

We would like to thank Bevelacqua et al. for their interest in our article [1] and their comments [2]. We completely agree with the authors that the complex space environment cannot be adequately simulated on Earth. However, this was not our intention for these studies on tumor spheroids. The influence of space radiation would even be counterproductive in targeting a metastasis model to reflect the situation in cancer patients on Earth.

For many years, cells have been exposed to conditions of microgravity and scientists have investigated how these cells might sense or adapt to microgravity [3,4]. Growing three-dimensional cell aggregates in space and in laboratories using microgravity simulators is a typical application of microgravity research [5]. The NASA-developed rotating wall vessel bioreactor was specially applied to engineer such tissue constructs. The random positioning machine (RPM) works as a three-dimensional clinostat negating the directional influence of the *g*-vector providing simulated conditions of micro- or partial gravity (also referred as "time-averaged microgravity" by some authors). It should be noted and we are aware that simulated microgravity differs from real (space) microgravity in several aspects. Nevertheless, the RPM has become a well-established device for gravitational research [6,7], which is also recommended by international space agencies (like NASA [8] or the European Space Agency, ESA [9]) for experiments in reduced gravity. Although it is known to produce additional shear stress and fluid dynamics within the cell culture flasks [10,11], Wuest et al. also evaluated the RPM as a "reliable tool supporting ground-based microgravity studies" and an "ideal tool for preliminary microgravity tests, screening studies in which simulated microgravity effects are checked on various organisms" [12]. This is caused by the fact that cellular effects that were observed in real microgravity could be reproduced with good agreement on RPMs [13–18]. In this study, however, these questions were not of importance, as the RPM merely served as a tool for the generation of tumor spheroids. We believe that the RPM is superior to other methods of spheroid generation such as liquid overlay techniques or spinner flasks, as in this system it is possible to induce cells to detach from an already established cellular network and to form spheroids suspended in the culture medium. These spheroids, which were also found after long-term cultures in space [19], are an important model system to mimic micrometastases or microregions of solid tumors [20]. Despite their simplicity, tumor spheroids are often used for drug testing in pharmacology today [21–23] and are still state of the art. More complex tumor organoids co-cultured with different cell types are in development and will replace the homogenous spheroids in the future, better mimicking tumor biology in vivo. In addition, the investigation of microgravity-induced cell detachment and spheroid formation delivers valuable information about processes like metastasis and in vivo cancer progression [22]. The RPM-based metastasis model can be effectively used to study in vitro whether drugs can favor or inhibit spheroid formation. Of course, mechanical forces have to be taken into consideration while designing experiments with the RPM and analyzing their results. However, this can be done by using appropriate controls. Moreover, cells inside the human body, especially those in the process of forming metastases, are subjected to a multitude of forces besides gravity; therefore, the implementation of a completely shear-force-free experimental environment would not improve the results.

In summary, the arguments of Bevelacqua et al. [2] may result from misunderstandings of the actual purpose of the experiments. Not only has it never been the intention of this study to simulate space conditions in complete detail, it was never intended to simulate space at all and we never claimed to do so. Our aim was to investigate molecular mechanisms of metastasis development and the identification of possible target molecules for potential antimetastatic pharmacological interventions using a device that, by simulating certain aspects of microgravity, induces the formation of spheroids from originally adherently growing tumor cells, something that cannot be achieved as easily by other established methods. Some years ago Becker and Souza wrote: "Combination of the resources available in the unique environment of microgravity with the tools and advanced technologies that exist in laboratories across Earth may inform new research approaches to expand the knowledge necessary for improving treatment options, and enhancing the quality of life for those affected by this illness." [24]. In our study, we used space-derived investigations to fight cancer on Earth.

We thank the editors for giving us the opportunity to provide a reply to the letter.

Author Contributions: Writing, M.K., M.W., D.G., S.K., and T.J.C.; project administration, D.G.; funding acquisition, D.G. All authors have read and agreed to the published version of the manuscript.

Funding: This research was funded by Deutsches Zentrum für Luft- und Raumfahrt (DLR), grant number 50WB1924.

Conflicts of Interest: The authors declare no conflict of interest.

References

1. Melnik, D.; Sahana, J.; Corydon, T.J.; Kopp, S.; Nassef, M.Z.; Wehland, M.; Infanger, M.; Grimm, D.; Krüger, M. Dexamethasone Inhibits Spheroid Formation of Thyroid Cancer Cells Exposed to Simulated Microgravity. *Cells* **2020**, *9*, 367. [CrossRef] [PubMed]

2. Bevelacqua, J.; Welsh, J.; Mortazavi, S. Comment on "Dexamethasone Inhibits Spheroid Formation of Thyroid Cancer Cells Exposed to Simulated Microgravity". *Cells* **2020**, *9*, 1738. [CrossRef]

3. Ingber, D. How cells (might) sense microgravity. *FASEB J.* **1999**, *13*, S3–S15. [CrossRef] [PubMed]

4. Cogoli, A.; Tschopp, A.; Fuchs-Bislin, P. Cell sensitivity to gravity. *Science* **1984**, *225*, 228–230. [CrossRef]

5. Unsworth, B.R.; Lelkes, P.I. Growing tissues in microgravity. *Nat. Med.* **1998**, *4*, 901–907. [CrossRef]

6. Herranz, R.; Anken, R.; Boonstra, J.; Braun, M.; Christianen, P.C.; de Geest, M.; Hauslage, J.; Hilbig, R.; Hill, R.J.; Lebert, M.; et al. Ground-based facilities for simulation of microgravity: Organism-specific recommendations for their use, and recommended terminology. *Astrobiology* **2013**, *13*, 1–17. [CrossRef]

7. Borst, A.G.; van Loon, J.J.W.A. Technology and Developments for the Random Positioning Machine, RPM. *Microgravity Sci. Technol.* **2008**, *21*, 287. [CrossRef]

8. NASA. Microgravity Simulation Support Facility. Available online: https://www.nasa.gov/sites/default/files/atoms/files/microgravity_simulation_support_facility_mssf_one_pager.pdf (accessed on 3 June 2020).

9. ESA. Ground-Based Facilities Programme. Available online: https://esamultimedia.esa.int/docs/HRE/ESA-CORA-GBF_final_rev.6.pdf (accessed on 3 June 2020).

10. Wuest, S.L.; Stern, P.; Casartelli, E.; Egli, M. Fluid Dynamics Appearing during Simulated Microgravity Using Random Positioning Machines. *PLoS ONE* **2017**, *12*, e0170826. [CrossRef]

11. Hauslage, J.; Cevik, V.; Hemmersbach, R. Pyrocystis noctiluca represents an excellent bioassay for shear forces induced in ground-based microgravity simulators (clinostat and random positioning machine). *NPJ Microgravity* **2017**, *3*, 12. [CrossRef]

12. Wuest, S.L.; Richard, S.; Kopp, S.; Grimm, D.; Egli, M. Simulated Microgravity: Critical Review on the Use of Random Positioning Machines for Mammalian Cell Culture. *Biomed. Res. Int.* **2015**, *2015*, 971474. [CrossRef]

13. Benavides Damm, T.; Walther, I.; Wüest, S.L.; Sekler, J.; Egli, M. Cell cultivation under different gravitational loads using a novel random positioning incubator. *Biotechnol. Bioeng.* **2014**, *111*, 1180–1190. [CrossRef]

14. Wuest, S.L.; Richard, S.; Walther, I.; Furrer, R.; Anderegg, R.; Sekler, J.; Egli, M. A Novel Microgravity Simulator Applicable for Three-Dimensional Cell Culturing. *Microgravity Sci. Technol.* **2014**, *26*, 77–88. [CrossRef]

15. Cogoli-Greuter, M. The Lymphocyte Story—An Overview of Selected Highlights on the in Vitro Activation of Human Lymphocytes in Space. *Microgravity Sci. Technol.* **2014**, *25*, 343–352. [CrossRef]

16. Schwarzenberg, M.; Pippia, P.; Meloni, M.A.; Cossu, G.; Cogoli-Greuter, M.; Cogoli, A. Signal transduction in T lymphocytes—A comparison of the data from space, the free fall machine and the random positioning machine. *Adv. Space Res.* **1999**, *24*, 793–800. [CrossRef]

17. Walther, I.; Pippia, P.; Meloni, M.A.; Turrini, F.; Mannu, F.; Cogoli, A. Simulated microgravity inhibits the genetic expression of interleukin-2 and its receptor in mitogen-activated T lymphocytes. *FEBS Lett.* **1998**, *436*, 115–118. [CrossRef]

18. Villa, A.; Versari, S.; Maier, J.A.; Bradamante, S. Cell behavior in simulated microgravity: A comparison of results obtained with RWV and RPM. *Gravit. Space Biol. Bull.* **2005**, *18*, 89–90.

19. Pietsch, J.; Ma, X.; Wehland, M.; Aleshcheva, G.; Schwarzwälder, A.; Segerer, J.; Birlem, M.; Horn, A.; Bauer, J.; Infanger, M.; et al. Spheroid formation of human thyroid cancer cells in an automated culturing system during the Shenzhou-8 Space mission. *Biomaterials* **2013**, *34*, 7694–7705. [CrossRef]

20. Kunz-Schughart, L.A. Multicellular tumor spheroids: Intermediates between monolayer culture and in vivo tumor. *Cell Biol. Int.* **1999**, *23*, 157–161. [CrossRef]

21. Krüger, M.; Melnik, D.; Kopp, S.; Buken, C.; Sahana, J.; Bauer, J.; Wehland, M.; Hemmersbach, R.; Corydon, T.J.; Infanger, M.; et al. Fighting Thyroid Cancer with Microgravity Research. *Int. J. Mol. Sci.* **2019**, *20*, 2553. [CrossRef]

22. Nunes, A.S.; Barros, A.S.; Costa, E.C.; Moreira, A.F.; Correia, I.J. 3D tumor spheroids as in vitro models to mimic in vivo human solid tumors resistance to therapeutic drugs. *Biotechnol. Bioeng.* **2019**, *116*, 206–226. [CrossRef]

23. Sant, S.; Johnston, P.A. The production of 3D tumor spheroids for cancer drug discovery. *Drug Discov. Today Technol.* **2017**, *23*, 27–36. [CrossRef] [PubMed]

24. Becker, J.L.; Souza, G.R. Using space-based investigations to inform cancer research on Earth. *Nat. Rev. Cancer* **2013**, *13*, 315–327. [CrossRef] [PubMed]

MDPI

St. Alban-Anlage 66

4052 Basel

Switzerland

Tel. +41 61 683 77 34

Fax +41 61 302 89 18

www.mdpi.com

Cells Editorial Office

E-mail: cells@mdpi.com

www.mdpi.com/journal/cells